Introduction to Wireless
Digital Communication

Introduction to Wireless Digital Communication

A Signal Processing Perspective

Robert W. Heath, Jr.

PEARSON

Library of Congress Control Number: 2016961387

ISBN-13: 978-0-13-443179-6
ISBN-10: 0-13-443179-0

To my family

—RWH

Contents

Preface

I wrote this book to make the principles of wireless communication more accessible. Wireless communication is the dominant means of Internet access for most people, and it has become the means by which our devices connect to the Internet and to each other. Despite the ubiquity of wireless, the principles of wireless communication have remained out of reach for many engineers. The main reason seems to be that the technical concepts of wireless communication are built on the foundations of digital communication. Unfortunately, digital communication is normally studied at the end of an undergraduate program in electrical engineering, leaving no room for a course on wireless communication. In addition, this puts wireless communication out of reach for students in related areas like computer science or aerospace engineering, where digital communication may not be offered. This book provides a means to learn wireless communication together with the fundamentals of digital communication.

The premise of this book is that wireless communication can be learned with only a background in digital signal processing (DSP). The utility of a DSP approach stems from the following fact: wireless communication signals (at least ideally) are bandlimited. Thanks to Nyquist's theorem, it is possible to represent bandlimited continuous-time signals from their samples in discrete time. As a result, discrete time can be used to represent the continuous-time transmitted and received signals in a wireless system. With this connection, channel impairments like multipath fading and noise can be written in terms of their discrete-time equivalents, creating a model for the received signal that is entirely in discrete time. In this way, a digital communication system can be viewed as a discrete-time system.

Many classical signal processing functions have a role to play in this discrete-time equivalent of the digital communication system. Linear time-invariant systems, which are characterized by convolution with an impulse response, model multipath wireless channels. Deconvolution is used to equalize the effects of the channel. Upsampling, downsampling, and multirate identities find application in the efficient implementation of pulse shaping at the transmitter and matched filtering at the receiver. Fast Fourier transforms are the foundation of two important modulation/demodulation techniques: orthogonal frequency-division multiplexing and single-carrier frequency-domain equalization. Linear estimation and least squares become the basis of algorithms for channel estimation (estimating an unknown filter response) and equalization (finding a deconvolution filter). Algorithms for estimating the parameters of an unknown sinusoid in noise find application in carrier frequency offset estimation. In short, signal processing has always been a part of communication; leveraging this fact, digital communication can be learned based on connections to signal processing.

I begin this book with an introduction to wireless communication and signal processing in Chapter 1, providing some historical context. A highlight of the chapter is the discussion of different applications of wireless communication, including broadcast radio and television, cellular communication, local area networking, personal area networks, satellite systems, ad hoc networks, sensor networks, and even underwater communications. The review of applications gives context for subsequent examples and homework problems in the book, which often draw on developments in wireless local area networks, personal area networks, or cellular communication systems.

In the next two chapters, I establish a fundamental background in digital communication and signal processing. I start with an overview of the typical block diagram of digital communication systems in Chapter 2, with an explanation of each block at the transmitter and the receiver. Important functions are described, including source coding, encryption, channel coding, and modulation, along with a discussion of the wireless channel. The remainder of the chapter focuses on a subset of these functions: modulation, demodulation, and the channel. To provide a proper mathematical background, I provide an extensive overview of important concepts from signal processing in Chapter 3, including deterministic and stochastic signals, passband and multirate signal processing, and linear estimation. This chapter gives tools that are used to describe the operations of the digital communication transmitter and receiver from a signal processing perspective.

With the fundamentals at hand, I continue with a more thorough treatment of modulation and demodulation in Chapter 4. Instead of considering all possible modulation formats as would be done in a deep-dive treatment, I focus on strategies described by complex pulse-amplitude modulation. This is general enough to describe most waveforms used in commercial wireless systems. The demodulation procedure is derived assuming an additive white Gaussian noise channel, including pulse shaping, maximum likelihood detection, and the probability of symbol error. The key parts of the transmitter and the receiver are described using multirate signal processing concepts. This chapter forms the basis of a classical introduction to digital communication but with a signal processing flair.

The full specialization of the receiver algorithms to the wireless environment comes in Chapter 5. Specific impairments are described, including symbol timing offset, frame timing offset, carrier frequency offset, and frequency-selective channels. Several methods for mitigation are also described, including algorithms for estimating unknown parameters based on least squares estimation and equalization strategies that leverage alternatively the time or frequency domain. I focus on algorithms for dealing with impairments in the simplest ways possible, to set the stage for more advanced algorithms that might be encountered in the future. The chapter concludes with a description of large- and small-scale channel models, and a discussion about how to characterize the time and frequency selectivity of a channel. The knowledge in this chapter is essential for the design and implementation of any wireless system.

The final chapter, Chapter 6, provides a generalization of the concepts from Chapters 4 and 5 to systems that use multiple transmit and/or multiple receive antennas, generally referred to as MIMO communication. In this chapter, I define different MIMO modes of operation and explore them in further detail, including receiver diversity, transmitter diversity, and spatial multiplexing. In essence, most of the descriptions in prior channels generalize to the MIMO case with suitable vector and matrix notation, and additional complexity in the receiver algorithms. While most of the chapter focuses on the flat-fading channel model, some generalizations to frequency-selective channels are provided at the end. This chapter provides important connections between wireless fundamentals and the types of communication systems now widely used in commercial wireless systems.

I developed this book as part of a course at The University of Texas at Austin (UT Austin) called the Wireless Communications Lab, which served both senior undergraduates and new graduate students. The lecture portion of the course materials was based on drafts of this book. The lab portion uses my laboratory manual *Digital Wireless Communication: Physical Layer Exploration Lab Using the NI USRP,* published by the National Technology and Science Press in 2012. This laboratory manual is bundled with a USRP hardware package available from National Instruments. In the lab, students implement quadrature amplitude modulation and demodulation. They must deal with a succession of more sophisticated impairments, including noise, multipath channels, symbol timing offset, frame timing offset, and carrier frequency offset. The lab provides a way to see the concepts in this book in action on real wireless signals. For self-study or courses without a laboratory component, I have included several computer problems that involve simulating key pieces of the communication link. To see the concepts in action, you can implement the described algorithms over an audio channel using a microphone and speaker. I have used this approach in the past to prototype a High Frequency Near Vertical Incidence Skywave wireless communication link, operating in amateur radio bands.

This book may be used in several different ways, with or without a laboratory component. In a junior-level undergraduate course, I would cover Chapters 1 through 5 and spend some extra time at the beginning on the mathematical fundamentals. For a senior-level undergraduate course, I would cover the material from the entire book. For a graduate course, I would cover the entire book with an additional implementation or research project. At UT Austin, the course I teach for both undergraduates and graduates covers the entire book and the first several laboratory experiments in the aforementioned lab manual. The graduate students also perform a project. It is an intense but rewarding course.

The textbook is highly recommended as independent study, as a comprehensive review of the fundamentals is provided in Chapters 2 and 3. I encourage careful review of the mathematical fundamentals before proceeding to the main content in Chapters 4 through 6. To digest the material, I suggest working out all of the example problems with pen and paper, to help visualize the key ideas. Attempting select homework problems is also recommended. I believe that this book will provide a good foundation for further studies in wireless communications.

As wireless communication continues to integrate into our lives, there continues to be a constant evolution of wireless technologies. My hope is that this book offers a fresh perspective on wireless and acts as a launching point for further developments to come.

Register your copy of *Introduction to Wireless Digital Communication* at informit.com for convenient access to downloads, updates, and corrections as they become available. To start the registration process, go to informit.com/register and log in or create an account. Enter the product ISBN (9780134431796) and click Submit. Once the process is complete, you will find any available bonus content under "Registered Products."

Acknowledgments

This book was developed as part of my course EE 371C / EE 371C / EE 387K-17 Wireless Communications Lab taught at the University of Texas at Austin, along with the laboratory manual *Digital Wireless Communication: Physical Layer Exploration Lab Using the NI USRP,* published by the National Technology and Science Press. Many of my teaching assistants provided valuable feedback during the development of my course and books, including Roopsha Samanta, Sachin Dasnurkar, Ketan Mandke, Hoojin Lee, Josh Harguess, Caleb Lo, Robert Grant, Ramya Bhagavatula, Omar El Ayache, Harish Ganapathy, Zheng Li, Tom Novlan, Preeti Kumari, Vutha Va, Jianhua Mo, and Anum Anumali. All the teaching assistants ran the labs, worked with the students, collected feedback, and made suggestions and revisions, including examples and homework problems. Hundreds of students have taken the course and provided valuable feedback on various drafts of the book. They also asked thought-provoking questions that inspired revisions and additions in the text. In particular, I would like to thank Bhavatharini Sridharan for her positive feedback to get the book published with the material that I had. I would also like to thank Kien T. Truong for his assistance with the digital communication background chapter. I sincerely thank all my past students and teaching assistants for their support.

I would also like to acknowledge the support of my colleagues in developing this book. In particular, Professor Nuria González-Prelcic provided a detailed review of all the chapters, including many suggestions and edits for the figures. Her overall help was invaluable. The Department of Electrical and Computer Engineering was helpful in supporting my teaching related to this book. I also appreciate the many colleagues who recommended my course to their students. Sam Shearman and Erik Luther at National Instruments provided much encouragement and support to get the book published, especially after they so graciously agreed to publish the accompanying laboratory manual.

The development of the course on which this book is based was supported by several groups. National Instruments (NI) made the course possible through its establishment of the Truchard Wireless Lab at the University of Texas at Austin. NI provided summer support that funded the endeavor and contributed countless hours of employee time. Several groups from UT at Austin, including the Department of Electrical Engineering and the Cockrell School of Engineering, provided teaching assistant support to develop and maintain the course materials. Working on the labs has inspired my research. I am pleased to acknowledge research support from the National Science Foundation, DARPA, the Office of Naval Research, the Army Research Labs, Huawei, National Instruments, Samsung, Semiconductor Research Corporation, Freescale, Andrew, Intel, Cisco, Verizon, AT& T, CommScope, Mitsubishi Electric Research Laboratories, Nokia, Toyota ITC, and Crown Castle. Their support has allowed me to train many graduate students who share my appreciation of theory and practice.

About the Author

Robert W. Heath, Jr., received B.S. and M.S. degrees from The University of Virginia in 1996 and 1997, respectively, and a Ph.D. from Stanford University in 2002, all in electrical engineering. From 1998 to 2001, he was a senior member of the technical sta then a senior consultant at Iospan Wireless, Inc., where he worked on the design and implementation of the physical and link layers of the rst commercial MIMO-OFDM communication system. He is a Distinguished Professor in the Department of Electrical and Computer Engineering at North Carolina State University. He is also president and CEO of MIMO Wireless Inc. His research interests include several aspects of wireless communication and signal processing—5G cellular systems, MIMO communication, millimeter wave communication, adaptive video transmission, and manifold signal processing—as well as applications of wireless communication to automotive, aerial vehicles, and wearable networks. He is a coauthor of the book *Millimeter Wave Wireless Communications* (Prentice Hall, 2015) and author of *Digital Wireless Communication: Physical Layer Exploration Lab Using the NI USRP* (National Technology and Science Press, 2012).

Dr. Heath is a coauthor of several best-paper award recipients, including most recently the 2010 and 2013 EURASIP Journal on Wireless Communications and Networking best-paper awards, the 2012 Signal Processing Magazine best paper award, a 2013 Signal Processing Society best-paper award, 2014 EURASIP Journal on Advances in Signal Processing best-paper award, the 2014 Journal of Communications and Networks best-paper award, the 2016 IEEE Communications Society Fred W. Ellersick Prize, and the 2016 IEEE Communications and Information Theory Societies Joint Paper Award. He was a distinguished lecturer in the IEEE Signal Processing Society and is an ISI Highly Cited Researcher. He is also an elected member of the Board of Governors for the IEEE Signal Processing Society, a licensed amateur radio operator, a private pilot, a registered professional engineer in Texas, and a fellow of the IEEE.

Chapter 1

Introduction

1.1 Introduction to Wireless Communication

During the last 100 years, wireless communication has invaded every aspect of our lives. Wireless communication, though, has existed for much longer than the wire it is replacing. Speech is a prehistoric example of a wireless system, though it is predated by gestures such as beating on one's chest to display authority (still common with gorillas). Sadly, the distance over which speech is effective is limited because of the constraints of human acoustic power and the natural reduction in power as a function of distance. Early attempts at engineering a wireless communication system include smoke signals, torch signals, signal flares, and drums. One of the more successful of these was the heliotrope, which used reflections from the sun in a small mirror to convey digital signals.

The modern notion of wireless communication relies on the transmission and reception of electromagnetic waves. The concept was theorized by Maxwell and demonstrated in practice by Hertz in 1888 [151]. Others contributed to the early demonstration of wireless communication, including Lodge, Bose, and de Moura.

The earliest examples of wireless communication used what is now known as digital communication. The term *digital* comes from *digitus* in Latin, which refers to a finger or toe. Digital communication is a form of communication that involves conveying information by selecting one symbol from a set at any given time. For example, by extending just one finger, a hand can convey one of five symbols. Extending two fingers at a time, a hand can convey one of $5 \times 4 = 20$ symbols. Repeating the hand gestures quickly allows multiple symbols to be sent in succession. This is the essence of digital communication.

Digital communication using electromagnetic waves involves varying the parameters of continuous-time signals (or analog signals) to send a sequence of binary information, or bits. The most common kind of wireline communication system in the 1800s was the telegraph, which used Morse code to send digital messages consisting of letters, numbers, stops, and spaces across the country, and even the ocean, over wires. A wireless telegraph was patented by Marconi in 1896, which is generally accepted as the first wireless (electromagnetic) digital communication system. The first transatlantic wireless Morse code message was sent by Marconi in 1901 [48]. The history of wireless digital communication is as old as wireless itself.

Though interest in wireless telegraphy continued, digital communication gave way to analog communication as the primary modulation method used in wireless applications until the 1980s. Using analog communication, the parameters of a waveform are varied continuously based on a continuous-time signal at the input. An early example of analog communication is the original telephone system, developed in the late 1870s [55], in which acoustic speech waves were converted via a microphone to electrical signals that could be amplified and propagated on a wire. Early examples of wireless analog communication, still in use today, include AM (amplitude modulation) and FM (frequency modulation) radio, and older broadcast TV (television). Analog communication has been widely used in wireless communication systems, but it is now being replaced by digital communication.

The primary reasons that digital communication has now overtaken analog communication are the prevalence of digital data and advancements in semiconductor technologies. Digital data was not common before the development of computers and computer networks. Nowadays, everything stored on a computer or exchanged over the Internet is digital, including e-mail, voice calls, music streaming, videos, and Web browsing among others. Advances in integrated circuits have led to increasing numbers of transistors in a given amount of semiconductor area, which has increased the potential of digital signal processing. While not required for digital communication, leveraging digital signal processing allows for much better transmitter and receiver algorithms. In wireline telephony, digital communication circuits began to completely replace analog circuits in the network backbone in the 1960s, in part because of the noise resilience of digital signals when transmitted over long distances (repeaters are less sensitive to noise than amplifiers). Similar developments in wireless communication, however, did not start in earnest until the 1980s. The reason, it seems, is that it was only in the 1980s that integrated circuit technology had developed to the point where it could be considered for use in portable wireless devices. About the same time, the compact disc started replacing the tape and vinyl record.

Digital communication is now a fundamental part of wireless communication. In fact, almost all current and next-generation wireless communication systems (actually all developing standards as well) make use of digital communication. Wherever there is currently a wire, there is a proposal to eliminate that wire via wireless. There are a multitude of commercial, military, and consumer applications of wireless digital communication.

1.2 Wireless Systems

This section reviews common applications of wireless communication and introduces key terminology that facilitates the discussion of wireless communication in practice. Several topics are addressed, including broadcast radio and broadcast television, cellular communication, wireless local area networks, personal area networks, satellite networks, ad hoc networks, sensor networks, and finally underwater communication. The key concepts and the connections to digital communication are highlighted along the way.

1.2.1 Broadcast Radio

Broadcasting music was one of the first applications of wireless communication. A typical broadcast radio or television architecture is illustrated in Figure 1.1. Until recently, radio

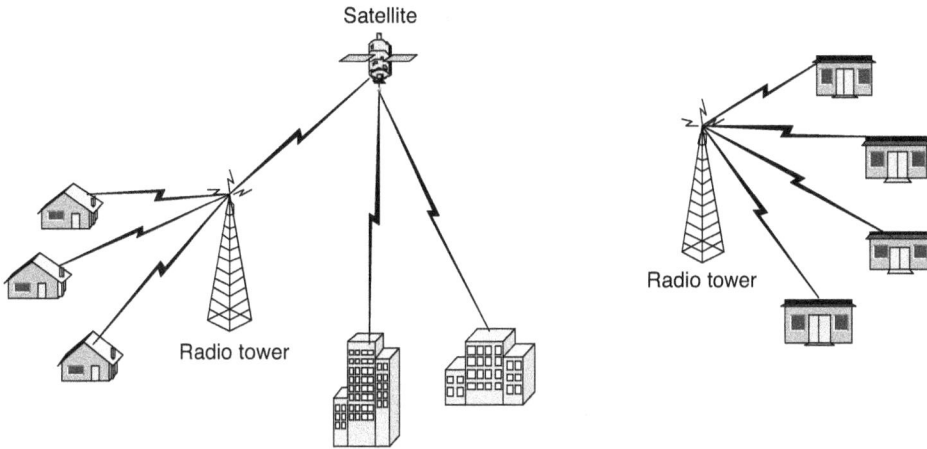

Figure 1.1 In a radio or television network, signals can be broadcast by a high radio/television tower on the ground or by a satellite in the sky.

was still analog, being transmitted in the usual AM and FM bands and taking advantage of technology developed in the 1920s and 1940s, respectively [243]. AM radio, the process of radio broadcasting using amplitude modulation, was the dominant method of radio broadcasting during the first 80 years of the twentieth century. Because of its susceptibility to atmospheric and electrical interference, AM radio now is mainly used for talk radio and news programming. In the 1970s, radio broadcasting shifted to FM radio, which uses frequency modulation to provide high-fidelity sound, especially for music radio and public radio.

In the 1990s, there was a transition of broadcast radio from analog to digital technology. In 1995, the digital audio broadcasting (DAB) standard, also known as Eureka 147, was developed [333]. DAB is used in Europe and other parts of the world, coexisitng in some cases with traditional AM and FM emissions. It uses a digital modulation technique known as COFDM (coded orthogonal frequency-division multiplexing) to broadcast multiple digital radio streams [304]. COFDM is a particular form of OFDM, which is discussed extensively in this book.

The United States uses a different digital method known as HD Radio (a trademarked name), which was approved by the FCC (Federal Communications Commission) in 2002 as the AM and FM digital broadcasting system to transmit digital audio and data along the existing analog radio signals [317, 266]. HD Radio uses a proprietary transmission technique, which also uses OFDM but fits in the gaps between existing FM broadcast stations. HD Radio started rolling out in force in 2007 in the United States. Digital coding and modulation techniques permit compact-disc-quality stereo signals to be broadcast from either satellites or terrestrial stations. In addition to audio quality improvement, digital audio broadcasting can provide other advantages: additional data services, multiple audio sources, and on-demand audio services. Just like today's analog AM and FM radio, HD Radio requires no subscription fee. HD Radio receivers are factory installed in most vehicles at present. Therefore, owners of new cars immediately have access to the HD Radio audio and data services offered [317].

1.2.2   Broadcast Television

Broadcasting television, after radio, is the other famous application of wireless. Analog television broadcasting began in 1936 in England and France, and in 1939 in the United States [233]. Until recently, broadcast TV was still using one of several analog standards developed in the 1950s: NTSC, named after the National Television System Committee, in the United States, Canada, and some other countries; PAL (phase alternating line) in much of Europe and southern Asia; and SECAM (*séquentiel couleur à mémoire*) in the former Soviet Union and parts of Africa. In addition to fundamental quality limitations, analog television systems, by their nature, are rigidly defined and constrained to a narrow range of performance that offers few choices. The move to digital television enabled a higher level of signal quality (high-definition pictures with high-quality surround-sound audio) as well as a wider range of services.

In the 1990s, the DVB (digital video broadcasting) suite of standards was initiated for digital and high-definition digital television broadcasting [274, 275]. DVB standards are deployed throughout much of the world, except in the United States. Like DAB, DVB also uses an OFDM digital modulation technique. There are several different flavors of DVB specified for terrestrial, satellite, cable, and handheld applications [104].

The United States chose to follow a different approach for high-definition digital broadcasting that produces a digital signal with a similar spectrum to the analog NTSC signal. The ATSC (Advanced Television Systems Committee) digital standard employs 8-VSB (vestigial sideband) modulation and uses a special trellis-encoder (one of the few examples of trellis-coded modulation in wireless systems) [86, 276, 85]. ATSC systems require directional antennas to limit the amount of multipath, since equalization is relatively more difficult compared with the OFDM modulation used in the DVB standard. In 2009, after over a half-century of use, the analog NTSC systems were replaced by ATSC in the United States.

1.2.3   Cellular Communication Networks

Cellular communication uses networks of base stations to provide communication with mobile subscribers over a large geographic area. The term *cell* is used to refer to the area covered by a single base station. The base stations are placed such that the cells overlap, to provide mobile users with coverage, as shown in Figure 1.2. Clusters of cells share a set of radio frequencies, which are reused geographically, to make the most use of limited radio spectrum. Cellular systems support handoff, where a connection is transferred from one base station to another as a mobile user moves. The base stations are networked together, typically with a wireline network, with several functional components to provide services such as roaming and billing. Cellular networks are typically connected to the public switched telephone network (the network used for making telephone calls) and the Internet.

The first generation of cellular communication devices used analog communication, in particular FM modulation, for the wireless link between mobile users and the base stations. The technology for these systems was conceived in the 1960s and deployed in the late 1970s and early 1980s [243, 366, 216]. The use of analog technology gave little security (it was possible to eavesdrop on a call with the right radio gear), and limited

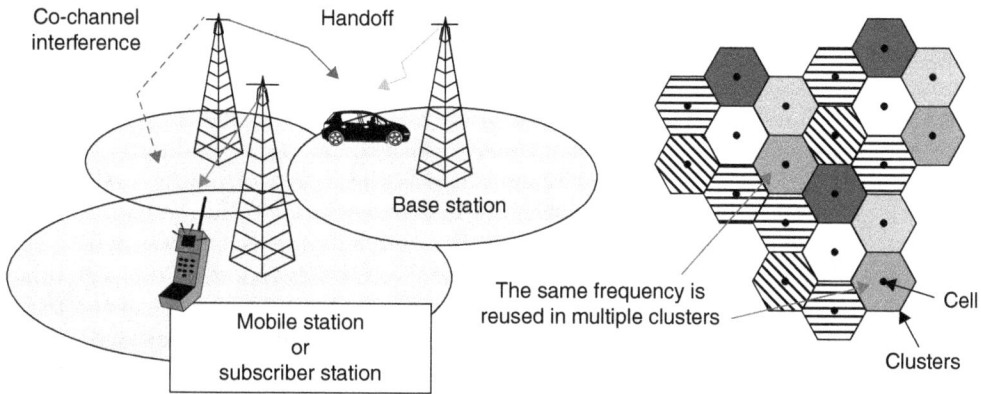

Figure 1.2 The components of a typical cellular system. Each cell has a base station serving multiple mobiles/users/subscribers. A backhaul network connects the base stations together and permits functions like handoff. Frequencies are reused in clusters of cells.

data rates were supported. Many similar, but not compatible, first-generation systems were deployed around the same time, including AMPS (Advanced Mobile Phone System) in the United States, NMT (Nordic Mobile Telephony) in Scandinavia, TACS (Total Access Communication System) in some countries in Europe, Radiocom 2000 in France, and RTMI (Radio Telefono Mobile Integrato) in Italy. Japan had several different analog standards. The plurality of standards deployed in different countries made international roaming difficult.

The second and subsequent generations of cellular standards used digital communication. Second-generation systems were conceived in the 1980s and deployed in the 1990s. The most common standards were GSM (Global System for Mobile Communications) [123, 170], IS-95 (Interim Standard 1995, also known as TIA-EIA-95) [139, 122], and the combination IS-54/IS-136 (known as Digital AMPS). GSM was developed in a collaboration among several companies in Europe as an ETSI (European Telecommunications Standards Institute) standard. It was adopted eventually throughout the world and became the first standard to facilitate global roaming. The IS-95 standard was developed by Qualcomm and used a new (at the time) multiple access strategy called CDMA (Code Division Multiple Access) [137]; therefore, IS-95 was also known as cdmaOne. IS-95 was deployed in the United States, South Korea, and several other countries. The IS-54/IS-136 standard was developed to provide a digital upgrade to the AMPS system and maintain a certain degree of backward compatibility. It was phased out in the 2000s in favor of GSM and third-generation technologies. The major enhancements of second-generation systems were the inclusion of digital technology, security, text messaging, and data services (especially in subsequent enhancements).

The third generation (3G) of cellular standards, deployed in the 2000s, was standardized by 3GPP (3rd Generation Partnership Project) and 3GPP2 (3rd Generation Partnership Project 2). UMTS (Universal Mobile Telecommunications System) was specified by 3GPP as the 3G technology based on the GSM standard [193, 79]. It used a similar network infrastructure and a higher-capacity digital transmission technology. The evolution of cdmaOne led to CDMA2000, which was standardized by 3GPP2 [364, 97]. Notably, both UMTS and CDMA2000 employ CDMA. The major advance of third-generation standards over the second generation was higher voice capacity (the ability to support more voice users), broadband Internet access, and high-speed data.

The fourth generation of cellular standards was the object of much development, and much debate (even over the definition of "fourth generation"). In the end, two systems were officially designated as fourth-generation cellular systems. One was 3GPP LTE (Long Term Evolution) Advanced release 10 and beyond [93, 253, 299, 16]. The other was WiMAX (Worldwide Interoperability for Microwave Access), a subset of the IEEE 802.16 m standard [194, 12, 98, 260]. Though WiMAX was deployed earlier, 3GPP LTE became the de facto 4G standard. A major departure from third-generation systems, fourth-generation systems were designed from the ground up to provide wireless Internet access in a large area. 3GPP LTE is an evolution of 3GPP that supports larger-bandwidth channels and a new physical layer based on OFDMA (orthogonal frequency-division multiple access) where subcarriers can be dynamically assigned to different users. OFDMA is a multiple-access version of OFDM (orthogonal frequency-division multiplexing), which is discussed in Chapter 5. 3GPP LTE Advanced adds other new capabilities, including more support for MIMO (multiple input multiple output) communication enabled by multiple antennas at the base station and handset, and thus supports higher data rates. WiMAX is based on the IEEE 802.16 standard. Essentially, the WiMAX Forum (an industry consortium) is specifying a subset of functions for implementation, and appropriate certification and testing procedures will ensure interoperability. WiMAX also employs OFDMA, though note that earlier versions used a slightly different access technique based on OFDM. Fourth-generation systems make more use of multiple antennas via MIMO communication, which is discussed in Chapter 6. The fourth generation of cellular systems promises higher data rates than previous systems along with network enhancements such as simplified backhaul architectures.

Research on the fifth generation of cellular standards has begun in 3GPP. At the time of the writing of this book, various technologies are being considered to further improve throughput and quality and reduce latency and costs. There is great interest in continuing to push MIMO communication to its limits [321]. Massive MIMO promises hundreds of antennas at the base station to support more users simultaneously [223, 185, 42], and full-dimensional MIMO uses horizontal and vertical beamforming to support more users [238]. Millimeter wave MIMO systems making use of spectrum above 30 GHz are also being considered for the fifth generation of cellular systems [268, 262, 269]. Research on all of these topics is ongoing [45, 11].

1.2.4 Wireless Local Area Networks (WLANs)

Wireless local area networks are a wireless counterpart to Ethernet networks, whose initial objective was to deliver data packets from one computer to another. A wireless local area network is illustrated in Figure 1.3. All WLANs use digital communication. The original objective of WLANs was simply to implement a local area network; in current deployments WLANs are seen as a primary means for wireless Internet access. Compared with cellular networks that use expensive licensed spectrum, WLANs are implemented in

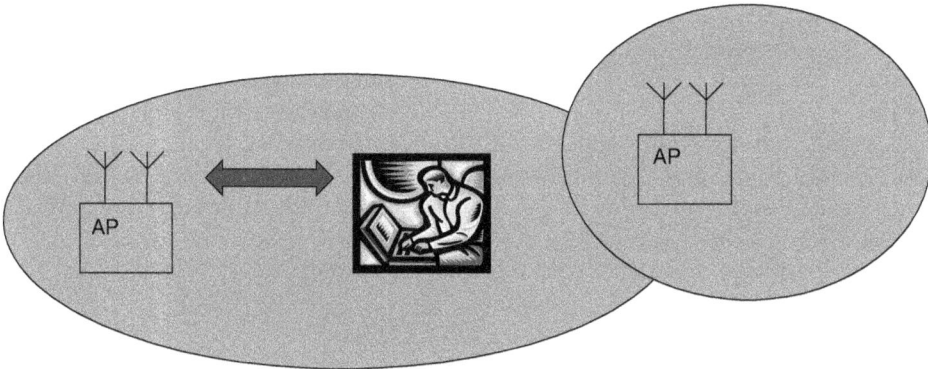

Figure 1.3 A wireless local area network. Access points (APs) serve clients. Unlike in cellular systems, handoff is generally not supported.

unlicensed bands like the ISM (Industrial, Scientific, and Medical) and U-NII (Unlicensed National Information Infrastructure) radio bands in the United States. This means they can be installed by anyone with approved equipment but cannot provide guaranteed service. WLANs are philosophically different from cellular networks. While both may be used for wireless Internet access, WLANs are primarily an extension of a wired network and are not designed to provide seamless large-area coverage, like a cellular network, for example. Most WLANs implement only basic forms of handoff, if any handoff is implemented at all.

The most widely used WLAN standards are developed within the IEEE 802.11 working group [279, 290]. IEEE 802 is a group that develops LAN and MAN (metropolitan area network) standards, focusing on the physical (PHY), media access control (MAC), and radio link protocol (link) layers, considered Layer 1 and Layer 2 in typical networking literature [81]. The IEEE 802.11 working group focuses on WLANs. The Wi-Fi Alliance is an organization for certifying IEEE 802.11 products to guarantee interoperability (often Wi-Fi is used interchangeably with IEEE 802.11, though they are not exactly the same). Different subgroups of IEEE 802.11 are associated with different letters, such as IEEE 802.11b, IEEE 802.11a, IEEE 802.11g, and IEEE 802.11n.

The original IEEE 802.11 standard supported 0.5Mbps (megabit-per-second) data rates with a choice of two different physical layer access techniques, either frequency-hopping spread spectrum or direct-sequence spread spectrum in the 2.4GHz ISM band. IEEE 802.11b provides data rates of 11bps by using Complementary Code Keying modulation, extending the direct-sequence spread spectrum mode. IEEE 802.11a and IEEE 802.11g provide data rates of 54Mbps in the 5.8GHz and 2.4GHz bands, respectively, using OFDM modulation, which is discussed in Chapter 5.

IEEE 802.11n is a high-throughput extension of IEEE 802.11g and IEEE 802.11a that uses MIMO communication, combined with OFDM, to provide even higher data rates [360, 257]. MIMO enables a new class of modulation techniques, some of which can be used to send multiple data streams in parallel, and others that provide higher reliability as described further in Chapter 6. More advanced high-throughput extensions of IEEE 802.11 were developed as IEEE 802.11ac and IEEE 802.11ad. Two letters are used since single letters have been exhausted through other extensions of the standard. IEEE 802.11ac focuses on sub-6GHz solutions [29], and IEEE 802.11ad focuses on higher-frequency, in particular the 60GHz millimeter wave unlicensed band, solutions [258, 268]. Compared with IEEE 802.11n, IEEE 802.11ac supports more advanced MIMO capability (up to eight antennas) and multiuser MIMO communication, where the access point communicates with several users at the same time. IEEE 802.11ad is the first WLAN solution at millimeter wave, providing gigabit-per-second (Gbps) peak throughputs. The next generation of WLAN is currently in development under the name IEEE 802.11ay; it will support multiuser operation, targeting 100Gbps data rates and an extended transmission distance of 300–500m.

1.2.5 Personal Area Networks (PANs)

Personal area networks (PANs) are digital networks intended for short-range connectivity, typically on the order of 10m in all directions, especially for wire replacement. An example of a PAN is illustrated in Figure 1.4. One of the most appropriate applications of a WPAN (wireless PAN) is to connect devices in the user's personal space, that is, the devices an individual carries on or near the person, such as keyboards, headphones, displays, audio/video players, tablets, or smartphones [353]. According to the standards, a PAN can be viewed as a "personal communication bubble" around a person. All PANs use digital communication. PANs have a major architectural difference from WLANs—they expect communication in an ad hoc fashion. This means that devices can set up ad hoc peer-to-peer networks without the aid of a central controller (or access point). PANs are also implemented in unlicensed spectrum.

Most PANs are developed within the IEEE 802.15 working group [308]. The Bluetooth standard, IEEE 802.15.1a and later extensions, is perhaps the most familiar protocol. It is

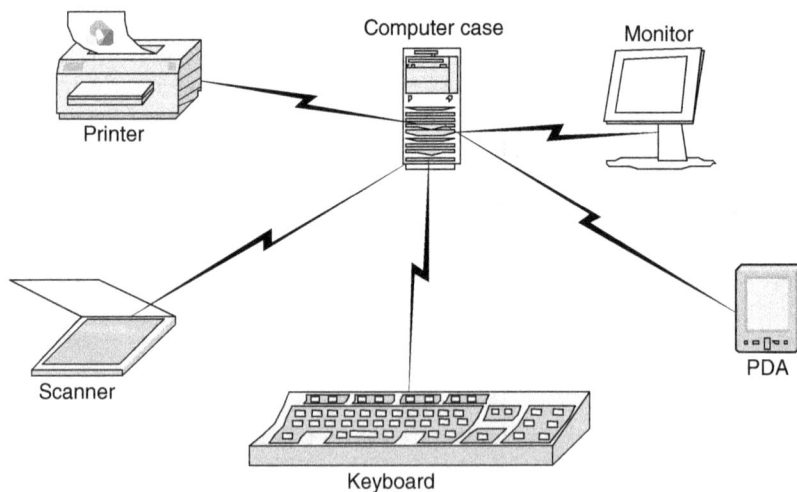

Figure 1.4 A wireless personal area network formed on an office desk. A computer connects with all the other devices, namely, a monitor, a keyboard, a PDA, a scanner, and a printer, via wireless links. The typical distance between devices is 10m.

most commonly used for wireless headset connections to cell phones, wireless keyboards, and wireless computer mice. Another PAN standard is IEEE 802.15.4, known as ZigBee, intended for low-power embedded applications like sensor networks, home monitoring and automation, and industry controls [227]. IEEE 802.15.3c was a high-data-rate extension of 802.15 to the millimeter wave unlicensed band (around 57GHz to 64GHz), which was not as successful as WirelessHD [356], which was developed by an industry consortium [268]. These systems provide high-bandwidth connections in excess of 2Gbps for applications such as wireless HDMI (High-Definition Multimedia Interface) and wireless video display connections. The boundaries between WLAN and PAN are starting to blur, with IEEE 802.11ad taking over many of the functions offered by 60GHz PAN. It is likely that such developments will continue with IEEE 802.11ay.

1.2.6 Satellite Systems

Satellite systems use space-based transceivers at very high altitudes over the Earth's surface to provide coverage over large geographic areas, as illustrated in Figure 1.5. They are an alternative to terrestrial communication networks, where the infrastructure equipment is located on the ground. The idea of telecommunication satellites originated from a paper by Arthur C. Clarke, a science fiction writer, in *Wireless World* magazine in 1945 [74]. That paper proposed the use of the orbital configuration of a constellation of three satellites in the geo-stationary Earth orbit (GEO) at 35,800km to provide intercontinental communication services. Other orbits, namely, LEO (low Earth orbit) between 500km and 1700km and MEO (medium Earth orbit) between 5000km and 10,000km and over 20,000km, are now employed as well [222]. The higher orbit provides more coverage,

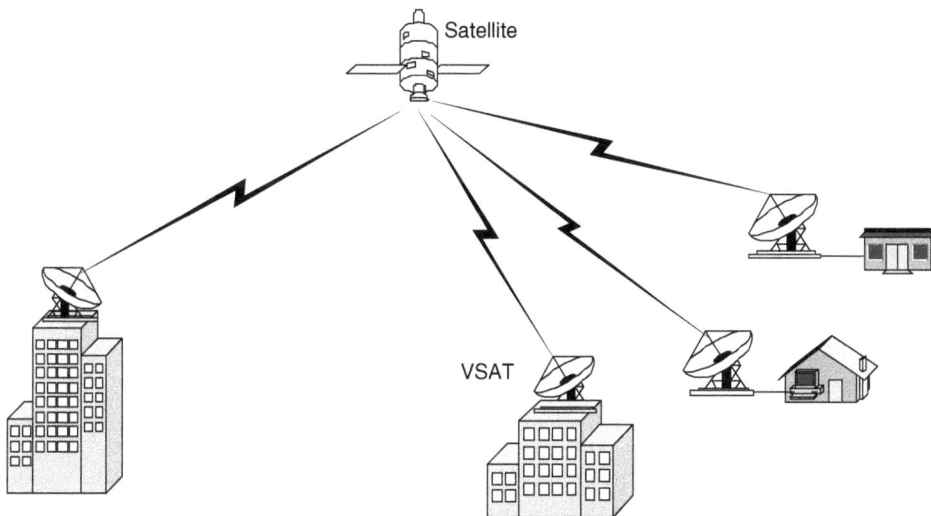

Figure 1.5 The components of a satellite system. A satellite located at a high altitude over the Earth's surface acts as a repeater to help point-to-point and point-to-multipoint transmissions between VSATs on the ground.

that is, fewer satellites, but at the cost of larger propagation delay and free space loss. Until the 1960s, though, satellites were not actually for telecommunications in practice, but for observation and probes. Project SCORE, launched in 1958, was the world's first communications satellite, providing a successful test of a space communication relay system. Since that time, the number of launched communication satellites has increased: 150 satellites during 1960–1970, 450 satellites during 1970–1980, 650 satellites during 1980–1990, and 750 satellites during 1990–2000 [221].

Satellites, in the context of telecommunications, act as repeaters to help both point-to-point and point-to-multipoint transmissions of signals. Traditionally, communication satellites provide a wide range of applications, including telephony, television broadcasting, radio broadcasting, and data communication services [94]. Compared to other systems, communication satellite systems stand out because of their broad coverage, especially their ability to provide services to geographically isolated regions or difficult terrains. For example, mobile satellite services would target land mobile users, maritime users [180], and aeronautical users [164].

Satellites provide both long-distance (especially intercontinental) point-to-point or trunk telephony services as well as mobile telephony services. In 1965, Intelsat launched the first commercial satellite, named Early Bird, to provide intercontinental fixed telephony services. Communication satellite systems are able to provide worldwide mobile telephone coverage, also via digital communication technology [286]. The first GEO satellite providing mobile services, Marisat, was launched into orbit in 1976. Other examples of systems include Iridium, Inmarsat, and Globalstar. Satellite phones are inherently more expensive because of the high cost of putting a satellite in orbit and their low capacity. Satellite phones are useful in remote areas and for sea-based communication; their use in populated areas has been eclipsed by cellular networks.

Television accounts for about 75% of the satellite market for communication services [221]. Early satellite TV systems used analog modulation and required a large receiving dish antenna. In 1989, TDF 1 was launched as the first television direct broadcasting satellite. Now most satellite TV programming is delivered via direct broadcast satellites, which use digital communication technology. Some examples of current communications satellites used for TV broadcasting applications are Galaxy and EchoStar satellites in the United States, Astra and Eutelsat Hot Bird in Europe, INSAT in India, and JSAT satellites in Japan.

A recent application of satellite broadcast is high-definition radio. In the last 20 years, satellite radio has taken off in many areas [355]. The initial applications of satellites in radio were to provide high-fidelity audio broadcast services to conventional AM or FM broadcast radio stations. Now they are widely used for transmitting audio signals directly to the users' radio sets. In satellite radio systems like SiriusXM [88], based on Sirius and XM technology [247, 84], digital communication is used to multicast digital music to subscribers. Other information may also be bundled in the satellite radio transmissions such as traffic or weather information.

A final application of satellites is for data communication. Satellite systems provide various data communication services, including broadcast, multicast, and point-to-point unidirectional or bidirectional data services [105]. Example services include messaging, paging, facsimile, data collection from sensor networks, and of course wireless Internet access. Unidirectional or broadcast data communication services are often provided by VSAT (very small aperture terminal) networks [4, 66, 188], using GEO satellites. VSAT networks work well for centralized networks with a central host and a number of geographically dispersed systems. Typical examples are small and medium-size businesses with a central office and banking institutions with branches in different locations. VSAT networks are also used for wireless Internet access in rural areas.

High-altitude platform (HAP) stations are a hybrid technology that combines the benefits of terrestrial and satellite communication systems. Examples of HAP are unmanned airships and manned/unmanned aircraft flying in the stratosphere just above the troposphere, at an altitude of about 17km or higher [76, 18, 103]. HAP stations may fill the gap between satellite-based communication systems, which are expensive and put high demands on the subscriber units because of the large distance to the satellites, and the terrestrial transmitters, which suffer from limited coverage. They may also be an alternative to cellular systems for telephony and wireless Internet access in parts of the world that lack cellular infrastructure.

1.2.7 Wireless Ad Hoc Networks

Ad hoc networks are characterized by their lack of infrastructure. Whereas users in cellular networks normally communicate with fixed base stations, users in ad hoc networks communicate with each other; all users transmit, receive, and relay data. A fantastic use case for ad hoc networks is by emergency services (police, search and rescue). Disasters, such as Hurricane Katrina, the earthquake in Haiti, or the typhoon in the Philippines, destroy the cellular infrastructure. Collaboration of rescue crews, communication with loved ones, and coordination of aid delivery are drastically hindered by the devastation. A mobile ad hoc network can transform a smartphone into both a cell tower and a cell phone. In this way, data can be transmitted throughout the disaster area. Ad hoc networks are also important in the military where there is high mobility and an inability to rely on existing fixed infrastructure. The soldiers of the future will require reliable, easily deployable, decentralized high-speed wireless communication networks for high-quality video, imagery, voice, and position data to ensure an information advantage in combat. There are many practical applications of ad hoc networks.

Ad hoc networking capability is a core part of most PANs. With Bluetooth, for example, devices self-organize into a piconet with one device acting as the master and the other devices slaved to that master. The master coordinates transmissions among the various devices. WLANs also support ad hoc capability for communication between devices, and also a more formal mesh capability in IEEE 802.11s [61]. Cellular networks are starting to support device-to-device communication where devices can exchange data directly without going through the base station [89, 110]. This is not a completely self-organized ad hoc operation, though, because the devices may coordinate key network operations like device discovery through the base station.

A recent application of mobile ad hoc networking is vehicular ad hoc networking (commonly known as VANETs in the literature) [329]. As illustrated in Figure 1.6, VANETs involve both vehicle-to-vehicle communication and vehicle-to-infrastructure communication and are a key ingredient in connected and automated vehicles. A difference between

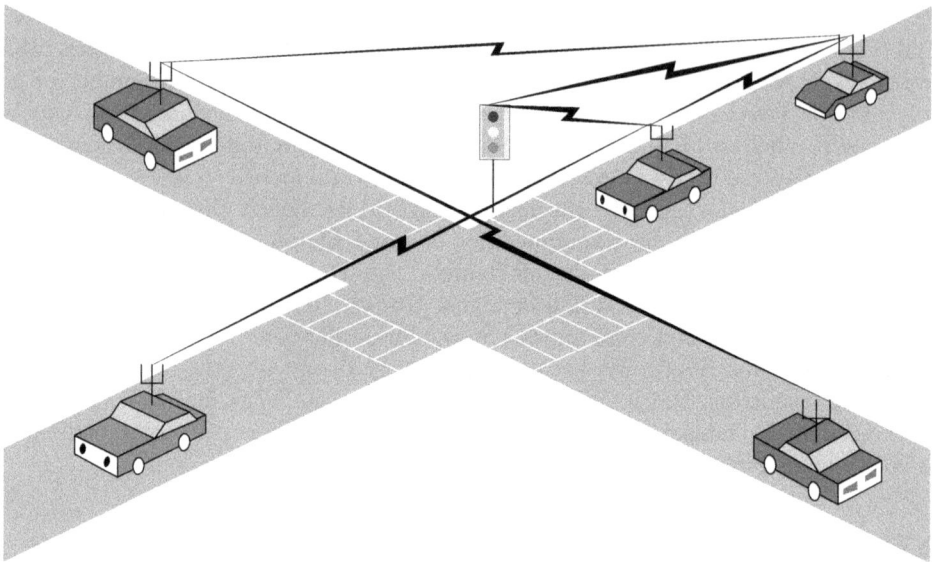

Figure 1.6 VANETs consist of vehicles, each of which is capable of communicating with other vehicles or infrastructure within its radio range for a variety of purposes, including, for example, collision avoidance.

VANETs and other ad hoc networks is in the overlying applications. Safety is a primary application of VANETs. For example, the dedicated short-range communication protocol [41, 232, 177] allows vehicles to exchange messages with position and velocity information for applications such as forward collision warning. Next-generation connected vehicles will exchange even more information. For example, sharing perceptual data among neighboring vehicles can extend a vehicle's perception range beyond its visual line of sight [69]. This data can be fused to create a bird's eye view of neighboring traffic, which can assist both automated and human drivers in difficult driving tasks such as overtaking and lane changing [234]. VANETs, especially at millimeter wave, continue to be an active area of research [337].

1.2.8   Wireless Sensor Networks

A wireless sensor network is a form of an ad hoc wireless network, where wirelessly connected sensors relay information to some selected nodes at appropriate times. Advances in wireless communication, signal processing, and electronics have enabled the development of low-cost, low-power, multifunctional sensor nodes that are small in size and capable of sensing, data processing, and communicating [7]. The most important factor in the design of wireless sensor networks is the short network lifetime due to finite-capacity batteries [5].

Energy networks provide another potential application of wireless communication in the form of sensor networks. The electric power grid is based on hundred-year-old technology where power is sent into the network and consumption is measured by electric meters, which are read infrequently. Sensors can be used to enable what is called a smart grid, which supports features like demand-based pricing and distributed power generation [58, 293]. Many aspects of the smart grid are enabled through wireless meters. Smart grids can be implemented with a host of different wireline or wireless technologies. There are many research challenges in smart grid technology, including control, learning, and system-level issues.

RFID (radio frequency identification) is a special type of communication that is used in applications such as manufacturing, supply chain management, inventory control, personal asset tracking, and telemedicine [361, 183, 62]. An RFID system consists of RFID tags, which are given to products and objects for identification purposes, and RFID readers. Readers broadcast queries to tags in their radio range for information control, and tags reply with stored identification information, typically using energy from the broadcast query to power the RFID circuit and transmitter [64]. Since no active transmission is involved, the power consumption for communication is very low [5]. RFID may be used in a sensor network as both a sensor and a means of communication to detect, for example, if the RFID tag (or the object that is tagged) is physically present in a given location. RFID has been standardized by EPCglobal and the ISO (International Organization for Standardization). The battery-free design of the typical RFID tag makes its design different from that of conventional communication systems.

1.2.9 Underwater Communication

Underwater communication is another niche application of wireless communication. Some applications of underwater communication are illustrated in Figure 1.7. The major difference from other forms of communication discussed in this chapter is that underwater communication is most often conceived with acoustic propagation versus electromagnetic waves in radio frequency wireless systems. The high conductivity in seawater, induced by salinity, causes large attenuation in electromagnetic radiation methods, making electromagnetic waves incapable of propagating over long distances. Acoustic methods have their own limitations, mainly a very limited bandwidth. Generally speaking, acoustic methods are used for low-rate long-distance transmission, whereas electromagnetic methods may be used for high-rate short-range transmission [168].

Modern underwater communication systems use digital transmission [311, 315, 265]. From a signal processing perspective, underwater communication requires the use of sophisticated adaptive receiver techniques [311]. The reason is that, relatively speaking, the underwater propagation channel changes and presents many multipaths. Most radio frequency wireless systems are designed with a kind of block invariance where time variation can be neglected in short processing intervals.

This assumption may not be appropriate for underwater communication due to the rapid channel variations. The main applications of underwater communication are found in the military, for example, ship-to-ship, ship-to-shore, and ship-to-sub, though there are commercial applications in the petroleum industry, such as autonomous underwater vehicles. Communicating underwater is a growth industry for the United States Navy. Two-way underwater digital communication between submarines and the AUTEC (Atlantic Undersea Test and Evaluation Center) range-control station in the Bahamas has been successfully demonstrated [144]. Sensor networks are also applied underwater

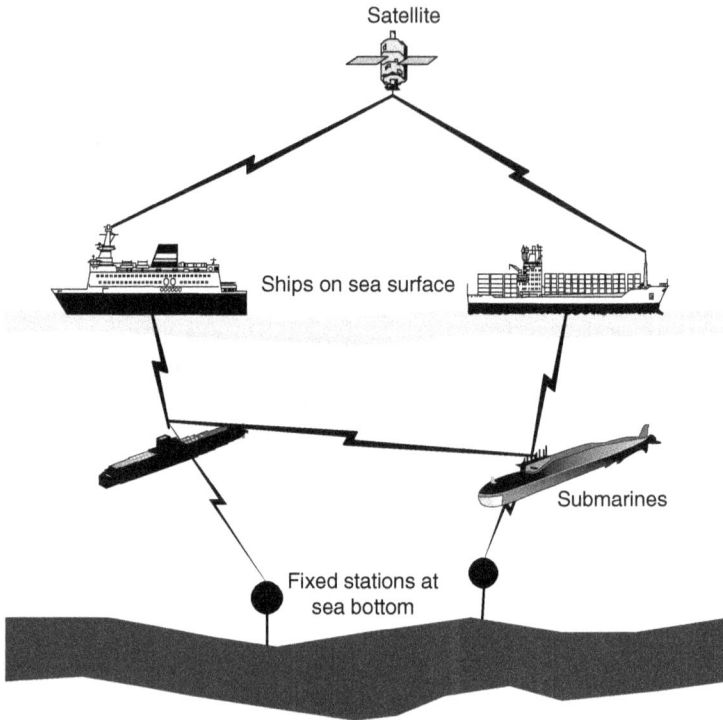

Figure 1.7 An underwater communication system. Submarines can communicate with fixed stations located at the sea bottom or with ships on the surface.

for oceanographic data collection, environment monitoring, explorations, and tactical surveillance [9]. Many of the concepts developed in this book can be applied to underwater communication systems, with some modifications to account for variability of the propagation channel.

1.3 Signal Processing for Wireless Communication

A signal is a function that describes how a physical or a nonphysical variable changes over time and/or space. Signals are usually acquired by sensors and transformed by a transducer into an appropriate form to be stored, processed, or transmitted. For example, a microphone contains a diaphragm to capture the audio signal and a transducer to convert that signal into a voltage. In a wireless communication system, typical signals are the currents and the electromagnetic fields used to carry data from a transmitter to a receiver through a wireless channel. There are many other types of signals besides audio and communications signals: speech, image, video, medical signals like an electrocardiogram, or financial signals measuring, for example, the evolution of stock prices. Signal processing is a relatively new engineering discipline that studies how to manipulate signals to extract information or to change the characteristics of the signal with a given purpose.

Though signal processing includes digital and analog techniques, DSP dominates most of the application scenarios. Therefore, an analog signal to be processed is discretized and quantized before manipulation. For example, the receiver in a wireless communication system has to apply some processing to the received signal to remove noise, cancel interference, or eliminate the distortion due to the propagation through the wireless channel; at the transmitter side, signal processing is used to generate the waveform to be transmitted and maximize the range or the amount of information per time unit that can be sent. The current trend is to perform all these operations in digital, placing an analog-to-digital converter (ADC) or a digital-to-analog converter (DAC) as close as possible to the receive or transmit antenna respectively. Figure 1.8 shows an example of a basic communication system using a signal processing approach, making use of analog and digital techniques.

Signal processing has many applications in other fields such as:

- Speech and audio, for speaker recognition, text-to-speech conversion, speech recognition, speech or audio compression, noise cancellation, or room equalization.

- Image and video, for image and video compression, noise reduction, image enhancement, features extraction, motion compensation, or tracking of objects.

- Medicine, for monitoring and analysis of biosignals.

- Genomics, for interpretation of genomic information.

- Finance, to analyze financial variables mainly for prediction purposes.

- Radar, for detecting targets and estimating their position and velocity.

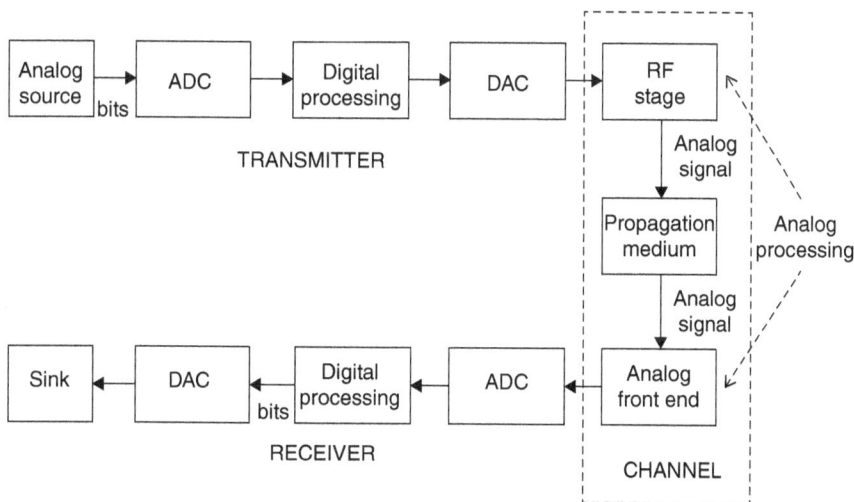

Figure 1.8 Basic block diagram of a digital communication system making use of analog and digital signal processing

Signal processing is a discipline at the intersection of signal processing and applied mathematics. It did not emerge as an independent field of study until the mid-twentieth century [239]. By that time Norbert Wiener had proposed a random process model for the information source. He also invented the Wiener filter, which provides a statistical estimate of an unknown process from an observed noisy process. The landmark paper "A Mathematical Theory of Communication," written by Claude Shannon in 1948 [302], established the foundations of communication theory by analyzing a basic digital communication system from a signal processing perspective, using Wiener's idea to model information signals. The sampling theorem proposed by Harry Nyquist in 1928 and proved by Shannon in 1949 in his paper "Communication in the Presence of Noise" addressed the problem of sampling and reconstruction of continuous signals, a milestone in DSP. For subsequent years, however, analog signal processing continued to dominate signal processing applications, from radar signal processing to audio engineering [239]. The publication in 1965 by Cooley and Tukey of an algorithm for the fast implementation of the Fourier transform (now known as FFT) led to the explosion of DSP, making it possible to implement convolution much more efficiently. Speech coding for telephone transmission was at that time a very active signal processing area, which started to benefit from adaptive algorithms and contributed to the success of DSP. Since that time, DSP algorithms have continued to evolve, leading to better performance and the expansion of the range of applications that benefit from them. Wireless communication is not an exception; the incredible increase in performance and data rates experienced in recent years in many communication systems was made possible by the increased complexity of DSP techniques.

A signal processing approach tackles problems from a system perspective, including models for the input and output signals at every block in the system. The different blocks represent the different processing stages, which can be realized with an analog device or a numerical algorithm implemented in a digital processor, as can be seen in Figure 1.8. There exists a trade-off between the complexity and the performance of the models used for the signals and the analog components of the system: more accurate models provide an excellent tool for the simulation and practical evaluation of the system, but they increase complexity and simulation time and make the theoretical analysis of the problems difficult. Statistical characterization of signals using random process theory and probability provides useful models for the signal carrying the information and also for the noise and the interfering signals that appear in a wireless communication system.

Signal processing theory also provides mathematical tools to relate the different signals in a system, using concepts from calculus, linear algebra, and statistics. Chapter 3 reviews in detail the fundamental signal processing results that can be used in the design and analysis of wireless communication systems. Linear time-invariant systems are used extensively in wireless communication to model different devices in the system such as filters or equalizers. Many of the features of a communication system are better understood in the frequency domain, so Fourier analysis is also a basic tool for wireless engineers. Digital communication systems leverage multirate theory results as well, since multirate filters lead to efficient implementations of many of the operations usually performed in a digital transmitter or receiver. Finally, fundamental results in linear algebra are the basis for many signal processing algorithms used for different tasks at the receiver such as channel equalization.

A digital signal processing approach to wireless communications, the so-called software-defined radio (SDR) concept, makes sense for many reasons, such as ease of reconfigurability (software download) or simultaneous reception of different channels and standards, as shown in Figure 1.9. Digitizing the communication signal at the output of the receive antenna may not be feasible, however, because of technical (a very high sampling frequency) or cost (too-high power consumption at the ADC) reasons. Therefore, a trade-off between analog signal processing and DSP is usually found in practical communication systems, which usually include an analog stage to downconvert the signal followed by a digital stage, as illustrated in Figure 1.9. Later chapters of this book provide several examples of functional block diagrams corresponding to current communication systems that make use of this approach.

1.4   Contributions of This Book

This book presents the fundamentals of wireless digital communication from a signal processing perspective. It makes three important contributions. First, it provides a foundation in the mathematical tools that are required for understanding wireless digital communication. Second, it presents the fundamentals of digital communication from a signal processing perspective, focusing on the most common modulations rather than the most general description of a communication system. Third, it describes specific receiver algorithms, including synchronization, carrier frequency offset estimation, channel estimation, and equalization. This book can be used in conjunction with the codeveloped laboratory course [147] or independently on its own.

There are already a number of textbooks on related topics of wireless communication and digital communication. Most other textbooks on wireless communication are targeted toward graduate students in communications, building on the foundations of graduate

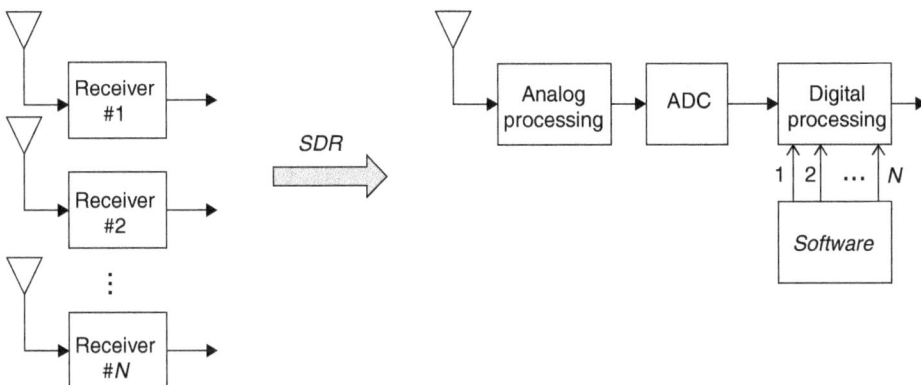

Figure 1.9 A current digital communication receiver based on the SDR concept. It allows simultaneous reception of channels using different standards with a single piece of hardware for the receiver.

courses in random processes and digital communication. Unfortunately, undergraduate students, graduate students in other areas, and practicing engineers may not have taken the typical graduate prerequisites for those textbooks. Other textbooks on digital communication are targeted toward one- or two-semester graduate courses, attempting to present digital communication in its most general form. This book, however, focuses on a subset of digital communication known as complex pulse-amplitude modulation, which is used in most commercial wireless systems. In addition, this book describes in detail important receiver signal processing algorithms, which are required to implement a wireless communication link. While most concepts are presented for a communication system with a single transmit and single receive antenna, they are extended at the end of the book to MIMO communication systems, which are now widely deployed in practice.

For communications engineers, this book provides background on receiver algorithms like channel estimation and synchronization, which are often not explained in detail in other textbooks. It also provides an accessible introduction to the principles of MIMO communication. For signal processing engineers, this book explains how to view a communication link through a signal processing lens. In particular, input-output relationships are built upon principles from digital signal processing so that the entire system can be represented in terms of discrete-time signals. Critical background on communication system impairments and their models is provided, along with an approachable introduction to the principles of wireless channel modeling. For analog, mixed-signal, and circuit designers, this book provides an introduction into the mathematical principles of wireless digital communication. The formulations are simplified from what is found in other textbooks, yet what is presented is immediately practical and can be used to prototype a wireless communication link [147].

Coverage in this book is intentionally narrow. No attempt is made to present a framework that includes every possible kind of digital communication. The focus is specifically on complex pulse-amplitude modulated systems. Nor is any attempt made to provide optimum receiver signal processing algorithms for all different channel impairments. Rather the emphasis is on using simpler estimators like linear least squares, which work in practice. The foundations developed in this book are a great platform for further work in wireless communication.

1.5 Outline of This Book

This book is organized to allow students, researchers, and engineers to build a solid foundation in key physical-layer signal processing concepts. Each chapter begins with an introduction that previews the material in each section and ends with a summary of key points in bullet form. Examples and numerous homework problems are provided to help readers test their knowledge.

This chapter serves as an introduction to wireless communication, providing a detailed overview of the myriad applications. It also provides some historical background on signal processing and makes the case for using signal processing to understand wireless communications.

Chapter 2 provides an overview of digital communication. The review is built around a canonical block diagram for a digital communication system to provide context for developments in subsequent chapters. Then the components of that diagram are discussed in more detail. First the types of distortion introduced by the wireless channel are reviewed, including additive noise, interference, path loss, and multipath. The presence of the wireless channel introduces many challenges in the receiver signal processing. Then a brief overview is provided of source coding and decoding, with examples of lossless and lossy coding. Source coding compresses data, reducing the number of bits that need to be sent. Next, some background is provided on secret-key and public-key encryption, which is used to keep wireless links secure from eavesdroppers. Then channel coding and decoding are reviewed. Channel coding inserts structured redundancy that can be exploited by the decoder to correct errors. The chapter concludes with an introduction to modulation and demodulation, including baseband and passband concepts, along with a preview of the impact of different channel impairments. Subsequent chapters in the book focus on modulation and demodulation, correcting channel impairments, modeling the channel, and extending the exposition to multiple antennas.

Chapter 3 provides a review of signal processing fundamentals, which are leveraged in subsequent parts of the book. It starts with an introduction to the relevant continuous-time and discrete-time signal notation, along with background on linear time-invariant systems, the impulse response, and convolution. Linear time-invariant systems are used to model multipath wireless channels. The chapter continues with a review of several important concepts related to probability and random processes, including stationarity, ergodicity, and Gaussian random processes. Next, some background is provided on the Fourier transform in both continuous and discrete time, as well as on signal power and bandwidth, as it is useful to view communication signals in both the time and frequency domains. The chapter continues with derivation of the complex baseband signal representation and complex baseband equivalent channel, both of which are used to abstract out the carrier frequency of a communication signal. It then provides a review of some multirate signal processing concepts, which can be developed for efficient digital implementation of pulse shaping. The chapter concludes with background on critical concepts from linear algebra, especially the least squares solution to linear equations.

Chapter 4 introduces the main principles of complex pulse-amplitude modulation. The main features of the modulation are provided, including symbol mapping, constellations, and the modulated signal's bandwidth. Then the most basic impairment of additive white Gaussian noise is introduced. To minimize the effects of additive noise, the optimal pulse-shaping design problem is formulated and solved by Nyquist pulse shapes. Assuming such pulse shapes are used, then the maximum likelihood symbol detector is derived and the probability of symbol error is analyzed. The topics in this chapter form a basic introduction to digital communication using pulse-amplitude modulation with perfect synchronization and only the most basic impairment of additive noise.

Chapter 5 describes other impairments introduced in wireless communication. It starts with an overview of symbol synchronization and frame synchronization for flat-fading channels. This involves knowing when to sample and the location of the beginning of a frame of data. It then proceeds to present a linear time-invariant model for the effects of multiple propagation paths called frequency selectivity. Several mitigation strategies are described, including linear equalization. Because the distortion introduced by the frequency-selective channel varies over time, the chapter then describes approaches for channel estimation. The channel estimate is used to compute the coefficients of the equalizer. Alternative modulation strategies that facilitate equalization are then introduced: single-carrier frequency-domain equalization (SC-FDE) and OFDM. Specific channel estimation and carrier frequency offset correction algorithms are then developed for single-carrier and OFDM systems. Most of the algorithms in this chapter are developed by formulating a linear system and taking the least squares solution. The chapter concludes with an introduction to propagation and fading-channel models. These statistical models are widely used in the design and analysis of wireless systems. A review is provided of both large-scale models, capturing channel variations over hundreds of wavelengths, and small-scale models, incorporating variations over fractions of a wavelength. Ways to quantify frequency selectivity and time selectivity are introduced. The chapter concludes with a description of common small-scale fading-channel models for both flat and frequency-selective channels.

Chapter 6 concludes the book with a concise introduction to MIMO communication. Essentially, the key concepts developed in this book are re-examined, assuming a plurality of transmit and/or receive antennas. Most of the development is built around flat-fading channels, with extensions to frequency selectivity through MIMO-OFDM provided at the end. The chapter starts with an introduction to the different configurations of multiple antennas in SIMO (single input multiple output), MISO (multiple input single output), and MIMO configurations. It then describes the basics of receiver diversity for SIMO systems, including antenna selection and maximum ratio combining, including their impact on the probability of vector symbol error. Next it explains some approaches for extracting diversity in MISO communication systems, including beamforming, limited feedback, and space-time coding. The chapter then introduces the important MIMO technique known as spatial multiplexing. Extensions to precoding, limited feedback, and channel estimation are also described. The chapter concludes with an overview of MIMO-OFDM, which combines MIMO spatial multiplexing with the ease of equalization in OFDM systems. Key ideas like equalization, precoding, channel estimation, and synchronization are revisited in this challenging setting of MIMO with frequency-selective channels. MIMO and MIMO-OFDM are used in many commercial wireless systems.

The concepts developed in this book are ideally suited for practical implementation in software-defined radio. The author has developed a companion laboratory manual [147], which is sold as part of a package with the National Instruments Universal Software Radio Peripheral. That laboratory manual features seven experiments that cover the main topics in Chapter 4 and Chapter 5, along with a bonus experiment that explores the benefits of error control coding. Of course, the concepts can be demonstrated in practice in other ways, even using a speaker as a transmit antenna and a microphone as a receive antenna. Readers are encouraged to simulate algorithms when possible, along with working example and homework problems.

1.6 Symbols and Common Definitions

We use the notation in Table 1.1 and assign specific definitions to the variables in Table 1.2 throughout this book.

Table 1.1 Generic Notation Used in This Book

$*$	Convolution operator
$\mathbf{a}$	Bold lowercase is used to denote column vectors
$\mathbf{A}$	Bold uppercase is used to denote matrices
a, A	Non-bold letters are used to denote scalar values
$\lvert a \rvert$	Magnitude of scalar a
$\lVert \mathbf{a} \rVert$	Vector 2-norm of $\mathbf{a}$
$\lVert \mathbf{A} \rVert_F$	Frobenius norm of $\mathbf{A}$
$\mathcal{A}$	Calligraphic letters denote sets
$\lvert \mathcal{A} \rvert$	Cardinality of set $\mathcal{A}$
$\mathbf{A}^{\mathrm{T}}$	Matrix transpose
$\mathbf{A}^*$	Conjugate transpose
$\mathbf{A}^{\mathrm{c}}$	Conjugate
$\mathbf{A}^{1/2}$	Matrix square root
$\mathbf{A}^{-1}$	Matrix inverse
$\mathbf{A}^{\dagger}$	Moore-Penrose pseudo-inverse
$\mathbf{a}_i$	i^{th} entry of vector $\mathbf{a}$
$[\mathbf{A}]_{i,j}$	Scalar entry of $\mathbf{A}$ in i^{th} row j^{th} column
$[\mathbf{A}]_{:,k}$	k^{th} column of matrix $\mathbf{A}$
$[\mathbf{A}]_{:,k:m}$	Column consisting of rows $k, k+1, \ldots, m$ of matrix $\mathbf{A}$
$(\cdot)$	Used to index a continuous signal
$a(t)$	Continuous scalar signal and value at t
$\mathbf{a}(t)$	Continuous vector signal and value at t
$\mathbf{A}(t)$	Continuous matrix signal and value at t
$[\cdot]$	Used to index a discrete-time signal
$a[n]$	Discrete-time scalar signal and value at n
$\mathbf{a}[n]$	Discrete-time vector signal and value at n
$\mathbf{A}[n]$	Discrete-time matrix signal and value at n
$\mathbf{a}[k]$	Discrete-time vector signal in frequency domain at subcarrier n
$\mathbf{A}[k]$	Discrete-time matrix signal in frequency domain at subcarrier n
$\log$	$\log_{10}$ unless otherwise mentioned
$\mathrm{phase}(\cdot)$	Principle phase of the argument

Table 1.2 Common Definitions Used in This Book

E_{s}	Signal energy
N_{o}	Noise energy
L	Channel order
$\{h[\ell]\}_{\ell=0}^{L}$	Discrete-time ISI (intersymbol interference) channel impulse response with $(L+1)$ taps

continues

Table 1.2 Common Definitions Used in This Book (*continued*)

$h[k] = \sum_{\ell=0}^{L} h[\ell] e^{-j2\pi k\ell/N}$	Frequency-domain channel transfer function
$(\cdot)_{\mathrm{p}}$	Notation used to explicitly denote a passband signal
$y[n]$	Symbol sampled received signal
$s[n]$	Symbols to be transmitted prior to scaling
$x[n]$	Symbol sampled transmitted signal, often a scaled version of $s[n]$
N_{tr}	Number of training symbols
N_{t}	Number of transmit antennas, assumed to be 1 except when dealing with MIMO communication
N_{r}	Number of receive antennas, assumed to be 1 except when dealing with MIMO communication
N_{s}	Number of MIMO data streams, usually equal to N_{t} unless precoding is applied
$\mathbf{y}[n]$	Symbol sampled received signal of dimension $N_{\mathrm{r}} \times 1$
$\mathbf{s}[n]$	Transmit symbol vector of dimension $N_{\mathrm{s}} \times 1$
$\mathbf{x}[n]$	Symbol sampled transmitted signal of dimension $N_{\mathrm{t}} \times 1$, often related to $\mathbf{s}[n]$ through linear precoding
$\mathbf{I}_N$	$N \times N$ identity matrix
$\mathbf{0}_{N,M}$	$N \times M$ all zeros matrix
$\mathbf{1}_{N,M}$	$N \times M$ all ones matrix
j	Imaginary number $\mathrm{j} = \sqrt{-1}$
$\mathbb{E}[\cdot]$	Expectation operator
$x \sim \mathcal{N}\left(m, \sigma^2\right)$	Means that x is a Gaussian random variable with mean m and variance σ^2
$x \sim \mathcal{N}_{\mathbb{C}}\left(m, \sigma^2\right)$	Means that x is a circularly symmetric complex Gaussian random variable with complex mean m, total variance σ^2, the real and imaginary parts of x are independent, and the variance of the real and imaginary parts is $\sigma^2/2$ each
f_{c}	Carrier frequency
c	Speed of light

1.7 Summary

- Wireless communication has a large number of applications, which are different from each other in the propagation environment, transmission range, and underlying technologies.

- Most major wireless communication systems use digital communication. Advantages of digital over analog include its suitability for use with digital data, robustness to noise, ability to more easily support multiple data rates and multiple users, and easier implementation of security.

- Digital signal processing is well matched with digital communication. Digital signal processing makes use of high-quality reproducible digital components. It also leverages Moore's law, which leads to more computation and reduced power consumption and cost.

- This book presents the fundamentals of wireless digital communication as seen through a signal processing lens. It focuses on complex pulse-amplitude modulation and the most common challenges faced when implementing a wireless receiver: additive noise, frequency-selective channels, symbol synchronization, frame synchronization, and carrier frequency offset synchronization.

Problems

1. **Wireless Devices/Networks in Practice**   This problem requires some research on the technical specifications of wireless networks or wireless devices.

    (a) Choose a current cell phone from three of these manufacturers: Nokia, Samsung, Apple, LG, Huawei, Sony, Blackberry, Motorola, or another of your choosing. Describe the wireless and cellular technologies and the frequency bands supported by each one.

    (b) Name at least three mobile service providers in your country. Which cellular technologies are currently supported by the networks?

    (c) Which of those three mobile service providers charge for data and what are the charges for a typical consumer plan (not business)? Why do you think some providers have stopped offering unlimited data plans?

2. **Wireless Device Comparison**   Fill in the following table for three cellular devices manufactured by the three companies in the table:

Manufacturer	Device #1	Device #2	Device #3
Device			
What wireless technologies are supported (i.e., Wi-Fi, cellular, etc.)?			
Which cellular standards are supported?			
Which frequency bands are used for communication?			
What is the maximum (claimed) data rate supported by the device using cellular?			
What is the maximum (claimed) data rate supported by the device using Wi-Fi?			
What operating system is employed?			

3. **Visible Light Communication (VLC)**   Do some research on VLC, an alternative to wireless communication using RF (radio frequency) signals. This is a topic that we do not cover much in the book, but the topics from the book can be used to understand the basic principles. Be sure to cite your sources in the answers below. Note: You should find some trustworthy sources that you can reference (e.g., there may be mistakes in the Wikipedia article or it may be incomplete).

   (a) Which part of the IEEE 802 LAN/MAN Standards Committee deals with VLC?

   (b) What is the concept of VLC?

   (c) What is the bandwidth of a typical VLC application?

   (d) Explain how VLC might be used for secure point-to-point communication.

   (e) Explain how VLC might be used for indoor location-based services.

   (f) Explain why VLC might be preferred on aircraft for multimedia delivery.

   (g) Explain how VLC could be used in intelligent transportation systems.

4. **Sensor Networking**   There are many kinds of wireless networks, such as wireless sensor networks, that have important applications in manufacturing. They are often classified as low-rate wireless personal area networks. This is a topic that we do not cover much in the book, but the topics from the book can be used to understand the basic principles. Be sure to cite your sources in the answers below. Note: You should find some trustworthy sources that you can reference (e.g., there may be mistakes in the Wikipedia article or it may be incomplete).

   (a) What is a wireless sensor network?

   (b) What is IEEE 802.15.4?

   (c) What is ZigBee?

   (d) How are IEEE 802.15.4 and ZigBee related?

   (e) What communication bands are supported by IEEE 802.15.4 in the United States?

   (f) What is the bandwidth of the communication channel specified by IEEE 802.15.4? Note: This is bandwidth in hertz, not the data rate.

   (g) What is the typical range of an IEEE 802.15.4 device?

(h) How long should the battery last in an IEEE 802.15.4 device?

(i) How are sensor networks being used to monitor highway bridges?

5. **Wireless and Intellectual Property**   The wireless industry has been plagued with lawsuits over intellectual property. Identify a recent case of interest and describe the parties and their positions. Then describe in at least half a page your opinion about the role of intellectual property in wireless communications.

Chapter 2

An Overview of Digital Communication

2.1 Introduction to Digital Communication

Communication is the process of conveying information from a transmitter to a receiver, through a channel. The transmitter generates a signal containing the information to be sent. Upon transmission, the signal propagates to the receiver, incurring various types of distortion captured together in what is called the channel. Finally, the receiver observes the distorted signal and attempts to recover the transmitted information. The more side information the receiver has about the transmitted signal or the nature of the distortion introduced by the channel, the better the job it can do at recovering the unknown information.

There are two important categories of communication: analog and digital. In analog communication, the source is a continuous-time signal, for example, a voltage signal corresponding to a person talking. In digital communication, the source is a digital sequence, usually a binary sequence of ones and zeros. A digital sequence may be obtained by sampling a continuous-time signal using an analog-to-digital converter, or it may be generated entirely in the digital domain using a microprocessor. Despite the differences in the type of information, both analog and digital communication systems send continuous-time signals. Analog and digital communication are widely used in commercial wireless systems, but digital communication is replacing analog communication in almost every new application.

Digital communication is becoming the de facto type of communication for a number of reasons. One reason is that it is fundamentally suitable for transmitting digital data, such as data found on every computer, cellular phone, and tablet. But this is not the only reason. Compared to analog systems, digital communication systems offer higher quality, increased security, better robustness to noise, reductions in power usage, and easy integration of different types of sources, for example, voice, text, and video. Because the majority of the components of a digital communication system are implemented digitally using integrated circuits, digital communication devices take advantage of the reductions in cost and size resulting from Moore's law. In fact, the majority of the public switched telephone network, except perhaps the connection from the local exchange to the home, is digital. Digital communication systems are also easy to reconfigure, using the concept of software-defined radio, though most digital communication systems still perform a majority of the processing using application-specific integrated circuits (ASICs) and field programmable gate arrays (FPGAs).

This chapter presents an overview of the main components in a single wireless digital communication link. The purpose is to provide a complete view of the key operations found in a typical wireless communication link. The chapter starts with a review of the canonical transmitter and receiver system block diagram in Section 2.2. A more detailed description of the input, the output, and the functions of each component is provided in the following sections. In Section 2.3, the impairments affecting the transmitted signal are described, including noise and intersymbol interference. Next, in Section 2.4, the principles of source coding and decoding are introduced, including lossless and lossy compression. In Section 2.5, encryption and decryption are reviewed, including secret-key and public-key encryption. Next, in Section 2.6, the key ideas of error control coding are introduced, including a brief description of block codes and convolutional codes. Finally, in Section 2.7, an overview of modulation and demodulation concepts is provided. Most of the remainder of this book will focus on modulation, demodulation, and dealing with impairments caused by the channel.

2.2   Overview of a Wireless Digital Communication Link

The focus of this book is on the point-to-point wireless digital communication link, as illustrated in Figure 2.1. It is divided into three parts: the transmitter, the channel, and the receiver. Each piece consists of multiple functional blocks. The transmitter processes a bit stream of data for transmission over a physical medium. The channel represents the physical medium, which adds noise and distorts the transmitted signal. The receiver attempts to extract the transmitted bit stream from the received signal. Note that the pieces do not necessarily correspond to physical distinctions between the transmitter, the propagation channel, and the receiver. In our mathematical modeling of the channel, for example, we include all the analog and front-end effects because they are included in the total distortion experienced by the transmitted signal. The remainder of this section summarizes the operation of each block.

Now we describe the components of Figure 2.1 in more detail based on the flow of information from the source to the sink. The first transmitter block is devoted to source encoding. The objective of source encoding is to transform the source sequence into an information sequence that uses as few bits per unit time as possible, while still enabling the source output to be retrieved from the encoded sequence of bits. For purposes of explanation, we assume a digital source that creates a binary sequence $\{b[n]\} = \ldots, b[-1], b[0], b[1], \ldots$. A digital source can be obtained from an analog source by sampling. The output of the source encoder is $\{i[n]\}$, the information sequence. Essentially, source encoding compresses the source as much as possible to reduce the number of bits that need to be sent.

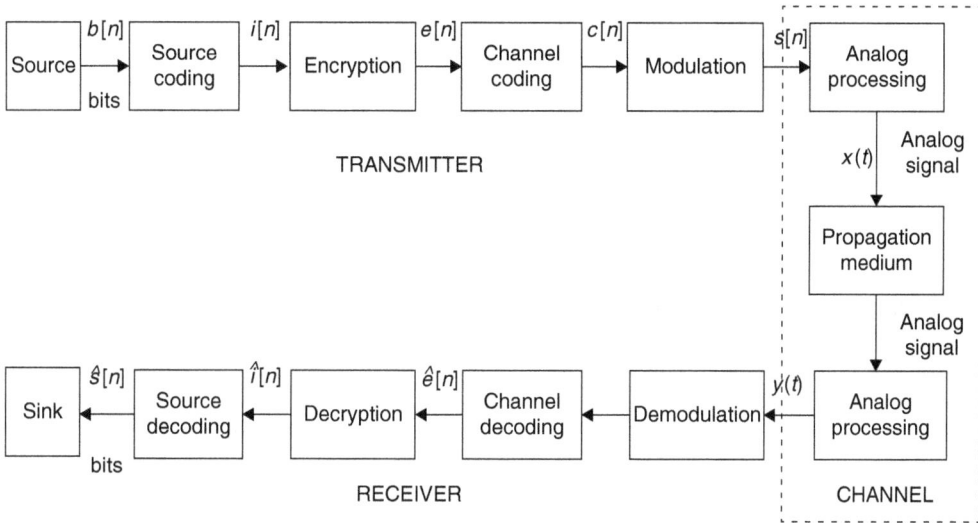

Figure 2.1 The components of a typical digital communication system

Following source encoding is encryption. The purpose of encryption is to scramble the data to make it difficult for an unintended receiver to interpret. Generally encryption involves applying a lossless transformation to the information sequence $\{i[n]\}$ to produce an encrypted sequence $e[n] = p(\{i[n]\})$. Unlike source coding, encryption does not compress the data; rather, it makes the data appear random to an uninformed receiver.

The next block is the channel coder. Channel coding, also called error control coding or forward error correction, adds redundancy to the encrypted sequence $\{e[n]\}$ to create $\{c[n]\}$ in a controlled way, providing resilience to channel distortions and improving overall throughput. Using common coding notation, for every k input bits, or information bits, there is an additional redundancy of r bits. The total number of bits is $m = k + r$; the coding rate is defined as k/m. This redundancy may enable errors to be detected, and even corrected, at the receiver so that the receiver can either discard the data in error or request a retransmission.

Following channel coding, the coded bits $\{c[n]\}$ are mapped to *signals* by the modulator. This is the demarcation point where basic transmitter-side digital signal processing for communication ends. Typically the bits are first mapped in groups to *symbols* $\{s[n]\}$. Following the symbol mapping, the modulator converts the digital symbols into corresponding analog signals $x(t)$ for transmission over the physical link. This can be accomplished by sending the digital signal through a DAC, filtering, and mixing onto a higher-frequency carrier. Symbols are sent at a rate of R_s symbols per second, also known as the baud rate; the symbol period $T_s = 1/R_s$ is the time difference between successive symbols.

The signal generated by the transmitter travels to the receiver through a propagation medium, known as the propagation channel, which could be a radio wave through a wireless environment, a current through a telephone wire, or an optical signal through a fiber. The propagation channel includes electromagnetic propagation effects such as reflection, transmission, diffraction, and scattering.

The first block at the receiver is the analog front end (AFE) which, at least, consists of filters to remove unwanted noise, oscillators for timing, and ADCs to convert the data into the digital regime. There may be additional analog components like analog gain control and automatic frequency control. This is the demarcation point for the beginning of the receiver-side digital signal processing for digital communication.

The channel, as illustrated in Figure 2.1, is the component of the communication system that accounts for all the noise and distortions introduced by the analog processing blocks and the propagation medium. From a modeling perspective, we consider the channel as taking the transmitted signal $x(t)$ and producing the distorted signal $y(t)$. In general, the channel is the part of a digital communication system that is determined by the environment and is not under the control of the system designer.

The first digital communication block at the receiver is the demodulator. The demodulator uses a sampled version of the received signal, and perhaps knowledge of the channel, to infer the transmitted symbol. The process of demodulation may include equalization, sequence detection, or other advanced algorithms to help in combatting channel distortions. The demodulator may produce its best guess of the transmitted bit or symbol (known as hard detection) or may provide only a tentative decision (known as a soft decision).

Following the demodulator is the decoder. Essentially the decoder uses the redundancy introduced by the channel coder to remove errors generated by the demodulation block as a result of noise and distortion in the channel. The decoder may work jointly with the demodulator to improve performance or may simply operate on the output of the demodulator, in the form of hard or soft decisions. Overall, the effect of the demodulator and the decoder is to produce the best possible guess $\widehat{e}[n]$ of the encrypted signal given the observations at the receiver.

After demodulation and detection, decryption is applied to the output of the demodulator. The objective is to descramble the data to make it intelligible to the subsequent receiver blocks. Generally, decryption applies the inverse transformation $p^{-1}(\cdot)$ corresponding to the encryption process to produce the transmitted information sequence $\{\widehat{i}[n]\}$ that corresponds to the encrypted sequence $\{\widehat{e}[n]\}$.

The final block in the diagram is the source decoder, which reinflates the data back to the form in which it was sent: $\widehat{s}[n] = g(\widehat{i}[n])$. This is essentially the inverse operation of the source encoder. After source decoding, the digital data is delivered to higher-level communication protocols that are beyond the scope of this book.

The processes at the receivers are in the reverse direction to the corresponding processes at the transmitter. Specifically, each block at the receiver attempts to reproduce the input of the corresponding block at the transmitter. If such an attempt fails, we say that an error occurs. For reliable communication, it is desirable to reduce errors as much as possible (have a low bit error rate) while still sending as many bits as possible per unit of time (having a high information rate). Of course, this must be done while making sensible trade-offs in algorithm performance and complexity.

This section provided a high-level picture of the operation of a typical wireless digital communication link. It should be considered as one embodiment of a wireless digital communication system. At the receiver, for example, there are often benefits to combining the operations. In the following discussion as well as throughout this book, we assume a point-to-point wireless digital communication system with one transmitter, one channel, and one receiver. Situations with multiple transmitters and receivers are practical, but the digital communication is substantially more complex. Fortunately, the understanding of the point-to-point wireless system is an outstanding foundation for understanding more complex systems.

The next sections in this chapter provide more detail on the functional blocks in Figure 2.1. Since the blocks at the receiver mainly reverse the processing of the corresponding blocks at the transmitter, we first discuss the wireless channel and then the blocks in transmitter/receiver pairs. This treatment provides some concrete examples of the operation of each functional pair along with examples of their practical application.

2.3   Wireless Channel

The wireless channel is the component that mainly distinguishes a digital wireless communication system from other types of digital communication systems. In wired communication systems, the signal is guided from the transmitter to the receiver through a physical connection, such as a copper wire or fiber-optic cable. In a wireless communication system, the transmitter generates a radio signal that propagates through a more open physical medium, such as walls, buildings, and trees in a cellular communication system. Because of the variability of the objects in the propagation environment, the wireless channel changes relatively quickly and is not under the control of wireless communication system designers.

The wireless channel contains all the impairments affecting the signal transmission in a wireless communication system, namely, noise, path loss, shadowing, multipath fading, intersymbol interference, and external interference from other wireless communication systems. These impairments are introduced by the analog processing at both the transmitter and the receiver and by the propagation medium. In this section, we briefly introduce the concepts of these channel effects. More detailed discussions and mathematical models for the effects of the wireless channel are provided in subsequent chapters.

2.3.1   Additive Noise

Noise is the basic impairment present in all communication systems. Essentially, the term *noise* refers to a signal with many fluctuations and is often modeled using a random process. There are many sources of noise and several ways to model the impairment created by noise. In his landmark paper in 1948, Shannon realized for the first time that noise limits the number of bits that can be reliably distinguished in data transmission, thus leading to a fundamental limitation on the performance of communication systems.

The most common type of noise in a wireless communication system is additive noise. Two sources of additive noise, which are almost impossible to eliminate by filters at the AFE, are thermal noise and quantization noise. Thermal noise results from the material properties of the receivers, or more precisely the thermal motion of electrons in the dissipate components like resistors, wires, and so on. Quantization noise is caused by the conversion between analog and digital representations at the DAC and the ADC due to quantization of the signal into a finite number of levels determined by the resolution of the DAC and ADC.

The noise is superimposed on the desired signal and tends to obscure or mask the signal. To illustrate, consider a simple system with only an additive noise impairment. Let $x(t)$ denote the transmitted signal, $y(t)$ the received signal, and $v(t)$ a signal corresponding to the noise. In the presence of additive noise, the received signal is

$$y(t) = x(t) + v(t). \tag{2.1}$$

For the purpose of designing a communication system, the noise signal $v(t)$ is often modeled as a Gaussian random process, which is explained in more detail in Chapter 3. Knowledge of the distribution is important in the design of receiver signal processing operations like detection.

An illustration of the effects of additive noise is provided in Figure 2.2. The performance of a system inflicted with additive noise is usually a function of the ratio between the signal power and the noise power, called the signal-to-noise ratio (SNR). A high SNR usually implies a good communication link, where exact determination of "high" depends on system-specific details.

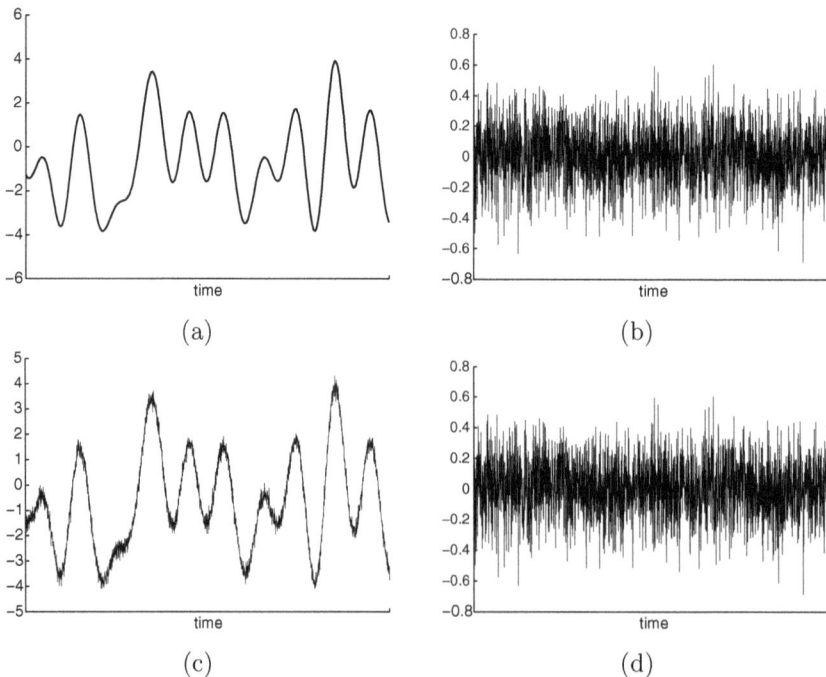

Figure 2.2 (a) An example of a pulse-amplitude modulated communication signal; (b) a realization of a noise process; (c) a noisy received signal with a large SNR; (d) a noisy received signal with a small SNR

2.3.2 Interference

The available frequency spectrum for wireless communication systems is limited and hence scarce. Therefore, users often share the same bandwidth to increase the spectral efficiency, as in cellular communication systems discussed in Chapter 1. This sharing, however, causes co-channel interference between signals of different pairs of users, as illustrated in Figure 2.3. This means that besides its desired signal, a receiver also observes additional additive signals that were intended for other receivers. These signals may be modeled as additive noise or more explicitly modeled as interfering digital communication signals. In many systems, like cellular networks, external interference may be much stronger than noise. As a result, external interference becomes the limiting factor and the system is said to be interference limited, not noise limited.

There are many other sources of interference in a wireless communication system. Adjacent channel interference comes from wireless transmissions in adjacent frequency bands, which are not perfectly bandlimited. It is strongest when a transmitter using another band is proximate to a receiver. Other sources of interference result from non-linearity in the AFE, for example, intermodulation products. The analog parts of the radio are normally designed to meet certain criteria, so that they do not become the performance-limiting factor unless at a very high SNR. This book focuses on additive thermal noise as the main impairment; other types of interference can be incorporated if interference is treated as additional noise.

2.3.3 Path Loss

During the course of propagation from the transmitter to the receiver in the physical medium, the signal power decays (typically exponentially) with the transmission distance. This distance-dependent power loss is called *path loss*. An example path loss function is illustrated in Figure 2.4.

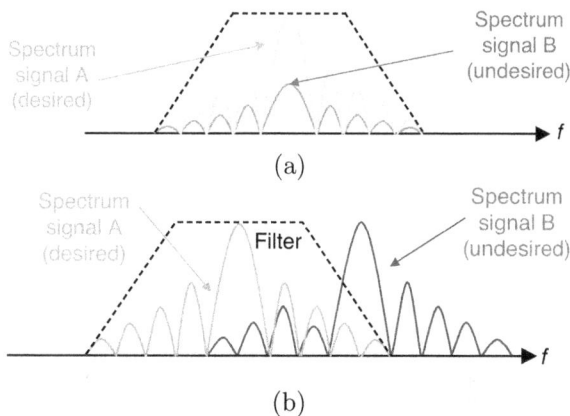

Figure 2.3 Interference in the frequency domain: (a) co-channel interference; (b) adjacent channel interference

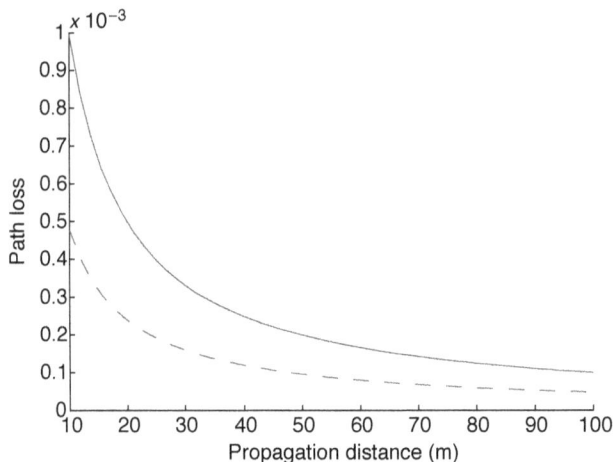

Figure 2.4 Path loss as a function of distance, measured as the ratio of receive power to transmit power, assuming a free-space path-loss model with isotropic antennas. More details about this channel model are found in Chapter 5.

There are a variety of models for path loss. In free space, path loss is proportional to the square of the frequency carrier, meaning that the higher-frequency carrier causes a higher path loss, assuming the receive antenna aperture decreases with antenna frequency. Path loss also depends on how well the transmitter can "see" the receiver. Specifically, the decay rate with distance is larger when the transmitter and receiver are obstructed in what is called non-line-of-sight (NLOS) propagation and is smaller when they experience line-of-sight (LOS) propagation. Several parameters like terrain, foliage, obstructions, and antenna heights are often taken into account in the computation of path loss. The path loss can even incorporate a random component, called shadowing, to account for variability of the received signal due to scattering objects in the environment.

A simple signal processing model for path loss is to scale the received signal by $\sqrt{G}$ where the square root is to remind us that path loss is usually treated as a power. Modifying (2.1), the received signal with path loss and additive noise is

$$y(t) = \sqrt{G}x(t) + v(t). \tag{2.2}$$

Path loss reduces the amplitude of the received signal, leading to a lower SNR.

2.3.4 Multipath Propagation

Radio waves traverse the propagation medium using mechanisms like diffraction, reflection, and scattering. In addition to the power loss described in the previous section, these phenomena cause the signal to propagate to the receiver via multiple paths. When there are multiple paths a signal can take between the transmitter and the receiver, the propagation channel is said to support multipath propagation, which results in *multipath fading*. These paths have different delays (due to different path lengths) and different attenuations, resulting in a smearing of the transmitted signal (similar to the effect of blurring an image).

There are different ways to model the effects of multipath propagation, depending on the bandwidth of the signal. In brief, if the bandwidth of the signal is small relative to a quantity called the coherence bandwidth, then multipath propagation creates an additional multiplicative impairment like the $\sqrt{G}$ in (2.2), but it is modeled by a random process and varies over time. If the bandwidth of the signal is large, then multipath is often modeled as a convolution between the transmitted signal and the impulse response of the multipath channel. These models are described and the choice between models is explored further in Chapter 5.

To illustrate the effects of multipath, consider a propagation channel with two paths. Each path introduces an attenuation α and a time delay τ. Modifying (2.2), the received signal with path loss, additive noise, and two multipaths is

$$y(t) = \sqrt{G}\alpha_1 x(t - \tau_1) + \sqrt{G}\alpha_2 x(t - \tau_2) + v(t). \tag{2.3}$$

The path loss $\sqrt{G}$ is the same for both paths because it is considered to be a locally averaged quantity; additional differences in received power are captured in differences in α_1 and α_2. The smearing caused by multipath creates intersymbol interference, which must be mitigated at the receiver.

Intersymbol interference (ISI) is a form of signal distortion resulting from multipath propagation as well as the distortion introduced by the analog filters. The name *intersymbol interference* means that the channel distortion is high enough that successively transmitted symbols interfere at the receiver; that is, a transmitted symbol arrives at the receiver during the next symbol period. Coping with this impairment is challenging when data is sent at high rates, since in these cases the symbol time is short and even a small delay can cause intersymbol interference. The effects of intersymbol interference are illustrated in Figure 2.5. The presence of ISI is usually analyzed using eye diagrams, as shown in this figure. The eye pattern overlaps many samples of the received signal. The effects of receiving several delayed and attenuated versions of the transmitted signal can be seen in the eye diagram as a reduction in the eye opening and loss of definition. Intersymbol interference can be mitigated by using equalization, as discussed in Chapter 5, or by using joint detection.

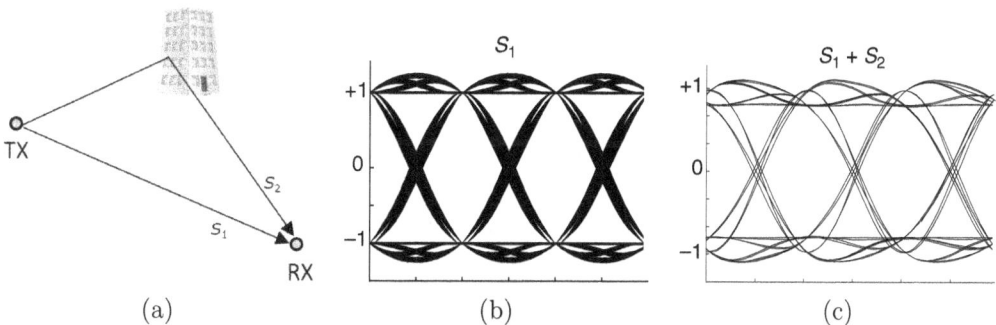

Figure 2.5 (a) Multipath propagation; (b) eye diagram corresponding to single-path propagation; (c) eye diagram corresponding to multipath propagation, showing ISI effects

2.4   Source Coding and Decoding

The main objective of a digital communication system is to transfer information. The source, categorized as either continuous or discrete, is the origin of this information. A continuous source is a signal with values taken from a continuum. Examples of continuous-valued sources include music, speech, and vinyl records. A discrete source has a finite number of possible values, usually a power of 2 and represented using binary. Examples of discrete sources include the 7-bit ASCII characters generated by a computer keyboard, English letters, and computer files. Both continuous and discrete sources may act as the origin of information in a digital communication system.

All sources must be converted into digital form for transmission by a digital communication system. To save the resources required to store and transport the information, as few bits as possible should be generated per unit time, with some distortion possibly allowed in the reconstruction. The objective of source coding is to generate a compressed sequence with as little redundancy as possible. Source decoding, as a result, reconstructs the original source from the compressed sequence, either perfectly or in a reasonable approximation.

Source coding includes both lossy and lossless compression. In lossy compression, some degradation is allowed to reduce the amount of resulting bits that need to be transmitted [35]. In lossless compression, redundancy is removed, but upon inverting the encoding algorithm, the signal is exactly the same [34, 345]. For discrete-time sources—for example, if f and g are the source encoding and decoding processes—then $\widehat{s}[n] = g(i[n]) = g(f(s[n]))$. For lossy compression, $s[n] \cong \widehat{s}[n]$, while for lossless compression, $s[n] = \widehat{s}[n]$. Both lossy encoding and lossless encoding are used in conjunction with digital communication systems.

While source coding and decoding are important topics in signal processing and information theory, they are not a core part of the physical-layer signal processing in a wireless digital communication system. The main reason is that source coding and decoding often happen at the application layer, for example, the compression of video in a Skype call, which is far removed from the physical-layer tasks like error control coding and modulation. Consequently, we provide a review of the principles of source coding only in this section. Specialized topics like joint source-channel coding [78] are beyond the scope of this exposition. In the remainder of this book, we assume that source coding has already been performed. Specifically, we assume that ideal source coding has been accomplished such that the output of the source encoder is an independent identically distributed sequence of binary data.

2.4.1   Lossless Source Coding

Lossless source coding, also known as zero-error source coding, is a means of encoding a source such that it can be perfectly recovered without any errors. Many lossless source codes have been developed. Each source code is composed of a number of codewords. The codewords may have the same number of bits—for example, the ASCII code for computer characters, where each character is encoded by 7 bits. The codewords may also have different numbers of bits in what is called a variable-length code. Morse code is a classic variable-length code in which letters are represented by sequences of dots and dashes of different lengths. Lossless coding is ideal when working with data files where even a 1-bit error can corrupt the file.

The efficiency of a source-coding algorithm is related to a quantity called *entropy*, which is a function of the source's alphabet and probabilistic description. Shannon showed that this source entropy is exactly the minimum number of bits per source symbol required for source encoding such that the corresponding source decoding can retrieve uniquely the original source information from the encoded string of bits. For a discrete source, the entropy corresponds to the average length of the codewords in the best possible lossless data compression algorithm. Consider a source that can take values from an alphabet with m distinct elements, each with probability p_1, p_2, $\cdots$, p_m. Then the entropy is defined as

$$H = -\sum_{i=1}^{m} p_i \log_2(p_i), \tag{2.4}$$

where in the special case of a binary source

$$H = -p_1 \log_2(p_1) - p_2 \log_2(p_2). \tag{2.5}$$

The entropy of a discrete source is maximized when all the symbols are equally likely, that is, $p_i = 1/m$ for $i = 1, 2, \ldots, m$. A more complicated entropy calculation is provided in Example 2.1.

Example 2.1   A discrete source generates an information sequence using the quaternary alphabet $\{a, b, c, d\}$ with probabilities $\mathbb{P}(s = a) = 1/8$, $\mathbb{P}(s = b) = 1/8$, $\mathbb{P}(s = c) = 1/2$, and $\mathbb{P}(s = d) = 1/8$. The entropy of the source is

$$\begin{aligned}
H(s) &= -\left(\frac{1}{8}\log_2\frac{1}{8} + \frac{1}{8}\log_2\frac{1}{8} + \frac{1}{4}\log_2\frac{1}{4} + \frac{1}{2}\log_2\frac{1}{2}\right) \tag{2.6}\\
&= 1.75. \tag{2.7}
\end{aligned}$$

This means that lossless compression algorithms for encoding the source need at least 1.75 bits per alphabet letter.

There are several types of lossless source-coding algorithms. Two widely known in practice are Huffman codes and Lempel-Ziv (ZIP) codes. Huffman coding takes as its input typically a block of bits (often a byte). It then assigns a prefix-free variable-length codeword in such a way that the more likely blocks have smaller lengths. Huffman coding is used in fax transmissions. Lempel-Ziv coding is a type of arithmetic coding where the data is not segmented a priori into blocks but is encoded to create one codeword. Lempel-Ziv encoding is widely used in compression software like LZ77, Gzip, LZW, and UNIX *compress*. An illustration of a simple Huffman code is provided in Example 2.2.

Example 2.2   Consider the discrete source given in Example 2.1. The Huffman coding procedure requires the forming of a tree where the branch ends correspond to the

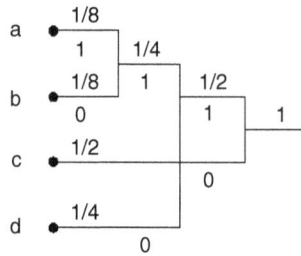

Figure 2.6 Huffman coding tree for the discrete source in Example 2.1

letters in the alphabet. Each branch is assigned a weight equal to the probability of the letter at the branch. Two branches with the lowest weights form a new branch whose weight is equal to the sum of the weights of the two branches. The procedure repeats with the new set of branches until forming the root with the weight of one. Each branch node is now labeled with a binary 1 or 0 decision to distinguish the two branches forming the node. The codeword for each letter is found by tracing the tree path from the root to the branching end where the letter is located. The procedure is illustrated in Figure 2.6.

The codewords for the letters in the alphabet are as follows: 111 for a, 110 for b, 0 for c, and 10 for d. Notice that the letters with a higher probability correspond to the shorter codewords. Moreover, the average length of the codewords is 1.75 bits, exactly equal to the source entropy $H(s)$.

2.4.2 Lossy Source Coding

Lossy source coding is a type of source coding where some information is lost during the encoding process, rendering perfect reconstruction of the source impossible. The purpose of lossy source coding is to reduce the average number of bits per unit time required to represent the source, at the expense of some distortion in the reconstruction. Lossy source coding is useful in cases where the end receiver of the information is human, as humans have some tolerance for distortion.

One of the most common forms of lossy source coding found in digital communication systems is quantization. Since continuous-time signals cannot be transmitted directly in a digital communication system, it is common for the source coding to involve sampling and quantization to generate a discrete-time signal from a continuous-time source. Essentially, the additional steps transform the continuous-time source into a close-to-equivalent digital source, which is then possibly further encoded. Note that these steps are the sources of errors in lossy source coding. In general, it is impossible to reconstruct uniquely and exactly the original analog sources from the resulting discrete source.

The sampling operation is based on the Nyquist sampling theorem, which is discussed in detail in Chapter 3. It says that no information is lost if a bandlimited continuous-time signal is periodically sampled with a small-enough period. The resulting discrete-time signal (with continuous-valued samples) can be used to perfectly reconstruct the bandlimited signal. Sampling, though, only converts the continuous-time signal to a discrete-time signal.

Quantization of the amplitudes of the sampled signals limits the number of possible amplitude levels, thus leading to data compression. For example, a B bit quantizer would represent a continuous-valued sample with one of 2^B possible levels. The levels are chosen to minimize some distortion function, for example, the mean squared error. An illustration of a 3-bit quantizer is provided in Example 2.3.

Example 2.3   Consider a 3-bit uniform quantizer. The dynamic range of the source, where most of the energy is contained, is assumed to be from -1 to 1. The interval $[-1, 1]$ is then divided into $2^3 = 8$ segments of the same length. Assuming the sampled signal is rounded to the closest end of the segment containing it, each sample is represented by the 3-bit codeword corresponding to the upper end of the segment containing the sample as illustrated in Figure 2.7. For example, a signal with a sample value of 0.44 is quantized to 0.375, which is in turn encoded by the codeword 101. The difference between 0.44 and 0.375 is the quantization error. Signals with values greater than 1 are mapped to the maximum value, in this case 0.875. This results in a distortion known as clipping. A similar phenomenon happens for signals with values less than -1.

There are several types of lossy coding beyond quantization. In other applications of lossy coding, the encoder is applied in a binary sequence. For example, the Moving Picture Experts Group (MPEG) defines various compressed audio and video coding standards used for digital video recording, and the Joint Photographic Experts Group (JPEG and JPEG2000) describes lossy still image compression algorithms, widely used in digital cameras [190].

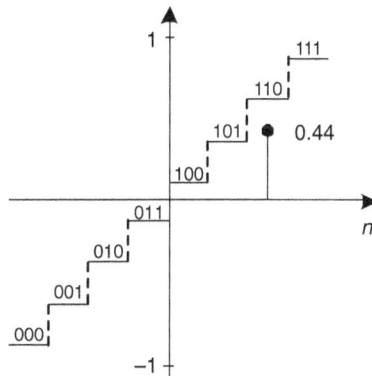

Figure 2.7 An example of pulse-code modulation using 3-bit uniform quantization

2.5 Encryption and Decryption

Keeping the information transmitted by the source understandable by only the intended receivers is a typical requirement for a communication system. The openness of the wireless propagation medium, however, provides no physical boundaries to prevent unauthorized users from getting access to the transmitted messages. If the transmitted messages are not protected, unauthorized users are able not only to extract information from the messages (eavesdropping) but also to inject false information (spoofing). Eavesdropping and spoofing are just some implications of an insecure communication link.

In this section, we focus on encryption, which is one means to provide security against eavesdropping. Using encryption, the bit stream after source encoding is further encoded (encrypted) such that it can be decoded (decrypted) only by an intended receiver. This approach allows sensitive information to travel over a public network without compromising confidentiality.

There are several algorithms for encryption, and correspondingly for decryption. In this section, we focus on *ciphers*, which are widely used in communication. Ciphers work by transforming data using a known algorithm and a *key* or pair of keys. In secret-key encryption, an example of a symmetric cipher, a single secret key is used by both the encryption and the decryption. In public-key encryption, an example of an asymmetric cipher, a pair of keys is used: a public one for encryption and a private one for decryption. In a good encryption algorithm, it is difficult or ideally impossible to determine the secret or private key, even if the eavesdropper has access to the complete encryption algorithm and the public key. Informing the receiver about the secret or private key is a challenge in itself; it may be sent to the receiver by a separate and more secure means. Figure 2.8 illustrates a typical secret-key communication system [303].

Now we describe a typical system. Let k_1 denote the key used for encryption at the transmitter and k_2 the key for decryption at the receiver. The mathematical model described in Section 2.2 for encryption should be revised as $e[n] = p(i[n], k_1)$, where $p(\cdot)$ denotes the cipher. At the receiver, the decrypted data can be written as $\widehat{i}[n] = q(\widehat{e}[n], k_2)$, where $q(\cdot)$ is the corresponding decipher (i.e., the decrypting algorithm). In the case of public-key encryption, k_1 and k_2 are different from each other; k_1 may be known to everyone, but k_2 must be known by only the intended receiver. With secret-key encryption, $k_1 = k_2$ and the key remains unknown to the public. Most wireless communication systems use secret-key encryption because it is less computationally intensive; therefore we focus on secret-key encryption for the remainder of this section.

Secret-key ciphers can be divided into two groups: block ciphers and stream ciphers [297]. Block ciphers divide the input data into nonoverlapping blocks and use the key

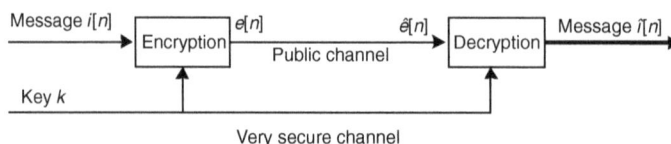

Figure 2.8 A typical secret-key communication system

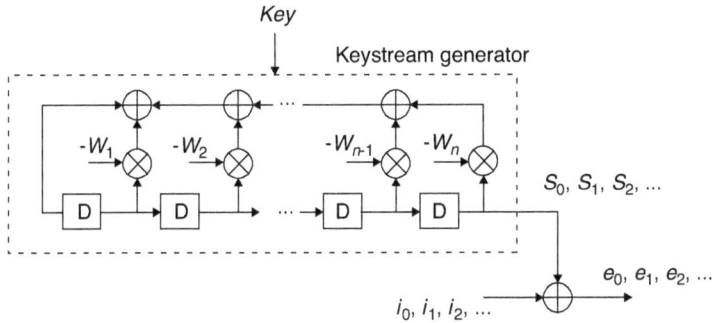

Figure 2.9 Stream ciphering with a linear feedback shift register used to generate the keystream

to encrypt the data block by block to produce blocks of encrypted data of the same length. Stream ciphers generate a stream of pseudo-random key bits, which are then bitwise XORed (i.e., the eXclusive OR [XOR] operation) with the input data. Since stream ciphers can process the data byte after byte, and even bit after bit, in general they are much faster than block ciphers and more suitable to transmission of continuous and time-sensitive data like voice in wireless communication systems [297]. Moreover, a single bit error at the input of the block decrypter can lead to an error in the decryption of other bits in the same block, which is called error propagation. Stream ciphers are widely used in wireless, including GSM, IEEE 802.11, and Bluetooth, though the usage of block ciphers is also increasing with their incorporation into the 3GPP third- and fourth-generation cellular standards.

The major challenge in the design of stream ciphers is in the generation of the pseudo-random key bit sequences, or keystreams. An ideal keystream would be long, to foil brute-force attacks, but would still appear as random as possible. Many stream ciphers rely on linear feedback shift registers (LFSRs) to generate a fixed-length pseudo-random bit sequence as a keystream [297]. In principle, to create an LFSR, a feedback loop is added to a shift register of length n to compute a new term based on the previous n terms, as shown in Figure 2.9. The feedback loop is often represented as a polynomial of degree n. This ensures that the numerical operation in the feedback path is linear. Whenever a bit is needed, all the bits in the shift register are shifted right 1 bit and the LFSR outputs the least significant bit. In theory, an n-bit LFSR can generate a pseudo-random bit sequence of length up to $(2^n - 1)$ bits before repeating. LFSR-based keystreams are widely used in wireless communication, including the A5/1 algorithm in GSM [54], E0 in Bluetooth [44], the RC4 and AES algorithms [297] used in Wi-Fi, and even in the block cipher KASUMI used in 3GPP third- and fourth-generation cellular standards [101].

2.6 Channel Coding and Decoding

Both source decoding and decryption require error-free input data. Unfortunately, noise and other channel impairments introduce errors into the demodulation process. One way of dealing with errors is through channel coding, also called forward error correction or error control coding. Channel coding refers to a class of techniques that aim to improve communication performance by adding redundancy in the transmitted signal to provide resilience to the effects of channel impairments. The basic idea of channel coding is to add some redundancy in a controlled manner to the information bit sequence. This redundancy can be used at the receiver for error detection or error correction.

Detecting the presence of an error gives the receiver an opportunity to request a retransmission, typically performed as part of the higher-layer radio link protocol. It can also be used to inform higher layers that the inputs to the decryption or source decoding blocks had errors, and thus the outputs are also likely to be distorted. Error detection codes are widely used in conjunction with error correction codes, to indicate whether the decoded sequence is highly likely to be free of errors. The cyclic redundancy check (CRC) reviewed in Example 2.4 is the most widely used code for error detection.

Example 2.4   CRC codes operate on blocks of input data. Suppose that the input block length is k and that the output length is n. The length of the CRC, the number of parity bits, is $n - k$ and the rate of the code is k/n. CRC operates by feeding the length k binary sequence into an LFSR that has been initialized to the zero state. The LFSR is parameterized by $n - k$ weights $w_1, w_2, \ldots, w_{n-k}$. The operation of the LFSR is shown in Figure 2.10. The multiplies and adds are performed in binary. The D notation refers to a memory cell. As the binary data is clocked into the LFSR, the data in the memory cells is clocked out. Once the entire block has been input into the LFSR, the bits remaining in the memory cells form the length $n - k$ parity. At the receiver, the block of bits can be fed into a similar LFSR and the resulting parity can be compared with the received parity to see if an error was made. CRC codes are able to detect about $1 - 2^{-(n-k)}$ errors.

Error control codes allow the receiver to reconstruct an error-free bit sequence, if there are not too many errors. In general, a larger number of redundant bits allows the receiver to correct for more errors. Given a fixed amount of communication resources, though, more redundant bits means fewer resources are available for transmission of the desired information bits, which means a lower transmission rate. Therefore, there exists a trade-off between the reliability and the rate of data transmission, which depends

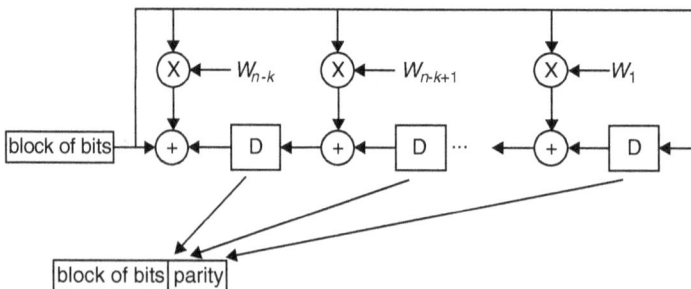

Figure 2.10 The process of generating a CRC using an LFSR

on the available resources and the channel characteristics. This trade-off is illustrated through the simplest error control code, known as repetition coding, in Example 2.5. The example shows how a repetition code, which takes 1 bit and produces 3 output bits (a rate 1/3 code), is capable of correcting one error with a simple majority vote decoder. The repetition code is just one illustration of an error control code. This section reviews other types of codes, with a more sophisticated mathematical structure, that are more efficient than the repetition code.

Example 2.5 Consider a rate 1/3 repetition code. The code operates according to the mapping in Table 2.1. For example, if the bit sequence to be sent is 010, then the encoded sequence is 000111000.

Suppose that the decoder uses the majority decoding rule for decoding. This leads to the decoder mapping in Table 2.2. Based on the decoding table, if there is only one bit error, then the decoder outputs the correctly sent bit. If there are two or three bit errors, then the decoder decides in favor of the incorrectly sent bit and an error is made.

The performance of the code is often measured in terms of the probability of bit error. Suppose that a bit error occurs with probability p and is memoryless. The probability that 2 or 3 bits are in error out of a group of 3 bits is $1-(1-p)^3+3p(1-p)^2$. For $p = 0.01$, the uncoded system sees an error probability of 0.01, which means on average one in 100 bits is in error. The coded system, though, sees an error probability of 0.000298, which means that on average about 3 bits in 10,000 are in error.

Table 2.1 The Encoding Rule for a Rate 1/3 Repetition Code

Input	Output
0	000
1	111

Table 2.2 Majority Decoding for a Rate 1/3 Repetition Code

Input	Output
000	0
001	0
010	0
011	1
100	0
101	1
110	1
111	1

While repetition coding provides error protection, it is not necessarily the most efficient code. This raises an important issue, that is, how to quantify the efficiency of a channel code and how to design codes to achieve good performance. The efficiency of a channel code is usually characterized by its code rate k/n, where k is the number of information bits and n is the sum of the information and redundant bits. A large code rate means there are fewer redundant bits but generally leads to worse error performance. A code is not uniquely specified by its rate as there are many possible codes with the same rate that may possess different mathematical structures. Coding theory is broadly concerned with the design and analysis of error control codes.

Coding is fundamentally related to information theory. In 1948, Shannon proposed the concept of *channel capacity* as the highest information rate that can be transmitted through a channel with an arbitrarily small probability of error, given that power and bandwidth are fixed [302]. The channel capacity can be expressed as a function of a channel's characteristics and the available resources. Let C be the capacity of a channel and R the information rate. The beauty of the channel capacity result is that for any noisy channel there exist channel codes (and decoders) that make it possible to transmit data reliably with the rate $R < C$. In other words, the probability of error can be as small as desired as long as the rate of the code (captured by R) is smaller than the capacity of the channel (measured by C). Furthermore, if $R > C$, there is no channel code that can achieve an arbitrarily small probability of error. The main message is that there is a bound on how large R can be (remember that since $R = k/n$, larger R means smaller $n - k$), which is determined by a property of the channel given by C. Since Shannon's paper, much effort has been spent on designing channel codes that can achieve the limits he set out. We review a few strategies for error control coding here.

Linear block codes are perhaps the oldest form of error control coding, beyond simply repetition coding. They take a block of data and add redundancy to create a longer block. Block codes generally operate on binary data, performing the encoding either in the binary field or in a higher-order finite field. Codes that operate using a higher-order finite field have burst error correction capability since they treat groups of bits as an element of the field. Systematic block codes produce a codeword that contains the original input data plus extra parity information. Nonsystematic codes transform the input data into a new block. The well-known Hamming block code is illustrated in Example 2.6.

Various block codes are used in wireless communication. Fire codes [108], cyclic binary codes capable of burst error correction, are used for signaling and control in GSM [91]. Reed-Solomon codes [272] operate in a higher-order finite field and are typically used for correcting multiple burst errors. They have been widely used in deep-space communication [114] and more recently in WiMAX [244] and IEEE 802.11ad [162].

Example 2.6   In this example we introduce the Hamming (7,4) code and explain its operation. The Hamming (7,4) code encodes blocks of 4 information bits into blocks of

7 bits by adding 3 parity/redundant bits into each block [136]. The code is characterized by the code generator matrix

$$\mathbf{G} = \begin{bmatrix} 1 & 0 & 0 & 0 \\ 0 & 1 & 0 & 0 \\ 0 & 0 & 1 & 0 \\ 0 & 0 & 0 & 1 \\ 1 & 1 & 0 & 1 \\ 1 & 0 & 1 & 1 \\ 0 & 1 & 1 & 1 \end{bmatrix}, \tag{2.8}$$

which takes an input vector of 4 bits $\mathbf{e}$ to produce an output vector of 7 bits $\mathbf{c} = \mathbf{Ge}$, where the multiplication and addition happen in the binary field. Note that it is conventional in coding to describe this operation using row vectors, but to be consistent within this book we explain it using column vectors. The code generator contains the 4×4 identity matrix, meaning that this is a systematic code.

The parity check matrix for a linear block code is a matrix $\mathbf{H}$ such that $\mathbf{H}^{\mathrm{T}}\mathbf{G} = \mathbf{0}$. For the Hamming code

$$\mathbf{H} = \begin{bmatrix} 1 & 1 & 0 \\ 1 & 0 & 1 \\ 0 & 1 & 1 \\ 1 & 1 & 1 \\ 1 & 0 & 0 \\ 0 & 1 & 0 \\ 0 & 0 & 1 \end{bmatrix}. \tag{2.9}$$

Now suppose that the observed data is $\widehat{\mathbf{c}} = \mathbf{c} + \mathbf{v}$ where $\mathbf{v}$ is a binary error sequence (being all zero means there is no error). Then

$$\mathbf{H}^{\mathrm{T}}\widehat{\mathbf{c}} = \mathbf{H}^{\mathrm{T}}\mathbf{v}. \tag{2.10}$$

If $\mathbf{H}^{\mathrm{T}}\widehat{\mathbf{c}}$ is not zero, then the received signal does not pass the parity check. The remaining component is called the syndrome, based on which the channel decoder can look for the correctable error patterns. The Hamming code is able to correct 1-bit errors, and its rate of 4/7 is much more efficient than the rate 1/3 repetition code that performs the same task.

Another major category of channel coding found in wireless communications is convolutional coding. A convolutional code is in general described by a triple (k, n, K), where k and n have the same meaning as in block codes and K is the constraint length defining the number of k-tuple stages stored in the encoding shift register, which is used to perform the convolution. In particular, the n-tuples generated by a (k, n, K) convolutional code are a function not only of the current input k-tuple but also of the previous $(K-1)$ input k-tuples. K is also the number of memory elements needed in the convolutional encoder and plays a significant role in the performance of the code and the resulting decoding complexity. An example of a convolutional encode is shown in Example 2.7.

Convolutional codes can be applied to encode blocks of data. To accomplish this, the memory of the encoder is initialized to a known state. Often it is initialized to zero and the sequence is then zero padded (an extra K zeros are added to the block) so that it ends in the zero state. Alternatively, tail biting can be used where the code is initialized using the end of the block. The choice of initialization is explained in the decoding algorithm.

Example 2.7   In this example we illustrate the encoding operation for a 64-state rate 1/2 convolutional code used in IEEE 802.11a as shown in Figure 2.11. This encoder has a single input bit stream ($k = 1$) and two output bit streams ($n = 2$). The shift register has six memory elements corresponding to $2^6 = 64$ states. The connections of the feedback loops are defined by octal numbers as 133_8 and 171_8, corresponding to the binary representations 001011011 and 001111001. Two sets of output bits are created,

$$c_1[n] = e[n] \oplus e[n-2] \oplus e[n-3] \oplus e[n-5] \oplus e[n-6] \qquad (2.11)$$
$$c_2[n] = e[n] \oplus e[n-1] \oplus e[n-2] \oplus e[n-3] \oplus e[n-6], \qquad (2.12)$$

where $\oplus$ is the XOR operation. The encoded bit sequence is generated by interleaving the outputs, that is, $c[2n] = c_1[n]$ and $c[2n+1] = c_2[n]$.

Decoding a convolutional code is challenging. The optimum approach is to search for the sequence that is closest (in some sense) to the observed error sequence based on some metric, such as maximizing the likelihood. Exploiting the memory of convolutional codes, the decoding can be performed using a forward-backward recursion known as the Viterbi algorithm [348]. The complexity of the algorithm is a function of the number of states, meaning that the error correction capability that results from longer constraint lengths comes at a price of more complexity.

Convolutional coding is performed typically in conjunction with bit interleaving. The reason is that convolutional codes generally have good random error correction capability but are not good with burst errors. They may also be combined with block codes that

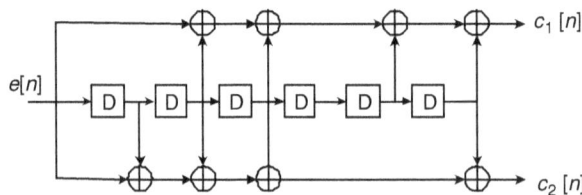

Figure 2.11 The 64-state rate 1/2 convolutional code used in IEEE 802.11a

have good burst error correction capability, like the Reed-Solomon code, in what is known as concatenated coding. Decoding performance can be further improved by using soft information, derived from the demodulation block. When soft information is used, the output of the demodulator $\widehat{c}[n]$ is a number that represents the likelihood that $\widehat{c}[n]$ is a zero or a one, as opposed to a hard decision that is just a binary digit. Most modern wireless systems use bit interleaving and soft decoding.

Convolutional codes with interleaving have been widely used in wireless systems. GSM uses a constraint length $K = 5$ code for encoding data packets. IS-95 uses a constraint length $K = 9$ code for data. IEEE 802.11a/g/n use a constraint length $K = 7$ code as the fundamental channel coding, along with puncturing to achieve different rates. 3GPP uses convolutional codes with a constraint length $K = 9$ code.

Other types of codes are becoming popular for wireless communication, built on the foundations of iterative soft decoding. Turbo codes [36] use a modified recursive convolutional coding structure with interleaving to achieve near-optimal Shannon performance in Gaussian channels, at the expense of long block lengths. They are used in the 3GPP cellular standards. Low-density parity check (LDPC) codes [217, 314] are linear block codes with a special structure that, when coupled with efficient iterative decoders, also achieve near-optimal Shannon performance. They are used in IEEE 802.11ac and IEEE 802.11ad [259]. Finally, there has been interest in the recently developed polar codes [15], which offer options for lower-complexity decoding with good performance, and will be used for control channels in 5G cellular systems.

While channel coding is an important part of any communication system, in this book we focus on the signal processing aspects that pertain to modulation and demodulation. Channel coding is an interesting topic in its own right, already the subject of many textbooks [200, 43, 352]. Fortunately, it is not necessary to become a coding expert to build a wireless radio. Channel encoders and decoders are available in simulation software like MATLAB and LabVIEW and can also be acquired as intellectual property cores when implementing algorithms on an FPGA or ASIC [362, 331].

2.7 Modulation and Demodulation

Binary digits, or bits, are just an abstract concept to describe information. In any communication system, the physically transmitted signals are necessarily analog and continuous time. Digital modulation is the process at the transmit side by which information bit sequences are transformed into signals that can be transmitted over wireless channels. Digital demodulation is then the process at the receive side to extract the information bits from the received signals. This section provides a brief introduction to digital modulation and demodulation. Chapter 4 explores the basis of modulation and demodulation in the presence of additive white Gaussian noise, and Chapter 5 develops extended receiver algorithms for working with impairments.

There are a variety of modulation strategies in digital communication systems. Depending on the type of signal, modulation methods can be classified into two groups: baseband modulation and passband modulation. For baseband modulation, the signals are electrical pulses, but for passband modulation they are built on radio frequency carriers, which are sinusoids.

2.7.1   Baseband Modulation

In baseband modulation, the information bits are represented by electric pulses. We assume the information bits require the same duration of transmission, called a bit time slot. One simple method of baseband modulation is that 1 corresponds to bit time slots with the presence of a pulse and 0 corresponds to bit time slots without any pulse. At the receiver, the determination is made as to the presence or absence of a pulse in each bit time slot. In practice, there are many pulse patterns. In general, to increase the chance of detecting correctly the presence of a pulse, the pulse is made as wide as possible at the expense of a lower bit rate. Alternatively, signals can be described as a sequence of transitions between two voltage levels (bipolar). For example, a higher voltage level corresponds to 1 and a lower voltage level corresponds to 0. Alternate pulse patterns may have advantages including more robust detection or easier synchronization.

Example 2.8   We consider two examples of baseband modulation: the non-return-to-zero (NRZ) code and the Manchester code. In NRZ codes, two nonzero different voltage levels H and L are used to represent the bits 0 and 1. Nevertheless, if the voltage is constant during a bit time slot, then a long sequence of all zeros or all ones can lead to loss of synchronization since there are no transitions in voltage to demarcate bit boundaries. To avoid this issue, the Manchester code uses the pulse with transitions that are twice as fast as the bit rate. Figure 2.12 illustrates the modulated signals for a bit sequence using the NRZ code and the Manchester code.

In principle, a pulse is characterized by its amplitude, position, and duration. These features of a pulse can be modulated by the information bits, leading to the corresponding modulation methods, namely, pulse-amplitude modulation (PAM), pulse-position modulation (PPM), and pulse-duration modulation (PDM). In this book we focus on complex pulse-amplitude modulations, which consist of two steps. In the first step, the bit sequence is converted into sequences of symbols. The set of all possible symbols is called the signal constellation. In the second step, the modulated signals are synthesized based on the sequence of symbols and a given pulse-shaping function. The entire process is illustrated in Figure 2.13 for the simplest case of real symbols. This is the well-known baseband modulation method called M-ary PAM.

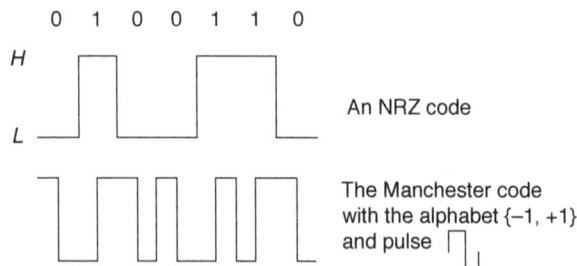

Figure 2.12 Examples of baseband modulation: an NRZ code and the Manchester code

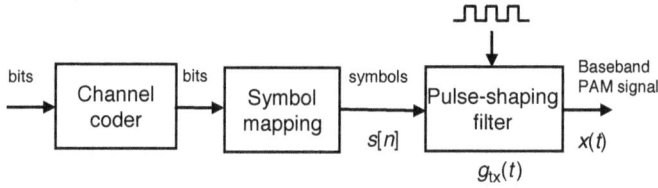

Figure 2.13 Block diagram for generating M-ary PAM signals

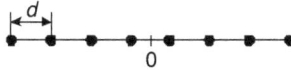

Figure 2.14 An 8-PAM signal constellation. d is the normalization factor.

Example 2.9 A pulse-amplitude modulated received signal at baseband may be written in the binary case as

$$x(t) = \sum_n (-1)^{i[n]} g_{\mathrm{tx}}(t - nT) + v(t), \qquad (2.13)$$

where $i[n]$ is the binary information sequence, $g_{\mathrm{tx}}(t)$ is the transmit pulse shape, T is the symbol duration, and $v(t)$ is additive noise (typically modeled as random). In this example, $i[n] = 0$ becomes a "1" symbol, and $i[n] = 1$ becomes a "-1" symbol.

In M-ary PAM one of M allowable amplitude levels is assigned to each of the M symbols of the signal constellation. A standard M-ary PAM signal constellation $\mathcal{C}$ consists of M d-spaced real numbers located symmetrically around the origin, that is,

$$\mathcal{C}_{\mathrm{PAM}} = \left\{ \frac{-d(M-1)}{2}, \cdots, \frac{-d}{2}, \frac{d}{2}, \cdots, \frac{d(M-1)}{2} \right\}, \qquad (2.14)$$

where d is the normalization factor. Figure 2.14 illustrates the positions of the symbols for an 8-PAM signal constellation. The modulated signal corresponding to the transmission of a sequence of M real symbols using PAM modulation can be written as the summation of uniform time shifts of the pulse-shaping filter multiplied by the corresponding symbol

$$x(t) = \sum_{k=1}^{M} s_k g_{\mathrm{tx}}(t - kT), \qquad (2.15)$$

where T is the interval between the successive symbols and s_k is the kth transmitted symbol, which is selected from $\mathcal{C}_{\mathrm{PAM}}$ based on the information sequence $i[n]$.

The transmitter outputs the modulated signals to the wireless channel. Baseband signals, which are produced by baseband modulation methods, are not efficient because the transmission through the space of the electromagnetic fields corresponding to low baseband frequencies requires very large antennas. Therefore, all practical wireless systems use passband modulation methods.

Example 2.10   Binary phase-shift keying (BPSK) is one of the simplest forms of digital modulation. Let $i[n]$ denote the sequence of input bits, $s[n]$ a sequence of symbols, and $x(t)$ the continuous-time modulated signal. Suppose that bit 0 produces A and bit 1 produces $-A$.

- Create a bit-to-symbol mapping table.

  Answer:

$i[n]$	$s[n]$
0	A
1	$-A$

- BPSK is a type of pulse-amplitude modulation. Let $g_{\mathrm{tx}}(t)$ denote the pulse shape and let

$$x(t) = \sum_{n=-\infty}^{\infty} s[n] g_{\mathrm{tx}}(t - nT) \tag{2.16}$$

denote the modulated baseband signal. Let the pulse $g_{\mathrm{tx}}(t) = \frac{1}{\sqrt{T}}$ for $t \in [0, T]$ and $g_{\mathrm{tx}}(t) = 0$ otherwise. Draw a graph of $x(t)$ for the input $i[0] = 1$, $i[1] = 1$, $i[2] = 0$, $i[3] = 1$, with $A = 3$ and $T = 2$. Illustrate for $t \in [0, 8]$.

Answer: The illustration is found in Figure 2.15.

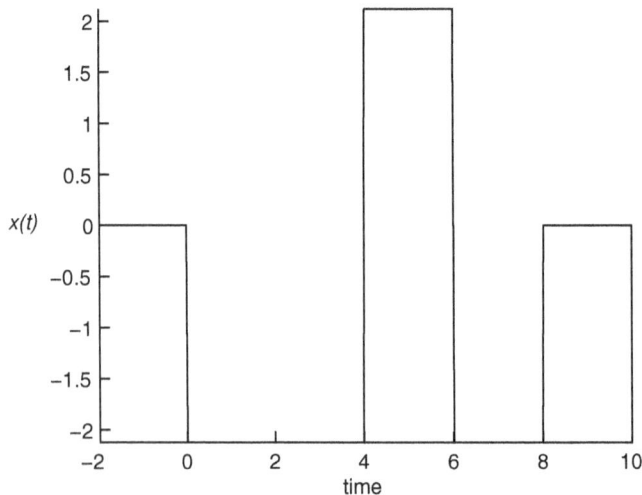

Figure 2.15 Output of a BPSK modulator when the transmit pulse-shaping filter is a rectangular function

2.7.2 Passband Modulation

The signals used in passband modulation are RF carriers, which are sinusoids. Sinusoids are characterized by their amplitude, phase, and frequency. Digital passband modulation can be defined by the process whereby any features, or any combination of features, of an RF carrier are varied in accordance with the information bits to be transmitted. In general, a passband modulated signal $x_{\mathrm{p}}(t)$ can be expressed as follows:

$$x_{\mathrm{p}}(t) = A(t)\cos(2\pi f_{\mathrm{c}}t + \phi(t)), \tag{2.17}$$

where $A(t)$ is the time-varying amplitude, f_{c} is the frequency carrier in hertz, and $\phi(t)$ is the time-varying phase. Several of the most common digital passband modulation types are amplitude-shift keying (ASK), phase-shift keying (PSK), frequency-shift keying (FSK), and quadrature-amplitude modulation (QAM).

In ASK modulation methods, only the amplitude of the carrier is changed according to the transmission bits. The ASK signal constellation is defined by different amplitude levels:

$$\mathcal{C}_{\mathrm{ASK}} = \{A_1, A_2, \cdots, A_M\}. \tag{2.18}$$

Therefore, the resulting modulated signal is given by

$$x_{\mathrm{p}}(t) = A(t)\cos(2\pi f_{\mathrm{c}}t + \phi_0), \tag{2.19}$$

where f_{c} and ϕ_0 are constant; T is the symbol duration time, and

$$A(t) = \sum_{n=-\infty}^{\infty} s[n]g_{\mathrm{tx}}(t - nT),$$

where $s(n)$ is drawn from $\mathcal{C}_{\mathrm{ASK}}$. Figure 2.16 shows the block diagram of a simple ASK modulator.

The PSK modulation methods transmit information by changing only the phase of the carrier while the amplitude is kept constant (we can assume the amplitude is 1). The modulation is done by the phase term as follows:

$$x_{\mathrm{p}}(t) = A\cos(2\pi f_{\mathrm{c}}t + \phi(t)), \tag{2.20}$$

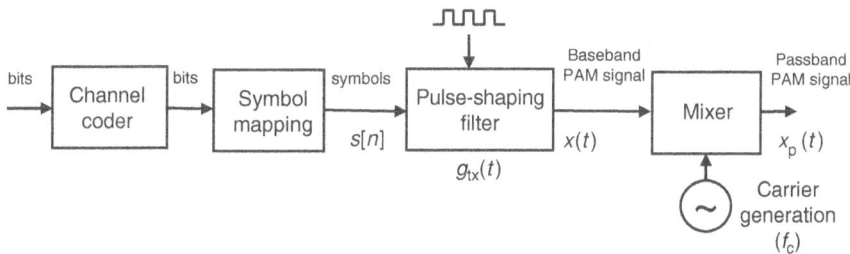

Figure 2.16 Block diagram for generating ASK signals

where $\phi(t)$ is a signal that depends on the sequence of bits. Quadrature PSK (QPSK) and 8-PSK are the best-known PSK modulations. 8-PSK modulation is often used when there is a need for a 3-bit constellation, for example, in the GSM and EDGE (Enhanced Data rates for GSM Evolution) standards.

In FSK modulation, each vector symbol is represented by a distinct frequency carrier f_k for $k = 1, 2, \cdots, M$. For example, it may look like

$$x_\mathrm{p}(t) = \cos(2\pi f_k t), \quad 0 \le t \le T. \tag{2.21}$$

This modulation method thus requires the availability of a number of frequency carriers that can be distinguished from each other.

The most commonly used passband modulation is M-QAM modulation. QAM modulation changes both the amplitude and the phase of the carrier; thus it is a combination of ASK and PSK. We can also think of an M-QAM constellation as a two-dimensional constellation with complex symbols or a Cartesian product of two $M/2$-PAM constellations. QAM modulated signals are often represented in the so-called IQ form as follows:

$$
\begin{aligned}
x_\mathrm{p}(t) &= \sum_{k=-\infty}^{\infty} \mathrm{Re}\{s[k]\} g_\mathrm{tx}(t - kT) \cos(2\pi f_\mathrm{c} t) \\
&\quad - \sum_{k=-\infty}^{\infty} \mathrm{Im}\{s[k]\} g_\mathrm{tx}(t - kT) \sin(2\pi f_\mathrm{c} t), \tag{2.22} \\
&= x_\mathrm{I}(t) \cos(2\pi f_\mathrm{c} t) - x_\mathrm{Q}(t) \sin(2\pi f_\mathrm{c} t), \tag{2.23}
\end{aligned}
$$

where $x_I(t)$ and $x_Q(t)$ are the $M/2$-PAM modulated signals for the inphase and quadrature components. M-QAM is described more extensively in Chapter 4.

Figure 2.17 shows the block diagram of a QAM modulator. M-QAM modulation is used widely in practical wireless systems such as IEEE 802.11 and IEEE 802.16.

Example 2.11   This example considers BPSK with passband modulation, continuing Example 2.10. For BPSK, the passband signal is denoted as

$$x_\mathrm{p}(t) = x(t) \cos(2\pi f_\mathrm{c} t) \tag{2.24}$$

where $x(t)$ is defined in (2.16). BPSK is a simplification of M-QAM with only an inphase component. The process of creating $x_\mathrm{p}(t)$ from $x(t)$ is known as upconversion.

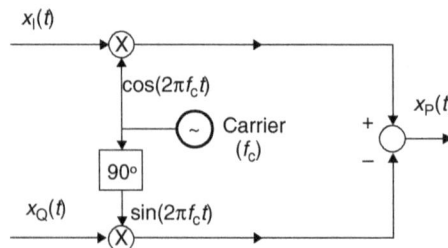

Figure 2.17 Block diagram for generating QAM signals

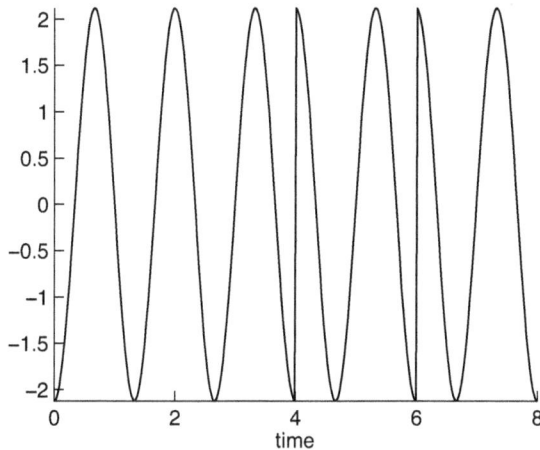

Figure 2.18 Output of a BPSK modulator with a square pulse shape from Example 2.11 when the carrier frequency is very small (just for illustration purposes)

- Plot $x_p(t)$ for $t \in [0,8]$ and interpret the results. Choose a reasonable value of f_c so that it is possible to interpret your plot.

  Answer: Using a small value of f_c shows the qualitative behavior of abrupt changes in Figure 2.18. The signal changes phase at the points in time when the bits change. Essentially, information is encoded in the phase of the cosine, rather than in its amplitude. Note that with more complicated pulse-shaping functions, there would be more amplitude variation.

- The receiver in a wireless system usually operates on the baseband signal. Neglecting noise, suppose that the received signal is the same as the transmit signal. Show how to recover $x(t)$ from $x_p(t)$ by multiplying by $\cos(2\pi f_c t)$ and filtering.

  Answer: After multiplying $x_p(t)$ by $\cos(2\pi f_c t)$,

  $$x_p(t)\cos(2\pi f_c t) = x(t)\cos^2(2\pi f_c t) \tag{2.25}$$

  $$= \frac{x(t)}{2}(1 + \cos(4\pi f_c t)). \tag{2.26}$$

The carrier at $2f_c$ can be removed by a lowpass-filtering (LPF) operation, so that the output of the filter is a scaled version of $x(t)$.

2.7.3 Demodulation with Noise

When the channel adds noise to the passband modulated signal that passes through it, the symbol sequence can be recovered by using a technique known as matched filtering, as shown in Figure 2.19. The receiver filters the received signal $y(t)$ with a filter whose shape is "matched" to the transmitted signal's pulse shape $g_{tx}(t)$. The matched filter limits the amount of noise at the output, but the frequencies containing the data signal

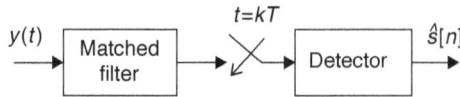

Figure 2.19 The demodulation procedure based on matched filtering

are passed. The output of the matched filter is then sampled at multiples of the symbol period. Finally, a decision stage finds the symbol vector closest to the received sample. This demodulation procedure and the mathematical definition of the matched filter are covered in detail in Chapter 4. In practical receivers this is usually implemented in digital, using sampled versions of the received signal and the filter.

Example 2.12 In this example we consider the baseband BPSK signal from Example 2.10. A simple detector is based on the matched filter, which in this example is the transmit rectangular pulse $g_{\text{tx}}(t)$ in Example 2.10.

- Show that $\int_{kT}^{kT+T} x(t)g_{\text{tx}}(t - kT)\mathrm{d}t = s[k]$ where $x(t)$ is given in (2.16).

  Answer:

$$\int_{kT}^{kT+T} x(t)g_{\text{tx}}(t - kT)\mathrm{d}t = \int_{kT}^{kT+T} \sum_{n=-\infty}^{\infty} s[n]g_{\text{tx}}(t - nT)g_{\text{tx}}(t - kT)\mathrm{d}t \quad (2.27)$$

$$= \sum_{n=-\infty}^{\infty} \int_{kT}^{kT+T} s[n]g_{\text{tx}}(t - nT)g_{\text{tx}}(t - kT)\mathrm{d}t \quad (2.28)$$

$$= \int_{kT}^{kT+T} s[k]g_{\text{tx}}(t - kT)g_{\text{tx}}(t - kT)\mathrm{d}t \quad +$$

$$\sum_{n\neq k} \int_{kT}^{kT+T} s[n]g_{\text{tx}}(t - nT)g_{\text{tx}}(t - kT)\mathrm{d}t \quad (2.29)$$

$$= s[k] \int_{kT}^{kT+T} \mathrm{d}t \quad (2.30)$$

$$= Ts[k] \quad (2.31)$$

  The original signal can be recovered by dividing by T.

- Plot the signal at the output of the matched filter, which is also a rectangular pulse, in $t \in [-2, 10]$, when $x(t)$ is the baseband BPSK signal from Example 2.10 corrupted by noise. Plot the sampled received signal at $t = 1, 3, 5, 7$. Interpret the results.

  Answer: Figure 2.20(a) shows the output of the matched filter for the signal $x(t)$ in Example 2.10 corrupted by additive noise. The samples of the received signal at multiples of the symbol period are represented in Figure 2.20(b) to show the output of the sampling stage. Then, the detector has to find the closest symbol to this received vector. Despite the noise, the detector provides the right symbol sequence $[-1, -1, 1, -1]$.

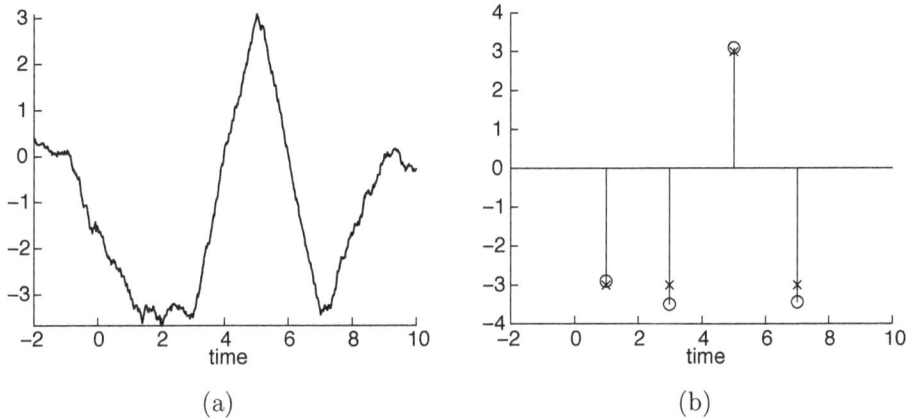

(a)   (b)

Figure 2.20 (a) Output of the matched filter for the signal in Example 2.10 corrupted by noise; (b) output samples of the sampling stage after the matched filter (circles) and transmitted symbols (x-marks)

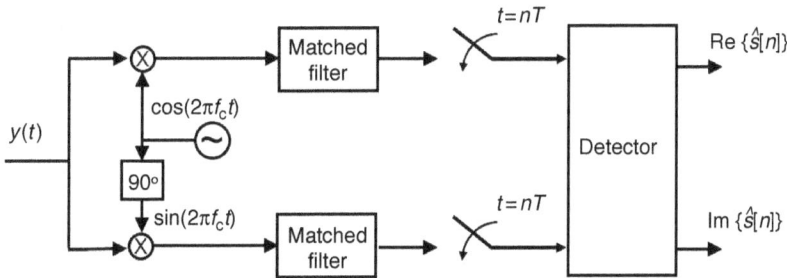

Figure 2.21 Block diagram of an M-QAM demodulator

When demodulating a general passband signal, we need to extract the symbol information from the phase or the amplitude of an RF carrier. For M-QAM, the demodulator operates with two basic demodulators in parallel to extract the real and imaginary parts of the symbols. This demodulator can be implemented as shown in Figure 2.21. The passband received signal $y(t)$ is corrupted by noise as in the baseband case, and the matched filters remove noise from both components.

In practice, however, the channel output signal is not equal to the modulated signal plus a noise term; signal distortion is also present, thus requiring more complicated demodulation methods, which are briefly described in Section 2.7.4 and discussed in detail in Chapter 5.

2.7.4 Demodulation with Channel Impairments

Real wireless channels introduce impairments besides noise in the received signal. Further, other functional blocks have to be added to the previous demodulation structures to recover the transmitted symbol. Moreover, practical realizations of the previous demodulators require the introduction of additional complexity in the receiver.

Consider, for example, the practical implementation of the demodulation scheme in Figure 2.19. It involves sampling at a time instant equal to multiples of the symbol duration, so it is necessary to find somehow the phase of the clock that controls this sampling. This is the so-called symbol timing recovery problem. Chapter 5 explains the main approaches for estimating the optimal sampling phase in systems employing complex PAM signals.

In a practical passband receiver some additional problems appear. The RF carriers used at the modulator are not exactly known at the receiver in Figure 2.21. For example, the frequency generated by the local oscillator at the receiver may be slightly different than the transmit frequency, and the phase of the carriers will also be different. Transmission through the wireless channel may also introduce a frequency offset in the carrier frequency due to the Doppler effect. Figure 2.22 shows the effect of a phase offset in a QPSK constellation. Because of these variations, a practical M-QAM receiver must include carrier phase and frequency estimation algorithms to perform the demodulation. Chapter 5 describes the main approaches for phase and frequency offset correction for single-carrier and OFDM systems.

Example 2.13   Consider BPSK with passband modulation, continuing Example 2.11. Suppose that during the downconversion process, the receiver multiplies by $\cos(2\pi(f_c + \epsilon)t)$, where $\epsilon \neq 0$ represents the frequency offset. Show how the received signal, in the absence of noise, is distorted.

Answer:

$$x_r(t) = \text{LPF}\{x_p(t)\cos(2\pi(f_c + \epsilon)t)\} \tag{2.32}$$

$$= \text{LPF}\{x(t)\cos(2\pi f_c t)\cos(2\pi(f_c + \epsilon)t)\} \tag{2.33}$$

$$= \text{LPF}\{x(t)\frac{1}{2}(\cos(2\pi f_c t - 2\pi(f_c + \epsilon)t) + \cos(2\pi f_c t + 2\pi(f_c + \epsilon)t))\} \tag{2.34}$$

$$= \text{LPF}\{x(t)\frac{1}{2}(\cos(2\pi\epsilon t) + \cos(2\pi(2f_c + \epsilon)t))\} \tag{2.35}$$

$$= x(t)\frac{1}{2}\cos(2\pi\epsilon t). \tag{2.36}$$

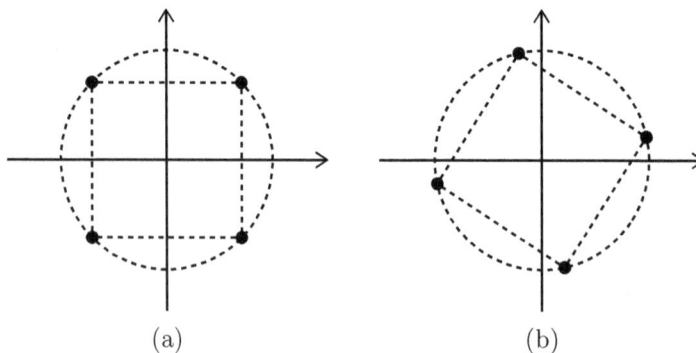

(a)   (b)

Figure 2.22 (a) QPSK constellation; (b) effect of phase offset in a QPSK constellation

Because of the term $\cos(2\pi\epsilon t)$ in the demodulated signal, reconstruction by simple filtering no longer works, even for small values of ϵ.

As described in Section 2.3, the wireless channel also introduces multipath propagation. The input of the demodulator is not the modulated signal $x(t)$ plus a noise term, but it includes a distorted version of $x(t)$ because of multipath. This filtering effect introduced by the channel has to be compensated at the receiver using an equalizer. Chapter 5 introduces the basis for the main channel equalization techniques used in current digital communication systems.

2.8 Summary

- Both analog and digital communication send information using continuous-time signals.

- Digital communication is well suited for digital signal processing.

- Source coding removes the redundancy in the information source, reducing the number of bits that need to be transmitted. Source decoding reconstructs the uncompressed sequence either perfectly or with some loss.

- Encryption scrambles data so that it can be decrypted by the intended receiver, but not by an eavesdropper.

- Channel coding introduces redundancy that the channel decoder can exploit to reduce the effects of errors introduced in the channel.

- Shannon's capacity theorem gives an upper bound on the rate that can be supported by a channel with an arbitrarily low probability of error.

- Significant impairments in a wireless communication system include additive noise, path loss, interference, and multipath propagation.

- Physically transmitted signals are necessarily analog. Digital modulation is the process by which bit sequences containing the information are transformed into continuous signals that can be transmitted over the channel.

- Demodulating a signal consists of extracting the information bits from the received waveform.

- Sophisticated signal processing algorithms are used at the receiver side to compensate for the channel impairments in the received waveform.

Problems

1. **Short Answer**

    (a) Why should encryption be performed after source coding?

(b) Why are digital communication systems implemented in digital instead of completely in analog?

2. Many wireless communication systems support adaptive modulation and coding, where the code rate and the modulation order (the number of bits per symbol) are adapted over time. Periodic puncturing of a convolutional code is one way to adapt the rate. Do some research and explain how puncturing works at the transmitter and how it changes the receiver processing.

3. A discrete source generates an information sequence using the ternary alphabet $\{a, b, c\}$ with probabilities $\mathbb{P}(s = a) = 1/4$, $\mathbb{P}(s = b) = 1/3$, and $\mathbb{P}(s = c) = 5/12$. How many bits per alphabet letter are needed for source encoding?

4. Look up the stream ciphering used in the original GSM standard. Draw a block diagram that corresponds to the cipher operation and also give a linear feedback shift register implementation.

5. Determine the CRC code used in the WCDMA (Wideband Code Division Multiple Access) cellular system, the first release of the third-generation cellular standard.

   (a) Find the length of the code.

   (b) Give a representation of the coefficients of the CRC code.

   (c) Determine the error detection capability of the code.

6. **Linear Block Code**   Error control coding is used in all digital communication systems. Because of space constraints, we do not cover error control codes in detail in this book. Linear block codes refer to a class of parity-check codes that independently transform blocks of k information bits into longer blocks of m bits. This problem explores two simple examples of such codes.

   Consider a Hamming (6,3) code (6 is the codeword length, and 3 is the message length in bits) with the following generator matrix:

$$\mathbf{G} = \begin{bmatrix} 1 & 0 & 0 \\ 0 & 1 & 0 \\ 0 & 0 & 1 \\ 1 & 1 & 0 \\ 0 & 1 & 1 \\ 1 & 0 & 1 \end{bmatrix}. \tag{2.37}$$

   (a) Is this a systematic code?

   (b) List all the codewords for this (6,3) code. In other words, for every possible binary input of length 3, list the output. Represent your answer in table form.

   (c) Determine the minimum Hamming distance between any two codewords in the code. The Hamming distance is the number of bit locations that differ between two locations. A code can correct c errors if the Hamming distance between any two codewords is at least $2c + 1$.

(d) How many errors can this code correct?

(e) Without any other knowledge of probability or statistics, explain a reasonable way to correct errors. In other words, there is no need for any mathematical analysis here.

(f) Using the (6,3) code and your reasonable method, determine the number of bits you can correct.

(g) Compare your method and the rate 1/3 repetition code in terms of efficiency.

7. **HDTV**   Do some research on the ATSC HDTV broadcast standard. Be sure to cite your sources in the answers. Note: You should find some trustworthy sources that you can reference (e.g., there may be mistakes in the Wikipedia article or it may be incomplete).

(a) Determine the types of source coding used for ATSC HDTV transmission. Create a list of different source coding algorithms, rates, and resolutions supported.

(b) Determine the types of channel coding (error control coding) used in the standard. Provide the names of the different codes and their parameters, and classify them as block code, convolution code, trellis code, turbo code, or LDPC code.

(c) List the types of modulation used for ATSC HDTV.

8. **DVB-H**   Do some research on the DVB-H digital broadcast standard for handheld devices. Be sure to cite your sources in the answers. Note: You should find some trustworthy sources that you can reference (e.g., there may be mistakes in the Wikipedia article or it may be incomplete).

(a) Determine the types of source coding used for DVB-H transmission. Create a list of different source coding algorithms, rates, and resolutions supported.

(b) Determine the types of channel coding (error control coding) used in the standard. Provide the names of the different codes and their parameters, and classify them as block code, convolution code, trellis code, turbo code, or LDPC code.

(c) List the types of modulation used for DVB-H.

(d) Broadly speaking, what is the relationship between DVB and DVB-H?

9. On-off keying (OOK) is another form of digital modulation. In this problem, you will illustrate the operation of OOK. Let $i[n]$ denote the sequence of input bits, $s[n]$ a sequence of symbols, and $x(t)$ the continuous-time modulated signal.

(a) Suppose that bit 0 produces 0 and bit 1 produces A. Fill in the bit mapping table.

$i[n]$	$s[n]$

(b) OOK is a type of amplitude modulation. Let $g(t)$ denote the pulse shape and let

$$x(t) = \sum_n s[n]g(t - nT) \tag{2.38}$$

denote the modulated baseband signal. Let $g(t)$ be a triangular pulse shape function defined as

$$g(t) = \begin{cases} 1 - 2\frac{|t - \frac{T}{2}|}{T}, & \text{if } t \in [0, T], \\ 0, & \text{otherwise.} \end{cases} \tag{2.39}$$

Draw a graph of $x(t)$ for the input $i[0] = 1$, $i[1] = 1$, $i[2] = 0$, $i[3] = 1$, with $A = 4$ and $T = 2$. You can illustrate for $t \in [0, 8]$. Please draw this by hand.

(c) Wireless communication systems use passbands signals. Let the carrier frequency $f_c = 1\text{GHz}$. For OOK, the passband signal is

$$x_p(t) = \cos(2\pi f_c t)x(t). \tag{2.40}$$

Plot $x(t)$ for the same range as the previous plot. Be careful! You might want to plot using a computer program, such as MATLAB or LabVIEW. Explain how the information is encoded in this case versus the previous case. The process of going from $x(t)$ to $x_p(t)$ is known as upconversion. Illustrate the qualitative behavior with a smaller value of f_c.

(d) The receiver in a wireless system usually operates on the baseband signal. Neglect to noise, suppose that the received signal is the same as the transmit signal. Show how to recover $x(t)$ from $x_p(t)$ by multiplying $x_p(t)$ by $\cos(2\pi f_c t)$ and filtering.

(e) What happens if you multiply instead by $\cos(2\pi(f_c + \epsilon)t)$ where $\epsilon \neq 0$? Can you still recover $x(t)$?

(f) A simple detector is the matched filter. Show that $\int_{kT}^{kT+T} x(t)g(t - kT)dt = \frac{T}{3}s[k]$.

(g) What happens if you compute $\int_{kT+\tau}^{kT+T+\tau} x(t)g(t - kT - \tau)dt$, where $\tau \in (0, T)$?

10. M-ary PAM is a well-known baseband modulation method in which one of M allowable amplitude levels is assigned to each of the M vector symbols of the

signal constellation. A standard 4-PAM signal constellation consists of four real numbers located symmetrically around the origin, that is,

$$\mathcal{C}_{4\text{-PAM}} = \left\{ -\frac{3}{2}, -\frac{1}{2}, \frac{1}{2}, \frac{3}{2} \right\}. \tag{2.41}$$

Since there are four possible levels, one symbol corresponds to 2 bits of information. Let $i[n]$ denote the sequence of input bits, $s[n]$ a sequence of symbols, and $x(t)$ the continuous-time modulated signal.

(a) Suppose we want to map the information bits to constellation symbols by their corresponding numeric ordering. For example, among the four possible binary sequences of length 2, 00 is the smallest, and among the constellation symbols, $-\frac{3}{2}$ is the smallest, so we map 00 to $-\frac{3}{2}$. Following this pattern, complete the mapping table below:

$i[n]$	$s[n]$
00	$-\frac{3}{2}$

(b) Let $g(t)$ denote the pulse shape and let

$$x(t) = \sum_n s[n]g(t - nT) \tag{2.42}$$

denote the modulated baseband signal. Let $g(t)$ be a triangular pulse shape function defined as

$$g(t) = \begin{cases} 1 - 2\frac{|t - \frac{T}{2}|}{T}, & \text{if } t \in [0, T], \\ 0, & \text{otherwise.} \end{cases} \tag{2.43}$$

By hand, draw a graph of $x(t)$ for the input bit sequence 11 10 00 01 11 01, with $T = 2$. You can illustrate for $t \in [0, 12]$.

(c) In wireless communication systems, we send passband signals. The concept of passband signals is covered in Chapter 3. Let the carrier frequency be f_c; then the passband signal is given by

$$x_p(t) = \cos(2\pi f_c t)x(t). \tag{2.44}$$

For cellular systems, the range of f_c is between 800MHz and around 2GHz. For illustration purposes here, use a very small value, $f_c = 2$Hz. You might want to plot using a computer program such as MATLAB or LabVIEW. Explain how the information is encoded in this case versus the previous case.

(d) The receiver in a wireless system usually operates on the baseband signal. At this point we neglect noise and suppose that the received signal is the same as the transmit signal. Show how to recover $x(t)$ from $x_\mathrm{p}(t)$ by multiplying $x_\mathrm{p}(t)$ by $\cos(2\pi f_c t)$ and filtering. The process of recovering a baseband signal from a passband signal is known as downconversion.

(e) What happens if you multiply instead by $\cos(2\pi(f_\mathrm{c} + \epsilon)t)$, where $\epsilon \neq 0$? Can you still recover $x(t)$? This is called carrier frequency offset and is covered in more detail in Chapter 5.

(f) A simple detector is the matched filter. Show that $\int_{kT}^{kT+T} x(t)g(t - kT)\mathrm{d}t = \frac{T}{3}s[k]$.

(g) What happens if $\int_{kT+\tau}^{kT+T+\tau} x(t)g(t - kT - \tau)\mathrm{d}t$, where $\tau \in (0, T)$? This is called symbol timing offset and is covered in more detail in Chapter 5.

Chapter 3

Signal Processing Fundamentals

A key feature of this book is the signal processing approach to wireless communications. This chapter reviews the fundamentals of signal processing, providing critical mathematical background that is used in subsequent chapters. We start with a review of concepts from signals and systems, including continuous-time and discrete-time signals, linear time-invariant systems, Fourier transforms, bandwidth, the sampling theorem, and discrete-time processing of continuous-time signals. These basic signal concepts are used extensively in this book. Then we explain concepts from statistical signal processing, providing background on probability, random processes, wide-sense stationarity, ergodicity, power spectrum, filtered random processes, and extensions to multivariate processes. The exposition focuses on just the main items from statistical signal processing that are needed in later chapters. Next we introduce passband signals and the associated concepts of the complex baseband equivalent, the complex envelope, and the complex baseband equivalent system. This leads to an important discrete-time abstraction of wireless communication channels. Then we provide some mathematical background on two additional topics. We review key results on multirate signal processing, which are used in subsequent chapters to implement pulse shaping in discrete time. Finally, we provide some background on estimation, including linear least squares, maximum likelihood, and minimum mean squared error estimators. We make extensive use of linear least squares to solve problems related to channel estimation and equalization.

3.1 Signals and Systems

In this section we introduce key concepts from signals and systems, which are essential for a signal processing approach to wireless communication. First we introduce continuous-time and discrete-time signal notations. Then we review linear time-invariant systems, which are used to model the effects of filtering and multipath in the wireless channel. Next we summarize the continuous-time, discrete-time, and discrete Fourier transforms. There are many cases where looking at the frequency domain is useful, including to determine the bandwidth of a signal or finding an easy way to equalize a channel. We use the Fourier transform to define different notions of bandwidth. Next we review the Nyquist sampling theorem, which forms the fundamental connection between bandlimited continuous-time and discrete-time signals. Finally, we use the sampling theorem to develop a discrete-time linear time-invariant equivalent of a continuous-time linear time-invariant system. This last result is used to develop a discrete-time equivalent of the continuous-time communication system model, leading to the wide use of digital signal processing in communications.

3.1.1   Types of Signals and Notation

At different points in the digital communication system there are different kinds of signals. We take some time here to describe some of these signals and also to introduce notation that is used in the rest of the book. A simple taxonomy of signals is based on the discreteness of a signal:

- $x(t)$: A continuous-time signal where t is a real variable that indicates time, and $x(t)$ is the value of the signal at that time

- $x[n]$: A discrete-time signal where n is an integer that indicates discrete time, and $x[n]$ is the value of the signal at that discrete time

- $x_Q[n]$: A digital signal where n is an integer that indicates discrete time, and $x_Q[n]$ is the quantized value of the signal at that discrete time

Quantization maps a continuous variable to one of a finite number of levels. These signal levels can be represented using binary notation from which we get the notion of digital information. Figure 3.1 shows examples of all these types of signals.

In all the cases mentioned above, the signal may be real or complex. A complex signal can be thought of as the sum of two real signals, for example, $x[n] = a[n] + jb[n]$. The signals transmitted in any communication system are always real. Complex signals, though, are used to more efficiently represent passband signals, resulting in more compact expressions for the signal and channel. Complex discrete-time signals are used to represent bandlimited signals by their sampled equivalents, allowing the ready application of DSP.

From a mathematical perspective, this book deals primarily with discrete-time signals. In most DSP, FPGA, or ASIC implementations, however, the signals are actually digital. This requires designing special algorithms to account for the format in which the quantized signals are stored and manipulated, which is important but beyond the scope of this book [273, Section 7.3.3].

3.1.2   Linear Time-Invariant Systems

Many of the challenges in wireless communications are a result of temporal dispersion of the transmitted signal due to multipath propagation in the wireless channel and the frequency selectivity of analog filters in the analog front end. Multipath propagation results from the plurality of paths between transmitter and receiver resulting from reflection, penetration, diffraction, and scattering. Depending on the carrier frequency and the bandwidth of the signal, the signals on these paths can arrive with different delays, attenuations, and phase shifts, resulting in dispersive distortion of the transmitted signal.

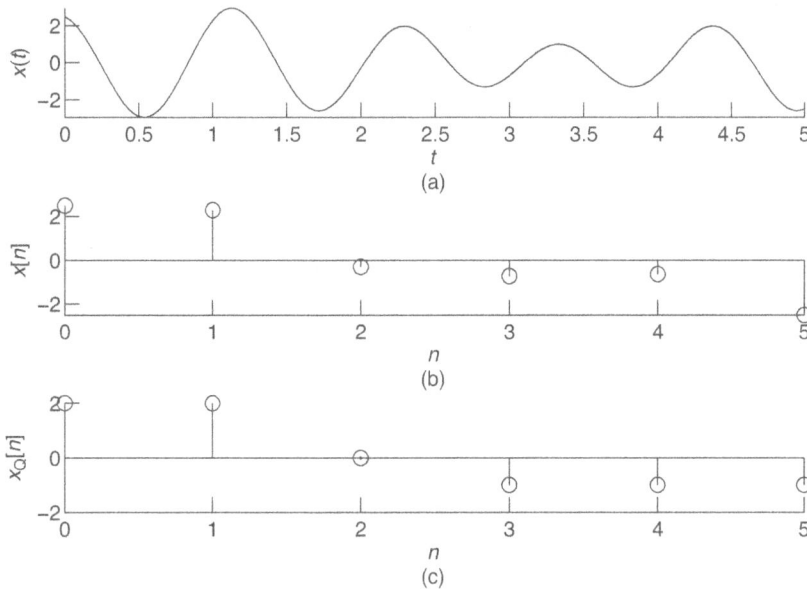

Figure 3.1 Different types of signals according to their discreteness: (a) continuous-time signal; (b) discrete-time signal; (c) digital signal quantized to four levels of $[-1, 0, 1, 2]$ corresponding to 2 bits of quantization

Analog filters are widely used in the analog front end in both the transmitter and the receiver, to help meet the transmit spectral mask, correct for imperfect digital-to-analog conversion, remove intermodulation products, and eliminate noise and interference from other bands. Digital filters may be used to supplement analog filtering. These aggregate effects are known collectively as the channel.

Linear time-varying systems are good models for wireless systems. Linearity is reasonable to describe far-field propagation. The model is time varying because sources of multipath can change over time because of the motion of the transmitter, receiver, or objects in the environment. Given the linear time-varying assumption, and in the absence of noise, the received continuous-time signal $y(t)$ is related to the transmitted continuous-time signal $x(t)$ through a superposition integral between the time-varying impulse response $h_c(t, \tau)$ via

$$y(t) = \int_{-\infty}^{\infty} h_c(t, \tau) x(\tau) d\tau. \tag{3.1}$$

The time-varying impulse response $h_c(t, \tau)$ captures all the effects of multipath and filtering in the transmission path as well as the variation of the transmission path over time.

Most wireless communication systems are designed so that the time variation of the channel can be neglected in the design of signal processing algorithms, at least during a short interval of time. The main reason is that dealing with time-varying channels $h_c(t, \tau)$ is challenging for the receiver: they can be difficult to estimate, track, and equalize.

Consequently, most systems are designed so that information is sent in short bursts, which are shorter than the coherence time of the channel (a measure of how fast the channel varies, as discussed further in Chapter 5). Therefore, for purposes of building a mathematical model, wireless communication channels are modeled as linear time-invariant (LTI) systems, which is known as the *time-invariance assumption*. All the models considered in this book are accurate for linear time-invariant systems where the channel is approximately constant over a period long enough that the channel can be estimated and compensated at the receiver.

Under the LTI assumption, the channel is characterized by its impulse response $h_c(t)$. For an LTI system, the input and output in (3.1) are related through the well-known convolution integral

$$y(t) = \int_{-\infty}^{\infty} h_c(\tau)x(t-\tau)\mathrm{d}\tau. \tag{3.2}$$

This gives a mathematically tractable relationship between the input and output, which has been widely studied [249, 186]. In discrete time, the convolution integral becomes a sum. Given input $x[n]$, output $y[n]$, and an LTI system with impulse response $h[n]$, the input and output are related as

$$y[n] = \sum_{\ell=-\infty}^{\infty} h[\ell]x[n-\ell]. \tag{3.3}$$

Discrete-time LTI systems are used to model the entire communication system, by virtue of the bandlimited property of complex baseband signals (see Section 3.3), the sampling theorem (see Section 3.1.5), and an equivalence between continuous-time and discrete-time processing of signals (see Section 3.1.6).

3.1.3 The Fourier Transform

The frequency domain is incredibly useful when engineering wireless communication systems. For example, our entire notion of bandwidth is built from understanding signals and systems in the frequency domain. It is also natural to look into the frequency domain, since the convolution becomes multiplication. Real-world signals can be viewed in the frequency domain through a spectrum analyzer.

In this section, we review some relevant Fourier transform relationships, including transforms of both continuous-time and discrete-time signals. Tables of transforms of common signals and properties are provided for each case. We also provide several examples involving both the direct calculation of transforms and applications of their properties.

First, we consider the continuous-time Fourier transform (CTFT) of a well-behaved signal.[1] A continuous-time function $x(t)$ and its Fourier transform $\mathsf{x}(f)$ are related by the synthesis and analysis equations,

---

1. There are several conditions for a Fourier transform to exist: that the integral of $|f(t)|$ from $-\infty$ to ∞ exists, there are a finite number of discontinuities in $f(t)$, and $f(t)$ has bounded variations [50]. All physically realizable signals have Fourier transforms. Discussion of more technical aspects, such as Lipschitz conditions, is beyond the scope of this book.

$$\text{Analysis} \quad \mathsf{x}(f) = \int_{-\infty}^{\infty} x(t)e^{-\mathsf{j}2\pi ft}\mathrm{d}t \tag{3.4}$$

$$\text{Synthesis} \quad x(t) = \int_{-\infty}^{\infty} \mathsf{x}(f)e^{\mathsf{j}2\pi ft}\mathrm{d}f. \tag{3.5}$$

We often use the shorthand $\mathsf{x}(f) = \mathcal{F}\{x(t)\}$ and $x(t) = \mathcal{F}^{-1}\{\mathsf{x}(f)\}$.

While the synthesis and analysis equations may be applied directly, it is convenient to leverage the common properties of the Fourier transform in Table 3.1 and the basic

Table 3.1 Continuous-Time Fourier Transform Relationships

Property	Aperiodic Signal	Fourier Transform
	$x(t)$	$\mathsf{x}(f)$
	$y(t)$	$\mathsf{y}(f)$
Linearity	$ax(t) + by(t)$	$a\mathsf{x}(f) + b\mathsf{y}(f)$
Time shifting	$x(t - t_0)$	$e^{-\mathsf{j}2\pi f_c t_0}\,\mathsf{x}(f)$
Frequency shifting	$e^{\mathsf{j}2\pi f_0 t}x(t)$	$\mathsf{x}(f - f_0)$
Conjugation	$x^*(t)$	$\mathsf{x}^*(-f)$
Time reversal	$x(-t)$	$\mathsf{x}(-f)$
Time scaling	$x(at)$	$\frac{1}{|a|}X\left(\frac{f}{a}\right)$
Convolution	$x(t) * y(t) = \int_{-\infty}^{\infty} x(\tau)y(t - \tau)\mathrm{d}\tau$	$\mathsf{x}(f)\mathsf{y}(f)$
Autocorrelation	$x(t) * x^*(-t)$	$|\mathsf{x}(f)|^2$
Multiplication	$x(t)y(t)$	$\mathsf{x}(f) * \mathsf{y}(f) = \int_{-\infty}^{\infty} X(\theta)Y(f - \theta)\mathrm{d}\theta$
Differentiation in time	$\frac{d^n x(t)}{dt^n}$	$(\mathsf{j}2\pi f)^n\,\mathsf{x}(f)$
Integration	$\int_{-\infty}^{t} x(\tau)\mathrm{d}\tau$	$\frac{1}{\mathsf{j}2\pi f}\mathsf{x}(f) + \frac{1}{2}X(0)\delta(f)$
Differentiation in frequency	$t^n x(t)$	$\left(\frac{\mathsf{j}}{2\pi}\right)^n \frac{d^n X(f)}{df^n}$
Modulation (1)	$x(t)e^{\mathsf{j}2\pi f_0 t}$	$\mathsf{x}(f - f_0)$
Modulation (2)	$x(t)\cos(2\pi f_0 t)$	$\frac{1}{2}\left[\mathsf{x}(f - f_0) + X(f + f_0)\right]$
Modulation (3)	$x(t)\sin(2\pi f_0 t)$	$\frac{1}{\mathsf{j}2}\left[\mathsf{x}(f - f_0) - X(f + f_0)\right]$
Conjugate symmetry for real signals	$x(t)$ is real	$\begin{cases} \mathsf{x}(f) = \mathsf{x}^*(-f) \\ \mathrm{Re}\{\mathsf{x}(f)\} = \mathrm{Re}\{\mathsf{x}(-f)\} \\ \mathrm{Im}\{\mathsf{x}(f)\} = -\mathrm{Im}\{\mathsf{x}(-f)\} \\ |\mathsf{x}(f)| = |\mathsf{x}(-f)| \\ \angle\mathsf{x}(f) = -\angle\mathsf{x}(-f) \end{cases}$
Symmetry for real and even signals	$x(t)$ real and even	$\mathsf{x}(f)$ real and even
Symmetry for real and odd signals	$x(t)$ real and odd	$\mathsf{x}(f)$ purely imaginary and odd
Even-odd decomposition for real signals	$x_e(t) = \mathrm{Ev}\{x(t)\} \quad [x(t)\text{real}]$ $x_o(t) = \mathrm{Od}\{x(t)\} \quad [x(t)\text{real}]$	$\mathrm{Re}\{\mathsf{x}(f)\}$ $\mathsf{j}\mathrm{Im}\{\mathsf{x}(f)\}$
Duality	$x(t) \longleftrightarrow \mathsf{x}(f)$	$\mathsf{x}(t) \longleftrightarrow x(-f)$
Parseval's theorem	$\int_{-\infty}^{\infty} |x(t)|^2\,\mathrm{d}t = \int_{-\infty}^{\infty} |\mathsf{x}(f)|^2\,\mathrm{d}f$ $\int_{-\infty}^{\infty} x(t)y^*(t)\mathrm{d}t = \int_{-\infty}^{\infty} \mathsf{x}(f)\mathsf{y}^*(f)\mathrm{d}f$	
$(x(t)$ and $y(t)$ real$)$	$\int_{-\infty}^{\infty} x(t)y(-t)\mathrm{d}t = \int_{-\infty}^{\infty} \mathsf{x}(f)\mathsf{y}(f)\mathrm{d}f$	

Table 3.2 Continuous-Time Fourier Transform Pairs

Function Name	Time-Domain Signal $x(t)$	Frequency-Domain Signal $\mathsf{x}(f)$
Impulse	$\delta(t)$	1
DC	1	$\delta(f)$
Complex exponential	$e^{j2\pi f_0 t}$	$\delta(f - f_0)$
Cosine	$\cos(2\pi f_0 t + \theta)$	$\frac{1}{2}\left[e^{j\theta}\delta(f - f_0) + e^{-j\theta}\delta(f + f_0)\right]$
Sine	$\sin(2\pi f_0 t + \theta)$	$\frac{1}{2j}\left[e^{j\theta}\delta(f - f_0) - e^{-j\theta}\delta(f + f_0)\right]$
Unit step	$u(t) = \begin{cases} 1 & t \geqslant 0 \\ 0 & t < 0 \end{cases}$	$\frac{1}{2}\delta(f) + \frac{1}{j2\pi f}$
Sign	$\text{sgn}(t) = \begin{cases} 1 & t \geqslant 0 \\ -1 & t < 0 \end{cases}$	$\frac{1}{j\pi f}$
Impulse train	$\text{III}(t/T) = \sum\limits_{k=-\infty}^{\infty} \delta(t - kT)$	$\frac{1}{T}\sum\limits_{n=-\infty}^{\infty} \delta(f - \frac{n}{T})$
Fourier series	$\sum\limits_{k=-\infty}^{\infty} a_k e^{j2\pi f_0 kt}$ where $a_k = \frac{1}{T}\int_T x(t)e^{-j2\pi f_0 kt}\text{d}t$	$\sum\limits_{n=-\infty}^{\infty} a_n\delta(f - nf_0)$
Rectangle pulse	$\text{rect}\left(\frac{t}{T}\right) = \begin{cases} 1 & \|t\| \leqslant \frac{T}{2} \\ 0 & \text{elsewhere} \end{cases}$	$T\text{sinc}\,(fT) = \frac{\sin(\pi fT)}{\pi f}$
Triangle pulse	$\Lambda\left(\frac{t}{W}\right) = \begin{cases} 1 - \frac{\|t\|}{W} & \|t\| \leqslant W \\ 0 & \text{elsewhere} \end{cases}$	$W\text{sinc}^2\,(fW)$
Sinc pulse	$\text{sinc}\,(Wt) = \frac{\sin(\pi Wt)}{\pi Wt}$	$\frac{1}{W}\text{rect}\left(\frac{f}{W}\right)$
Sinc^2 pulse	$\text{sinc}^2\,(Wt)$	$\frac{1}{W}\Lambda\left(\frac{f}{W}\right)$
Exponential pulse	$e^{-a\|t\|}$ with $a > 0$	$\frac{2a}{a^2+(2\pi f)^2}$
Decaying exponential	$e^{-at}u(t)$ with Re $\{a\} > 0$	$\frac{1}{a+j2\pi f}$
	$te^{-at}u(t)$ with Re $\{a\} > 0$	$\frac{1}{(a+j2\pi f)^2}$
	$\frac{t^{n-1}}{(n-1)!}e^{-at}u(t)$ with Re $\{a\} > 0$	$\frac{1}{(a+j2\pi f)^n}$
Linear decaying	$\frac{1}{t}$	$-j\pi\,\text{sgn}(f)$

Fourier transforms in Table 3.2. An example of the direct application of the analysis equation is provided in Example 3.1. The application of transform pairs and properties is shown in Example 3.2 and Example 3.3.

Example 3.1   Compute the Fourier transform of

$$\text{rect}\left(\frac{t}{T}\right) = \begin{cases} 1 & |t| \leq \frac{T}{2} \\ 0 & \text{elsewhere} \end{cases} \tag{3.6}$$

using the analysis equation to verify the result in Table 3.2.

Answer: Through direct computation,

$$\mathcal{F}\left\{\text{rect}\left(\frac{t}{T}\right)\right\} = \int_{-\infty}^{\infty} \text{rect}\left(\frac{t}{T}\right)e^{-j2\pi ft}\text{d}t \tag{3.7}$$

$$= \int_{-T/2}^{T/2} e^{-j2\pi ft}\text{d}t \tag{3.8}$$

$$= \left[\frac{-1}{\mathrm{j}2\pi f} e^{-\mathrm{j}2\pi ft} \right]_{-T/2}^{T/2} \tag{3.9}$$

$$= \frac{e^{\mathrm{j}2\pi f \frac{T}{2}} - e^{-\mathrm{j}2\pi f \frac{T}{2}}}{\mathrm{j}2\pi f} \tag{3.10}$$

$$= \frac{\sin(\pi fT)}{\pi f} \tag{3.11}$$

Example 3.2 Compute the Fourier transform of

$$\Lambda \left(\frac{t}{T} \right) = \begin{cases} 1 - \frac{|t|}{T} & |t| \leq T \\ 0 & \text{elsewhere} \end{cases} \tag{3.12}$$

using the fact that $\Lambda(\frac{t}{T}) = \frac{1}{T}\mathrm{rect}(\frac{t}{T}) * \mathrm{rect}(\frac{t}{T})$.

Answer: The Fourier transform is

$$\mathcal{F} \left\{ \Lambda \left(\frac{t}{T} \right) \right\} = \mathcal{F} \left\{ \frac{1}{T}\mathrm{rect} \left(\frac{t}{T} \right) * \mathrm{rect} \left(\frac{t}{T} \right) \right\} \tag{3.13}$$

$$= \frac{1}{T}\mathcal{F} \left\{ \mathrm{rect} \left(\frac{t}{T} \right) \right\} \mathcal{F} \left\{ \mathrm{rect} \left(\frac{t}{T} \right) \right\} \tag{3.14}$$

$$= \frac{\sin^2(\pi fT)}{(\pi f)^2 T}. \tag{3.15}$$

Example 3.3 Compute the inverse CTFT of $\mathsf{x}(2f)\cos(2\pi ft_0)$, where t_0 is constant.

Answer: Exploit the properties

$$\mathcal{F}^{-1}\left\{ \mathsf{x}(2f) \right\} = \frac{1}{2}x(t/2) \tag{3.16}$$

and

$$\mathcal{F}^{-1}\left\{ \mathsf{y}(f)\cos(2\pi ft_0) \right\} = \frac{1}{2}\left[\mathcal{F}^{-1}\left\{ \mathsf{y}(f)e^{\mathrm{j}2\pi ft_0} \right\} + \mathcal{F}^{-1}\left\{ \mathsf{y}(f)e^{-\mathrm{j}2\pi ft_0} \right\} \right] \tag{3.17}$$

$$= \frac{1}{2}\left[y(t - t_0) + y(t + t_0) \right]. \tag{3.18}$$

This gives

$$\mathcal{F}^{-1}\left\{ \mathsf{x}(2f)\cos(2\pi ft_0) \right\} = \frac{1}{4}\left[x((t - t_0)/2) + x((t + t_0)/2) \right]. \tag{3.19}$$

The CTFT's analysis and synthesis equations are strictly defined only for well-behaved functions. Without going into details, this applies to functions where the integrals in (3.5) and (3.4) exist. Periodic functions do not fall under the umbrella of well-behaved functions since they contain infinite energy. This means that an infinite integral over a periodic function is unbounded. Unfortunately, we send information on RF carriers, which are sines and cosines; thus periodic functions are a part of our wireless life. A way around this is to recall the definition of the Fourier series and to use our poorly defined friend the Dirac delta function.

Consider a periodic signal $x(t)$. The period $T > 0$ is the smallest real number such that $x(t) = x(t + T)$ for all t. The continuous-time Fourier series (CTFS) of a (well-behaved) periodic signal $x(t)$ can be defined as

$$\text{Analysis} \quad \mathsf{x}[n] = \frac{1}{T}\int_0^T x(t)e^{-j\frac{2\pi}{T}nt}\mathrm{d}t \tag{3.20}$$

$$\text{Synthesis} \quad x(t) = \sum_{n=-\infty}^{\infty} \mathsf{x}[n]e^{j\frac{2\pi}{T}nt}. \tag{3.21}$$

Note that the CTFS creates as an output a weighting of the fundamental frequency of the signal $\exp(j2\pi/T)$ and its harmonics. Using the CTFS, we can *define* the Fourier transform of a periodic signal as

$$\text{Analysis} \quad \mathsf{x}(f) = \sum_{n=-\infty}^{\infty} \mathsf{x}[n]\delta\left(f - \frac{1}{T}n\right) \tag{3.22}$$

$$\text{Synthesis} \quad x(t) = \int_f \mathsf{x}(f)e^{j2\pi ft}\mathrm{d}f. \tag{3.23}$$

We use the notion of "define" here because the δ function is a generalized function and not a proper function. But the intuition from (3.22) is accurate. In a spectrum analyzer, the Fourier spectrum of a periodic signal has a comblike structure.

A example application of the application of the CTFS to compute the Fourier transform of a periodic signal is provided in Example 3.4. Generalization of this example gives the transforms for the sine and cosine in Table 3.2.

Example 3.4   Compute the Fourier transform of $x(t) = \exp(j2\pi f_c t)$.
Answer: Note that the period of this signal is $T = 1/f_c$. From (3.20):

$$\mathsf{x}[n] = \frac{1}{T}\int_0^T e^{j2\pi f_c t}e^{-j\frac{2\pi}{T}nt}\mathrm{d}t \tag{3.24}$$

$$= \frac{1}{T}\int_0^T e^{j\frac{2\pi}{T}t(1-n)}\mathrm{d}t \tag{3.25}$$

$$= \delta[n - 1]. \tag{3.26}$$

Then from (3.22):

$$\mathsf{x}(t) = \sum_n \mathsf{x}[n]\delta\left(f - \frac{1}{T}n\right) \tag{3.27}$$

$$= \sum_n \delta[n-1]\delta\left(f - \frac{1}{T}n\right) \tag{3.28}$$

$$= \delta\left(f - \frac{1}{T}\right). \tag{3.29}$$

Note in (3.28) we use both the Kronecker delta $\delta(n)$ and the Dirac delta $\delta(f)$.

Because most of the signal processing in a communication system is performed in discrete time, we also need to use the discrete-time Fourier transform (DTFT). The DTFT analysis and synthesis pair (for well-behaved signals) is

$$\text{Analysis} \quad \mathsf{x}(e^{\mathrm{j}2\pi f}) = \sum_{n=-\infty}^{\infty} x[n]e^{-\mathrm{j}2\pi fn} \tag{3.30}$$

$$\text{Synthesis} \quad x[n] = \int_{-1/2}^{1/2} \mathsf{x}(e^{\mathrm{j}2\pi fn})e^{\mathrm{j}2\pi fn}\mathrm{d}f. \tag{3.31}$$

The DTFT is also often written in radians using $\omega = 2\pi f$, in other books. The DTFT $\mathsf{x}(e^{\mathrm{j}2\pi fn})$ is periodic in f. This implies that the DTFT is unique only on any finite interval of unit length, typically $f \in (-1/2, 1/2]$ or $f \in [0,1)$.

Transform properties are illustrated in Table 3.3, and relevant transforms are presented in Table 3.4.

Example 3.5   The frequency response of an ideal lowpass filter with cutoff frequency f_c is given by

$$h(e^{\mathrm{j}2\pi f}) = \begin{cases} 1 & |f| < f_c \\ 0 & f_c < f \le 1/2. \end{cases} \tag{3.32}$$

Determine its discrete-time impulse response.

Answer: Applying (3.31):

$$h[n] = \int_{-1/2}^{1/2} h(e^{\mathrm{j}2\pi f})e^{\mathrm{j}2\pi fn}\mathrm{d}f \tag{3.33}$$

$$= \int_{-f_c}^{f_c} e^{\mathrm{j}2\pi fn}\mathrm{d}f \tag{3.34}$$

$$= \left[\frac{1}{\mathrm{j}2\pi n}e^{\mathrm{j}2\pi fn}\right]_{-f_c}^{f_c} \tag{3.35}$$

$$= \frac{\sin(2\pi f_c n)}{\pi n}. \tag{3.36}$$

Note that since $h[n]$ is not zero for $n < 0$, the filter is noncausal.

Table 3.3 Discrete-Time Fourier Transform Relationships

Property	Sequence	Fourier Transform
	$x[n]$	$\mathsf{x}(e^{\mathrm{j}2\pi f})$
	$y[n]$	$\mathsf{y}(e^{\mathrm{j}2\pi f})$
Linearity	$ax[n]+by[n]$	$a\mathsf{x}(e^{\mathrm{j}2\pi f})+b\mathsf{y}(e^{\mathrm{j}2\pi f})$
Time shifting	$x[n-n_0]$	$e^{-\mathrm{j}2\pi f n_0}\mathsf{x}(e^{\mathrm{j}2\pi f})$
Frequency shifting	$e^{\mathrm{j}2\pi f_0 n}x[n]$	$\mathsf{x}(e^{\mathrm{j}2\pi(f-f_0)})$
Conjugation	$x^*[n]$	$\mathsf{x}^*(e^{-\mathrm{j}2\pi f})$
Time reversal	$x[-n]$	$\mathsf{x}(e^{-\mathrm{j}2\pi f})$
Conjugate symmetry	$x[n]$ real	$\mathsf{x}(e^{\mathrm{j}2\pi f})=\mathsf{x}^*(e^{-\mathrm{j}2\pi f})$
Even symmetry	$x[n]=x[-n]$	$\mathsf{x}(e^{\mathrm{j}2\pi f})=\mathsf{x}(e^{-\mathrm{j}2\pi f})$
Odd symmetry	$x[n]=-x[-n]$	$\mathsf{x}(e^{\mathrm{j}2\pi f})=-\mathsf{x}(e^{-\mathrm{j}2\pi f})$
Convolution	$x[n]*y[n]=\sum_{m=-\infty}^{\infty}x[m]y[n-m]$	$\mathsf{x}(e^{\mathrm{j}2\pi f})\mathsf{y}(e^{\mathrm{j}2\pi f})$
Autocorrelation	$x[n]*x^*[-n]$	$\lvert\mathsf{x}(e^{\mathrm{j}2\pi f})\rvert^2$
Multiplication	$x[n]y[n]$	$\int_{-1/2}^{1/2}\mathsf{x}(e^{\mathrm{j}2\pi\theta})Y(e^{\mathrm{j}2\pi(f-\theta)})\mathrm{d}\theta$
Multiplication by n	$nx[n]$	$\frac{1}{-\mathrm{j}2\pi}\frac{d}{df}\mathsf{x}(e^{\mathrm{j}2\pi f})$
Sum	$\sum_{n=-\infty}^{\infty}x[n]$	$\mathsf{x}(e^{\mathrm{j}2\pi 0})$
Value at origin	$x[0]$	$\int_{-1/2}^{1/2}\mathsf{x}(e^{\mathrm{j}2\pi f})\mathrm{d}f$
Modulation (1)	$x[n]e^{\mathrm{j}2\pi f_0 n}$	$\mathsf{x}(e^{\mathrm{j}2\pi(f-f_0)})$
Modulation (2)	$x[n]\cos(2\pi f_0 n)$	$\frac{1}{2}\left[\mathsf{x}(e^{\mathrm{j}2\pi(f-f_0)})+\mathsf{x}(e^{\mathrm{j}2\pi(f+f_0)})\right]$
Modulation (3)	$x[n]\sin(2\pi f_0 n)$	$\frac{1}{\mathrm{j}2}\left[\mathsf{x}(e^{\mathrm{j}2\pi(f-f_0)})-\mathsf{x}(e^{\mathrm{j}2\pi(f+f_0)})\right]$
Parseval's theorem	$\sum_{n=-\infty}^{\infty}\lvert x[n]\rvert^2=\int_{-1/2}^{1/2}\lvert\mathsf{x}(e^{\mathrm{j}2\pi f})\rvert^2\,\mathrm{d}f$	

Table 3.4 Discrete-Time Fourier Transform Pairs

Function Name	Sequence	Fourier Transform
Impulse	$\delta[n]$	1
	$\delta[n-n_0]$	$e^{-\mathrm{j}2\pi n_0 f}$
DC	$1\ (-\infty<n<\infty)$	$\sum_{k=-\infty}^{\infty}\delta(f-k)$
Complex exponential	$e^{\mathrm{j}2\pi f_0 n}$	$\sum_{k=-\infty}^{\infty}\delta(f-f_0-k)$
Cosine	$\cos(2\pi f_0 n+\theta)$	$\sum_{k=-\infty}^{\infty}\frac{1}{2}\left[e^{\mathrm{j}\theta}\delta(f-f_0-k)\right.$ $\left.+e^{-\mathrm{j}\theta}\delta(f+f_0-k)\right]$
Sine	$\sin(2\pi f_0 n+\theta)$	$\sum_{k=-\infty}^{\infty}\frac{1}{2\mathrm{j}}\left[e^{\mathrm{j}\theta}\delta(f-f_0-k)\right.$ $\left.-e^{-\mathrm{j}\theta}\delta(f+f_0-k)\right]$
Unit step	$u[n]$	$\frac{1}{1-e^{-\mathrm{j}2\pi f}}+\frac{1}{2}\sum_{k=-\infty}^{\infty}\delta(f-k)$
	$a^n u[n]$	$\frac{1}{1-ae^{-\mathrm{j}2\pi f}}$
	$(n+1)a^n u[n]$	$\frac{1}{(1-ae^{-\mathrm{j}2\pi f})^2}$
Window	$\mathrm{rect}\left(\frac{n}{M}\right)=\begin{cases}1 & \lvert n\rvert\leqslant M\\ 0 & \lvert n\rvert>M\end{cases}$	$\frac{\sin\left(2\pi(n+\frac{1}{2})f\right)}{\sin(\pi f)}$
Sinc	$2f_c\mathrm{sinc}\,(2f_c n)$	$\sum_{k=-\infty}^{\infty}\mathrm{rect}\left(\frac{f-k}{2f_c}\right)$
	$\frac{r^n\sin[2\pi f_{\mathrm{p}}(n+1)]}{\sin 2\pi f_{\mathrm{p}}}$ with $\lvert f_{\mathrm{p}}\rvert<1$	$\frac{1}{1-2r\cos 2\pi f_{\mathrm{p}}e^{-\mathrm{j}2\pi f}+r^2 e^{-\mathrm{j}4\pi f}}$

Example 3.6   Compute the DTFT of $x[n] \cos(2\pi f_1 n) \cos(2\pi f_2 n)$, where f_1 and f_2 are constant.

Answer: Using the properties in Table 3.3,

$$\mathcal{F}\left\{x[n] \cos(2\pi f_1 n) \cos(2\pi f_2 n)\right\} =$$

$$\frac{1}{2}\left[\mathcal{F}\left\{x[n] \cos(2\pi f_1 n)\right\}(e^{j2\pi(f-f_2)}) + \mathcal{F}\left\{x[n] \cos(2\pi f_1 n)\right\}(e^{j2\pi(f+f_2)})\right] \qquad (3.37)$$

and

$$\mathcal{F}\left\{x[n] \cos(2\pi f_1 n)\right\}(e^{j2\pi(f \pm f_2)}) = \frac{1}{2}\left[\mathsf{x}(e^{j2\pi(f \pm f_2 - f_1)}) + \mathsf{x}(e^{j2\pi(f \pm f_2 + f_1)})\right]. \qquad (3.38)$$

Therefore,

$$\mathcal{F}\left\{x[n] \cos(2\pi f_1 n) \cos(2\pi f_2 n)\right\} =$$

$$\frac{1}{4}\left[\mathsf{x}(e^{j2\pi(f - f_2 - f_1)}) + \mathsf{x}(e^{j2\pi(f - f_2 + f_1)}) + \mathsf{x}(e^{j2\pi(f + f_2 - f_1)}) + \mathsf{x}(e^{j2\pi(f + f_2 + f_1)})\right]. \qquad (3.39)$$

Example 3.7   Compute the inverse DTFT of $\mathsf{x}(e^{j2\pi(f-f_1)})(1 - je^{j2\pi(f-f_1)}) + \mathsf{x}(e^{j2\pi(f+f_1)})(1 + je^{j2\pi(f+f_1)})$, where f_1 is constant.

Answer: Using Table 3.3,

$$\mathcal{F}^{-1}\left\{\mathsf{x}(e^{j2\pi(f-f_1)}) + \mathsf{x}(e^{j2\pi(f+f_1)})\right\} = 2x[n] \cos(2\pi f_1 n), \qquad (3.40)$$

then it also follows that

$$\mathcal{F}^{-1}\left\{-j\mathsf{x}(e^{j2\pi(f-f_1)})e^{j2\pi(f-f_1)} + j\mathsf{x}(e^{j2\pi(f+f_1)})e^{j2\pi(f+f_1)}\right\}$$

$$= 2\mathcal{F}^{-1}\left\{\mathsf{x}(e^{j2\pi f})e^{j2\pi f}\right\} \sin(2\pi f_1 n) \qquad (3.41)$$

$$= 2x[n+1] \sin(2\pi f_1 n). \qquad (3.42)$$

Therefore, the answer is $x[n] \cos(2\pi f_1 n) + 2 x[n+1] \sin(2\pi f_1 n)$.

The DTFT provides a sound theoretical analysis of discrete-time signals. Unfortunately, it is difficult to employ in real-time systems because of the continuous frequency variable. An alternative is the discrete Fourier transform (DFT) and its easy-to-implement friend the fast Fourier transform (FFT) when the signal length is a power of 2 (and in some other cases as well). Of all the transforms, the DFT is the most useful from a wireless signal processing perspective. It is critical in enabling the low-complexity equalization strategies of SC-FDE and OFDM, which are discussed further in Chapter 5.

The DFT applies only to *finite-length* signals; the DTFT works for both finite and infinite-length signals. The DFT analysis and synthesis equations are

$$\text{Analysis} \quad \mathsf{x}[k] = \sum_{n=0}^{N-1} x[n] e^{-j\frac{2\pi}{N}kn} \quad k = 0, 1, ..., N-1 \tag{3.43}$$

$$\text{Synthesis} \quad x[n] = \frac{1}{N} \sum_{k=0}^{N-1} \mathsf{x}[k] e^{j\frac{2\pi}{N}kn} \quad n = 0, 1, ..., N-1. \tag{3.44}$$

Note the presence of discrete time and discrete frequency. The DFT equations are often written in an alternative form using $W_N = \exp(-j2\pi/N)$.

A list of relevant properties is provided in Table 3.5. There are some subtle differences relative to the DTFT. In particular, shifts are all given in terms of a modulo by N

Table 3.5 Discrete Fourier Transform Relationships

Finite-Length Sequence (Length N)	N-Point DFT (Length N)
$x[n]$	$\mathsf{x}[k]$
$x_1[n], x_2[n]$	$\mathsf{x}_1[k], \mathsf{x}_2[k]$
$ax_1[n] + bx_2[n]$	$a\,\mathsf{x}_1[k] + b\,\mathsf{x}_2[k]$
$\mathsf{x}[n]$	$Nx[((-k))_N]$
$x[((n-m))_N]$	$W_N^{km}\,\mathsf{x}[k]$
$W_N^{-\ell n}x[n]$	$\mathsf{x}[((k-\ell))_N]$
$\sum_{m=0}^{N-1} x_1[m]x_2[((n-m))_N]$	$\mathsf{x}_1[k]\,\mathsf{x}_2[k]$
$x_1[n]x_2[n]$	$\frac{1}{N}\sum_{l=0}^{N-1} \mathsf{x}_1[\ell]\,\mathsf{x}_2[((k-\ell))_N]$
$x^*[n]$	$\mathsf{x}^*[((-k))_N]$
$x^*[((-n))_N]$	$\mathsf{x}^*[k]$
$\text{Re}\{x[n]\}$	$X_{\text{ep}}[k] = \frac{1}{2}\left\{\mathsf{x}[((k))_N] + \mathsf{x}^*[((-k))_N]\right\}$
$j\text{Im}\{x[n]\}$	$X_{\text{op}}[k] = \frac{1}{2}\left\{\mathsf{x}[((k))_N] - \mathsf{x}^*[((-k))_N]\right\}$
$x_{\text{ep}}[k] = \frac{1}{2}\left\{x[n] + x^*[((-n))_N]\right\}$	$\text{Re}\{\mathsf{x}[k]\}$
$x_{\text{op}}[k] = \frac{1}{2}\left\{x[n] - x^*[((-n))_N]\right\}$	$j\text{Im}\{X[k]\}$
† $x[n]$ real	
Symmetry properties	$\begin{cases} \mathsf{x}[k] = \mathsf{x}^*[((-k))_N] \\ \text{Re}\left\{\mathsf{x}[k]\right\} = \text{Re}\left\{\mathsf{x}[((-k))_N]\right\} \\ \text{Im}\left\{\mathsf{x}[k]\right\} = -\text{Im}\left\{\mathsf{x}[((-k))_N]\right\} \\ \left|\mathsf{x}[k]\right| = \left|\mathsf{x}[((-k))_N]\right| \\ \measuredangle\mathsf{x}[k] = -\measuredangle\mathsf{x}[((-k))_N] \end{cases}$
$x_{\text{ep}}[k] = \frac{1}{2}\left\{x[n] + x[((-n))_N]\right\}$	$\text{Re}\{\mathsf{x}[k]\}$
$x_{\text{op}}[k] = \frac{1}{2}\left\{x[n] - x[((-n))_N]\right\}$	$j\text{Im}\{\mathsf{x}[k]\}$

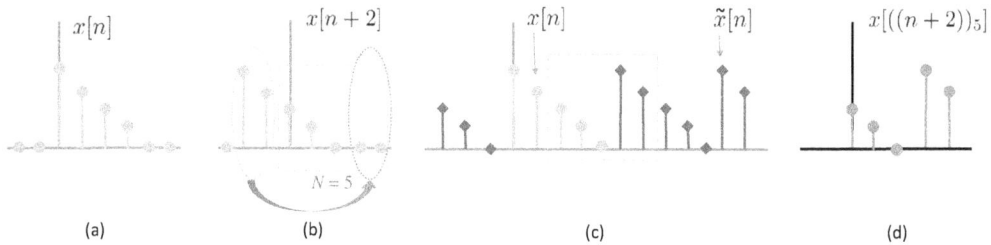

Figure 3.2 (a) Finite-length signal. (b) The circular shift can be computed from the linear shift of $x[n]$. (c) The circular shift can also be computed from a periodic version of the finite-length signal. (d) Circularly shifted version of $x[n]$ for $N = 5$

operation denoted as $(\cdot)_N$. This ensures that the argument falls in $0, 1, \ldots, N - 1$, for example, $(5)_4 = 1$ and $(-1)_4 = 3$. Figure 3.2 illustrates how to compute the circular shift $x[((n + 2))_5]$ of a finite-length signal $x[n]$ when $N = 5$. The circular shift modulo N can be computed from the linearly shifted version of $x[n]$ in Figure 3.2(a) by moving the two samples located at negative indices in the linearly shifted signal to the other end of the signal, as illustrated in Figure 3.2(b). It can also be computed by creating an N-periodic signal $\tilde{x}[n]$ from the original finite-length signal and extracting the first N samples of $\tilde{x}[n - 2]$, as shown in Figure 3.2(c). The result of the circular shift $x[((n + 2))_5]$ is shown in Figure 3.2(d).

The modulo shift shows up in the shift property as well as in the duality between multiplication in the frequency domain and convolution in the time domain. In particular, a new kind of convolution is used called the circular convolution, denoted by $\circledast$, which uses shifts of modulo N to compute the convolution sum:

$$x_1[n] \circledast x_2[n] = \sum_{m=0}^{N-1} x_1[m] x_2[((n - m))_N]. \qquad (3.45)$$

The DFT of the circular convolution of two finite-length signals is the product of the DFTs of both signals, as shown in Table 3.5. The linear and the circular convolutions are different operations, but the DFT can be used to efficiently compute linear convolutions because in some cases both operations are equivalent. Thus, if $x_1[n]$ of size N_1 and $x_2[n]$ of size N_2 are padded with zeros to length at least $N_1 + N_2 - 1$, the linear convolution of $x_1[n]$ and $x_2[n]$ can be computed as the inverse DFT of the product $x_1[k]x_2[k]$. For more details on the DFT and the circular convolution see [248]. The circular convolution plays a key role in the development of SC-FDE and OFDM, as explained in Chapter 5.

Now we provide several examples to illustrate computations with the DFT. Example 3.8 shows a basic DFT calculation. Example 3.9 gives an example of circular convolution. Example 3.10 shows how to perform the DFT by creating the input signal as a vector and defining a DFT matrix. Example 3.11 through Example 3.14 provide example calculations of the DFT.

Example 3.8   Calculate the DFT of $\{1, -1, 1\}$.
   Answer:

$$X[0] = \sum_{n=0}^{2} x[n] e^{-j2\pi n 0/3} = \sum_{n=0}^{2} x[n] = 1 + -1 + 1 = 1 \tag{3.46}$$

$$X[1] = \sum_{n=0}^{2} x[n] e^{-j2\pi n 1/3} = e^{-j2\pi \cdot 0 \cdot 1/3} + -e^{-j2\pi \cdot 1 \cdot 1/3} + e^{-j2\pi \cdot 2 \cdot 1/3} = 1 + j\sqrt{3} \tag{3.47}$$

$$X[2] = \sum_{n=0}^{2} x[n] e^{-j2\pi n 2/3} = e^{-j2\pi \cdot 0 \cdot 2/3} + -e^{-j2\pi \cdot 1 \cdot 2/3} + e^{-j2\pi \cdot 2 \cdot 2/3} = 1 - j\sqrt{3} \tag{3.48}$$

Example 3.9   Calculate the circular convolution between the sequences $\{1, -1, 2\}$ and $\{0, 1, 2\}$.
   Answer:

$$y[n] = \sum_{\ell=0}^{2} x[\ell] h[(n - \ell)_3] \tag{3.49}$$

$$y[0] = \sum x[\ell] h[(0 - \ell)_3] = x[0]h[0] + x[1]h[2] + x[2]h[1] = 0 \tag{3.50}$$

$$y[1] = \sum x[\ell] h[(1 - \ell)_3] = x[0]h[1] + x[1]h[0] + x[2]h[2] = 5 \tag{3.51}$$

$$y[2] = \sum x[\ell] h[(2 - \ell)_3] = x[0]h[2] + x[1]h[1] + x[2]h[0] = 1 \tag{3.52}$$

$$y[n] = \{0, 5, 1\}. \tag{3.53}$$

Example 3.10   It is useful to write the DFT analysis in matrix form for a length N sequence. In this example, write this for general N, then compute the DFT matrix (which transforms the time-domain vector to the frequency-domain vector) explicitly for $N = 4$.
   Answer: In the general case,

$$\begin{bmatrix} \mathsf{x}[0] \\ \mathsf{x}[1] \\ \vdots \\ \mathsf{x}[N-1] \end{bmatrix} = \begin{bmatrix} 1 & 1 & \cdots & 1 \\ 1 & e^{-j\frac{2\pi}{N}} & \cdots & e^{-j\frac{2\pi}{N}(N-1)} \\ \vdots & \vdots & \ddots & \vdots \\ 1 & e^{-j\frac{2\pi}{N}(N-1)} & \cdots & e^{-j\frac{2\pi}{N}(N-1)^2} \end{bmatrix} \begin{bmatrix} x[0] \\ x[1] \\ \vdots \\ x[N-1] \end{bmatrix}. \tag{3.54}$$

For $N = 4$, $e^{-j\frac{\pi}{2}} = -j$, so the DFT matrix $\mathbf{D}$ is

$$\mathbf{D} = \begin{bmatrix} 1 & 1 & 1 & 1 \\ 1 & -j & (-j)^2 & (-j)^3 \\ 1 & (-j)^2 & (-j)^4 & (-j)^6 \\ 1 & (-j)^3 & (-j)^6 & (-j)^9 \end{bmatrix} = \begin{bmatrix} 1 & 1 & 1 & 1 \\ 1 & -j & -1 & j \\ 1 & -1 & 1 & -1 \\ 1 & j & -1 & -j \end{bmatrix}. \tag{3.55}$$

Example 3.11   Find DFT{DFT{DFT{DFT{$x[n]$}}}}.
Answer: Use the reciprocity property of the DFT: $x[n] \leftrightarrow X[k] \iff X[n] \leftrightarrow Nx[(-k)_N]$. Therefore,

$$\text{DFT}\{x[n]\} = X[k] \tag{3.56}$$

$$\text{DFT}\{\text{DFT}\{x[n]\}\} = \text{DFT}\{X[k]\} \qquad\qquad = Nx[(-k)_N] \tag{3.57}$$

$$\text{DFT}\{\text{DFT}\{\text{DFT}\{x[n]\}\}\} = \text{DFT}\{Nx[(-k)_N]\} \qquad = NX[(-k)_N] \tag{3.58}$$

$$\text{DFT}\{\text{DFT}\{\text{DFT}\{\text{DFT}\{x[n]\}\}\}\} = \text{DFT}\{NX[(-k)_N]\} \tag{3.59}$$

$$= N^2 x[k] \tag{3.60}$$

Example 3.12   Assuming that N is even, compute the DFT of

$$x[n] = \begin{cases} 1 & n \text{ even} \\ 0 & n \text{ odd.} \end{cases} \tag{3.61}$$

Answer:

$$\mathsf{x}[k] = \sum_{n=0}^{N/2-1} e^{-j\frac{2\pi}{N}2kn} \tag{3.62}$$

If $k = 0$ or $k = N/2$, then $e^{-j\frac{2\pi}{N}2kn} = 1$, so $\mathsf{x}[k = 0] = \mathsf{x}[k = N/2] = \frac{N}{2}$. If $k \neq 0$ and $k \neq N/2$,

$$\sum_{n=0}^{N/2-1} e^{-j\frac{2\pi}{N}2kn} = \frac{1 - e^{-j2\pi k}}{1 - e^{j\frac{4\pi}{N}k}} = 0. \tag{3.63}$$

Therefore,

$$\mathsf{x}[k] = \frac{N}{2}(\delta[k] + \delta[k - N/2]). \tag{3.64}$$

Example 3.13   Assuming that N is even, compute the DFT of

$$x[n] = \begin{cases} 1 & 0 \le n \le N/2 - 1 \\ 0 & N/2 \le n \le N - 1. \end{cases} \tag{3.65}$$

Answer: From the definition,

$$\mathsf{x}[k] = \sum_{n=0}^{N/2-1} e^{-\mathrm{j}\frac{2\pi}{N}kn}. \tag{3.66}$$

It should be clear that $\mathsf{x}[k=0] = \frac{N}{2}$. For $k \ne 0$,

$$\mathsf{x}[k] = \frac{1 - e^{-\mathrm{j}k\pi}}{1 - e^{-\mathrm{j}\frac{2\pi}{N}k}} = \frac{e^{-\mathrm{j}\frac{k\pi}{2}}(e^{\mathrm{j}\frac{k\pi}{2}} - e^{-\mathrm{j}\frac{k\pi}{2}})}{e^{-\mathrm{j}\frac{k\pi}{N}}(e^{\mathrm{j}\frac{k\pi}{N}} - e^{-\mathrm{j}\frac{k\pi}{N}})} \tag{3.67}$$

$$= e^{-\mathrm{j}\frac{k\pi}{N}(N/2-1)}\frac{\sin(\frac{k\pi}{2})}{\sin(\frac{k\pi}{N})}. \tag{3.68}$$

For k even $\sin(\frac{k\pi}{2}) = 0$, and for k odd $\sin(\frac{k\pi}{2}) = \frac{1}{\mathrm{j}}e^{\mathrm{j}\frac{k\pi}{2}} = (\mathrm{j})^{k-1} = (-1)^{(k-1)/2}$, thus

$$\mathsf{x}[k] = \begin{cases} N/2 & k = 0 \\ e^{-\mathrm{j}\frac{k\pi}{N}(N/2-1)}\frac{(-1)^{(k-1)/2}}{\sin(\frac{k\pi}{N})} & k \text{ odd} \\ 0 & k \text{ even.} \end{cases} \tag{3.69}$$

Example 3.14   Compute the DFT of the length N sequence $x[n] = a^n$ using direct application of the analysis equation.

Answer: Inserting into the analysis equation:

$$\mathsf{x}[k] = \sum_{n=0}^{N-1} a^n e^{-\mathrm{j}\frac{2\pi}{N}kn} \tag{3.70}$$

$$= \frac{1 - a^N}{1 - e^{-\mathrm{j}\frac{2\pi}{N}ka}}. \tag{3.71}$$

Using the Fourier transform, we are now equipped to define a notion of bandwidth for signals.

3.1.4   Bandwidth of a Signal

The continuous-time signals transmitted in a wireless communication system are ideally bandlimited. This results because of the desire to achieve bandwidth efficiency (send as much data for as little bandwidth as possible) and also the way governments allocate spectrum. The classic definition of the bandwidth of a signal $x(t)$ is the portion of the frequency spectrum $\mathsf{x}(f)$ for which $\mathsf{x}(f)$ is nonzero. This is known as the absolute bandwidth. Unfortunately, signals in engineering settings are finite in duration, and thus with this proper definition their bandwidth is infinite per Example 3.15. Consequently, other definitions are used to define an operational notion of bandwidth (an approximation of the bandwidth of a signal that is treated as if it were the absolute bandwidth).

Example 3.15   Show that finite-duration signals are in general not bandlimited.

   Answer: Let $x(t)$ denote a signal. We take a snapshot of this signal from $[-T/2, T/2]$ by *windowing* with a rectangle function to form $x_T(t) = \text{rect}(t/T)x(t)$. The term windowing reminds us that we are "looking" at only a piece of the infinite-duration signal. From the multiplication property in Table 3.1:

$$\mathsf{x}_T(f) = \mathsf{x}(f) * T\text{sinc}\,(fT). \tag{3.72}$$

Since $\text{sinc}(fT)$ is an infinite-duration function, the convolution is also an infinite-duration function for general $\mathsf{x}(f)$.

   The bandwidth of a signal is defined based on the energy spectral density (ESD) or power spectral density (PSD). For nonrandom signals, which are not periodic, the ESD is simply the magnitude squared of the Fourier transform of the signal (assuming it exists). For example, the ESD of $x(t)$ is given as $|\mathsf{x}(f)|^2$ and the ESD of $x[n]$ is $|\mathsf{x}(e^{j2\pi f})|^2$. For a periodic signal, with period T_0, the PSD is defined as

$$S_x(f) = \sum_{n=-\infty}^{\infty} |x[n]|^2 \delta(f - nf_0), \tag{3.73}$$

where $x[n]$ are the CTFS of $x(t)$, and $f_0 = \frac{1}{T_0}$. For a WSS (wide-sense stationary) random process, discussed in Section 3.2, the PSD comes from the Fourier transform of the autocorrelation function. We now define ways of calculating bandwidth assuming that a PSD is given (similar calculations apply for an ESD).

Half-Power Bandwidth   The half-power bandwidth, or 3dB bandwidth, is defined as the value of the frequency over which the power spectrum is at least 50% of its maximum value, that is,

$$B = f_{3\text{dB}}^{(h)} - f_{3\text{dB}}^{(\ell)}, \tag{3.74}$$

   where $f_{3\text{dB}}^{(h)}$ and $f_{3\text{dB}}^{(\ell)}$ are as shown in Figure 3.3.

XdB Bandwidth   The XdB bandwidth is defined as the difference between the largest frequency that suffers XdB of attenuation and the smallest frequency. Similar to the half-power bandwidth, this can be expressed as

$$B = f_{X\text{dB}}^{(h)} - f_{X\text{dB}}^{(\ell)}, \tag{3.75}$$

   where $f_{X\text{dB}}^{(h)}$ and $f_{X\text{dB}}^{(\ell)}$ are defined in the same manner as $f_{3\text{dB}}^{(h)}$ and $f_{3\text{dB}}^{(\ell)}$.

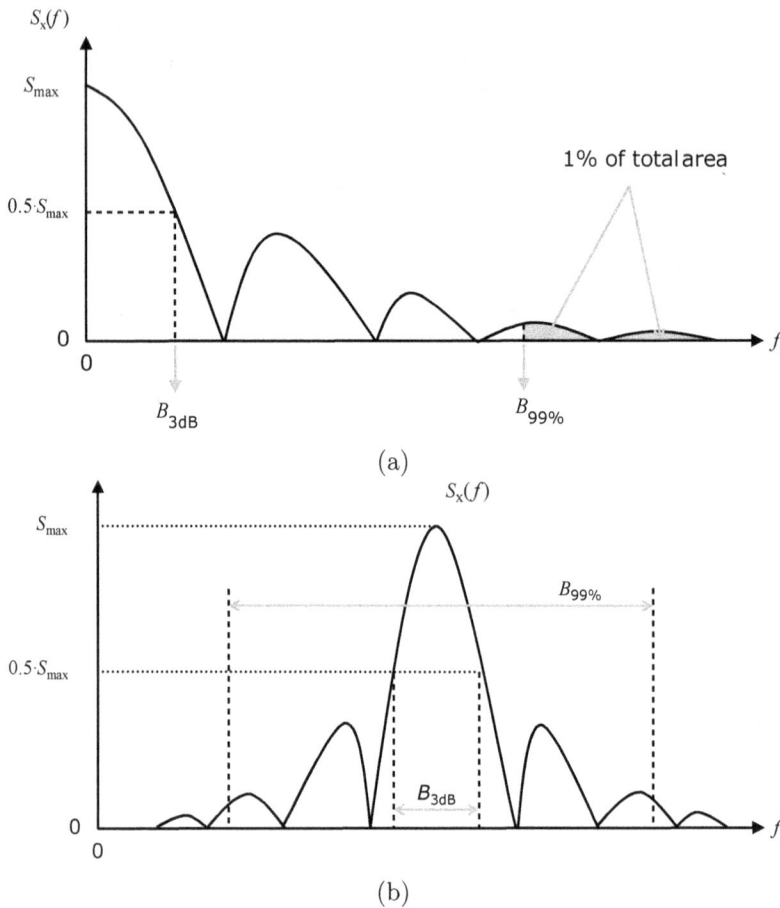

Figure 3.3 The different ways of defining bandwidth for (a) baseband signals and (b) passband signals

Noise Equivalent Bandwidth The noise equivalent bandwidth is defined as

$$B = \frac{1}{P_x(f_c)} \int_f P_x(f) \mathrm{d}f. \tag{3.76}$$

Fractional Power Containment Bandwidth One of the most useful notions of bandwidth, often employed by the FCC, is the fractional power containment bandwidth. The fractional containment bandwidth is

$$\int_0^{B/2} P_x(f)\mathrm{d}f = \alpha \int_{-\infty}^{\infty} P_x(f)\mathrm{d}f, \tag{3.77}$$

where α is the fraction of containment. For example, if $\alpha = 0.99$, then the bandwidth B would be defined such that the signal has $\alpha\%$ of its bandwidth.

The bandwidth depends on whether the signal is *baseband* or *passband*. Baseband signals have energy that is nonzero for frequencies near the origin. They are transmitted at DC, and bandwidth is measured based on the single-sided spectrum of the signal. Passband signals have energy that is nonzero for a frequency band concentrated about $f = \pm f_c$ where f_c is the carrier frequency and $f_c \gg 0$. Passband signals have a spectrum that is centered at some carrier frequency, with a mirror at $-f_c$ because the signal is real, and do not have frequencies that extend to DC. It is common to use a notion of bandwidth for passband signals, which is measured about the carrier frequency as illustrated in Figure 3.3. The aforementioned definitions of bandwidth all extend in a natural way to the passband case. Passband and baseband signals are related through upconversion and downconversion, as discussed in Sections 3.3.1 and 3.3.2.

Example 3.16   Consider a simplified PSD as shown in Figure 3.4, and compute the half power bandwidth, the noise equivalent bandwidth, and the fractional power containment bandwidth with $\alpha = 0.9$.

Answer: From Figure 3.4, we have $f_{3\text{dB}}^{(\ell)} = f_c - 1.5\text{MHz}$, and $f_{3\text{dB}}^{(h)} = f_c + 1.5\text{MHz}$; thus

$$B_{3\text{dB}} = (f_c + 1.5) - (f_c - 1.5) = 3\text{MHz}. \tag{3.78}$$

We have $P_x(f_c) = 1$, and the integral can be computed as the area of the trapezoid

$$B_{\text{noise equivalent}} = \frac{1}{P(f_c)} \int_f P(f)\mathrm{d}f = 3\text{MHz}. \tag{3.79}$$

The total area of the PSD is 3, so for $\alpha = 0.9$, the area that should be left out is 0.3 on both sides or 0.15 for one side of f_c. This corresponds to the point $f_c - 2 + \sqrt{0.3}$ on the left. Therefore,

$$B_{\text{fractional}} = (f_c + 2 - \sqrt{0.3}) - (f_c - 2 + \sqrt{0.3}) = 2.90\text{MHz}. \tag{3.80}$$

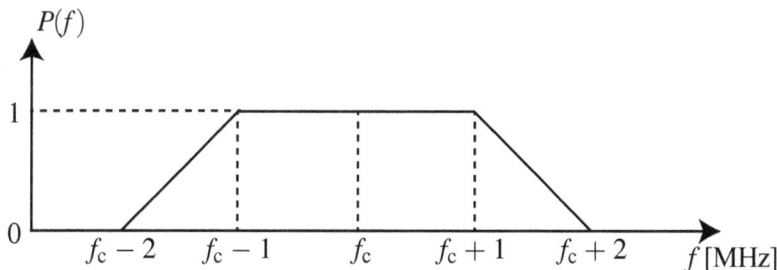

Figure 3.4 An example PSD for purposes of the bandwidth calculations in Example 3.16

3.1.5 Sampling

The fundamental connection between digital communication and DSP is through the sampling theorem. This results from the following fact: *communication systems are band-limited.* Even ultra-wideband systems are bandlimited, just with a very large bandwidth. The bandlimited property means that the transmitted signal is bandlimited. Therefore, it is sufficient for the transmitter to generate a bandlimited signal. Because the channel is assumed to be LTI, and LTI systems do not expand the bandwidth of the input signals, the received signal can also be treated as bandlimited. This means that the receiver has to process only the signals found in the bandwidth of interest.

As a consequence of the bandlimited property, it is possible to exploit the connection between bandlimited signals and their periodically sampled counterparts established by Nyquist's theorem. The essence of this is described in the following (paraphrased from [248]):

Nyquist Sampling Theorem Let $x(t)$ be a bandlimited signal, which means that $X(f) = 0$ for $f \geq f_N$. Then $x(t)$ is uniquely determined by its samples $\{x[n] = x(nT)\}_{n=-\infty}^{\infty}$ if the sampling frequency satisfies

$$f_s = \frac{1}{T} > 2f_N, \tag{3.81}$$

where f_N is the *Nyquist frequency* and $2f_N$ is generally known as the *Nyquist rate.* Furthermore, the signal can be reconstructed from its samples through the reconstruction equation

$$x(t) = \sum_{n=-\infty}^{\infty} x[n]\text{sinc}\left(\frac{t - nT}{T}\right). \tag{3.82}$$

Nyquist's theorem says that a bandlimited signal can be represented with no loss through its samples, provided the sampling period T is chosen to be sufficiently small. Additionally, Nyquist shows how to properly reconstruct $x(t)$ from its sampled representation.

An illustration of sampling is provided in Figure 3.5. The values of the continuous signal $x(t)$ taken at integer multiples of T are used to create the discrete-time signal $x[n]$.

Throughout this book, we use the block diagram notation found in Figure 3.6 to represent the C/D (continuous-to-discrete) converter and in Figure 3.7 to represent the D/C (discrete-to-continuous) converter. The sampling and reconstruction processes are assumed to be ideal. In practice, the C/D would be performed using an ADC and the D/C would be performed using a DAC. ADC and DAC circuits have practical performance limitations, and they introduce additional distortion in the signal; see [271] for further details.

Through sampling, the Fourier transforms of $x[n]$ and $x(t)$ are related through

$$X\left(e^{j2\pi f}\right) = \frac{1}{T} \sum_{n=-\infty}^{\infty} X\left(\frac{f}{T} - \frac{n}{T}\right). \tag{3.83}$$

While this expression might look odd, recall that the discrete-time Fourier transform is periodic in f with a period of 1. The summation preserves the periodicity with the result that there are multiple periodic replicas.

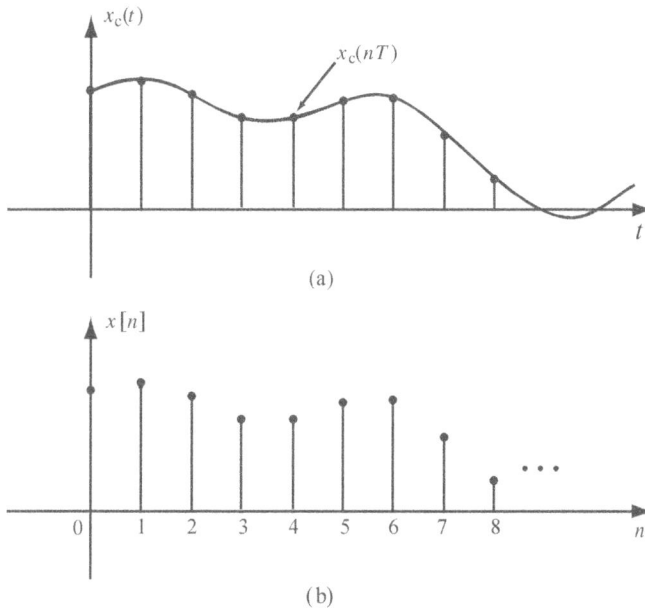

Figure 3.5 Sampling in the time domain. The periodic samples of $x(t)$ in (a) form the values for the discrete-time signal $x[n]$ in (b).

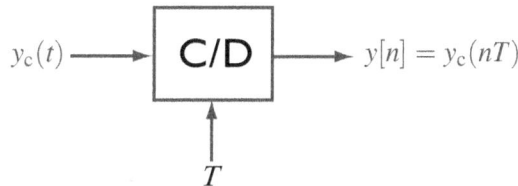

Figure 3.6 Sampling using a continuous-to-discrete-time converter

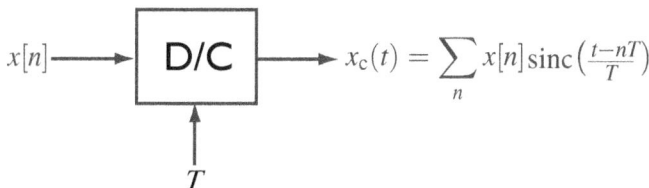

Figure 3.7 Generation of a bandlimited signal using a discrete-to-continuous converter

If care is not taken in the sampling operation, then the resulting received signal will be distorted. For example, suppose that $x(t)$ is bandlimited with maximum frequency f_N. Then $\mathsf{x}(f) = 0$ for $|f| > f_N$. To satisfy Nyquist, $T < 1/2f_N$. For $f \in [-1/2, 1/2)$ it follows that $f/T \in [-1/2T, 1/2T)$. Since $f_N < 1/2T$, then the sampled signal is fully

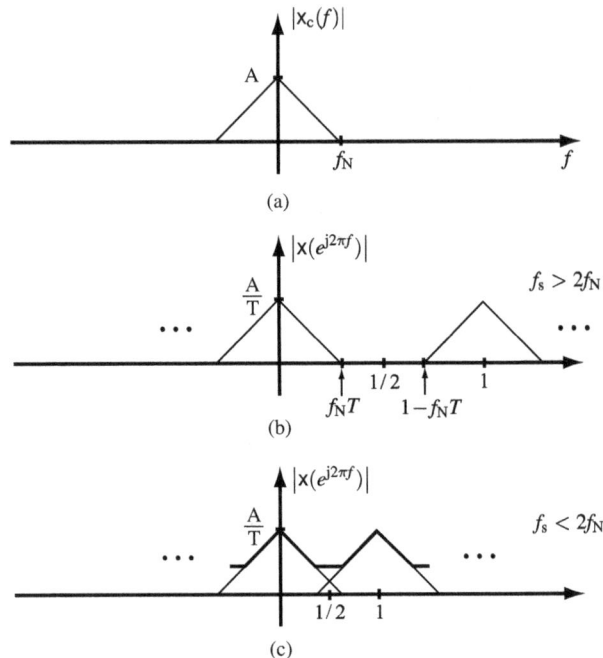

Figure 3.8 Sampling in the frequency domain. (a) Magnitude spectrum of a continuous-time signal. (b) Corresponding magnitude spectrum for a discrete-time signal assuming that the Nyquist criterion is satisfied. (c) Corresponding magnitude spectrum for a discrete-time signal assuming that the Nyquist criterion is not satisfied. In this example $f_s > f_N$ and $f_s < 2f_N$.

contained in one period of the DTFT and there is no overlap among replicas in (3.83). Zooming in on just one period,

$$\mathsf{x}\left(e^{j2\pi f}\right) = \frac{1}{T}\mathsf{x}\left(\frac{f}{T}\right) \quad \text{for } f \in [-1/2, 1/2). \tag{3.84}$$

Consequently, sampling of bandlimited signals by a suitable choice of T completely preserves the spectrum. If a small-enough T is not chosen, then according to (3.83) the summations will result in aliasing and (3.84) will not hold. An illustration of sampling in the frequency domain is provided in Figure 3.8. Examples with sampling are provided in Example 3.17 and Example 3.18. For more details refer to [248, Chapter 4].

Example 3.17

• Suppose that we wish to create the bandlimited signal

$$x(t) = 3\cos(2\pi 10^6 t + \pi/4) + \cos(6\pi 10^6 t + \pi/8). \tag{3.85}$$

Find the Nyquist frequency and Nyquist rate of $x(t)$.

Answer: The Nyquist frequency of $x(t)$ is $f_N = 3\text{MHz}$ because that is the largest frequency in $x(t)$. The Nyquist rate of $x(t)$ is twice the Nyquist frequency; thus $2f_N = 6\text{MHz}$.

- Suppose that we want to generate $x(t)$ using a discrete-to-continuous converter operating at five times the Nyquist rate. What function $x[n]$ do you need to input into the discrete-to-continuous converter to generate $x(t)$?

Answer: In this case $T = 1/(5 \times 3\text{MHz}) = (1/15) \times 10^{-6}\text{s}$. Using the reconstruction formula in (3.82) and substituting, the required input is

$$x[n] = x(n(1/15) \times 10^{-6}) \tag{3.86}$$
$$= 3\cos(\pi n/15 + \pi/4) \;+\; \cos(\pi n/5 + \pi/8). \tag{3.87}$$

- Suppose that we sample $x(t)$ with a sampling frequency of 4MHz. What is the largest frequency of the discrete-time signal $x[n] = x(nT)$?

Answer: In this case $T = 1/(4\text{MHz}) = 0.25 \times 10^{-6}\text{s}$. Substituting,

$$x[n] = x(n0.25 \times 10^{-6}) \tag{3.88}$$
$$= 3\cos(2\pi n/4 + \pi/4) \;+\; \cos(2\pi n3/4 + \pi/8). \tag{3.89}$$

The largest frequency in $[-1/2, 1/2)$ is $1/4$.

Example 3.18   Consider the signal

$$x(t) = 40^2 \text{sinc}^2(40\pi t). \tag{3.90}$$

Suppose that $x(t)$ is sampled with frequency $f_x = 60\text{Hz}$ to produce the discrete-time sequence $a[n] = x(nT)$ with Fourier transform $A(e^{j2\pi f})$. Find the maximum value of f_0 such that

$$A(e^{j2\pi f}) = \frac{1}{T} \times \left(\frac{f}{T}\right) \quad \text{for } |f| \le f_0, \tag{3.91}$$

where $\times(f)$ is the Fourier transform of $x(t)$.

Answer: The sampling period $T = \frac{1}{f_s}$. From Table 3.2 the Fourier transform of $x(t)$ is

$$\times(f) = \frac{1}{40}\Lambda\left(\frac{f}{40}\right). \tag{3.92}$$

The maximum frequency, that is, the Nyquist frequency, of $x(t)$ is $f_N = 40\text{Hz}$, and thus the Nyquist rate is $f_s = 80\text{Hz}$, which is larger than the sampling rate of 60Hz. Consequently, there will be aliasing in the discrete-time frequency range of $[-1/2, 1/2)$.

The sampled signal $a[n]$ has transform

$$A(e^{j2\pi f}) = \frac{1}{T} \sum_{k=-\infty}^{\infty} \times \left(\frac{f}{T} - \frac{k}{T}\right) \tag{3.93}$$

$$= 60 \sum_{k=-\infty}^{\infty} \times (60f - 60k). \tag{3.94}$$

To solve this problem, notice that there is overlap only between adjacent replicas, because $f_N < 1/T < f_s$. This means that inside the range $f \in [-1/2, 1/2)$,

$$A(e^{j2\pi f}) = 60 \times (60f + 60) + 60 \times (60f) + 60 \times (60f - 60). \tag{3.95}$$

Looking at positive frequencies, the replica $X(60f - 60)$ extends from $f = (f_s - f_N)/f_s = (60 - 40)/60 = 1/3$ to $(f_s/2)/f_s = 1/2$. Thus $f_0 = 1/3$ and

$$A(e^{j2\pi f}) = 60 \times (60f) \text{ for } |f| \leq 1/3. \tag{3.96}$$

To visualize the sampling operation, it often helps to examine the proxy signal $x_s(t) = x(t)\frac{1}{T}\text{III}(t/T)$, where $\text{III}(t) = \sum_k \delta(t - k)$ is the comb function with unit period. The unit period comb function satisfies $\text{III}(f) = \mathcal{F}\{\text{III}(t)\}$, so the same function is used in both time and frequency domains. Example 3.19 shows how the proxy signal can be used to provide intuition about aliasing in the sampling process.

Example 3.19   Consider a continuous-time signal $x(t)$ with transform $\mathsf{x}(f)$ as indicated in Figure 3.9. The transform is real to make illustration easier.

- What is the largest value of T that can be used to sample the signal and still achieve perfect reconstruction?

  Answer: From Figure 3.9, the highest frequency is 1000Hz; therefore $T = 1/2000$.

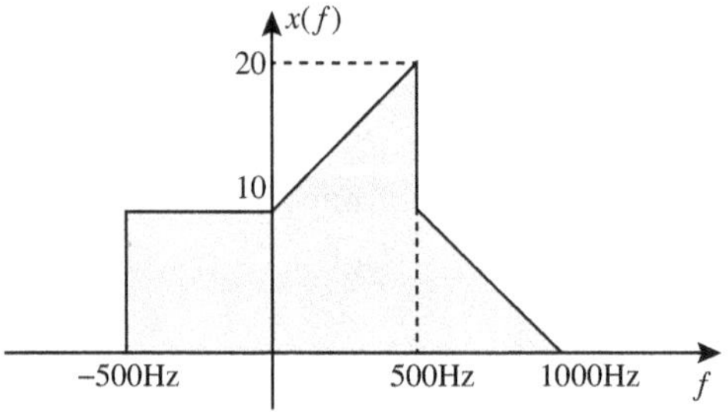

Figure 3.9 The Fourier transform of a signal. The transform is real. See Example 3.19.

- Suppose we sample with $T = 1/3000$. Illustrate $\mathsf{x}(f)$ and $\mathsf{x}_{\mathsf{s}}(f)$ in the interval $-1/2T$ to $1/2T$ and $\mathsf{x}(e^{\mathsf{j}2\pi f})$ in the interval $-1/2$ to $1/2$.

  Answer: The transform is computed as

  $$\mathsf{x}_{\mathsf{s}}(f) = \mathcal{F}\left\{ x(t) \sum_k \delta(t - kT) \right\} = \mathsf{x}(f) * \frac{1}{T} \sum_n \delta\left(f - \frac{n}{T}\right) = \sum_n \frac{1}{T}\mathsf{x}\left(f - \frac{n}{T}\right)$$

  $$(3.97)$$

  and is plotted in Figure 3.10.

- Suppose we sample with $T = 1/1500$. Illustrate $\mathsf{x}(f)$ and $\mathsf{x}_{\mathsf{s}}(f)$ in the interval $-1/2T$ to $1/2T$ and $\mathsf{x}(e^{\mathsf{j}2\pi f})$ in the interval $-1/2$ to $1/2$.

  Answer: The transform is plotted in Figure 3.11.

- Write an equation for $\mathsf{x}(f)$.

  Answer: From Figure 3.9, $\mathsf{x}(f)$ is a superposition of a rectangular pulse and a triangular pulse. Based on the figure,

  $$\mathsf{x}(f) = 10\left[\mathrm{rect}\left(\frac{f}{1000}\right) + \Lambda\left(\frac{f - 500}{500}\right)\right].$$

  $$(3.98)$$

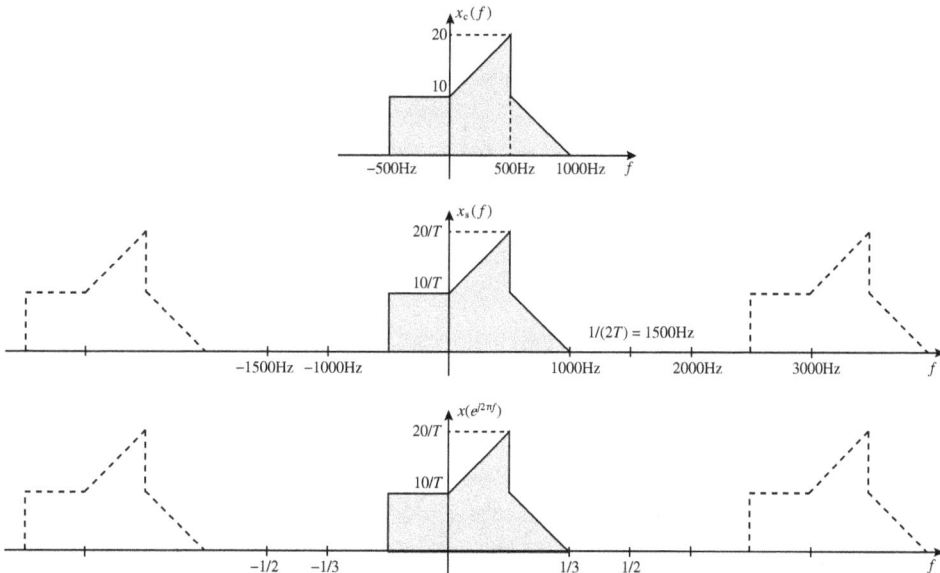

Figure 3.10 Connecting $\mathsf{x}(f)$, $\mathsf{x}_{\mathsf{s}}(f)$, and $\mathsf{x}\left(e^{\mathsf{j}2\pi f}\right)$ for $T = 1/3000$ in Example 3.19

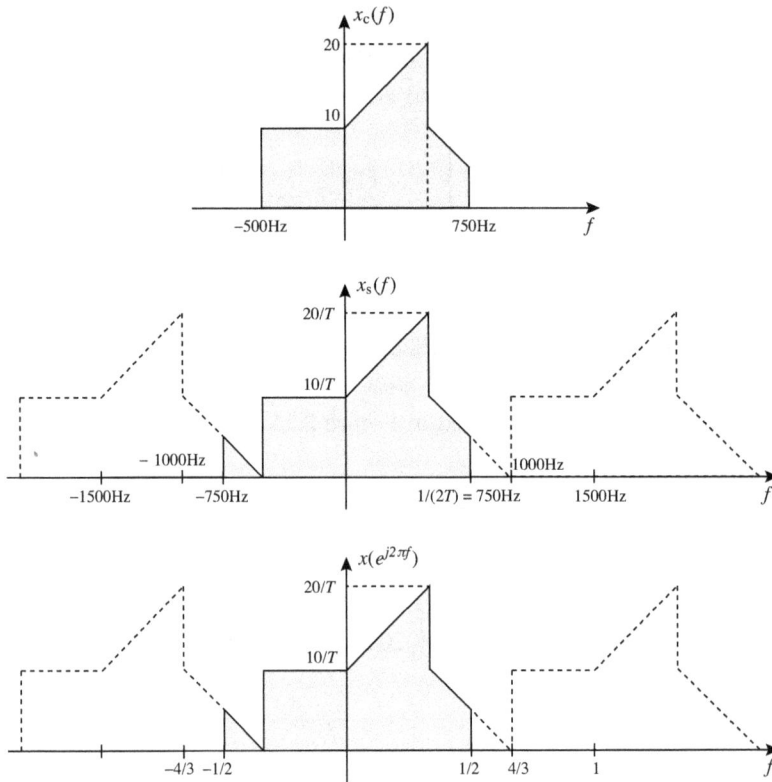

Figure 3.11 Connecting $\mathsf{x}(f)$, $\mathsf{x}_s(f)$, and $\mathsf{x}\left(e^{j2\pi f}\right)$ for $T = 1/1500$ in Example 3.19

- Determine $x(t)$ given $\mathsf{x}(f)$ using the Fourier transform table and properties.
  Answer:
  $$x(t) = 10 \left[1000 \operatorname{sinc}(1000t) + 500 \operatorname{sinc}^2(500t)\, e^{j2\pi 500 t}\right]. \tag{3.99}$$

- Determine $x[n]$ for $T = 1/3000$.
  Answer: Substituting $T = 1/3000$ into $x[n] = x(nT)$, we have
  $$x[n] = 5000 \left[2 \operatorname{sinc}\left(\frac{n}{3}\right) + \operatorname{sinc}^2\left(\frac{n}{6}\right) e^{j\frac{\pi}{3}n}\right]. \tag{3.100}$$

- Determine $x[n]$ for $T = 1/1500$.
  Answer: Substituting $T = 1/1500$ into $x[n] = x(nT)$, we have
  $$x[n] = 5000 \left[2 \operatorname{sinc}\left(\frac{2n}{3}\right) + \operatorname{sinc}^2\left(\frac{n}{3}\right) e^{j\frac{2\pi}{3}n}\right]. \tag{3.101}$$

3.1.6   Discrete-Time Processing of Bandlimited Continuous-Time Signals

Digital signal processing is the workhorse of digital communication systems as explained in Chapter 1. This section presents an important result on filtering continuous-time signals using a discrete-time system. This observation is exploited in Section 3.3.5 to develop a fully discrete-time signal model for a communication system. The steps of the derivation are summarized in Figure 3.12.

We start by explaining the implications of processing a bandlimited signal by an LTI system. Let $x(t)$ be a bandlimited signal input into an LTI system with impulse response $h_c(t)$, which is not necessarily bandlimited, and let the output of the system be $y(t)$. The input, output, and system are in general complex because of complex envelope notation (see Sections 3.3.1 and 3.3.2) and the concept of the baseband equivalent channel (see Section 3.3.3). The fact that $x(t)$ is bandlimited has three important implications:

1. The signal $x(t)$ can be generated in discrete time using a discrete-to-continuous converter operating at an appropriate sampling frequency.

2. The signal $y(t)$ is also bandlimited since $\mathsf{x}(f)$ is bandlimited and $\mathsf{y}(f) = \mathsf{h}_c(f)\mathsf{x}(f)$. Therefore, $y(t)$ can be processed in discrete time using a continuous-to-discrete converter operating at an appropriate sampling frequency.

3. Only the part of the channel $\mathsf{h}_c(f)$ that lives in the bandwidth of $\mathsf{x}(f)$ is important to $\mathsf{y}(f)$.

These facts can be used with additional results on sampling to obtain a relationship between the sampled input, the output, and a sampled filtered version of the system.

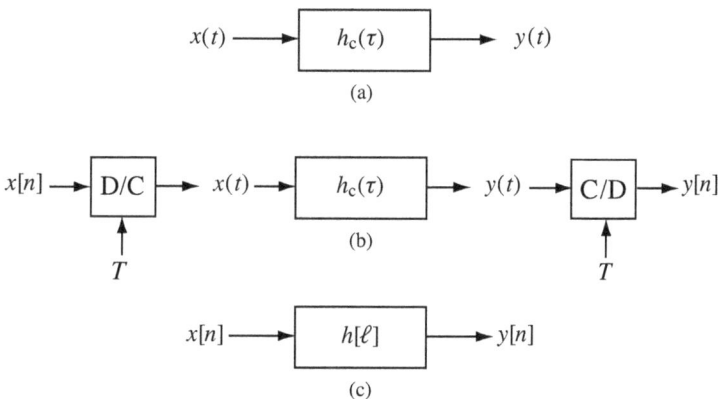

Figure 3.12 (a) A continuous-time LTI system with a bandlimited input; (b) generating a bandlimited input using a discrete-to-continuous converter, followed by processing with an LTI system, followed by sampling with a continuous-to-discrete converter; (c) the discrete-time equivalent system

Suppose that $x(t)$ has a bandwidth of $f_N = B/2$. For this derivation, absolute bandwidth is assumed. The sample period T is chosen to satisfy $T < 1/B$. Define the portion of the channel that lives in the bandwidth of $x(t)$ as

$$h_{low}(f) = \text{rect}(f/B)h_c(f), \tag{3.102}$$

where $\text{rect}(f/B)$ is an ideal lowpass filter with bandwidth of $B/2$. Using Table 3.2, in the time domain,

$$h_{low}(t) = B \int_{-\infty}^{\infty} \text{sinc}(\tau B)h_c(t - \tau)d\tau. \tag{3.103}$$

With these definitions, because $x(t)$ and $y(t)$ are bandlimited with bandwidth $B/2$,

$$y(f) = h_{low}(f)x(f). \tag{3.104}$$

The filtered impulse response $h_{low}(t)$ contains only the part of the system that is relevant to $x(t)$ and $y(t)$.

Now consider the spectrum of $y[n]$. Referring to (3.84),

$$y(e^{j2\pi f}) = \frac{1}{T}y\left(\frac{f}{T}\right) \quad \text{for } f \in [-1/2, 1/2) \tag{3.105}$$

$$= \frac{1}{T}h_{low}\left(\frac{f}{T}\right)x\left(\frac{f}{T}\right) \quad \text{for } f \in [-1/2, 1/2). \tag{3.106}$$

Substituting for $x(e^{j2\pi f})$ from (3.84),

$$y(e^{j2\pi f}) = h_{low}\left(\frac{f}{T}\right)x(e^{j2\pi f}) \quad \text{for } f \in [-1/2, 1/2) \tag{3.107}$$

$$= \frac{1}{T}Th_{low}\left(\frac{f}{T}\right)x(e^{j2\pi f}) \quad \text{for } f \in [-1/2, 1/2). \tag{3.108}$$

Now suppose that

$$h(f) = Th_{low}(f). \tag{3.109}$$

Then

$$y(e^{j2\pi f}) = \frac{1}{T}h\left(\frac{f}{T}\right)x(e^{j2\pi f}) \quad \text{for } f \in [-1/2, 1/2). \tag{3.110}$$

Again using (3.84),

$$y(e^{j2\pi f}) = h(e^{j2\pi f})x(e^{j2\pi f}) \quad \text{for } f \in [-1/2, 1/2). \tag{3.111}$$

Therefore, it follows that the discrete-time system that generates $y[n]$ from $x[n]$ is

$$h[n] = Th_{low}(nT) \tag{3.112}$$

$$= TB \int \mathrm{sinc}(\tau B) h_{\mathrm{c}}(nT - \tau)\mathrm{d}\tau \tag{3.113}$$

$$= \int \mathrm{sinc}(\tau B) h_{\mathrm{c}}(nT - \tau)\mathrm{d}\tau. \tag{3.114}$$

The discrete-time equivalent $h[n]$ is a smoothed and sampled version of the original impulse response $h(t)$. We refer to $h[n]$ as the discrete-time equivalent system. An example calculation of the discrete-time equivalent of an LTI system that delays as input is provided in Example 3.20.

Example 3.20   Suppose that an LTI system delays the input by an amount τ_d. Determine the impulse response of this system and its discrete-time equivalent assuming that the input signal has a bandwidth of $B/2$.

 Answer: This system corresponds to a delay; therefore, $h_{\mathrm{c}}(t) = \delta(t - \tau_d)$. Applying (3.114), the discrete-time equivalent system is

$$h[n] = \int \mathrm{sinc}(\tau B) h_{\mathrm{c}}(nT - \tau)\mathrm{d}\tau \tag{3.115}$$

$$= \int \mathrm{sinc}(\tau B)\delta(nT - \tau_d - \tau)\mathrm{d}\tau \tag{3.116}$$

$$= \mathrm{sinc}(nBT - B\tau_d) \tag{3.117}$$

$$= \mathrm{sinc}(n - B\tau_d). \tag{3.118}$$

It may seem surprising that this is not a delta function, but recall that τ_d can take any value; if τ_d is an integer fraction of T, then $B\tau_d$ will be an integer and $h[n]$ will become a Dirac delta function.

3.2   Statistical Signal Processing

Randomness is a component of every communication system. The transmitted information $i[n]$, the effects of the propagation channel $h(t)$, and the noise $n(t)$ may all be modeled as random signals. For purposes of system design and analysis it is useful to mathematically characterize the randomness in a signal using tools from probability theory. While this is a deep area, the class of random signals required for wireless communication is sufficiently basic and is summarized in this section, with an emphasis on application of probability tools, leaving the appropriate proofs to the references.

3.2.1   Some Concepts from Probability

Consider an experiment with outcomes that are generated from the sample space $\mathcal{S} = \{\omega_1, \omega_2, \ldots\}$ whose cardinality is finite or infinite according to some probability distribution $\mathbb{P}[\omega_i]$. To each outcome ω_i from this space is associated a real number $x(\omega)$, which is called a random variable. A random variable can be continuous, discrete, or mixed in nature and is completely characterized by its cumulative distribution function (CDF). For continuous-valued random variables let us use the notation $F_x(\alpha) = \int_{-\infty}^{\alpha} f_x(z)dz$ to denote the cumulative distribution function where the probability that x is less than or equal to α is $\mathbb{P}(x \leq \alpha) = F_x(\alpha)$, and $f_x(z)$ denotes the probability distribution function (PDF) of x. The PDF should satisfy $f_x(z) \geq 0$ and $\int_z f_x(z)dz = 1$. For discrete-valued random variables, let $\mathcal{X}$ denote the ordered set of outcomes of the random variable indexed from $-\infty$ to ∞, and let $p_x[m]$ denote the probability mass function (PMF) of x where the index m corresponds to the index of the outcome in $\mathcal{X}$. Then $p_x[m]$ is the probability that the value with the mth index in $\mathcal{X}$ is chosen. Note that $p_x[m] \geq 0$ and $\sum_m p_x[m] = 1$. It is possible to define joint distributions of random variables in both the continuous and discrete cases.

The expectation operator $\mathbb{E}_x[\cdot]$ is used in various characterizations of a random variable. In its most general form, for a continuous-valued random variable and given function $g(x)$,

$$\mathbb{E}_x[g(x)] = \int_{-\infty}^{\infty} g(z)f_x(z)dz. \tag{3.119}$$

The expectation operator satisfies linearity; for example, $\mathbb{E}_x[g(x) + h(x)] = \mathbb{E}_x[g(x)] + \mathbb{E}_x[h(x)]$. The mean of a random variable is

$$m_x = \mathbb{E}_x[x] \tag{3.120}$$

$$= \int_{-\infty}^{\infty} z f_x(z)dz. \tag{3.121}$$

The variance of a random variable is

$$\sigma_x^2 = \mathbb{E}\left[|x - m_x|^2\right] \tag{3.122}$$

$$= \mathbb{E}\left[|x|^2\right] - |m_x|^2. \tag{3.123}$$

The variance is nonnegative.

Random variables x and y are said to be independent if $F_{x,y}(\alpha, \beta) = F_x(\alpha)F_y(\beta)$. Essentially the outcome of x does not depend on the outcome of y. Random variables x and y are said to be uncorrelated if $\mathbb{E}_{x,y}[xy] = 0$. Be careful: being uncorrelated does not imply independence except in certain cases, for example, if x and y are Gaussian.

Example 3.21   Suppose that $f_x(x) = 1/2$ for $x \in [0, 2)$ and is zero elsewhere. Compute the mean and variance of x.

Answer: The mean is

$$\mathbb{E}_x[x] = \int_0^2 z f_x(z)dz \tag{3.124}$$

$$= \int_0^2 0.5z dz \tag{3.125}$$

$$= 0.25z^2|_0^2 \tag{3.126}$$

$$= 0.25(4 - 0) \tag{3.127}$$

$$= 1. \tag{3.128}$$

Now computing the variance:

$$\sigma_x^2 = \mathbb{E}_x\left[|x|^2\right] - |m_x|^2 \tag{3.129}$$

$$= \int_0^2 z^2 f_x(z)\mathrm{d}z - 1^2 \tag{3.130}$$

$$= \int_0^2 \frac{1}{2}z^2\mathrm{d}z - 1 \tag{3.131}$$

$$= \frac{1}{6}z^3\Big|_0^2 - 1 \tag{3.132}$$

$$= \frac{1}{6}(8-0) - 1 \tag{3.133}$$

$$= 8/6 - 6/6 \tag{3.134}$$

$$= 1/3. \tag{3.135}$$

3.2.2   Random Processes

A random or stochastic process is a mathematical concept that is used to model a signal, sequence, or function with random behavior. In brief, a random process is a probabilistic model for a random signal. It is a characterization of an *ensemble* of random signals, not just a single realization. Random processes are used in diverse areas of science, including non-wireless applications such as video denoising and stock market prediction. This book primarily considers discrete-time random processes, though it does require description of the continuous-time case as well for the purpose of defining the bandwidth of the transmitted signal.

Now we extend the definition of random variable to a process. Suppose that each outcome is associated with an entire sequence $\{x[n, \omega_i]\}_{n=-\infty}^{\infty}$, according to some rule. A discrete-time random process is constituted by the sample space $\mathcal{S}$, the probability distribution $\mathbb{P}[\omega_i]$, and the sequences $\{x[n, \omega_i]\}_{n=-\infty}^{\infty}$. Each sequence $\{x[n, \omega_i]\}_{n=-\infty}^{\infty}$ is called a *realization* or *sample path* and the set of all possible sequences $\{x[n, \omega_i]\}_{n=-\infty}^{\infty}$, $\omega_i \in \mathcal{S}$ is called an *ensemble*. The ensemble description of a random process is given in Figure 3.13.

Random processes can also be visualized as collections of random variables at different points in time. We say that $x[n, \omega]$ is a random process if for a fixed value of n, say $n = \bar{n}$, $x[\bar{n}, \omega]$ is a random variable. Having understood the connection with the sample space $\mathcal{S}$, we eliminate the explicit dependence on ω from the notation and pursue the notion that a discrete-time random process is essentially a collection of indexed random variables where n is the index and $x[n]$ is the value of the random variable at index n.

To simplify the exposition, suppose that the random process $x[n]$ is continuous valued. In this way, we suppose that $x[n]$ has a well-defined PDF.

Random processes are fully characterized by knowledge of their joint probability distributions for all orders. When only some joint distributions are known, this leads to a partial characterization. For example, the first-order characterization of $x[n]$ is knowing the CDF $F_{x[n]}(\alpha)$, or equivalently the PDF $f_{x[n]}(\alpha)$, for all n. The second-order characterization of a discrete-time random process is knowing the joint distribution between $x[n_1]$

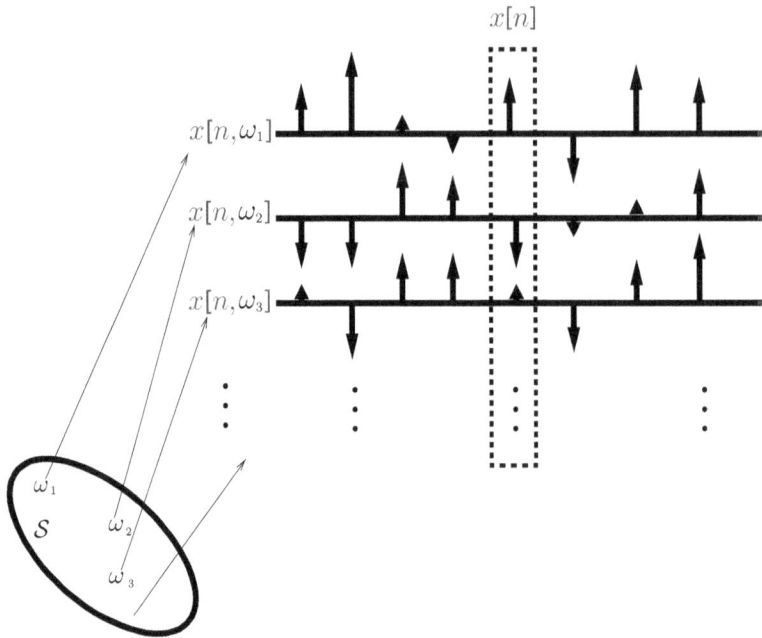

Figure 3.13 Ensemble description of a random process in discrete time. Each outcome corresponds to the realization of a signal.

and $x[n_2]$ for any choice of n_1 and n_2: $F_{x[n_1],x[n_2]}(\alpha_1, \alpha_2)$. Of course, the second-order characterization includes the first-order characterization as a special case. Higher-order characterizations can be defined in a similar fashion.

An important class of random processes consists of those that are independent identically distributed (IID). Such processes satisfy two important properties. First, all of the collections of random variables $(x[n_1], x[n_2], \ldots, x[n_k])$ for $k > 1$ are independent. This means that for any $m \leq k$,

$$F_{x[n_1],x[n_2],\ldots,x[n_m]}(\alpha_1, \alpha_2, \ldots, \alpha_m) = F_{x[n_1]}(\alpha_1) F_{x[n_2]}(\alpha_2) \cdots F_{x[n_m]}(\alpha_m). \quad (3.136)$$

Second, the random variables are identically distributed, thus

$$F_{x[n_k]}(\alpha_k) = F_x(\alpha_k) \quad (3.137)$$

for all $x[n_1], x[n_2], \ldots, x[n_k]$, and

$$F_{x[n_1],x[n_2],\ldots,x[n_m]}(\alpha_1, \alpha_2, \ldots, \alpha_m) = F_x(\alpha_1) F_x(\alpha_2) \cdots F_x(\alpha_m). \quad (3.138)$$

We often model the transmitted data and additive noise as IID.

Example 3.22   Find a random process that models an independent sequence of equally likely bits.

Answer: Let $i[n]$ denote the sequence of bits. Because the process is independent, each $i[n]$ is generated independently. As the distribution is the same for each bit, it is also identically distributed. Let $i[n]$ be a discrete random variable with two possible outcomes: $\mathbb{P}[i = 0] = 1/2$ and $\mathbb{P}[i = 1] = 1/2$. Denoting the set of outcomes $\mathcal{X} = \{0, 1\}$ with index set $\{0, 1\}$, the random process is completely characterized by the first-order probability mass function $p_i[0] = 1/2$ and $p_i[1] = 1/2$.

3.2.3   Moments of a Random Process

The concept of expectation can be extended to provide another characterization of random processes based on their moments. The mean of a random process is

$$m_x[n] = \mathbb{E}_{x[n]}\left[x[n]\right] \tag{3.139}$$

$$= \int_{-\infty}^{\infty} z f_{x[n]}(z) \mathrm{d}z. \tag{3.140}$$

The mean is in general a function of n since the distribution may vary as a function of n. The correlation of a random process is

$$R_{xx}[n_1, n_2] = \mathbb{E}_{x[n_1], x[n_2]}\left[x[n_1]x^*[n_2]\right] \tag{3.141}$$

$$= \int_{-\infty}^{\infty} \int_{-\infty}^{\infty} z_1 z_2^* f_{x[n_1], x[n_2]}(z_1, z_2) \mathrm{d}z_1 \mathrm{d}z_2. \tag{3.142}$$

The covariance of a random process generalizes the concept of variance:

$$C_{xx}[n_1, n_2] = \mathbb{E}_{x[n_1], x[n_2]}\left[x[n_1]x^*[n_2]\right] - m_{x[n_1]}m_{x[n_2]}^*. \tag{3.143}$$

If the process is zero mean, $C_{xx}[n_1, n_2] = R_{xx}[n_1, n_2]$. Most processes in this book are zero mean. The covariance function is conjugate symmetric, that is, $C_{xx}[n_1, n_2] = C_{xx}^*[n_2, n_1]$. This is also true of the correlation function. The variance is

$$\sigma_x^2[n] = C_{xx}[n, n] \tag{3.144}$$

and is a measure of the power of a random signal.

A random process is called *uncorrelated* if

$$C_{xx}[n_1, n_2] = \begin{cases} \sigma_x^2[n_1], & n_1 = n_2 \\ 0, & n_1 \neq n_2. \end{cases} \tag{3.145}$$

Note that the uncorrelated term is applied to the covariance function and not (surprisingly) to the correlation function.

A random process is called *orthogonal* if

$$R_{xx}[n_1, n_2] = \begin{cases} \mathbb{E}_{x[n_1]}|x[n_1]]|^2, & n_1 = n_2 \\ 0, & n_1 \neq n_2. \end{cases} \tag{3.146}$$

An uncorrelated zero-mean process is also an orthogonal process.

The cross-correlation between two stochastic processes $x[n]$ and $y[n]$ is

$$R_{xy}[n_1, n_2] = \mathbb{E}_{x[n_1], y[n_2]}\left[x[n_1]y^*[n_2]\right]. \tag{3.147}$$

Analogously, the covariance between $x[n]$ and $y[n]$ is

$$C_{xy}[n_1, n_2] = \mathbb{E}_{x[n_1], y[n_2]}\left[(x[n_1] - m_x[n_1])(y[n_2] - m_y[n_2])^*\right]. \tag{3.148}$$

3.2.4 Stationarity

Stationarity is a characterization of a random process. A random process is *stationary of order N* if

$$F_{x[n_1], x[n_2], \ldots, x[n_N]}(\alpha_1, \alpha_2, \ldots, \alpha_N) = F_{x[n_1+k], x[n_2+k], \ldots, x[n_N+k]}(\alpha_1, \alpha_2, \ldots, \alpha_N) \tag{3.149}$$

for all k. A process that is stationary for all orders $N, 1, 2, \ldots$ is called *strict-sense stationary* (SSS). Perhaps the most important class of random processes that we deal with are WSS random processes.

WSS is a particular type of stationary that is useful for random systems. A WSS random process $x[n]$ satisfies these criteria:

- The mean is constant $m_{x[n]}[n] = m_x$.

- The covariance (or equivalently the correlation) is only a function of the difference

$$C_{xx}[n_1, n_2] = C_{xx}[n_2 - n_1]. \tag{3.150}$$

No assumptions are made about joint probability distributions associated with a WSS random process. By construction, *an IID random process is also WSS* and any random process that is stationary of order $N \geq 2$ is WSS.

A similar definition of WSS applies for continuous-time random processes. A WSS random process $x(t)$ satisfies these criteria:

- The mean is constant:

$$m_{x(t)}(t) = \int_{-\infty}^{\infty} z f_{x(t)}(z)\mathrm{d}z \tag{3.151}$$

$$= m_x. \tag{3.152}$$

- The covariance (or equivalently the correlation since the mean is constant) is only a function of the difference:

$$C_{xx}(t_1, t_2) = C_{xx}(t_2 - t_1). \tag{3.153}$$

Estimation of the mean and covariance for continuous-time random processes in practice is challenging because a realization of $x(t)$ for all values of t is not necessarily available. Consequently, $x(t)$ is usually sampled and treated as a discrete-time random process (DTRP) for experimental purposes.

There are several useful properties of autocorrelation function $R_{xx}[n] = R_{xx}[n_2 - n_1]$ WSS random processes:

1. **Symmetry**: The autocorrelation sequence of a WSS random process $x[n]$ is a conjugate symmetric function:

$$R_{xx}[n] = R_{xx}^*[-n]. \tag{3.154}$$

When the process is real, the autocorrelation sequence is symmetric: $R_{xx}[n] = R_{xx}[-n]$.

2. **Maximum Value**: The magnitude of the autocorrelation sequence of a WSS random process $x[n]$ is maximum at $n = 0$:

$$R_{xx}[n] \leq R_{xx}[0], \forall n. \tag{3.155}$$

3. **Mean Square Value**: The autocorrelation sequence of a WSS random process at $n = 0$ is equal to the mean square value of the process:

$$R_{xx}[0] = \mathbb{E}[|x[n]|^2]. \tag{3.156}$$

We conclude the discussion of WSS with some example calculations involving computing the mean and covariance of signals and checking if they are WSS. We present classic examples, including a moving average function in Example 3.23, an accumulator in Example 3.24, an autoregressive process in Example 3.25, and a harmonic process in Example 3.26.

Example 3.23   Suppose that $s[n]$ is a real, IID WSS random process with zero mean and covariance $R_{ss}[k] = \sigma_s^2 \delta[k]$. Suppose that $x[n] = s[n] + 0.5s[n - 1]$, which is an example of what is called a moving average process. Compute the mean and covariance of $x[n]$, and determine if it is WSS.

Answer: The mean is

$$\mathbb{E}_s[x[n]] = \mathbb{E}_s\left[s[n] + 0.5s[n - 1]\right] \tag{3.157}$$
$$= 0 \tag{3.158}$$

since $s[n]$ is zero mean and the expectation operator is linear.

The covariance is the same as the correlation function, since the process is zero mean. Computing the correlation function,

$$R_{xx}[n_1, n_2] = \mathbb{E}_s\left[x[n_1]x^*[n_2]\right] \tag{3.159}$$
$$= \mathbb{E}_s\left[(s[n_1] + 0.5s[n_1 - 1])(s[n_2] + 0.5s[n_2 - 1])^*\right] \tag{3.160}$$
$$= \mathbb{E}_s\left[s[n_1]s^*[n_2] + 0.5s[n_1]s^*[n_2 - 1]\right.$$
$$\left. + 0.5s[n_1 - 1]s[n_2] + 0.25s[n_1 - 1]s^*[n_2 - 1]\right] \tag{3.161}$$
$$= R_{ss}[n_2 - n_1] + 0.5R_{ss}[n_2 - n_1 - 1]$$
$$+ 0.5R_{ss}[n_2 - n_1 + 1] + 0.25R_{ss}[n_2 - n_1] \tag{3.162}$$
$$= 1.25R_{ss}[n_2 - n_1] + 0.5R_{ss}[n_2 - n_1 - 1] + 0.5R_{ss}[n_2 - n_1 + 1]. \tag{3.163}$$

The correlation is only a function of the difference $n_2 - n_1$; thus $x[n]$ is WSS and the correlation can be simplified by substituting $k = n_2 - n_1$:

$$R_{xx}[k] = 1.25R_{ss}[k] + 0.5R_{ss}[k-1] + 0.5R_{ss}[k+1]. \tag{3.164}$$

Substituting the given value of $R_{ss}[k]$,

$$R_{xx}[k] = 1.25\sigma_s^2\delta[k] + 0.5\sigma_s^2\delta[k-1] + 0.5\sigma_s^2\delta[k+1]. \tag{3.165}$$

Example 3.24   Toss a fair coin at each n, $-\infty < n < \infty$ and let

$$w[n] = \begin{cases} +1, & \text{if heads is the outcome, } \mathbb{P}(H) = \frac{1}{2} \\ -1, & \text{if tails is the outcome, } \mathbb{P}(T) = \frac{1}{2}. \end{cases} \tag{3.166}$$

$w[n]$ is an IID random process. Define a new random process $x[n]$, $n \geq 0$ as

$$x[0] = w[0] \tag{3.167}$$
$$x[1] = w[0] + w[1] \tag{3.168}$$
$$\vdots \qquad \vdots$$
$$x[n] = \sum_{i=0}^{n} w[i]. \tag{3.169}$$

This is an example of an accumulator. Determine the mean and covariance of $x[n]$ and determine if it is WSS.

Answer: The mean of this process is

$$\mathbb{E}_w[x[n]] = \sum_{i=0}^{n} \mathbb{E}_w[w[i]] \tag{3.170}$$
$$= 0. \tag{3.171}$$

The covariance is the same as the correlation function, since the mean is zero. We can assume $n_2 \geq n_1$ without loss of generality and compute

$$R_{xx}[n_1, n_2] = \mathbb{E}_w[x[n_1]x^*[n_2]] \tag{3.172}$$
$$= \mathbb{E}_w\left[\left(\sum_{i=0}^{n_1} w[i] \right) \left(\sum_{k=0}^{n_2} w[k] \right)^* \right]. \tag{3.173}$$

To simplify the computation, we exploit the fact that $n_1 \leq n_2$ and break up the second summation:

$$R_{xx}[n_1, n_2] = \sum_{i=0}^{n_1}\sum_{k=0}^{n_1} \mathbb{E}_w[w[i]w^*[k]] + \sum_{i=0}^{n_1}\sum_{k=n_1+1}^{n_2} \mathbb{E}_w[w[i]w^*[k]]. \tag{3.174}$$

Note that the expectation of the second term is zero since the random process is independent and zero mean. To see this, note that for two independent random variables x and y, $\mathbb{E}_{x,y}[xy] = \mathbb{E}_x[x]\,\mathbb{E}_y[y]$. To compute the expectation of the first term, we factor out the square products:

$$R_{xx}[n_1, n_2] = \sum_{i=0}^{n_1} \mathbb{E}_w\left[|w[i]|^2\right] + \sum_{i=0}^{n_1} \sum_{k=0,i\neq k}^{n_1} \mathbb{E}_w\left[w[i]w^*[k]\right]. \tag{3.175}$$

Notice that again the second term is zero. Because $w[n]$ is IID,

$$R_{xx}[n_1, n_2] = (n_1 + 1)\mathbb{E}_w[|w[i]|^2] \tag{3.176}$$

$$= (n_1 + 1)\left(\frac{1}{2}(1)^2 + \frac{1}{2}(-1)^2\right) \tag{3.177}$$

$$= (n_1 + 1). \tag{3.178}$$

Since the correlation is not a function of the difference $n_2 - n_1$, the process is not WSS. This process is commonly referred to as a *random walk* or a *discrete-time Wiener process*.

Example 3.25   Let $x[n]$ be a random process generated by

$$x[n] = ax[n - 1] + w[n], \tag{3.179}$$

where $n \geq 0$, $x[-1] = 0$, and $w[n]$ is an IID (μ_w, σ_w^2) process. We can express the process more conveniently as

$$x[0] = ax[-1] + w[0] = w[0] \tag{3.180}$$
$$x[1] = ax[0] + w[1] = aw[0] + w[1] \tag{3.181}$$

$$\vdots \qquad \qquad \vdots$$

$$x[n] = a^n w[0] + a^{n-1}w[1] + a^{n-2}w[2] + \ldots + w[n] \tag{3.182}$$

$$= \sum_{i=0}^{n} a^{n-i}w[i]. \tag{3.183}$$

Determine the mean and covariance of $x[n]$ and determine if it is WSS.
    Answer: The mean of this process is

$$\mathbb{E}_w[x[n]] = \sum_{i=0}^{n} a^{n-i}\mathbb{E}_w[w[i]] \tag{3.184}$$

$$= \mu_w \sum_{i=0}^{n} a^{n-i} \tag{3.185}$$

$$= \mu_w a^n \sum_{i=0}^{n} a^{-i}. \tag{3.186}$$

For $a = 1$, $\mathbb{E}_w[x[n]] = \mu_w(n+1)$, whereas $a \neq 1$, $\mathbb{E}_w[x[n]] = \frac{\mu_w a^n(1-a^{-(n+1)})}{1-a^{-1}} = \frac{\mu_w(1-a^{n+1})}{1-a}$. Hence, for $\mu_w \neq 0$, the mean depends on n and the process is not stationary. For $\mu_w = 0$, however, the mean does not depend on n.

For $\mu_w \neq 0$ and $|a| < 1$, we observe an interesting phenomenon as $n \to \infty$. The asymptotic mean is

$$\lim_{n\to\infty} \mathbb{E}_w[x[n]] = \frac{\mu_w}{1-a}. \tag{3.187}$$

Random processes with this property are said to be asymptotically stationary of order 1. In fact, one can show that the process is asymptotically WSS by studying $\lim_{n_1\to\infty} R_{xx}[n_1, n_2]$ and showing that it is a function of only $n_2 - n_1$.

Example 3.26 Consider a harmonic random process $x[n] = \cos(2\pi f_o n + \phi)$ where ϕ is a random phase that is uniformly distributed on $[-\pi, \pi]$. This process is corrupted by additive white Gaussian noise $v[n] \sim \mathcal{N}(0, \sigma_v^2)$ that is uncorrelated with $x[n]$. The resulting signal $y[n] = x[n] + v[n]$ is available to the receiver. Compute the mean, covariance, and power spectral density of $y[n]$. Is $y[n]$ a WSS random process? Suppose that $x[n]$ is the desired signal and $v[n]$ is the noise. Compute the SNR, which is the ratio between the variance in $x[n]$ and the variance in $v[n]$.

Answer: The mean of $x[n]$ is

$$\mathbb{E}\left[\cos(2\pi f_o n + \phi)\right] = \int_{-\pi}^{\pi} \frac{1}{2\pi} \cos(2\pi f_o n + x) \mathrm{d}x \tag{3.188}$$

$$= 0. \tag{3.189}$$

The covariance of $x[n]$ is the same as the correlation function since $x[n]$ is zero mean:

$$R_{xx}[n, n+\ell] = \mathbb{E}\left[\cos(2\pi f_o n + \phi)\cos(2\pi f_o(n+\ell) + \phi)\right] \tag{3.190}$$

$$= \int_{-\pi}^{\pi} \frac{1}{2\pi} \cos(2\pi f_o n + x)\cos(2\pi f_o(n+\ell) + x)\mathrm{d}x \tag{3.191}$$

$$= \frac{\cos(2\pi f_o \ell)}{2} \tag{3.192}$$

which is not a function of n and thus can be written $R_{xx}[\ell]$. Hence $R_{yy}[\ell] = R_{xx}[\ell] + \sigma_v^2 \delta[\ell]$. The power spectral density of $y[n]$ is simply the DTFT of $R_{xx}[\ell]$ (discussed in more detail in Section 3.2.6):

$$P_y(e^{j2\pi f}) = \sum_{m=-\infty}^{\infty} \left(\frac{1}{2}\delta(f - f_o + m) + \frac{1}{2}\delta(f + f_o + m)\right) + \sigma_v^2. \tag{3.193}$$

Because it is zero mean and the covariance is only a function of the time difference, $y[n]$ is a WSS random process.

The SNR is

$$\frac{R_{xx}[0]}{R_{vv}[0]} = \frac{1}{2\sigma_v^2}. \tag{3.194}$$

The SNR is an important quantity for evaluating the performance of a digital communication system, as discussed further in Chapter 4.

3.2.5 Ergodicity

One challenge of working with measured signals is that their statistical properties may not be known a priori and must be estimated. Assuming a signal is a WSS random process, the key information is captured in the mean and the correlation function. Applying a conventional approach from estimation theory, the mean and correlation can be estimated by generating many realizations and using ensemble averages. Unfortunately, this strategy does not work if only one realization of the signal is available.

Most WSS random processes are *ergodic*. To a signal processing engineer, ergodicity means that one realization of the random process can be used to generate a reliable estimate of the moments. For WSS random processes, the main message is that the mean and correlation function can be estimated from one realization. With enough samples, the sample average converges to the true average of the process, called the ensemble average.

Ergodicity is a deep area of mathematics and physics, dealing with the long-term behavior of signals and systems. Necessary and sufficient conditions for a WSS random process to be mean or covariance ergodic are provided in various books—see, for example, [135, Theorems 3.8 and 3.9]—but discussion of these conditions is beyond the scope of this book. The WSS random processes considered here are (unless otherwise specified) assumed to be covariance ergodic, and therefore also mean ergodic.

The mean and covariance of a covariance-ergodic random process can be estimated from the sample average from *a realization of the process*. With enough samples, the sample average converges to the true average of the process. For example, given observations $x[0], x[1], \ldots, x[N-1]$, then the sample average

$$\widehat{m}_x = \lim_{N \to \infty} \frac{1}{N} \sum_{n=0}^{N-1} x[n] \tag{3.195}$$

converges to the ensemble average $\mathbb{E}[x[n]]$, and

$$\widehat{R}_{xx}[k] = \lim_{N \to \infty} \frac{1}{N-k} \sum_{n=0}^{N-1-k} x[n]x^*[n+k] \tag{3.196}$$

converges to the ensemble correlation function $\mathbb{E}[x[n]x^*[n+k]]$. Convergence is normally in the mean square sense, though other notions of convergence are possible.

Assuming that a WSS random process is ergodic, the sample estimators in (3.195) and (3.196) can be used in place of their ensemble equivalents in subsequent signal processing calculations.

Example 3.27   Suppose that $x[n]$ is real, independent and identically distributed with mean m_x and covariance $C_{xx}[k] = \sigma_x^2 \delta[k]$. Show that this process is ergodic in the mean, in the mean square sense. Essentially, prove that

$$\lim_{N \to \infty} \mathbb{E}_x \left[\left(\frac{1}{N} \sum_{n=0}^{N-1} x[n] \right) - m_x \right]^2 = 0. \tag{3.197}$$

Answer: Let $m_N = \frac{1}{N} \sum_{n=0}^{N-1} x[n]$. First we find the mean of the random variable m_N,

$$\mathbb{E}[m_N] = \mathbb{E} \left[\frac{1}{N} \sum_{n=0}^{N-1} x[n] \right] \tag{3.198}$$

$$= \frac{1}{N} \sum_{n=0}^{N-1} \mathbb{E}[x[n]] \tag{3.199}$$

$$= m_x. \tag{3.200}$$

Second, we find the mean of the squared sample average

$$\mathbb{E}\left[m_N^2 \right] = \mathbb{E} \left[\left(\frac{1}{N} \sum_{n=0}^{N-1} x[n] \right) \left(\frac{1}{N} \sum_{n=0}^{N-1} x[n] \right) \right] \tag{3.201}$$

$$= \frac{1}{N^2} \mathbb{E} \left[\sum_{m=0}^{N-1} \sum_{n=0}^{N-1} x[m]x[n] \right] \tag{3.202}$$

$$= \frac{1}{N^2} \left(\sum_{n=0}^{N-1} \mathbb{E}\left[x[n]^2 \right] + \sum_{n \neq m} \mathbb{E}\left[x[n]x[m] \right] \right) \tag{3.203}$$

$$= \frac{N}{N^2} \mathbb{E}\left[x[n]^2 \right] + \frac{N^2 - N}{N^2} m_x^2 \tag{3.204}$$

$$= \frac{1}{N} \mathbb{E}\left[x[n]^2 \right] - m_x^2 + m_x^2 \tag{3.205}$$

$$= \frac{\sigma_x^2}{N} + m_x^2. \tag{3.206}$$

Then the mean squared error is

$$\mathbb{E}\left[(m_N - m_x)^2 \right] = \mathbb{E}\left[m_N^2 - 2m_N m_x + m_x^2 \right] \tag{3.207}$$

$$= \mathbb{E}\left[m_N^2 \right] - 2m_x \mathbb{E}\left[m_N \right] + m_x^2 \tag{3.208}$$

$$= \frac{\sigma_x^2}{N} + m_x^2 - 2m_x^2 + m_x^2 \tag{3.209}$$

$$= \frac{\sigma_x^2}{N}. \tag{3.210}$$

As long as σ_x^2 is finite, the limit as $N \to \infty$ is zero. Therefore, any IID WSS random process with finite variance is ergodic in the mean, in the mean square sense.

3.2.6 Power Spectrum

The Fourier transform of a random process is not well defined. When it exists, which it normally does not, it is not informative. For WSS random processes, the preferred way to visualize the signal in the frequency domain is through the Fourier transform of the autocovariance function. This is called the power spectrum or the PSD. Loosely, it is a measure of the distribution of the power of a signal as a function of frequency. The power spectrum of a random process is used for all bandwidth calculations instead of the usual Fourier transform.

For continuous-time random processes, the power spectrum is defined using the CTFT as

$$P_x(f) = \int_{-\infty}^{\infty} C_{xx}(t) e^{-\mathrm{j}2\pi ft} \mathrm{d}t. \tag{3.211}$$

For discrete-time signals, it is defined in a similar way using the DTFT as

$$P_x(e^{\mathrm{j}2\pi f}) = \sum_{n=-\infty}^{\infty} C_{xx}[n] e^{-\mathrm{j}2\pi fn}. \tag{3.212}$$

In both cases the power spectrum is real and nonnegative. The real property follows from the conjugate symmetry of the covariance, for example, $C_{xx}[n] = C_{xx}^*[-n]$.

The total power in a random signal is the sum of the power per frequency over all frequencies. This is just the variance of the process. For continuous-time random processes,

$$\sigma_x^2 = C_{xx}(0) \tag{3.213}$$

$$= \int_{-\infty}^{\infty} P_x(f) \mathrm{d}f, \tag{3.214}$$

whereas for a discrete-time random process,

$$\sigma_x^2 = C_{xx}[0] \tag{3.215}$$

$$= \int_{-1/2}^{1/2} P_x\left(e^{\mathrm{j}2\pi f}\right) \mathrm{d}f. \tag{3.216}$$

Examples of PSD calculations are provided in Example 3.28, Example 3.29, and Example 3.30, using direct calculation of the Fourier transform.

Example 3.28 Find the power spectrum for a zero-mean WSS random process $x(t)$ with an exponential correlation function with parameter $\beta > 0$:

$$R_{xx}(t) = e^{-2\beta|t|} \qquad t \in (-\infty, \infty). \tag{3.217}$$

Answer: Because the process is zero mean, computing the power spectrum involves taking the Fourier transform of the autocorrelation function. The answer follows from calculus and simplification:

$$P_x(f) = \int_{-\infty}^{\infty} R_{xx(t)} e^{-j2\pi ft} dt \tag{3.218}$$

$$= \int_{-\infty}^{\infty} e^{-2\beta|t|} e^{-j2\pi ft} dt \tag{3.219}$$

$$= \int_{-\infty}^{0} e^{(2\beta - j2\pi f)t} dt + \int_{0}^{\infty} e^{-(2\beta - j2\pi f)t} dt \tag{3.220}$$

$$= \frac{1}{2\beta - j2\pi f} + \frac{1}{2\beta + j2\pi f} \tag{3.221}$$

$$= \frac{4\beta}{4\beta^2 + 4\pi^2 f^2} \tag{3.222}$$

$$= \frac{\beta}{\beta^2 + \pi^2 f^2}. \tag{3.223}$$

Example 3.29 Find the power spectrum of a zero-mean WSS process $x[n]$ with $C_{xx}[n] = a^{|n|}$, $|a| < 1$.

Answer: Because the process is zero mean, computing the power spectrum again involves taking the Fourier transform of the autocorrelation function. The answer follows by manipulating the summations as

$$P_x(e^{j2\pi f}) = \sum_{n=-\infty}^{\infty} C_{xx}[n] e^{-j2\pi fn} \tag{3.224}$$

$$= \sum_{n=-\infty}^{\infty} a^{|n|} e^{-j2\pi fn} \tag{3.225}$$

$$= \sum_{n=-\infty}^{0} a^{-n} e^{-j2\pi fn} + \sum_{n=1}^{\infty} a^n e^{-j2\pi fn} \tag{3.226}$$

$$= \sum_{m=0}^{\infty} a^m e^{j2\pi fm} + \sum_{n=0}^{\infty} a^n e^{-j2\pi fn} - 1 \tag{3.227}$$

$$= \frac{1}{1 - ae^{j2\pi f}} + \frac{1}{1 - ae^{-j2\pi f}} - 1 \tag{3.228}$$

$$= \frac{1 - a^2}{1 + a^2 - 2a\cos(2\pi f)}. \tag{3.229}$$

Example 3.30   A discrete-time harmonic process is defined by

$$x[n] = \sum_{k=1}^{M} a_k e^{\mathrm{j}(2\pi f_k n + \phi_k)} \tag{3.230}$$

where $\{(a_k, f_k)\}_{k=1}^{M}$ are given constants and $\{\phi_k\}_{k=1}^{M}$ are IID uniformly distributed random variables on $[0, 2\pi]$. Using $\mathbb{E}_\phi[\cdot]$ to denote $\mathbb{E}_{\{\phi_k\}_{k=1}^{M}}[\cdot]$, the mean of this function is

$$\mathbb{E}_\phi[x[n]] = \mathbb{E}_{x[n]}\left[\sum_{k=1}^{M} a_k e^{\mathrm{j}(2\pi f_k n + \phi_k)}\right] \tag{3.231}$$

$$= \sum_{k=1}^{M} a_k \mathbb{E}_{\phi_k}\left[e^{\mathrm{j}(2\pi f_k n + \phi_k)}\right] \tag{3.232}$$

$$= 0 \tag{3.233}$$

which follows from

$$\mathbb{E}_{\phi_k}\left[e^{\mathrm{j}(2\pi f_k n + \phi_k)}\right] = \int_0^{2\pi} e^{\mathrm{j}(2\pi f_k n + z)} \frac{1}{2\pi} \mathrm{d}z \tag{3.234}$$

$$= \frac{e^{\mathrm{j}(2\pi f_k n)}}{2\pi} \int_0^{2\pi} e^{\mathrm{j}z} \mathrm{d}z \tag{3.235}$$

$$= 0, \quad \forall k. \tag{3.236}$$

The autocorrelation function of the signal is given by

$$R_{xx}[\ell] = \mathbb{E}_\phi\left[\left(\sum_{i=1}^{M} a_i e^{\mathrm{j}(2\pi f_i n + \phi_i)}\right)\left(\sum_{k=1}^{M} a_k e^{\mathrm{j}(2\pi f_k (n+\ell) + \phi_k)}\right)^*\right] \tag{3.237}$$

$$= \mathbb{E}_\phi\left[\sum_{k=1}^{M} |a_k|^2 e^{-\mathrm{j}(2\pi f_k \ell)}\right]$$

$$+ \sum_{i=1}^{M}\sum_{k \neq i}^{M} a_i a_k^* \mathbb{E}_\phi\left[e^{\mathrm{j}(2\pi f_i n + \phi_i)} e^{-\mathrm{j}(2\pi f_k (n+\ell) + \phi_k)}\right] \tag{3.238}$$

$$= \sum_{k=1}^{M} |a_k|^2 e^{-\mathrm{j}(2\pi f_k \ell)} + 0 \tag{3.239}$$

$$= \sum_{k=1}^{M} |a_k|^2 e^{-\mathrm{j}(2\pi f_k \ell)}. \tag{3.240}$$

Since the autocorrelation is a sum of complex exponentials, the power spectrum is given by

$$P_x(e^{\mathrm{j}2\pi f}) = \sum_{k=1}^{M} |a_k|^2 \delta(f - f_k). \tag{3.241}$$

When the PSD of an ergodic WSS random process is unknown, it can be estimated
from one realization. This operation is performed by a spectrum analyzer in a lab. There
are several approaches to constructing the estimator. One approach is to use some variant
of a periodogram. The windowed periodogram involves directly estimating the power
spectrum $P_x\left(e^{j2\pi f}\right)$ from N samples of available data $\{x[n]\}_{n=0}^{N-1}$ with window function
$w[n]$, $n = 0, \ldots, N-1$ defined as

$$\widehat{P}_x\left(e^{j2\pi f}\right) = \frac{1}{N}\left|\sum_{n=0}^{N-1} w[n]x[n]e^{-j2\pi fn}\right|. \tag{3.242}$$

This method is simply referred to as the periodogram when the windowing function is
chosen to be a rectangular function. For ease of implementation, the DFT can be used to
compute the windowed periodogram $\widehat{P}_x(e^{j2\pi f})$ at frequencies $f_k = \frac{k}{N}$, $k = 0, \ldots, N-1$:

$$\widehat{P}_x[k] = \frac{1}{N}\left|\sum_{n=0}^{N-1} w[n]x[n]e^{-j2\pi\frac{kn}{N}}\right|. \tag{3.243}$$

The resolution of the resulting estimators is inversely proportional to N. Different window
functions can be used to achieve different trade-offs in main lobe width (which determines
the resolvability of close sinusoids) and sidelobe level (spectral masking).

The periodogram and the windowed periodogram are asymptotically unbiased but
are not consistent estimators. This means that the variance of the error does not go
to zero even with a large number of samples. A solution to this problem is to use
Bartlett's method (periodogram averaging) or Welch's method (averaging with overlap). With Bartlett's method, for example, a measurement is broken into several pieces,
the periodogram is computed in each piece, and the results are averaged together. The
resolution of the estimator decreases (since the transform is taken over a fraction of
N) but the consistency of the estimator is improved. Alternatively, an estimate of the
covariance function $\{\widehat{c}_{xx}[n]\}_{n=-L}^{L}$ for $L < N$ samples can be used (possibly with additional windowing) and the Fourier transform of those samples taken in what is called the
Blackman-Tukey method. All of these methods yield consistent estimators with somewhat more complexity than the periodogram. More information on spectrum estimation
is available in [143, Chapter 8].

3.2.7 Filtering Random Signals

WSS random processes will be used frequently in this book. Because filters are simply
LTI systems, it is of interest to establish the effects of an LTI system on WSS random
inputs. In this section, we characterize the mean and covariance of filtered WSS signals.

First consider an example that requires a brute-force calculation.

Example 3.31   Consider a zero-mean random process $x[n]$ with correlation function
$R_{xx}[n] = (1/2)^{|n|}$. Suppose that $x[n]$ is input into an LTI system with impulse response
$h[n] = \delta[n] + 0.5\delta[n-1] - 0.5\delta[n-2]$. The output is $y[n]$. Compute the mean and
correlation of $y[n]$.

Answer: The output of the system is

$$y[n] = x[n] * h[n] \tag{3.244}$$
$$= x[n] + 0.5x[n-1] - 0.5x[n-2]. \tag{3.245}$$

First, we compute the mean of $y[n]$,

$$\mathbb{E}[y[n]] = \mathbb{E}[x[n]] + 0.5\mathbb{E}[x[n-1]] - 0.5\mathbb{E}[x[n-2]] \tag{3.246}$$
$$= 0, \tag{3.247}$$

where the first equality follows from the linearity of expectation and the second equality follows from the fact that $x[n]$ is zero mean.

Second, we compute the correlation function,

$$R_{yy}[m, m+n] = \mathbb{E}[y[m]y^*[m+n]] \tag{3.248}$$
$$= \mathbb{E}[(x[m] + \frac{1}{2}x[m-1] - \frac{1}{2}x[m-2])(x^*[m+n]$$
$$+ \frac{1}{2}x^*[m+n-1] - \frac{1}{2}x^*[m+n-2])] \tag{3.249}$$
$$= \mathbb{E}[x[m]x[m+n]] + \frac{1}{2}\mathbb{E}[x[m]x^*[m+n-1]] - \frac{1}{2}\mathbb{E}[x[m]x^*[m+n-2]]$$
$$+ \frac{1}{2}\mathbb{E}[x[m-1]x^*[m+n]] + \left(\frac{1}{2}\right)^2 \mathbb{E}[x[m-1]x^*[m+n-1]]$$
$$- \left(\frac{1}{2}\right)^2 \mathbb{E}[x[m-1]x^*[m+n-2]] - \frac{1}{2}\mathbb{E}[x[m-2]x^*[m+n]]$$
$$- \left(\frac{1}{2}\right)^2 \mathbb{E}[x[m-2]x^*[m+n-1]] + \left(\frac{1}{2}\right)^2 \mathbb{E}[x[m-2]x^*[m+n-2]] \tag{3.250}$$
$$= \frac{3}{2}R_{xx}[n] + \frac{1}{4}R_{xx}[n+1] - \frac{1}{2}R_{xx}[n+2] + \frac{1}{4}R_{xx}[n-1] - \frac{1}{2}R_{xx}[n-2] \tag{3.251}$$
$$= \frac{3}{2}\frac{1}{2^{|n|}} + \frac{1}{2^{|n+1|+2}} + \frac{1}{2^{|n-1|+2}} - \frac{1}{2^{|n+2|+1}} - \frac{1}{2^{|n-2|+1}}. \tag{3.252}$$

Note that for $|n| \geq 2$, $|n \pm 1| = |n| \pm 1$ and $|n \pm 2| = |n| \pm 2$,

$$R_{yy}[n] = \frac{1}{2^{|n|}}\left\{1.5 + \frac{1}{2^3} + \frac{1}{2^1} - \frac{1}{2^3} - \frac{1}{2^{-1}}\right\} = 0, \tag{3.253}$$

for $|n| = 1$,

$$R_{yy}[n] = \frac{1.5}{2} + \frac{1}{2^4} + \frac{1}{2^2} - \frac{1}{2^4} - \frac{1}{2^2} = 0.75, \tag{3.254}$$

and

$$R_{yy}[0] = \frac{1.5}{2} + \frac{1}{2^3} + \frac{1}{2^3} - \frac{1}{2^3} - \frac{1}{2^3} = 1.5. \tag{3.255}$$

Therefore,

$$R_{yy}[n] = \begin{cases} 1.5 & \text{if } n = 0 \\ 0.75 & \text{if } |n| = 1 \\ 0 & \text{if } |n| \geq 2. \end{cases} \tag{3.256}$$

We make two observations from Example 3.31. First, the output in this case is WSS. Indeed, this is true in general as we will now show. Second, the brute-force calculation of the autocorrelation is cumbersome. We can simplify the calculations by connecting the output to the autocorrelation of the input and the cross-correlation between the input and the output.

Suppose that

$$y[n] = \sum_{\ell=-\infty}^{\infty} h[\ell]x[n - \ell] \tag{3.257}$$

and that $x[n]$ is a WSS random process with mean m_x, correlation $R_{xx}[k]$, and covariance $C_{xx}[k]$. A pair of random processes $x[n]$ and $y[n]$ are jointly WSS if each is individually WSS and their cross-correlation is only a function of the time difference

$$R_{xy}[n1, n2] = R_{xy}[n1 + k, n2 + k], \tag{3.258}$$

so that the cross-correlation can be described using a single variable:

$$R_{xy}[k] = \mathbb{E}_{xy}\left[x[n]y^*[n + k]\right]. \tag{3.259}$$

Similarly, the cross-covariance function $C_{xy}[k]$ of two jointly WSS random processes $x[n]$ and $y[n]$ is

$$C_{xy}[k] = \mathbb{E}_{xy}\left[x[n]y^*[n + k]\right] - m_x m_y^*. \tag{3.260}$$

Now we compute the mean, cross-correlation, and autocorrelation of the output of (3.257). The mean of the output signal is given by

$$\mathbb{E}_y\left[y[n]\right] = \sum_{\ell=-\infty}^{\infty} h[\ell]\mathbb{E}_x[x[n - \ell]] \tag{3.261}$$

$$= m_x \sum_{\ell=-\infty}^{\infty} h[\ell] \tag{3.262}$$

$$= m_x \mathsf{h}(e^{j2\pi 0}). \tag{3.263}$$

The last step follows from the definition of the DTFT, discussed further in Section 3.1.3. Among other things, notice that if the input is zero mean, then it follows from (3.263) that the output is also zero mean.

The cross-correlation between the input and the output of an LTI system can be computed as

$$R_{xy}[k] = \mathbb{E}_x\left[x[n]y^*[n+k]\right] \tag{3.264}$$

$$= \sum_{\ell=-\infty}^{\infty} h^*[\ell]\mathbb{E}_x\left[x[n]x^*[n+k-\ell]\right] \tag{3.265}$$

$$= \sum_{\ell=-\infty}^{\infty} h^*[\ell]R_{xx}[k-\ell] \tag{3.266}$$

$$= h^*[k] * R_{xx}[k]. \tag{3.267}$$

The cross-correlation can then be computed directly from the convolution between the impulse response and the autocorrelation of the system.

The autocorrelation of $y[n]$ can be calculated as

$$R_{yy}[k] = \mathbb{E}_x[y[n]y^*[n+k]] \tag{3.268}$$

$$= \sum_{\ell=-\infty}^{\infty} h[\ell]\mathbb{E}_x[x[n-\ell]y^*[n+k]] \tag{3.269}$$

$$= \sum_{\ell=-\infty}^{\infty} h[\ell]R_{xy}[k+\ell] \tag{3.270}$$

$$= h[-k] * R_{xy}[k]. \tag{3.271}$$

Substituting (3.267) into (3.271),

$$R_{yy}[k] = h[-k] * h^*[k] * R_{xx}[k]. \tag{3.272}$$

If the LTI system is bounded input bounded output stable, then the quantity $h[-k]*h^*[k]$ is well defined (is not infinity, for example). Then from (3.263) and (3.272), the output process is WSS.

Using these results, in Example 3.32 we reconsider the calculation in Example 3.31.

Example 3.32 Consider the same setup as in Example 3.31. Compute the mean and autocorrelation function.

Answer: From (3.262),

$$\mathbb{E}[y[n]] = \mathbb{E}[x[n]] \sum_{\ell=-\infty}^{\infty} h[\ell] \tag{3.273}$$

$$= 0 \tag{3.274}$$

since the process is zero mean.

To compute the autocorrelation function from (3.272) we need

$$h[-k] * h^*[k] = \sum_{\ell=0}^{2} h^*[\ell]h[\ell-k] \tag{3.275}$$

$$= -\frac{1}{2}\delta[k+2] + \frac{1}{4}\delta[k+1] + \frac{3}{2}\delta[k] + \frac{1}{4}\delta[k-1] - \frac{1}{2}\delta[k-2]. \tag{3.276}$$

Substituting into (3.272) gives

$$R_{yy}[k] = -\frac{1}{2}R_{xx}[k+2] + \frac{1}{4}R_{xx}[k+1] + \frac{3}{2}R_{xx}[k] + \frac{1}{4}R_{xx}[k-1] - \frac{1}{2}R_{xx}[k-2],$$

(3.277)

which is identical to (3.251). Similar simplifications lead to the final answer in (3.256).

Example 3.33   A random process $v[n]$ is white noise if the values $v[n_i]$, $v[n_j]$ are uncorrelated for every n_i and n_j, that is,

$$C_v[n_i, n_j] = 0, \forall n_i \neq n_j.$$

(3.278)

It is usually assumed that the mean value of white noise is zero. In that case, for a WSS white noise process

$$C_v[n_i, n_j] = R_v[n_i, n_j] = R_v[k] = \sigma_v^2 \delta[k],$$

(3.279)

with $k = n_j - n_i$. Suppose that a WSS white noise process $v[n]$ is the input to an LTI system with impulse response $h[n] = \delta[n] + 0.8\delta[n-1]$. Compute the autocorrelation of the output of the LTI system. Is the output an uncorrelated process?

Answer: From (3.272),

$$R_{yy}[k] = h[-k] * h * [k] * R_x x[k]$$

(3.280)

$$= (\delta[k] + 0.8\delta[k+1]) * (\delta[k] + 0.8\delta[k-1]) * \sigma_v^2 \delta[k]$$

(3.281)

$$= (0.8\delta[k+1] + 1.6\delta[k] + 0.8\delta[k-1]) * \sigma_v^2 \delta[k]$$

(3.282)

$$= \sigma_v^2 (0.8\delta[k+1] + 1.6\delta[k] + 0.8\delta[k-1]).$$

(3.283)

From the expression of the autocorrelation it is clear that filtered white noise samples become correlated.

3.2.8   Gaussian Random Processes

The most common model for additive noise is the AWGN (additive white Gaussian noise) distribution. This refers to an IID random process where each sample is chosen from a Gaussian distribution. The real Gaussian distribution is written in shorthand as $\mathcal{N}(m, \sigma^2)$ where for random variable v

$$f_v(x) = \frac{1}{\sqrt{2\pi\sigma^2}} e^{\frac{(x-m)^2}{2\sigma^2}}.$$

(3.284)

We will also encounter the circularly symmetric complex Gaussian distribution, written in shorthand as $\mathcal{N}_{\mathbb{C}}(m, \sigma^2)$. In this case, the real and imaginary parts are independent with distribution $\mathcal{N}(\text{Re}[m], \sigma^2/2)$ and $\mathcal{N}(\text{Im}[m], \sigma^2/2)$. It may be written as

$$f_v(x) = \frac{1}{\pi \sigma^2} e^{\frac{|x-m|^2}{\sigma^2}}, \tag{3.285}$$

where x is complex. The relevant moments for an IID Gaussian random process are

$$m_x = m \tag{3.286}$$

$$C_{xx}[n] = \sigma^2 \delta[n]. \tag{3.287}$$

An important fact about Gaussian random variables is that linear combinations of Gaussian random variables are Gaussian [280]. Because of this, it follows that a filtered Gaussian random process is also a Gaussian random process. This fact is illustrated in Example 3.34.

Example 3.34   Let x and y be two independent $\mathcal{N}(0,1)$ random variables. Show that $z = x + y$ is $\mathcal{N}(0,2)$.

Answer: The distribution of the sum of two independent random variables is given by the convolution of the distributions of the two random variables [252, Section 6.2]. Let $z = x + y$,

$$f_z(z) = f_x(z) * f_y(z) \tag{3.288}$$

$$= \frac{1}{2\pi} \int_{-\infty}^{\infty} e^{\frac{t^2 - (z-t)^2}{2}} \, dt \tag{3.289}$$

$$= \frac{1}{2\pi} e^{-z^2/4} \int_{-\infty}^{\infty} e^{-(t-z/2)^2} \, dt \tag{3.290}$$

$$= \frac{1}{2\sqrt{\pi}} e^{-z^2/4} \int_{-\infty}^{\infty} \frac{1}{\sqrt{2\pi \frac{1}{2}}} e^{-\frac{(t-z/2)^2}{2\frac{1}{2}}} \, dt \tag{3.291}$$

$$= \frac{1}{2\sqrt{\pi}} e^{-z^2/4} \tag{3.292}$$

$$= \frac{1}{\sqrt{2\pi \times 2}} e^{-\frac{(z-0)^2}{2 \times 2}}. \tag{3.293}$$

Thus, z is $\mathcal{N}(0,2)$. This can be generalized to show that if x is $\mathcal{N}(m_x, \sigma_x^2)$ and y is $\mathcal{N}(m_y, \sigma_y^2)$ and they are independent, then $z = x + y$ is $\mathcal{N}(m_x + m_y, \sigma_x^2 + \sigma_y^2)$.

3.2.9   Random Vectors and Multivariate Random Processes

The concepts of a random variable and a random process can be extended to vector-valued random variables and vector-valued random processes. This is useful for MIMO communication. In the case of random vectors, each point ω_i in the sample space $\mathcal{S}$ is associated with a vector $\mathbf{x} \in \mathbb{R}^M$ or $\mathbb{C}^M$ instead of a scalar value. In the case of a random vector process, each point ω_i in the sample space $\mathcal{S}$ is associated with a sequence of vectors $\mathbf{x}[n, \omega_i]$. Hence, for a fixed value of n, say $n = \bar{n}$, $\mathbf{x}[\bar{n}, \omega]$ is a vector random variable. This section defines some quantities of interest for vector random processes that will be useful in subsequent chapters.

The mean of a random vector process $\mathbf{x}[n]$ of length M is

$$\mathbf{m_x}[n] = \mathbb{E}_{\mathbf{x}[n]}[\mathbf{x}[n]] \tag{3.294}$$

$$= \begin{bmatrix} \mathbb{E}_{x_1[n]}[x_1[n]] \\ \mathbb{E}_{x_2[n]}[x_2[n]] \\ \vdots \\ \mathbb{E}_{x_M[n]}[x_M[n]] \end{bmatrix}. \tag{3.295}$$

The covariance matrix of a random vector process is

$$\mathbf{C_{xx}}[n_1, n_2] = \mathbb{E}_{\mathbf{x}[n_1], \mathbf{x}[n_2]} \left[(\mathbf{x}[n_1] - \mathbf{m_x}[n]) (\mathbf{x}[n_2] - \mathbf{m_x}[n_2])^* \right] \tag{3.296}$$

$$= \mathbb{E}_{\mathbf{x}[n_1], \mathbf{x}[n_2]} \left[\mathbf{x}[n_1] \mathbf{x}^*[n_2] \right] - \mathbf{m_x}[n_1] \mathbf{m_x^*}[n_2]. \tag{3.297}$$

A similar definition is possible for the correlation matrix $\mathbf{R_{xx}}[n_1, n_2]$.

WSS vector random processes have a constant mean $\mathbf{m_x}$ and a covariance that is only a function of the time difference $\mathbf{C_{xx}}[n] = \mathbf{C_{xx}}[n_2 - n_1]$. For IID WSS vector random processes, $\mathbf{C_{xx}}[n] = \mathbf{C}\delta[n]$ where $\mathbf{C}$ is known simply as the covariance matrix. If the entries of the vector are also independent, then $\mathbf{C}$ is a diagonal matrix. If they are further identically distributed, then $\mathbf{C} = \sigma^2 \mathbf{I}$. We deal primarily with IID WSS vector random processes in this book.

Covariance matrices have some interesting properties that have been exploited in statistical signal processing to develop fast algorithms [143]. They are conjugate symmetric with $\mathbf{C}^* = \mathbf{C}$. They are also positive semidefinite, meaning that

$$\mathbf{x}^* \mathbf{C} \mathbf{x} \geq 0, \text{ for all } \mathbf{x} \in \mathbb{C}^M. \tag{3.298}$$

From the *spectral theorem*, for all conjugate symmetric matrices, there exists a decomposition

$$\mathbf{C} = \mathbf{Q}^* \mathbf{\Lambda} \mathbf{Q}, \tag{3.299}$$

where $\mathbf{Q}$ is a unitary matrix and $\mathbf{\Lambda}$ is a diagonal matrix. The columns of $\mathbf{Q}$ contain the eigenvectors of $\mathbf{C}$, and $\mathbf{\Lambda}$ contains the eigenvalues of $\mathbf{C}$ on its diagonal. The eigenvectors and eigenvalues can be obtained from the eigendecomposition of $\mathbf{C}$. In many cases it is desirable that the eigenvalues be ordered from largest to smallest. This can be achieved using the singular value decomposition [316, 354].

Multivariate Gaussian distributions [182] are widely used in the analysis of MIMO communication systems, as discussed more in Chapter 6. The multivariate real Gaussian distribution is written as $\mathcal{N}(\mathbf{m_v}, \mathbf{C_v})$ for random variable $\mathbf{v}$ where

$$f_{\mathbf{v}}(x) = \frac{1}{(2\pi)^{\frac{M}{2}} |\mathbf{C_v}|^{\frac{1}{2}}} e^{-\frac{(\mathbf{x}-\mathbf{m_v})^{\mathrm{T}} \mathbf{C_v}^{-1}(\mathbf{x}-\mathbf{m_v})}{2}}. \tag{3.300}$$

The multivariate circularly symmetric complex Gaussian distribution is written as $\mathcal{N}_{\mathbb{C}}(\mathbf{m_v}, \mathbf{C_v})$ where

$$f_{\mathbf{v}}(x) = \frac{1}{\pi^M |\mathbf{C_v}|} e^{-(\mathbf{x}-\mathbf{m_v})^* \mathbf{C_v}^{-1}(\mathbf{x}-\mathbf{m_v})}. \tag{3.301}$$

Noise in a MIMO system is modeled as an IID multivariate random process with $\mathcal{N}_{\mathbb{C}}(\mathbf{m_v}, \mathbf{C_v})$.

The results of filtering a WSS random process can be extended to multivariate random processes by convolving $\mathbf{x}[n]$ with a multivariate impulse response $\mathbf{H}[n]$ as $\mathbf{y}[n] = \sum_{\ell} \mathbf{H}[\ell]\mathbf{x}[n-\ell]$. We provide an illustration for a special case in Example 3.35. The other results in Section 3.2.7 can be similarly extended with some care taken to preserve the proper order of vector and matrix operations.

Example 3.35 Suppose that $\mathbf{x}[n]$ is a WSS random process with mean $\mathbf{m_x}$ and covariance $\mathbf{C_x}[k]$. Suppose that $\mathbf{x}[n]$ is filtered by $\mathbf{H}[n] = \mathbf{H}\delta[n]$ to produce $\mathbf{y}[n]$. Determine the mean and covariance of $\mathbf{y}[n]$.

Answer: The output signal is

$$\mathbf{y}[n] = \sum_{\ell} \mathbf{H}[\ell]\mathbf{x}[n-\ell] \tag{3.302}$$

$$= \mathbf{H}\mathbf{x}[n]. \tag{3.303}$$

The mean is

$$\mathbb{E}\left[\mathbf{y}[n]\right] = \mathbb{E}\left[\mathbf{H}\mathbf{x}[n]\right] \tag{3.304}$$

$$\mathbb{E}\left[\mathbf{H}\mathbf{x}[n]\right] = \mathbf{H}\mathbb{E}\left[\mathbf{x}[n]\right] \tag{3.305}$$

$$= \mathbf{H}\mathbf{m_x}. \tag{3.306}$$

Given $\mathbf{R_{xx}}[k] = \mathbf{C_{xx}}[k] + \mathbf{m_x}\mathbf{m_x^*}$, the correlation matrix of $\mathbf{y}[n]$ is

$$\mathbf{R_{yy}}[k] = \mathbb{E}\left[\mathbf{y}[n]\mathbf{y}^*[n+k]\right] \tag{3.307}$$

$$= \mathbb{E}\left[\mathbf{H}\mathbf{x}[n]\mathbf{x}^*[n+k]\mathbf{H}^*\right] \tag{3.308}$$

$$= \mathbf{H}\mathbf{R_{xx}}[k]\mathbf{H}^*, \tag{3.309}$$

thus

$$\mathbf{C_{yy}}[k] = \mathbf{H R_{xx}}[k]\mathbf{H}^* - \mathbf{H m_x m_x^* H}^* \tag{3.310}$$
$$= \mathbf{H}\left(\mathbf{C_{xx}}[k] + \mathbf{m_x m_x^*}\right)\mathbf{H}^* - \mathbf{H m_x m_x^* H}^* \tag{3.311}$$
$$= \mathbf{H C_{xx}}[k]\mathbf{H}^*. \tag{3.312}$$

3.3 Signal Processing with Passband Signals

Wireless communication systems use *passband signals*, where the energy of the signal is concentrated around a carrier frequency f_c. In this section, we present the fundamentals of signal processing for passband signals. The key idea is to represent a passband signal through its *complex baseband signal representation*, which is also called its *complex envelope*. In this way, it is possible to decompose a passband signal into two terms, one that is a baseband signal without dependence on f_c and another a complex sinusoid, which is a function of f_c. A main advantage of the complex baseband signal representation is that it permits dealing with passband signals in a way that does not require explicit mention of the carrier frequency f_c. This is convenient because the carrier frequency is typically added in the RF hardware. As a result, complex baseband signal representations permit an important abstraction of critical analog portions of a wireless system.

This section reviews important concepts associated with passband signals. First, we describe upconversion, the process of creating a passband signal from a complex baseband signal. Then we describe downconversion, the process of extracting a complex baseband signal from a passband signal. In a wireless communication system, upconversion is performed at the transmitter whereas downconversion is performed at the receiver. Then we develop a complex baseband equivalent representation of a system that filters passband signals, and a related concept called the pseudo-complex baseband equivalent channel. We conclude with a discrete-time equivalent system, in which the transmitted signal, LTI system, and received signal are all represented in discrete time. This final formulation is used extensively in subsequent chapters to build signal processing algorithms for wireless communication systems.

3.3.1 Upconversion—Creating a Passband Signal

A real continuous-time signal $x_p(t)$ is a passband signal around carrier frequency f_c if its Fourier transform $x(f) = 0$ for $|f| \notin [f_c - f_{low}, f_c + f_{high}]$ where $f_{low} > 0$ and $f_{high} > 0$. For convenience, in this book double-sided signals are considered where $f_{high} = f_{low} = B/2$. The exposition can be generalized to remove this assumption. The bandwidth of the signal is B, where the bandwidth is computed according to the definitions in Section 3.1.4. For the mathematical derivation, the continuous-time signals are assumed to be ideally bandlimited.

Complex envelope notation is used to represent passband signals in a convenient mathematical framework. To develop the complex baseband signal representation, the bandwidth must be sufficiently smaller than the carrier frequency: $B \ll f_c$. This is known as the *narrowband assumption*. The narrowband assumption is not limiting as it applies to virtually every wireless communication system except for some ultra-wideband systems [281].

Now we provide a representation of a passband signal $x_p(t)$ and subsequently simplify it to obtain its baseband equivalent. Suppose that $x_p(t)$ is a narrowband passband signal with carrier f_c. A passband signal can be written as

$$x_p(t) = A(t)\cos(2\pi f_c t + \phi(t)), \qquad (3.313)$$

where $A(t)$ is an amplitude function and $\phi(t)$ is a phase function. In AM (amplitude modulation), information is contained in the $A(t)$, whereas in FM (frequency modulation), information is contained in the derivative of the phase term $\phi(t)$. The formulation in (3.313) is not convenient from a signal processing perspective since the $\phi(t)$ occurs in conjunction with the carrier inside the cosine term.

Another way to represent passband signals is to use inphase and quadrature notation. Recall the trigonometric identity $\cos(A+B) = \cos(A)\cos(B) - \sin(A)\sin(B)$. Simplifying (3.313) leads to

$$x_p(t) = \underbrace{A(t)\cos(\phi(t))}_{x_i(t)}\cos(2\pi f_c t) - \underbrace{A(t)\sin(\phi(t))}_{x_q(t)}\sin(2\pi f_c t). \qquad (3.314)$$

The signals $x_i(t)$ and $x_q(t)$ are called the inphase and quadrature components. They are both real by construction. Based on (3.314), the passband signal $x_p(t)$ can be generated from inphase $x_i(t)$ and quadrature $x_q(t)$ signals through multiplication by a carrier. This process is illustrated in Figure 3.14 and is known as *upconversion*. The multipliers are implemented using mixers in hardware. An oscillator generates a cosine wave, and a delay element would be used to generate a sine wave. More elaborate architectures may use multiple stages of mixing and filtering [271].

Using complex envelope notation, the inphase and quadrature signals are combined into a single complex signal. Let $x(t) = x_i(t) + jx_q(t)$ denote the complex baseband equivalent or complex envelope of $x_p(t)$. Recall from Euler's formula that $e^{j\theta} = \cos(\theta) + j\sin(\theta)$. Multiplying the complex baseband equivalent by the complex sinusoid $e^{j2\pi f_c t}$, then

$$\begin{aligned} x(t)e^{j2\pi f_c t} &= x_i(t)\cos(2\pi f_c t) - x_q(t)\sin(2\pi f_c t) \\ &\quad + j(x_i(t)\sin(2\pi f_c t) + x_q(t)\cos(2\pi f_c t)). \end{aligned} \qquad (3.315)$$

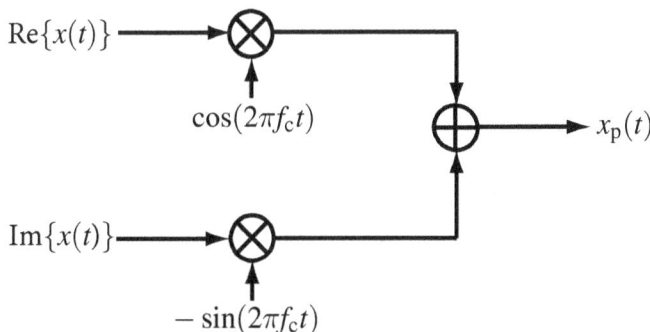

Figure 3.14 Mathematical block diagram model for direct upconversion of a baseband signal to a passband signal. The real analog RF hardware may not perform a direct conversion but rather may have a series of upconversions.

The real part corresponds to the passband signal in (3.314). The result can be written more compactly as

$$x_p(t) = \text{Re}\left[x(t)e^{j2\pi f_c t}\right].\tag{3.316}$$

The expression in (3.316) shows how $x_p(t)$ is separated into a term that depends on the complex baseband equivalent and a term that depends on the carrier f_c.

Now we establish why $x(t)$ in (3.316) is necessarily bandlimited with bandwidth $B/2$. First rewrite (3.316) as

$$x_p(t) = \frac{1}{2}x(t)e^{j2\pi f_c t} + \frac{1}{2}x^*(t)e^{-j2\pi f_c t}\tag{3.317}$$

to remove the real operation. Using the Fourier transform properties from Table 3.1,

$$\mathsf{x}_p(f) = \frac{1}{2}\left(\mathsf{x}(f - f_c) + \mathsf{x}^*(-f - f_c)\right).\tag{3.318}$$

Essentially, a scaled version of $\mathsf{x}(f)$ is shifted to f_c, and its mirror image $\mathsf{x}^*(-f)$ is shifted to $-f_c$. The mirroring of the spectrum in (3.318) preserves the conjugate symmetry in the frequency domain, which is expected since $x_p(t)$ is a real signal. The main implication of (3.318) is that if $x(t)$ is a baseband signal with bandwidth $B/2$, then $x_p(t)$ is a passband signal with bandwidth B. This process is illustrated in Figure 3.15.

Example 3.36   Let $x(t)$ be the complex envelope of a real-valued passband signal $x_p(t)$. Prove that $x(t)$ is generally a complex-valued signal and state the condition under which $x(t)$ is real valued.

Answer: Because $x_p(t)$ is a passband signal, it can be written as

$$x_p(t) = \text{Re}[x(t)e^{j2\pi f_c t}]\tag{3.319}$$
$$= \text{Re}[x(t)(\cos(2\pi f_c t) + j\sin(2\pi f_c t)]\tag{3.320}$$
$$= \text{Re}[x(t)]\cos(2\pi f_c t) - \text{Im}[x(t)]\sin(2\pi f_c t).\tag{3.321}$$

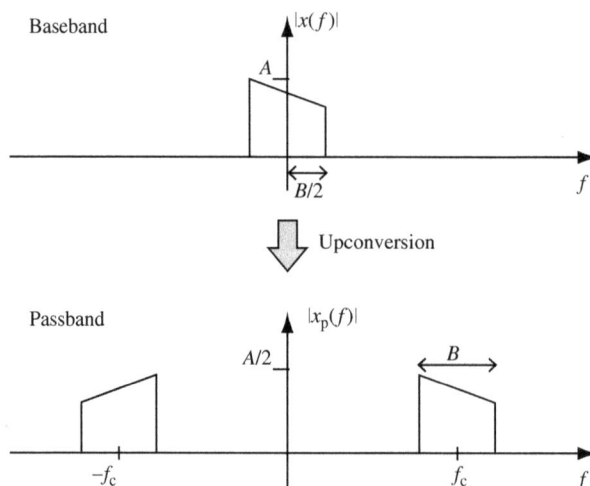

Figure 3.15 Upconversion in the frequency domain, which takes a baseband signal to a passband signal. The scaling factor A is used to show how the amplitude scales in each case.

By equating like terms in (3.321) and (3.314), we see that $\mathrm{Re}[x(t)] = x_{\mathrm{i}}(t)$ and $\mathrm{Im}[x(t)] = x_{\mathrm{q}}(t)$. Therefore, $x(t)$ is generally complex valued. It is real valued when $x_{\mathrm{q}}(t) = 0$, that is, $x_{\mathrm{p}}(t)$ does not have a quadrature component.

3.3.2 Downconversion—Extracting a Complex Baseband Signal from a Passband Signal

Downconversion is the reverse of upconversion. It is the process of taking a passband signal and extracting the complex baseband equivalent signal. In this section, we explain the process of downconversion.

Suppose that $y_{\mathrm{p}}(t)$ is a passband signal. Because it is passband, there exists an equivalent baseband $y(t)$ such that

$$y_{\mathrm{p}}(t) = \mathrm{Re}\left[y(t)e^{\mathrm{j}2\pi f_{\mathrm{c}}t}\right]. \tag{3.322}$$

To understand how downconversion works, recall the following trigonometric identities:

$$\sin(u)\sin(v) = \frac{1}{2}\left[\cos(u-v) - \cos(u+v)\right] \tag{3.323}$$

$$\cos(u)\cos(v) = \frac{1}{2}\left[\cos(u-v) + \cos(u+v)\right] \tag{3.324}$$

$$\sin(u)\cos(v) = \frac{1}{2}\left[\sin(u-v) + \sin(u+v)\right]. \tag{3.325}$$

Using (3.323)–(3.325):

$$y_{\mathrm{p}}(t)\cos(2\pi f_{\mathrm{c}}t) = \frac{1}{2}y_{\mathrm{i}}(t) + \frac{1}{2}y_{\mathrm{i}}(t)\cos(4\pi f_{\mathrm{c}}t) - \frac{1}{2}y_{\mathrm{q}}(t)\sin(4\pi f_{\mathrm{c}}t) \tag{3.326}$$

$$y_{\mathrm{p}}(t)\sin(2\pi f_{\mathrm{c}}t) = -\frac{1}{2}y_{\mathrm{q}}(t) + \frac{1}{2}y_{\mathrm{q}}(t)\cos(4\pi f_{\mathrm{c}}t) + \frac{1}{2}y_{\mathrm{i}}(t)\sin(4\pi f_{\mathrm{c}}t). \tag{3.327}$$

The $2f_{\mathrm{c}}$ components of the received signal are significantly higher than the baseband components. To extract the baseband components, therefore, it suffices to apply a lowpass filter to the outputs of (3.326) and (3.327) and correct for the scaling factor.

We can explain downconversion mathematically in a compact fashion. An ideal lowpass filter with unity gain and cutoff $B/2$ has a Fourier transform given by $\mathrm{rect}(f/B)$ or $B\mathrm{sinc}(tB)$ in the time domain from Table 3.2. The downconverted signal can be written as

$$y(t) = 2B\mathrm{sinc}(tB) * \left(y_{\mathrm{p}}(t)e^{-\mathrm{j}2\pi f_{\mathrm{c}}t}\right). \tag{3.328}$$

This expression captures the key ideas in a way that leads to a complex baseband signal.

Figure 3.16 shows a block diagram for downconversion. Effectively, the downconversion operation shifts the spectrum of the passband signal through multiplication by sine and cosine, then filters out the remaining higher-frequency replicas and corrects for the scaling factor. As is the case for upconversion, there are many practical hardware architectures to implement downconversion [271].

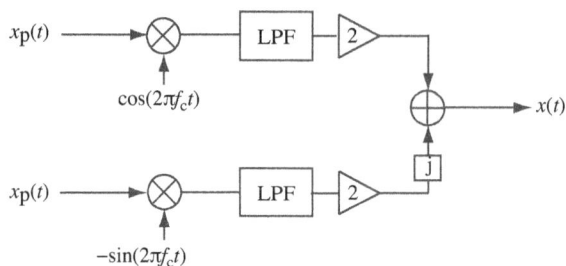

Figure 3.16 Mathematical block diagram for direct downconversion of a passband signal to a baseband signal. The real analog may not perform a direct conversion but rather may have a series of upconversions and downconversions. Additionally, there is typically a low-noise amplifier and a bandlimiting filter just after the antenna and a series of gain control circuits.

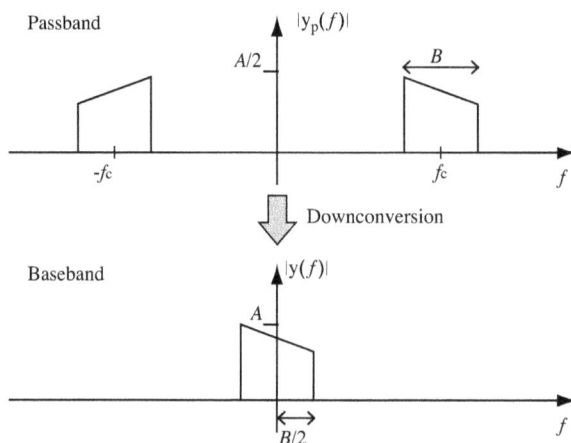

Figure 3.17 Downconversion in the frequency domain, which takes a passband signal to a baseband signal. The scaling factor A is used to show how the amplitude scales in each case.

It is useful to understand how downconversion works in the frequency domain. Taking the Fourier transform of (3.328), the downconversion process can be expressed mathematically as

$$y(f) = 2 \, \text{rect} \left(\frac{f}{B} \right) y_\text{p}(f + f_\text{c}). \tag{3.329}$$

Essentially, as illustrated in Figure 3.17, the entire spectrum of $y(t)$ is shifted down by f_c, and then a filter is applied to remove the high-frequency image.

Equations (3.329) and (3.328) assume an ideally lowpass filter exactly matched to the bandwidth of the signal, but a wider nonideal lowpass filter could be substituted, leading to the same mathematical result. In practice, the bandwidth is typically taken to be near B for considerations related to noise and adjacent channel interference rejection.

The connection between baseband and passband signals is important in viewing wireless communication systems through the lens of signal processing. Throughout this book, when dealing with wireless communication signals, only their complex baseband equivalents are considered. Issues related to imperfections in the upconversion and downconversion, due to differences between the carrier frequency at the transmitter and the receiver, are explored further in Chapter 5.

3.3.3 Complex Baseband Equivalent Channel

Thus far we have established a convenient representation for generating passband signals from complex baseband signals, and for extracting complex baseband signals from passband signals. Upconversion is used at the transmitter to create a passband signal, and downconversion is used at the receiver to extract a baseband signal from a passband signal. These are key operations performed by the analog front end in a wireless communication system.

From a signal processing perspective, it is convenient to work exclusively with complex baseband equivalent signals. A good model for the wireless channel is an LTI system as argued in Section 3.1.2. This system, though, is applied to the transmitted passband signal $x_p(t)$ to produce the received passband signal

$$y_p(t) = \int h_c(t-\tau)x_p(\tau)d\tau \tag{3.330}$$

as illustrated in Figure 3.18. Because it is at passband, the input-output relationship in (3.330) is a function of the carrier frequency. Based on the results in Section 3.3.2, it would be convenient to have an equivalent baseband representation that is just a function of the complex envelopes of the input signal and the output signal for some baseband channel $h(t)$. In other words, it would be nice to find a relationship

$$y(t) = \int h(t-\tau)x(\tau)d\tau \tag{3.331}$$

for some suitable $h(t)$. The complex baseband equivalent channel $h(t)$ thus acts on the input signal $x(t)$ to produce $y(t)$, all at baseband.

In this section, we summarize the steps in the derivation of the complex baseband equivalent channel $h(t)$ from the real channel $h_c(t)$ that is not bandlimited. The idea illustrated in Figure 3.18 is to recognize that only a portion of the channel response impacts the passband signal. Then an equivalent passband channel response can be considered as illustrated in Figure 3.19, without any loss. Finally, the passband channel is converted to its baseband equivalent in Figure 3.20.

Now we provide the mathematical steps for deriving the complex baseband equivalent channel. We work with the frequency domain, following Figure 3.20. The objective is to find the baseband equivalent channel $h(f)$ such that $y(f) = h(f) \times (f)$. Attention is needed to the scaling factors in the derivation.

Consider the received signal

$$y_p(f) = h_c(f)x_p(f). \tag{3.332}$$

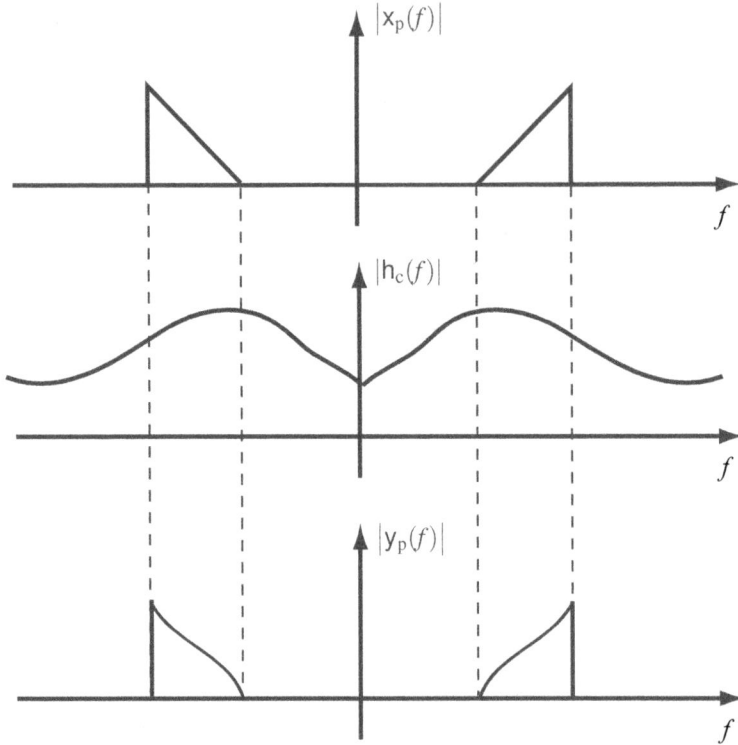

Figure 3.18 Convolution between the passband signal and the channel. In this case the total response of the channel $h_c(f)$ is illustrated.

Because $x_p(f)$ is a passband signal, only the part of $h_c(f)$ that is within the passband of $x_p(f)$ is important. Let us denote this portion of the channel as $h_p(f)$.

The passband channel $h_p(f)$ is derived by filtering $h(f)$ with an ideal passband filter. Assuming that $x_p(f)$ has a passband bandwidth of B, the corresponding ideal passband filter with bandwidth of B centered around f_c is

$$p(f) = \text{rect}\left(\frac{f - f_c}{B}\right) \; + \; \text{rect}\left(\frac{-(f + f_c)}{B}\right) \qquad (3.333)$$

$$= \text{rect}\left(\frac{f - f_c}{B}\right) \; + \; \text{rect}\left(\frac{f + f_c}{B}\right) \qquad (3.334)$$

or equivalently in the time domain is

$$p(t) = 2B \cos(2\pi f_c t)\text{sinc}(Bt). \qquad (3.335)$$

Since $x_p(f) = x_p(f)p(f)$, it follows that

$$y_p(f) = h_c(f)p(f)x_p(f) \qquad (3.336)$$

$$= h_p(f)x_p(f) \qquad (3.337)$$

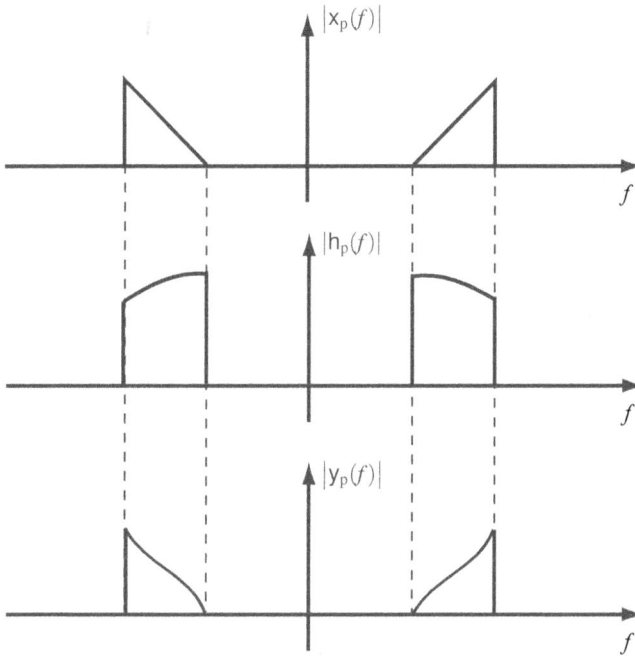

Figure 3.19 Convolution between the passband signal and the channel

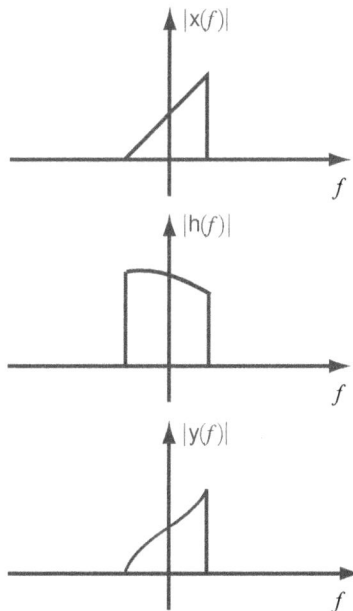

Figure 3.20 Convolution between the baseband signal and the baseband equivalent channel

where

$$h_p(f) = p(f)h_c(f) \tag{3.338}$$

is the passband filtered channel.

Now suppose that $h(f)$ is the complex baseband equivalent channel so that $y(f) = h(f) \times (f)$. The passband signal obtained using (3.318) is then

$$y_p(f) = \frac{1}{2}h(f - f_c) \times (f - f_c) + \frac{1}{2}h^*(-f - f_c) \times^*(-f - f_c). \tag{3.339}$$

But it is also true that

$$y_p(f) = h_p(f) \times_p(f). \tag{3.340}$$

Let $h_b(f)$ be the complex baseband equivalent of $h_p(f)$. Substituting for $h_p(f)$ and $x_p(f)$ again using (3.318),

$$y_p(f) = \left(\frac{1}{2}h_b(f - f_c) + \frac{1}{2}h_b^*(-f - f_c)\right)\left(\frac{1}{2}\times(f - f_c) + \frac{1}{2}\times^*(-f - f_c)\right) \tag{3.341}$$

$$= \frac{1}{4}h_b(f - f_c) \times (f - f_c) + \frac{1}{4}h_b^*(-f - f_c) \times^*(-f - f_c). \tag{3.342}$$

Equating terms in (3.342) and (3.339), it follows that

$$h(f) = \frac{1}{2}h_b(f). \tag{3.343}$$

The factor of $1/2$ arises because the passband channel was obtained by passband filtering a non-bandlimited signal, then downconverting. With similar calculations as in (3.329),

$$h(f) = \frac{1}{2}\, 2\, \text{rect}\left(\frac{f}{B}\right)h_p(f + f_c) \tag{3.344}$$

$$= \text{rect}\left(\frac{f}{B}\right)\left[\text{rect}\left(\frac{f}{B}\right)h_c(f + f_c) + \text{rect}\left(\frac{f + f_c + f_c}{B}\right)h_c(f - f_c)\right] \tag{3.345}$$

$$= \text{rect}\left(\frac{f}{B}\right)h_c(f + f_c). \tag{3.346}$$

Compared with (3.329), the factor of $1/2$ cancels the factor of 2 in (3.346). The entire process of generating the complex baseband equivalent channel is illustrated in Figure 3.21.

The time-domain response follows directly from the Fourier transform properties in Table 3.1:

$$h(t) = B\text{sinc}(Bt) * h_c(t)e^{j2\pi f_c t} \tag{3.347}$$

$$= B\int \text{sinc}(B(t - \tau))h_c(\tau)e^{j2\pi f_c \tau}d\tau. \tag{3.348}$$

Essentially, the complex baseband equivalent channel is a demodulated and filtered version of the continuous-time channel $h_c(t)$. The time-domain operations are illustrated in Figure 3.22. Several example calculations are provided in Examples 3.37 and 3.38.

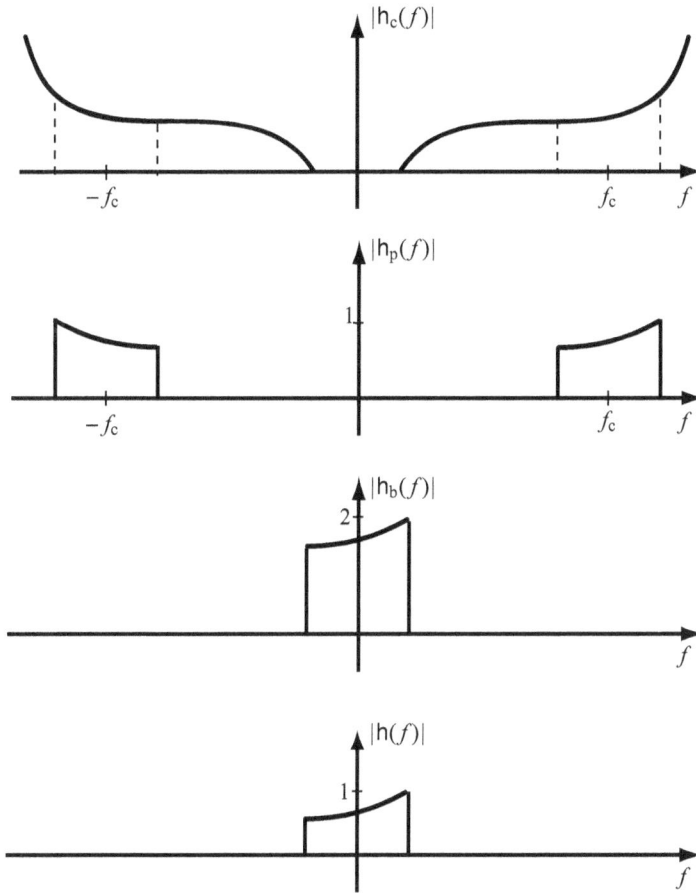

Figure 3.21 Frequency-domain illustration of the transformation from $h_c(f)$ to $h_p(f)$ to $h_b(f)$ to $h(f)$. An arbitrary scaling is selected in $h_p(f)$ just to illustrate subsequent rescalings.

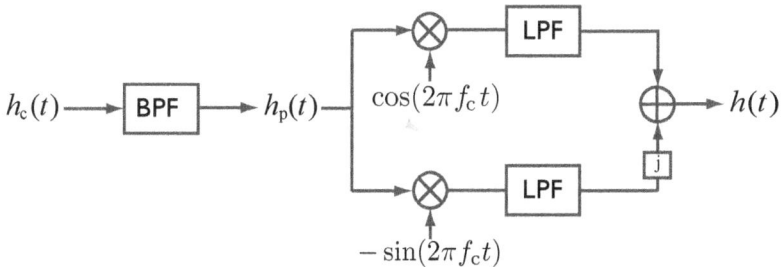

Figure 3.22 Creating the complex baseband equivalent by demodulating the channel

Example 3.37   Suppose that the signal has to propagate over a distance of 100m to arrive at the receiver and undergoes an attenuation of 0.1. Find a linear time-invariant model for this channel $h_c(t)$, then find the passband channel $h_p(t)$ and the baseband equivalent channel $h(t)$. The baseband signal bandwidth is 5MHz and the carrier frequency is $f_c = 2\text{GHz}$.

Answer: Consider the speed of propagation to be $c = 3 \times 10^8 m/s$. Then the delay between the transmitter and the receiver is $\tau_d = 100/c = 1/3\mu s = 1/3 \times 10^{-6}$s. Consequently, the channel can be modeled as

$$h_c(t) = 0.1\delta(t - 1/3 \times 10^{-6}). \tag{3.349}$$

The baseband signal bandwidth is 5MHz; thus the passband bandwidth is $B = 10\text{MHz}$. Now

$$h_p(t) = h_c(t) * p(t) \tag{3.350}$$

$$= \int p(t - \tau)h_c(\tau)d\tau \tag{3.351}$$

$$= 0.1 \int p(t - \tau)\delta(t - 1/3 \times 10^{-6})d\tau \tag{3.352}$$

$$= 0.1\, p(t - 1/3 \times 10^{-6}) \tag{3.353}$$

$$= 2 \times 0.1 \times 10^7 \cos(2\pi 2 \times 10^9(t - 1/3 \times 10^{-6}))\text{sinc}(10^7(t - 1/3 \times 10^{-6})). \tag{3.354}$$

Similarly, to find the baseband channel, we use (3.348):

$$h(t) = 0.1 \times 10^7 \int \text{sinc}(10^7(t - \tau))\delta(\tau - 1/3 \times 10^{-6})e^{-j2\pi \times 2 \times 10^9 \tau}d\tau \tag{3.355}$$

$$= 0.1 \times 10^7 \text{sinc}(10^7(t - 1/3 \times 10^{-6}))e^{-j2\pi\frac{2}{3} \times 10^3}. \tag{3.356}$$

Simplifying further given the numbers in the problem,

$$h(t) = 10^6 \text{sinc}(10^7(t - 1/3 \times 10^{-6}))e^{-j\pi\frac{4}{3}}. \tag{3.357}$$

Example 3.38   Consider a wireless communication system with the carrier frequency of $f_c = 900\text{MHz}$ and the absolute bandwidth of $B = 5\text{MHz}$. The two-path channel model is

$$h_c(t) = \delta(t - 2 \times 10^{-6}) + 0.5\delta(t - 5 \times 10^{-6}). \tag{3.358}$$

- Determine the Fourier transform of $h_c(t)$.

  Answer: Let $t_1 = 2 \times 10^{-6}$ and $t_2 = 5 \times 10^{-6}$ and rewrite the channel response in the time domain as

$$h_c(t) = \delta(t - t_1) + 0.5\delta(t - t_2). \tag{3.359}$$

In the frequency domain the channel response is

$$\mathsf{h}(f) = e^{-\mathsf{j}2\pi f_c t_1} + 0.5 e^{-\mathsf{j}2\pi f_c t_2} \tag{3.360}$$

and the magnitude (suitable for plotting) is

$$|\mathsf{h}(f)| = (1.25 + \cos(2\pi f(t_1 - t_2)))^{1/2} \tag{3.361}$$

$$= \left(1.25 + \cos(6 \times 10^{-6}\pi f)\right)^{1/2}. \tag{3.362}$$

- Find the passband channel $h_\mathrm{p}(t)$ in both time and frequency domains.

  Answer: In the time domain

$$h_\mathrm{p}(t) = h_\mathrm{c}(t) * p(t) \tag{3.363}$$

$$= p(t - t_1) + 0.5\, p(t - t_2). \tag{3.364}$$

In the frequency domain

$$\mathsf{h_p}(f) = \mathsf{h_c}(f)\mathsf{p}(f) \tag{3.365}$$

$$= \begin{cases} \mathsf{h}(f), & \text{for } f \in [-f_c - B/2, -f_c + B/2] \text{ or } [f_c - B/2, f_c + B/2] \\ 0, & \text{elsewhere} \end{cases}$$

$$\tag{3.366}$$

$$= (e^{-\mathsf{j}2\pi f_c t_1} + 0.5 e^{-\mathsf{j}2\pi f_c t_2}) \left[\mathrm{rect}\left(\frac{f - f_c}{B}\right) + \mathrm{rect}\left(\frac{f + f_c}{B}\right) \right]. \tag{3.367}$$

The magnitude of the channel is

$$|\mathsf{h_p}(f)| = \left[1.25 + \cos(6 \times 10^{-6}\pi f)\right]^{1/2} \left[\mathrm{rect}\left(\frac{f - f_c}{B}\right) + \mathrm{rect}\left(\frac{f + f_c}{B}\right) \right].$$

$$\tag{3.368}$$

- Find the complex baseband equivalent channel $h(t)$ in both time and frequency domains.

  Answer: Applying (3.348), in the time domain

$$h(t) = B \int \mathrm{sinc}(B(t - \tau))[\delta(t - t_1) + 0.5\,\delta(t - t_2)]e^{-\mathsf{j}2\pi f_c \tau} d\tau \tag{3.369}$$

$$= B \left[\mathrm{sinc}(B(t - t_1))e^{-\mathsf{j}2\pi f_c t_1} + 0.5\,\mathrm{sinc}(B(t - t_2))e^{-\mathsf{j}2\pi f_c t_2} \right] \tag{3.370}$$

$$= B \left[\mathrm{sinc}(B(t - t_1)) + 0.5\,\mathrm{sinc}(B(t - t_2)) \right] \tag{3.371}$$

since $e^{-\mathsf{j}2\pi f_c t_1} = e^{-\mathsf{j}2\pi f_c t_2} = 1$.

In the frequency domain

$$\mathsf{h}(f) = \left[e^{-\mathsf{j}2\pi f_c t_1} + 0.5 e^{-\mathsf{j}2\pi f_c t_2} \right] \mathrm{rect}\left(\frac{f}{B}\right). \tag{3.372}$$

The magnitude of the frequency response is

$$|\mathsf{h}(f)| = \frac{1}{2}\left[1.25 + \cos(6 \times 10^{-6}\pi f)\right]^{1/2}\mathrm{rect}\left(\frac{f}{10 \times 10^6}\right). \tag{3.373}$$

3.3.4 Pseudo-baseband Equivalent Channel

For the special case when the channel is a sum of delayed impulses, there is an alternative to the complex baseband equivalent channel. We refer to this as the pseudo-baseband equivalent channel; this terminology is not standard in the literature.

To derive the pseudo-baseband equivalent channel, consider the following channel with R rays or paths, each with amplitude α_r and delay τ_r:

$$h_{\mathrm{c}}(t) = \sum_{r=0}^{R-1} \alpha_r \delta(t - \tau_r). \tag{3.374}$$

Generalizing the results on the two-path channel models, it follows that the complex baseband equivalent is

$$h(t) = \sum_{r=0}^{R-1} \alpha_r e^{-\mathrm{j}2\pi f_c \tau_r} B\mathrm{sinc}(B(t - \tau_r)). \tag{3.375}$$

Note that this can be rewritten by factoring out the sinc function as

$$h(t) = B\mathrm{sinc}(Bt) * \sum_{r=0}^{R-1} \alpha_r e^{-\mathrm{j}2\pi f_c \tau_r} \delta(t - \tau_r). \tag{3.376}$$

We refer to the quantity $h_{\mathrm{pb}}(t) = \sum_{r=0}^{R-1} \alpha_r e^{-\mathrm{j}2\pi f_c \tau_r} \delta(t - \tau_r)$ as the pseudo-baseband equivalent channel. This channel has complex coefficients $\alpha_r e^{-\mathrm{j}2\pi f_c \tau_r}$ but otherwise is strikingly similar to $h_{\mathrm{c}}(t)$. Furthermore, since the signal is bandlimited, it follows that $y(t) = h(t) * x(t) = h_{\mathrm{pb}}(t) * x(t)$, so $h_{\mathrm{pb}}(t)$ can still be convolved with $x(t)$ to obtain the complex baseband equivalent signal $y(t)$. Unfortunately, $h_{\mathrm{pb}}(t)$ itself is not bandlimited and cannot be sampled; therefore, caution should be used when deriving the discrete-time model as described next.

Example 3.39   Find the pseudo-baseband equivalent channel for the case described in Example 3.37.

Answer: Since $h_{\mathrm{c}}(t) = 0.1\delta(t - 1/3 \times 10^{-6})$, it follows that

$$h_{\mathrm{pb}}(t) = 0.1e^{-\mathrm{j}2\pi \frac{2}{3} \times 10^3}\delta(t - 1/3 \times 10^{-6}) \tag{3.377}$$

$$= 0.1e^{-\mathrm{j}\pi \frac{4}{3}}\delta(t - 1/3 \times 10^{-6}). \tag{3.378}$$

Example 3.40   Find the pseudo-baseband equivalent channel for the case described in Example 3.38.

Answer: Since $h_c(t) = \delta(t - 2 \times 10^{-6}) + 0.5\delta(t - 5 \times 10^{-6})$, it follows that

$$
\begin{aligned}
h_{\text{pb}}(t) &= e^{-j2\pi 900 \times 10^6 \times 2 \times 10^{-6}} \delta(t - 2 \times 10^{-6}) \\
&\quad + 0.5 e^{-j2\pi 900 \times 10^6 \times 5 \times 10^{-6}} \delta(t - 5 \times 10^{-6}) \tag{3.379} \\
&= \delta(t - 2 \times 10^{-6}) + 0.5\delta(t - 5 \times 10^{-6}). \tag{3.380}
\end{aligned}
$$

In this case the pseudo-baseband equivalent channel is the same as the original channel since the complex phase effects vanish because of the choice of numbers in the problem.

3.3.5   The Discrete-Time Equivalent Channel

The bandlimited property of complex baseband equivalent signals allows us to develop a system model that depends entirely on the sampled complex baseband equivalent signals and a discrete-time equivalent channel. From this representation it will be clear how digital signal processing can be used to generate the transmitted signal and to process the received signal.

Suppose that $x(t)$ and $y(t)$ are bandlimited complex baseband signals with bandwidth $B/2$ and that the sampling period $T \geq 1/B$ satisfies Nyquist. The complex baseband signal $x(t)$ to be transmitted can be generated from its samples $\{x[n]\}$ using the reconstruction formula in (3.82) as

$$
x(t) = \sum_{n=-\infty}^{\infty} x[n]\operatorname{sinc}((t - nT)/T)). \tag{3.381}
$$

In practice, $x(t)$ would usually be generated using a pair of digital-to-analog converters (one for the inphase component and one for the quadrature component). The complex baseband signal $y(t)$ at the receiver can be periodically sampled without loss as long as Nyquist is satisfied to generate

$$
y[n] = y(nT), \tag{3.382}
$$

which would be performed with a pair of analog-to-digital converters (one for the inphase component and one for the quadrature component) in practice.

Applying the results from Section 3.1.6, the input-output relationship in (3.331) can be written equivalently in discrete time as

$$
y[n] = \sum_{k=-\infty}^{\infty} h[k]x[n - k] \tag{3.383}
$$

where $h[n] = Th(nT)$ since $h(t)$ is already bandlimited by construction in (3.346).

The sampled channel $h[n]$ is known as the discrete-time complex baseband equivalent channel. It is obtained from the continuous-time channel by combining (3.114) and (3.348) as

$$h[n] = TB \int \text{sinc}(B(nT - \tau))h_{\text{c}}(\tau)e^{-\text{j}2\pi f_{\text{c}}\tau}\text{d}\tau. \qquad (3.384)$$

In the special case of critical sampling where $T = 1/B$ (often used for examples and homework problems), the scaling factors disappear and

$$h[n] = \int \text{sinc}(n - B\tau)h_{\text{c}}(\tau)e^{-\text{j}2\pi f_{\text{c}}\tau}\text{d}\tau. \qquad (3.385)$$

The importance of discrete-time signal processing to digital communication systems results from the use of passband communication signals. Exploiting the passband property to develop complex baseband equivalents of the input and output signals and the resulting bandlimited property of those signals allows the development of a discrete-time signal model. The steps of abstraction are summarized in Figure 3.23. Throughout this book we will continue refining this model, adding additional impairments like additive noise and eliminating assumptions like perfect downconversion as we progress.

Example 3.41   Find the discrete-time complex baseband channel $h[n]$ obtained from $h(t)$ for Example 3.37 with sampling exactly at the Nyquist rate.

Answer: The baseband signal bandwidth is 5MHz; thus the passband bandwidth is $B = 10$MHz. The Nyquist frequency of the baseband signal is 5MHz; thus the Nyquist

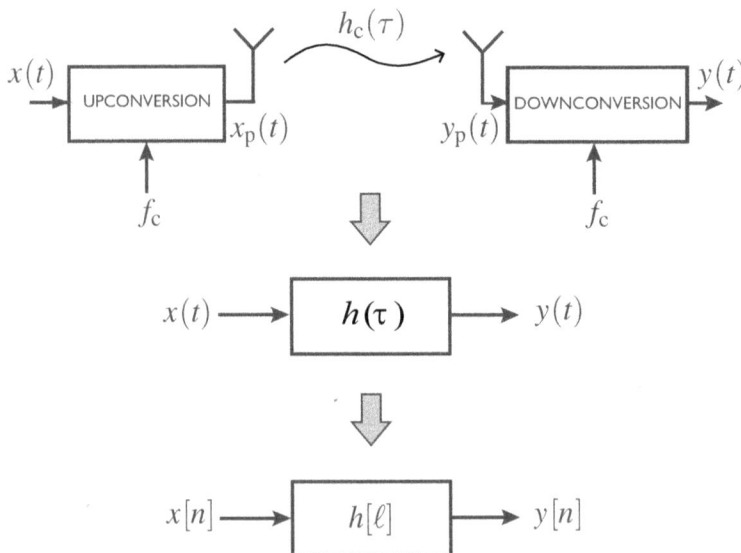

Figure 3.23 Refinement of the digital communication system

rate is 10MHz. Sampling at the Nyquist rate, $T = 1/10^7$s. Since $h(t)$ is already band-limited, the Nyquist sampling theorem can be applied directly:

$$h[n] = T\, h(nT) \tag{3.386}$$

$$= \frac{1}{10^7} 0.1 \times 10^7 \mathrm{sinc}(10^7(n/10^7 - 1/3 \times 10^{-6})) e^{-j2\pi \frac{2}{3} \times 10^3} \tag{3.387}$$

$$= 0.1\ \mathrm{sinc}(n - 10/3) e^{-j\pi \frac{4}{3}}. \tag{3.388}$$

The relatively simple channel that delays and attenuates the transmitted signal results in an impulse response that attenuates, phase shifts, and convolves the received signal with a sampled shifted sinc function. In special cases, further simplifications are possible as shown in the next example.

Example 3.42   Find the discrete-time complex baseband channel $h[n]$ obtained from $h(t)$ for Example 3.38 with sampling exactly at the Nyquist rate.

   Answer: From Example 3.38, $B = 5 \times 10^6$Hz; therefore, sampling at the Nyquist rate, $T = 1/B = 0.2 \times 10^{-6}$s. The discrete-time complex baseband channel is

$$h[n] = T\, h(nT). \tag{3.389}$$

Substituting for $h(t)$ from (3.371),

$$h[n] = TB\ \mathrm{sinc}(n - Bt_1) + 0.5\ \mathrm{sinc}(n - Bt_2) \tag{3.390}$$

where $t_1 = 2 \times 10^{-6}$s and $t_2 = 5 \times 10^{-6}$s. Substituting for all variables,

$$h[n] = \mathrm{sinc}(n - 10) + 0.5\ \mathrm{sinc}(n - 25) \tag{3.391}$$

$$= \delta[n - 10] + 0.5\delta[n - 25]. \tag{3.392}$$

The final simplification occurs because $\mathrm{sinc}(n) = \delta[n]$. The delays in this problem were specially selected relative to the bandwidth to create this final simplification.

3.4   Multirate Signal Processing

As the book progresses, there will be several opportunities to employ digital signal processing techniques at either the transmitter or the receiver to simplify implementation and improve overall radio flexibility. To that end, multirate signal processing will be useful. Multirate signal processing is a class of DSP techniques that facilitate changing the sampling rate after sampling has occurred. In this section, we review some important concepts from multirate signal processing, including downsampling, upsampling, polyphase sequences, filtering, and interpolation or resampling.

3.4.1   Downsampling

Downsampling, also called decimation, is a means for reducing the sample rate of a discrete-time sequence. Let M be a positive integer. Downsampling a signal by M involves creating a new signal where every $M-1$ samples in the original sequence are thrown away. The block diagram notation for downsampling is $\downarrow M$. Its application and a graphical example are provided in Figure 3.24.

To explain the operation of the downsampler, let $z[n]$ be the input and $y[n]$ the output. Mathematically the input and output are related as

$$y[n] = z[nM]. \tag{3.393}$$

Samples in $z[n]$ are discarded to produce $y[n]$; therefore, information is lost. The reduction in information creates an aliasing effect in the frequency domain:

$$y(e^{j2\pi f}) = \frac{1}{M} \sum_{m=0}^{M-1} z\left(e^{j2\pi f/M - j2\pi m/M}\right). \tag{3.394}$$

Because of aliasing, it is not possible in general to recover $z(e^{j2\pi f})$ from $y(e^{j2\pi f})$. Filtering the signal $z[n]$ prior to downsampling can eliminate aliasing in (3.394).

Example 3.43   Consider the samples of the function $x(t) = \sin(2\pi f_1 t) + \sin(2\pi f_2 t + \frac{pi}{3})$ at $t = n\frac{1}{f_s}$ for $f_1 = 18\text{KHz}$, $f_2 = 6\text{KHz}$, and a sampling frequency $f_s = 48\text{KHz}$. Some of these samples are shown in Figure 3.25(a) for $n = 1, \ldots, 20$. The magnitude spectrum of the resulting sequence can be seen in Figure 3.25(b), with the frequency axis in normalized units. The signal is decimated by $M = 2$, and part of the resulting sequence is shown in Figure 3.25(c). The spectrum of the decimated version of the signal is shown in Figure 3.25(d). Explain why the magnitude spectrum of the decimated signal consists of only two impulses.

Answer: The spectrum of the original sequence consists of impulses at $\pm \text{F}_1$ and $\pm \text{F}_2$, which correspond to the normalized frequencies $\text{F}_1 = \frac{f_1}{f_s} = \frac{1}{4}$ and $\text{F}_2 = \frac{f_2}{f_s} = \frac{3}{4}$. After

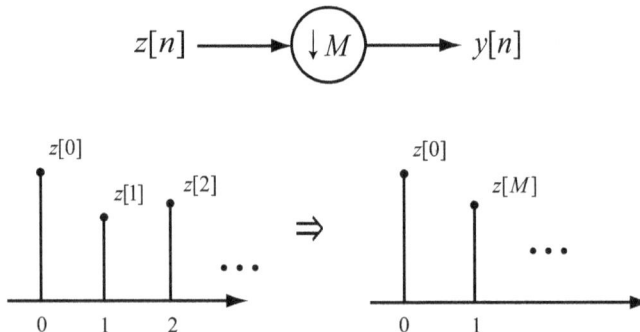

Figure 3.24 Downsampling of a discrete-time signal. The top figure illustrates the downsampling notation for use in the block diagram, and the bottom figures illustrate the downsampling operation.

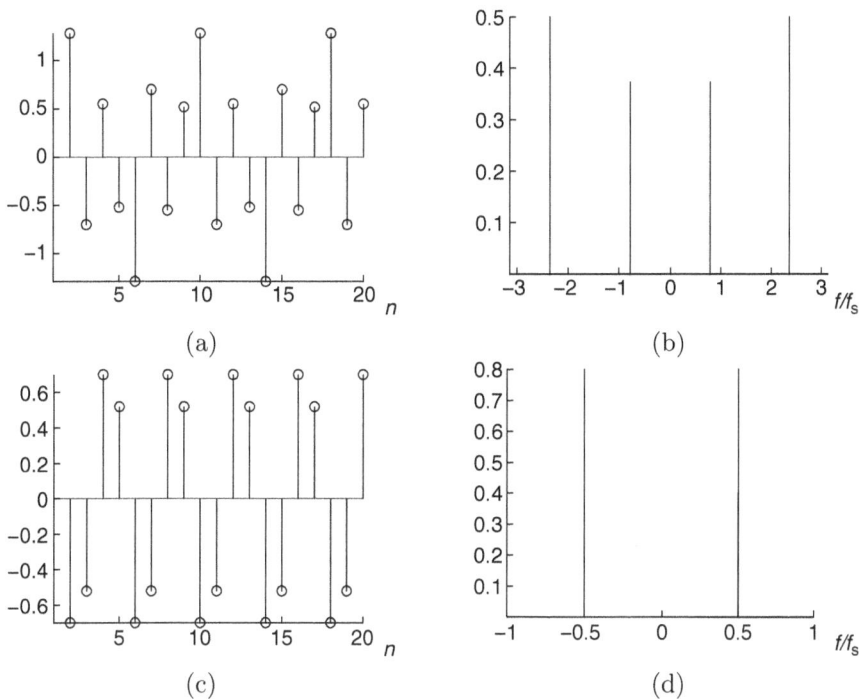

(a)

(b)

(c)

(d)

Figure 3.25 (a) Original sequence obtained by sampling the two sinusoids in Example 3.43; (b) magnitude spectrum of the sequence in (a); (c) sequence in (a) decimated by 2; (d) magnitude spectrum of the decimated signal

decimating the signal, the new sampling frequency is $f'_s = 24\text{KHz}$, and only two impulses can be observed at $\Omega = \frac{1}{2}$.

This result can be explained by building the spectrum of the decimated signal from (3.394). When decimating a signal by $M = 2$, every component at frequency Ω_j appears at $2\Omega_j$. The impulses in the original spectrum at $\Omega = \pm\frac{1}{4}$ appear now at $\Omega = \pm\frac{1}{2}$. The impulses at $\Omega = \pm\frac{3}{4}$ go to $\Omega = \pm\frac{6}{4} > |1|$ and can be observed also at $\pm(2 - \frac{6}{4}) = \pm\frac{1}{2}$. Thus, because of the aliasing effect, a single impulse resulting from the sum of both components is observed in the spectrum of the decimated signal.

3.4.2 Upsampling

Upsampling is a transformation of a discrete-time sequence that involves inserting zeros after every sample. Let L be a positive integer. Upsampling a signal by L means that a new signal is produced where each sample of the original signal is followed by $L - 1$ zeros. In block diagrams, the upsampler is written as $\uparrow L$. Its application and an example are illustrated in Figure 3.26.

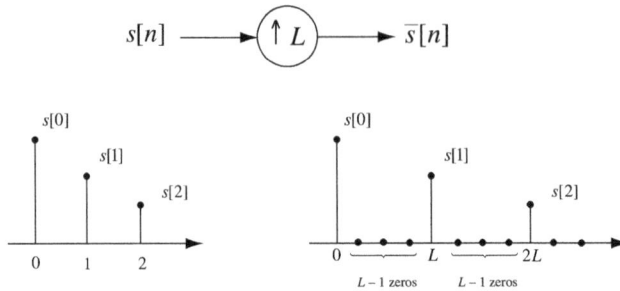

Figure 3.26 Upsampling of a discrete-time signal. The top figure illustrates the upsampling notation for use in a block diagram, and the bottom figures illustrate the effect of upsampling—inserting zeros in the signal.

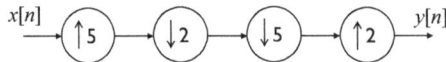

Figure 3.27 Block diagram of a system including a combination of downsampling and upsampling operations

Let $s[n]$ be the input to an upsampler, and let $\bar{s}_L[n]$ be the output. When there is no possibility of confusion, for simplicity we drop the subscript L. Mathematically, the input-output relationship of the upsampler can be written as

$$\bar{s}[n] = \sum_k s[k]\delta[n - kL].\tag{3.395}$$

Using Fourier theorems, the input and output signals are related in the frequency domain as

$$\bar{s}(e^{j2\pi f}) = s(e^{j2\pi fL}).\tag{3.396}$$

In the frequency domain, the upsampling operation has the effect of compressing the spectrum. There is no aliasing because samples are not lost in the upsampling process.

Example 3.44   Consider the system in Figure 3.27, including downsampling and upsampling operations. Compute the expression of $y[n]$ in terms of $x[n]$.

Answer: Using the fact that 5 and 3 are coprime numbers, the upsampler and the decimator commute, so we can switch the order of the first two blocks. Then, note that the 5-fold expander followed by the 5-fold decimator is equivalent to an identity system. Thus, the system is equivalent to a 5-fold decimator followed by a 5-fold upsampler, and consequently

$$y[n] = \begin{cases} x[n], & n = 5k, \quad k = 0, 1, 2, \ldots \\ 0, & \text{otherwise.} \end{cases}\tag{3.397}$$

3.4.3   Polyphase Decomposition

Multirate signal processing often involves decomposing a sequence into its polyphase components and processing each component separately. Let M be a positive integer. The signal $s_m[n] = s[nM + m]$ denotes the mth polyphase subsequence of $s[n]$.

A sequence can be decomposed based on its polyphase components. Writing $s[n]$ using the Kronecker delta function,

$$s[n] = \sum_{k=-\infty}^{\infty} s[k]\delta[n - k]. \tag{3.398}$$

Rewriting the sum over $Mr + m$ instead of k,

$$s[n] = \sum_{m=0}^{M-1} \sum_{r=-\infty}^{\infty} s[Mr + m]\delta[n - Mr - m] \tag{3.399}$$

$$= \sum_{m=0}^{M-1} \sum_{r=-\infty}^{\infty} s_m[r]\delta[n - Mr - m]. \tag{3.400}$$

There is no information lost when decomposing a signal into its polyphase components. The operation of decomposing $s[n]$ into its polyphase components is illustrated in Figure 3.28.

Rewriting (3.400) gives us a clue about how to reconstruct a sequence from its polyphase components using the upsampling operation. Recognizing the connection between (3.400) and (3.395), (3.400) can be rewritten as

$$s[n] = \sum_{m=0}^{M-1} \bar{s}_m[n - m] \tag{3.401}$$

where $\bar{s}_m[n]$ is the output of upsampling by M. The reconstruction of a signal from its polyphase components is illustrated in Figure 3.29.

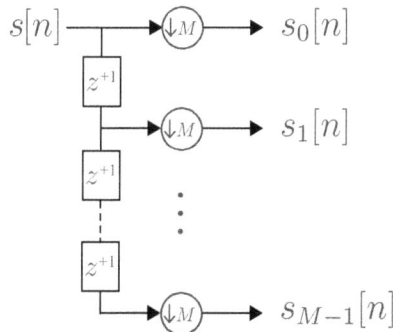

Figure 3.28 Decomposing a signal into its polyphase components. The notation z^{+1} is used to refer to an operation that advances the signal by one sample.

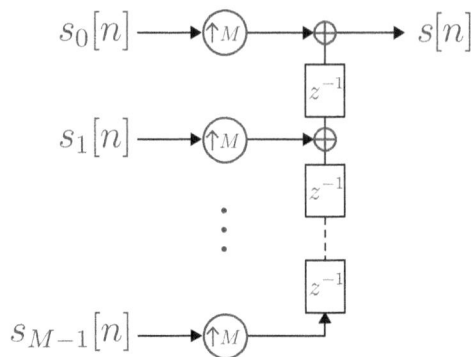

Figure 3.29 Reconstructing a signal into its polyphase components. The notation z^{-1} is used to refer to an operation that delays the signal by one sample.

Example 3.45 Consider the sequence $s[n] = \delta[n]+0.5\delta[n-1]+0.75\delta[n-2]-0.25\delta[n-3]$ and find its polyphase components when $M = 2$.

Answer: From (3.400) the polyphase components of a signal are $s_m[n] = s[nM + m]$, $\quad 0 \le m \le M - 1$. In this simple example there are two polyphase components:

$$s_0[n] = s[2n] = \delta[n] + 0.75\delta[n - 1], \tag{3.402}$$

$$s_1[n] = s[2n + 1] = 0.5\delta[n] - 0.25\delta[n - 1]. \tag{3.403}$$

3.4.4   Filtering with Upsampling and Downsampling

Multirate signal processing often occurs in conjunction with filtering. Multirate identities are used to exchange the order of operations between upsampling or downsampling and the filtering (convolution) operations. In software-defined radio, smart applications of multirate identities can reduce complexity [273].

The downsampling filtering identities are illustrated in Figure 3.30. The downsampling equivalence shows how to exchange filtering after downsampling with filtering (with a new filter) before the downsampling operation. In the time domain, the relationship is

$$y[n] = z[nM] * g[n] \tag{3.404}$$

$$= \sum_k z[kM]g[n - k] \tag{3.405}$$

$$= \sum_k z[k]\bar{g}[nM - k]. \tag{3.406}$$

By upsampling the filter $g[n]$ by M to create $\bar{g}[n]$, the convolution can be performed prior to the downsampling operation.

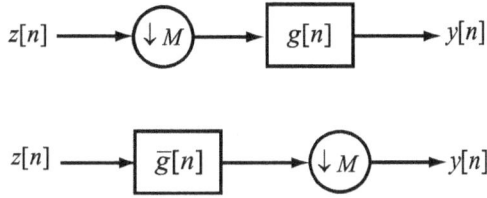

Figure 3.30 Downsampling equivalence. Filtering a downsampled signal can be equivalently performed by downsampling a signal filtered with an upsampled filter.

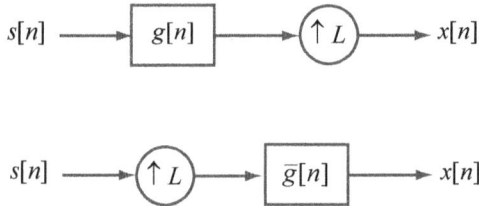

Figure 3.31 Upsampling equivalence. Filtering a signal and then upsampling is equivalent to upsampling the signal and then filtering with an upsampled filter.

The upsampling filtering identities are illustrated in Figure 3.31. The upsampling equivalence shows how to exchange filtering before the upsampler with filtering after the upsampler. In the time domain, the equivalence is

$$x[n] = \sum_k \left(\sum_m s[m]g[k-m] \right) \delta[n-kL] \tag{3.407}$$

$$= \sum_m \left(\sum_k s[k]\delta[m-kL] \right) \left(\sum_p g[p]\delta[n-m-pL] \right) \tag{3.408}$$

$$= \sum_m \bar{s}[m]\bar{g}[n-m]. \tag{3.409}$$

Essentially the upsampling equivalence states that a filter signal can be upsampled or the upsampled signal can be convolved with the upsampled filter.

3.4.5 Changing the Sampling Rate

The sampling rate at the transmitter is often determined by the rate of the digital-to-analog converter whereas the sampling rate of the receiver is determined by the rate of the analog-to-digital converter. It is useful when correcting for more sophisticated receiver impairments to increase the sampling rate to mimic the case where oversampling (sampling with a higher-rate analog-to-digital converter) is available. It may also be valuable in prototyping to design a system that uses very-high-rate converters (much more than the minimum needed to satisfy Nyquist), but the subsequent processing is performed at a lower rate to reduce complexity.

Increasing the Sampling Rate by an Integer Consider a bandlimited signal $x(t)$ that is sampled to produce $x[n] = x(nT)$ and the sampling period T is sufficient for the Nyquist sampling theorem to be satisfied. Suppose that we would like $z[n] = x(nT/L)$ where L is a positive integer. Since $x(t)$ is bandlimited,

$$x(t) = \sum_m x[m] \operatorname{sinc}((t - mT)/T), \tag{3.410}$$

therefore,

$$x(nT/L) = \sum_m x[m] \operatorname{sinc}((nT/L - mT)/T) \tag{3.411}$$

$$= \sum_m x[m] \operatorname{sinc}((n - mL)/L). \tag{3.412}$$

Increasing the sampling rate requires upsampling by L, then filtering with an ideal discrete-time lowpass filter with gain L and cutoff $1/L$ as illustrated in Figure 3.32.

Decreasing the Sampling Rate by an Integer Decreasing the sampling rate is slightly different from increasing the rate because in this case information is lost. Assume that we have a bandlimited signal $x(t)$ sampled at sampling period T such that the Nyquist sampling theorem is satisfied. Suppose that a signal with sampling period TM is required where M is a positive integer. If TM does not satisfy the Nyquist sampling theorem, then aliasing will destroy the integrity of the original signal. The typical solution is to lowpass filter the discrete-time signal prior to downsampling as illustrated in Figure 3.33. Mathematically this leads to

$$\widetilde{x}[n] = M \sum_m x[m] \operatorname{sinc}((nM - m)/M) \tag{3.413}$$

where the factor of M is required to maintain the appropriate scaling factor in the signal. Lowpass filtering before downsampling is equivalent to lowpass filtering the continuous-time signal corresponding to bandwidth $1/2MT$ prior to sampling.

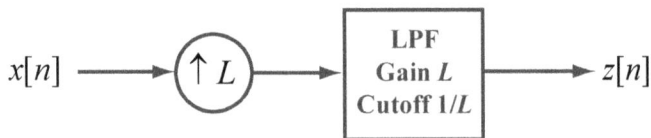

Figure 3.32 Increasing the sampling rate by an integer factor L, also known as interpolation

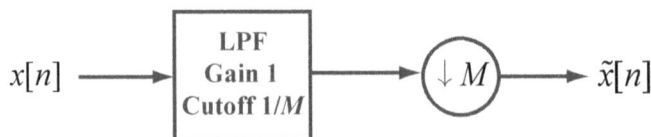

Figure 3.33 Decreasing the sampling rate by an integer factor M, also known as decimation

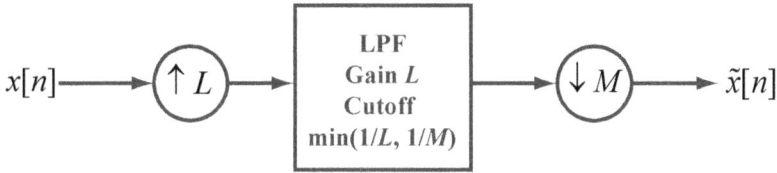

Figure 3.34 Changing the sampling rate by a factor of L/M, also known as resampling

Changing the Sampling Rate by a Rational Factor Given a signal sampled at rate $1/T$, suppose that a signal sampled at rate L/TM is desired where M and L are positive integers. It is possible to solve this problem by first increasing the sampling rate to L/T, then downsampling by M to produce the desired rate. Concatenating Figure 3.32 and Figure 3.33 produces the combined system in Figure 3.34. Let $R = \max(L, M)$. Mathematically, the new signal can be written as

$$\widetilde{x}[n] = \frac{L}{R} \sum_m x[m] \operatorname{sinc}((nM - m)/R). \tag{3.414}$$

If the sampling period is greater than that of the original signal, then effectively the signal is lowpass filtered to avoid aliasing. Changing the sampling rate is useful in communication systems to correct for mismatch between the available hardware and the signal being processed. (For example, in software-defined radio the analog-to-digital converter may operate at a very high sampling rate, which is unnecessary to process a low-bandwidth signal.)

3.5 Linear Estimation

In subsequent chapters it will be useful to find estimates of unknown parameters. In this section, we review some background on linear algebra, then introduce three important estimators from statistical signal processing: linear least squares estimators, maximum likelihood estimators, and linear minimum mean squared error estimators. Under some assumptions, linear least squares is also the maximum likelihood estimator. We make use of linear least squares extensively in Chapter 5. Many of the results can be extended to linear minimum mean squared error estimation.

3.5.1 Linear Algebra

In this book, we denote matrices with bold capital letters like $\mathbf{A}$ and vectors (always column) with lowercase bold letters like $\mathbf{b}$. We refer to the kth row and ℓ^{th} column of $\mathbf{A}$ as $[\mathbf{A}]_{k,\ell} = a_{k,\ell}$. We refer to $\mathbf{a}_\ell$ as the ℓth column of $\mathbf{A}$. The kth entry of $\mathbf{b}$ is b_k. The identity matrix $\mathbf{I}$ is square with 1 on its main diagonal and zero elsewhere. The vector $\mathbf{e}_k$ refers to the kth column of $\mathbf{I}$; thus it is a vector with a 1 in the kth row and zeros otherwise. We use $\mathbf{0}$ to denote a zero vector or matrix. The notation $\mathbb{C}^N$ is used to denote the space of N-dimensional complex vectors and $\mathbb{C}^{N \times M}$ to denote the space of $N \times M$ complex matrices. We similarly define $\mathbb{R}^N$ and $\mathbb{R}^{N \times M}$ for real vectors and matrices.

Let $\mathbf{A}$ denote an $N \times M$ matrix. If $N = M$, we say that the matrix is square. If $N > M$, we call the matrix tall, and if $M > N$, we say that the matrix is fat. We use the notation $\mathbf{A}^{\mathrm{T}}$ to denote the transpose of a matrix, $\mathbf{A}^*$ to denote the Hermitian or conjugate transpose, and $\mathbf{A}^c$ to denote the conjugate of the entries of $\mathbf{A}$. Let $\mathbf{b}$ be an $N \times 1$ vector. The 2-norm of the vector is given by $\|\mathbf{b}\| = \sqrt{\sum_{n=1}^{N} |b_n|^2}$. Other kinds of norms are possible, but we do not consider them in this book.

The inner product between two vectors $\mathbf{a}$ and $\mathbf{b}$ is given by $\langle \mathbf{a}, \mathbf{b} \rangle = \mathbf{a}^* \mathbf{b}$. A useful result is the Cauchy-Schwarz inequality: $|\langle \mathbf{a}, \mathbf{b} \rangle| \leq \|\mathbf{a}\| \|\mathbf{b}\|$ where equality holds when $\mathbf{b} = \beta \mathbf{a}$ for some scalar β (essentially when $\mathbf{a}$ is parallel to $\mathbf{b}$). The Cauchy-Schwarz inequality applies to functions in Hilbert space as well.

Consider a collection of vectors $\{\mathbf{x}_n\}_{n=1}^{K}$. We say that they are linearly independent if there does not exist a set of nonzero weights $\{\alpha_n\}_{n=1}^{K}$ such that $\sum_n \alpha_n \mathbf{x}_n = \mathbf{0}$. In $\mathbb{C}^N$ and $\mathbb{R}^N$, at most N vectors can be linearly independent.

We say that a square matrix $\mathbf{A}$ is invertible if the columns of $\mathbf{A}$ (or equivalently the rows) are linearly independent. If $\mathbf{A}$ is invertible, then a matrix inverse $\mathbf{A}^{-1}$ exists and satisfies the properties $\mathbf{A}\mathbf{A}^{-1} = \mathbf{A}^{-1}\mathbf{A} = \mathbf{I}$. If $\mathbf{A}$ is tall, we say that $\mathbf{A}$ is full rank if the columns of $\mathbf{A}$ are linearly independent. Similarly, if $\mathbf{A}$ is a fat matrix, we say that it is full rank if the rows are linearly independent.

Consider a system of linear equations written in matrix form

$$\mathbf{Ax} = \mathbf{b} \qquad (3.415)$$

where $\mathbf{A}$ is the known matrix of coefficients, sometimes called the data, $\mathbf{x}$ is a vector of unknowns, and $\mathbf{b}$ is a known vector, often called the observation vector. We suppose that $\mathbf{A}$ is full rank. If $N = M$, then the solution to (3.415) is $\mathbf{x} = \mathbf{A}^{-1}\mathbf{b}$. This solution is unique and can be computed efficiently using Gaussian elimination and its variations [131]. If $\mathbf{A}$ is square and low rank or fat, then an infinite number of exact solutions exist. It is common to take the solution that has the minimum norm. If $\mathbf{A}$ is tall and full rank, then an exact solution does not exist. This leads us to pursue an approximate solution.

3.5.2 Least Squares Solution to a System of Linear Equations

Consider again a system of linear equations as in (3.415). Now suppose that $N > M$ and that $\mathbf{A}$ is still full rank. In this case $\mathbf{A}$ is tall and the system is overdetermined in general. This means there are N equations but M unknowns; thus it is unlikely (except in special cases) that an exact solution exists. In this case we pursue an approximate solution known as linear least squares. Instead of solving (3.415) directly, we propose to find the solution that minimizes the squared error

$$\|\mathbf{Ax} - \mathbf{b}\|^2. \qquad (3.416)$$

There are other related least squares problems, including nonlinear least squares, weighted least squares, and total least squares, among others. We focus on linear least squares in this book and drop the term *linear* except where there is the possibility of confusion.

One direct approach to solving (3.416) is to expand the terms, differentiate, and solve. This is cumbersome and does not exploit the elegance of the linear system equations.

Alternatively, we exploit the following facts (see, for example, [143, 2.3.10] or the classic reference [51]). Let $f(\mathbf{x}, \mathbf{x}^c)$ be a real-valued function of complex vector $\mathbf{x}$ and possibly its conjugate $\mathbf{x}^c$. Then $\mathbf{x}$ and $\mathbf{x}^c$ may be treated as independent variables, and the direction of the maximum rate of change is given by $\frac{d}{d\mathbf{x}^c} f(\mathbf{x}) = [\partial/\partial x_1^c f(\mathbf{x}), \ldots, \partial/\partial x_N^c f(\mathbf{x}, \mathbf{x}^c)]^{\mathrm{T}}$. Furthermore, the stationary points of this function can be found by setting this vector derivative to zero. Many tables for vector derivatives exist [32, 220]. We find the following useful: $\frac{d}{d\mathbf{x}^c}\mathbf{a}^*\mathbf{x} = \mathbf{0}$, $\frac{d}{d\mathbf{x}^c}\mathbf{x}^*\mathbf{a} = \mathbf{a}$, and $\frac{d}{d\mathbf{x}^c}\mathbf{x}^*\mathbf{A}^*\mathbf{A}\mathbf{x} = \mathbf{A}^*\mathbf{A}\mathbf{x}$.

Now we can solve the problem in (3.416):

$$\frac{d}{d\mathbf{x}^c}\|\mathbf{A}\mathbf{x} - \mathbf{b}\|^2 = \frac{d}{d\mathbf{x}^c}\mathbf{x}^*\mathbf{A}^*\mathbf{A}\mathbf{x} - \frac{d}{d\mathbf{x}^c}\mathbf{x}^*\mathbf{A}^*\mathbf{b} - \frac{d}{d\mathbf{x}^c}\mathbf{b}^*\mathbf{A}\mathbf{x} + \frac{d}{d\mathbf{x}^c}\mathbf{b}^*\mathbf{b} \qquad (3.417)$$

$$= \mathbf{A}^*\mathbf{A}\mathbf{x} - \mathbf{A}^*\mathbf{b}. \qquad (3.418)$$

Setting (3.418) equal to zero gives the *orthogonality condition*

$$\mathbf{A}^*(\mathbf{A}\mathbf{x} - \mathbf{b}) = \mathbf{0} \qquad (3.419)$$

and then the *normal equations*

$$\mathbf{A}^*\mathbf{A}\mathbf{x} = \mathbf{A}^*\mathbf{b}. \qquad (3.420)$$

Note that $\mathbf{A}^*\mathbf{A}$ is a square and invertible matrix (invertible because $\mathbf{A}$ is full rank); thus it is possible to solve (3.420) for

$$\mathbf{x}_{\mathrm{LS}} = (\mathbf{A}^*\mathbf{A})^{-1}\mathbf{A}^*\mathbf{b} \qquad (3.421)$$

where the subscript LS denotes that this is the least squares solution to this set of linear equations.

We use the squared error achieved by $\mathbf{x}_{\mathrm{LS}}$ to measure the quality of the solution. To derive an expression for the squared error, note that

$$J(\mathbf{x}_{\mathrm{LS}}) = \|\mathbf{A}\mathbf{x}_{\mathrm{LS}} - \mathbf{b}\|^2 \qquad (3.422)$$

$$= \mathbf{x}_{\mathrm{LS}}^*\mathbf{A}^*(\mathbf{A}\mathbf{x}_{\mathrm{LS}} - \mathbf{b}) - \mathbf{b}^*(\mathbf{A}\mathbf{x}_{\mathrm{LS}} - \mathbf{b}) \qquad (3.423)$$

$$= \mathbf{b}^*\mathbf{b} - \mathbf{b}^*\mathbf{A}\mathbf{x}_{\mathrm{LS}} \qquad (3.424)$$

where the first term in (3.423) is zero, which follows from the orthogonality property in (3.419). Substituting for $\mathbf{x}_{\mathrm{LS}}$,

$$J(\mathbf{x}_{\mathrm{LS}}) = \mathbf{b}^*\mathbf{b} - \mathbf{b}^*\mathbf{A}(\mathbf{A}^*\mathbf{A})^{-1}\mathbf{A}^*\mathbf{b} \qquad (3.425)$$

$$= \mathbf{b}^*(\mathbf{I} - \mathbf{A}(\mathbf{A}^*\mathbf{A})^{-1}\mathbf{A}^*)\mathbf{b}. \qquad (3.426)$$

The special matrix $\mathbf{A}(\mathbf{A}^*\mathbf{A})^{-1}\mathbf{A}^*$ is known as a projection matrix for reasons that become clear when viewing least squares from a graphical perspective [172].

Now we present a classic visualization of least squares. Suppose that $N = 3$, $M = 2$, and $\mathbf{x}$, $\mathbf{A}$, and $\mathbf{b}$ are all real. The relevant quantities are illustrated in Figure 3.35. The observation vector $\mathbf{b}$ is unconstrained in $\mathbb{R}^3$. The objective is to find the best linear

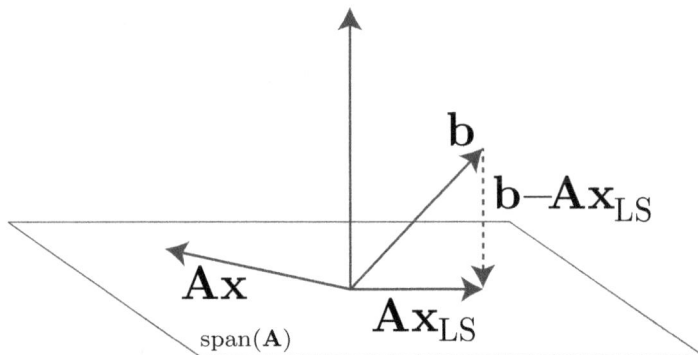

Figure 3.35 The typical illustration of a linear least squares problem. The objective is to find a vector $\mathbf{Ax}$ that is as close as possible to $\mathbf{b}$ in the minimum squared error sense. The resulting vector is the orthogonal projection of $\mathbf{b}$ on the space spanned by the columns of $\mathbf{A}$ and is given by $\mathbf{Ax}_{LS} = \mathbf{A}(\mathbf{A}^*\mathbf{A})^{-1}\mathbf{A}^*\mathbf{b}$.

combination of the columns of $\mathbf{A}$, in the sense that the squared error is minimized. The vector $\mathbf{Ax}$ is a linear combination of the columns of $\mathbf{A}$, which lives in the subspace denoted as span($\mathbf{A}$), in this case a plane. An exact solution to the set of equations is possible only when $\mathbf{b}$ can be exactly represented by a linear combination of the columns of $\mathbf{A}$, in this case being in the same plane defined by the span($\mathbf{A}$). The least squares solution gives $\mathbf{x}_{LS}$ such that $\mathbf{Ax}_{LS}$ is the orthogonal projection of $\mathbf{b}$ onto span($\mathbf{A}$). That projection is given by $\mathbf{A}(\mathbf{A}^*\mathbf{A})^{-1}\mathbf{A}^*\mathbf{b}$ where the matrix $\mathbf{A}(\mathbf{A}^*\mathbf{A})^{-1}\mathbf{A}^*$ is a projection matrix. The error vector is $\mathbf{b} - \mathbf{Ax}_{LS}$. It is orthogonal to $\mathbf{Ax}_{LS}$ according to the orthogonality condition in (3.419), and has the smallest 2-norm among all error vectors. The error can be written as $(\mathbf{I} - \mathbf{A}(\mathbf{A}^*\mathbf{A})^{-1}\mathbf{A}^*)\mathbf{b}$ where $(\mathbf{I} - \mathbf{A}(\mathbf{A}^*\mathbf{A})^{-1}\mathbf{A}^*)$ is a matrix that projects $\mathbf{b}$ onto the orthogonal complement of span($\mathbf{A}$). The length of the error is given in (3.426).

The least squares solution of a linear system is useful in many contexts in signal processing and communications. It can be used as an estimator (discussed further in Section 3.5.3) and as a way to choose parameters of a model. Least squares solutions also have deep connections with adaptive signal processing [172, 143, 294]. The least squares solution can be found using recursive least squares, where the least squares solution is updated as additional rows are added to $\mathbf{A}$ and additional entries in $\mathbf{b}$. Least squares is also related with linear minimum mean squared error (LMMSE) estimation (discussed further in Section 3.5.4).

An application of least squares is provided in Example 3.46. In this example, the coefficients of an affine signal model are selected. This example shows how to find the unknowns with and without linear algebra. Applications of least squares to wireless communications are found in Chapter 5.

Example 3.46   Consider a system that provides an output $y[n]$ for an input $x[n]$ at time n, where the input-output relationship is believed to have the following relationship:

$$y[n] = \beta_0 + \beta_1 x[n] \tag{3.427}$$

where β_0 and β_1 are unknown coefficients to be estimated. Let $e[n] = y[n] - \beta_0 - \beta_1 x[n]$. Suppose that there are $N \geq 3$ observations $y[0], y[1], \ldots, y[N-1]$ corresponding to N known inputs $x[0], x[1], \ldots, x[N-1]$.

- Assuming that $\beta_1 = 0$, find the least squares estimation of β_0 by expanding the sum, differentiating, and solving.

  Answer: The sum squares error is

$$\sum_{n=0}^{N-1} |e[n]|^2 = \sum_{n=0}^{N-1} |y[n] - \beta_0|^2 \tag{3.428}$$

$$= \sum_{n=0}^{N-1} (y^*[n] - \beta_0^*)(y[n] - \beta_0) \tag{3.429}$$

$$= \sum_{n=0}^{N-1} y[n]^2 - \left(\sum_{n=0}^{N-1} y^*[n] \right) \beta_0 - \left(\sum_{n=0}^{N-1} y[n] \right) \beta_0^* + N\beta_0^*\beta_0. \tag{3.430}$$

Taking the derivative of the sum squares error over β_0^* and setting the result to zero, we obtain the least squares estimation of β_0 as follows,

$$\widehat{\beta}_0 = \frac{\sum_{n=0}^{N-1} y[n]}{N}, \tag{3.431}$$

which is the mean of the observations $y[n]$.

- Assuming that $\beta_0 = 0$, find the linear least squares estimate of β_1 by expanding the sum, differentiating, and solving.

  Answer: The sum squares error is

$$\sum_{n=0}^{N-1} |e[n]|^2 = \sum_{n=0}^{N-1} |y[n] - \beta_1 x[n]|^2 \tag{3.432}$$

$$= \sum_{n=0}^{N-1} (y^*[n] - x^*[n]\beta_1^*)(y[n] - \beta_1 x[n]) \tag{3.433}$$

$$= \sum_{n=0}^{N-1} y[n]^2 - \left(\sum_{n=0}^{N-1} y^*[n]x[n] \right) \beta_1 -$$

$$\left(\sum_{n=0}^{N-1} y[n]x^*[n] \right) \beta_1^* + \left(\sum_{n=0}^{N-1} |x[n]|^2 \right) \beta_1^*\beta_1. \tag{3.434}$$

Taking the derivative of the sum squares error over β_1^* and setting the result to zero, we obtain the least squares estimation of β_1 as follows:

$$\widehat{\beta}_1 = \frac{\sum_{n=0}^{N-1} y[n]x^*[n]}{\sum_{n=0}^{N-1} |x[n]|^2}. \tag{3.435}$$

- Find the linear least squares estimation of β_0 and β_1 by expanding the sum, differentiating, and solving.

Answer: The sum squares error is

$$\sum_{n=0}^{N-1} |e[n]|^2 = \sum_{n=0}^{N-1} |y[n] - \beta_0 - \beta_1 x[n]|^2 \tag{3.436}$$

$$= \sum_{n=0}^{N-1} (y^*[n] - \beta_0^* - x^*[n]\beta_1^*)(y[n] - \beta_0 - \beta_1 x[n]) \tag{3.437}$$

$$= \left[\sum_{n=0}^{N-1} y^*[n](y[n] - \beta_0 - \beta_1 x[n]) \right] - \left[\sum_{n=0}^{N-1} (y[n] - \beta_0 - \beta_1 x[n]) \right] \beta_0^* -$$

$$- \left[\sum_{n=0}^{N-1} x^*[n](y[n] - \beta_0 - \beta_1 x[n]) \right] \beta_1^*. \tag{3.438}$$

Taking the derivative of the sum squares error over β_0^* and setting the result to zero, we obtain

$$\sum_{n=0}^{N-1} (y[n] - \beta_0 - \beta_1 x[n]) = 0. \tag{3.439}$$

Taking the derivative of the sum squares error over β_1^* and setting the result to zero, we obtain

$$\sum_{n=0}^{N-1} x^*[n](y[n] - \beta_0 - \beta_1 x[n]) = 0. \tag{3.440}$$

Solving the linear equations (3.439) and (3.440) gives us the least squares estimation of β_0 and β_1 as follows:

$$\widehat{\beta_0} = \frac{\left(\sum_{n=0}^{N-1} x^2[n] \right) \left(\sum_{n=0}^{N-1} y[n] \right) - \left(\sum_{n=0}^{N-1} x[n] \right) \left(\sum_{n=0}^{N-1} x^*[n]y[n] \right)}{N \left(\sum_{n=0}^{N-1} x^2[n] \right) - \left(\sum_{n=0}^{N-1} x^*[n] \right) \left(\sum_{n=0}^{N-1} x[n] \right)} \tag{3.441}$$

$$\widehat{\beta_1} = \frac{N \left(\sum_{n=0}^{N-1} x^*[n]y[n] \right) - \left(\sum_{n=0}^{N-1} x^*[n] \right) \left(\sum_{n=0}^{N-1} y[n] \right)}{N \left(\sum_{n=0}^{N-1} x^2[n] \right) - \left(\sum_{n=0}^{N-1} x^*[n] \right) \left(\sum_{n=0}^{N-1} x[n] \right)}. \tag{3.442}$$

- Use the least squares estimation method to find the coefficients β_0 and β_1 based on the known inputs and their corresponding observations using the matrix form.

Answer: In matrix form, the given model can be written as

$$\underbrace{\begin{bmatrix} y[0] \\ y[1] \\ \vdots \\ y[N-1] \end{bmatrix}}_{\mathbf{y}} = \underbrace{\begin{bmatrix} 1 & x[0] \\ 1 & x[1] \\ \vdots & \\ 1 & x[N-1] \end{bmatrix}}_{\mathbf{X}} \underbrace{\begin{bmatrix} \beta_0 \\ \beta_1 \end{bmatrix}}_{\mathbf{b}}. \tag{3.443}$$

The sum squares error is given by

$$J(\mathbf{b}) = \sum_{n=0}^{N-1} |y[n] - \beta_0 - \beta_1 x[n]|^2 \tag{3.444}$$

$$= |\mathbf{y} - \mathbf{X}\mathbf{b}|^2. \tag{3.445}$$

The formulation of the least squares problem for estimating β_0 and β_1 is

$$\widehat{\mathbf{b}}_{\text{LS}} = \underset{\mathbf{b} \in \mathbb{C}^2}{\arg\min} \, J(\mathbf{b}). \tag{3.446}$$

- Since $N \geq 3$, then $\mathbf{X}$ is tall and is likely to be full rank. Assuming this is the case, the least squares solution to this problem is given by

$$\mathbf{b}_{\text{LS}} = (\mathbf{X}^*\mathbf{X})^{-1}\mathbf{X}^*\mathbf{y}. \tag{3.447}$$

Because the matrices are 2×2, it is possible to simplify the answer even further:

$$\mathbf{b}_{\text{LS}} = \begin{bmatrix} N & \sum_{n=0}^{N-1} x[n] \\ \sum_{n=0}^{N-1} x^*[n] & \sum_{n=0}^{N-1} x[n]x^*[n] \end{bmatrix}^{-1} \begin{bmatrix} \sum_{n=0}^{N-1} y[n] \\ \sum_{n=0}^{N-1} x^*[n]y[n] \end{bmatrix} \tag{3.448}$$

$$= \frac{1}{N\sum_{n=0}^{N-1} x[n]x^*[n] - |\sum_{n=0}^{N-1} x[n]|^2}$$

$$\times \begin{bmatrix} \sum_{n=0}^{N-1} x[n]x^*[n] & -\sum_{n=0}^{N-1} x[n] \\ -\sum_{n=0}^{N-1} x^*[n] & N \end{bmatrix} \begin{bmatrix} \sum_{n=0}^{N-1} y[n] \\ \sum_{n=0}^{N-1} x^*[n]y[n] \end{bmatrix}. \tag{3.449}$$

The squared error given by the least squares solution is

$$J(\mathbf{b}_{\text{LS}}) = \mathbf{y}^*\mathbf{y} - \mathbf{y}^*\mathbf{X}\mathbf{b}_{\text{LS}} \tag{3.450}$$

$$= \sum_{n=0}^{N-1} |y[n]|^2 - \widehat{\beta_0} \sum_{n=0}^{N-1} y^*[n] - \widehat{\beta_1} \sum_{n=0}^{N-1} y^*[n]x[n]. \tag{3.451}$$

3.5.3   Maximum Likelihood Parameter Estimation in AWGN

The least squares solution is used extensively in this book as a way to estimate parameters when the observations are corrupted by noise. In this section we establish that under some assumptions, the least squares solution is also the maximum likelihood solution. This means that it performs well as an estimator.

Consider a system of equations where the observations are corrupted by additive noise:

$$\mathbf{y} = \mathbf{A}\mathbf{x} + \mathbf{v}. \tag{3.452}$$

We call $\mathbf{y}$ the $N \times 1$ observation vector, $\mathbf{A}$ the $N \times M$ data matrix, $\mathbf{x}$ the $M \times 1$ vector of unknowns, and $\mathbf{v}$ is an $N \times 1$ random vector called additive noise. The objective is to estimate $\mathbf{x}$ given an observation $\mathbf{y}$ and knowledge of $\mathbf{A}$ but with only statistical knowledge of $\mathbf{v}$—the instantaneous value of the noise is unknown. This is an estimation problem.

Now we focus on the special case where everything is complex and the additive noise is circularly symmetric complex Gaussian distributed with distribution $\mathcal{N}_{\mathbb{C}}(\mathbf{0}, \sigma^2 \mathbf{I})$. In this case the system is said to have AWGN.

There are different objective functions that can be used when performing estimation. A common estimator in signal processing and communications is based on the maximum likelihood objective function. The likelihood function corresponding to (3.452) is the conditional distribution of $\mathbf{y}$ given $\mathbf{A}$ and supposing that $\mathbf{x}$ the unknown takes the value $\bar{\mathbf{x}}$. Since $\mathbf{v}$ is circularly symmetric complex Gaussian, given $\mathbf{A}\bar{\mathbf{x}}$, then $\mathbf{y}$ is also circularly symmetric complex Gaussian with mean

$$\mathbb{E}_{\mathbf{y}|\mathbf{A},\mathbf{x}}[\mathbf{y}|\mathbf{A}, \bar{\mathbf{x}}] = \mathbf{A}\bar{\mathbf{x}} \tag{3.453}$$

and covariance

$$\mathbb{E}_{\mathbf{y}|\mathbf{A},\mathbf{x}}[(\mathbf{y} - \mathbf{A}\bar{\mathbf{x}})(\mathbf{y} - \mathbf{A}\bar{\mathbf{x}})^*] = \sigma^2 \mathbf{I}. \tag{3.454}$$

Therefore, the likelihood function is

$$f_{\mathbf{y}|\mathbf{A},\mathbf{x}}(\mathbf{y}|\mathbf{A}, \bar{\mathbf{x}}) = \frac{1}{\pi^N} e^{-(\mathbf{y}-\mathbf{A}\bar{\mathbf{x}})^*(\mathbf{y}-\mathbf{A}\bar{\mathbf{x}})}. \tag{3.455}$$

Taking derivatives with respect to $\bar{\mathbf{x}}$

$$\frac{d}{d\bar{\mathbf{x}}^c} f_{\mathbf{y}|\mathbf{A},\mathbf{x}}(\mathbf{y}|\mathbf{A}, \bar{\mathbf{x}}) = \frac{-1}{\pi^N} \mathbf{A}^*(\mathbf{y} - \mathbf{A}\bar{\mathbf{x}}) e^{-(\mathbf{y}-\mathbf{A}\bar{\mathbf{x}})^*(\mathbf{y}-\mathbf{A}\bar{\mathbf{x}})} \tag{3.456}$$

and setting the result equal to zero gives

$$\mathbf{A}^*(\mathbf{y} - \mathbf{A}\bar{\mathbf{x}}) = \mathbf{0}, \tag{3.457}$$

which is exactly the orthogonality condition in (3.419). Assuming that $N \geq M$ and $\mathbf{A}$ is full rank,

$$\mathbf{x}_{\mathrm{ML}} = (\mathbf{A}^*\mathbf{A})^{-1}\mathbf{A}^*\mathbf{y}. \tag{3.458}$$

The main conclusion is that linear least squares is also the maximum likelihood estimator when the input and output are linearly related and the observations are perturbed by AWGN. The maximum likelihood estimator is a good estimator in the sense that as the number of observations grows large, it converges in probability to its true value [175]. In this case it is also the best linear unbiased estimator. As a result, the specialization to least squares estimators in this book is not a major limitation, and the algorithms presented in subsequent chapters will be useful in practical wireless systems.

3.5.4   Linear Minimum Mean Squared Error Estimation

Another objective function that is widely used in statistical signal processing is minimizing the mean squared error [172, 143, 294]. It is perhaps more widely used than maximizing the likelihood because it requires fewer statistical assumptions. This section reviews the linear minimum mean squared error (MMSE) estimator. There are many different variations of the MMSE, but we focus specifically on the linear case, dropping the term *linear* unless there is possibility for confusion. We consider a vector case that allows us to relate the results to the least squares solution and also has applications to equalization in MIMO communication systems in Chapter 6. Most of the results using least squares estimators can be modified with some additional terms to become MMSE estimators.

Suppose that there is an unknown $M \times 1$ vector $\mathbf{x}$ with zero mean and covariance $\mathbf{C_{xx}}$ and an observation vector $\mathbf{y}$ with zero mean and covariance $\mathbf{C_{yy}}$. The vectors $\mathbf{x}$ and $\mathbf{y}$ are jointly correlated with correlation matrix $\mathbf{C_{yx}} = \mathbb{E}[\mathbf{y}\mathbf{x}^*]$. Both $\mathbf{x}$ and $\mathbf{y}$ are assumed to be zero mean, which can be relaxed with some additional notation. The objective of the linear MMSE estimator is to determine a linear transformation such that

$$\mathbf{G}_{\mathrm{MMSE}} = \arg\min_{\mathbf{G}} \mathbb{E}\left[\|\mathbf{x} - \mathbf{G}^*\mathbf{y}\|^2\right]. \tag{3.459}$$

Let $\mathbf{x}_m = [\mathbf{x}]_m$ and $\mathbf{g}_m = [\mathbf{G}]_{:,m}$. Then

$$\mathbf{G}_{\mathrm{MMSE}} = \arg\min_{\mathbf{G}} \mathbb{E}\left[\sum_{m=1}^{M} |\mathbf{x}_m - \mathbf{g}_m^*\mathbf{y}|^2\right]. \tag{3.460}$$

Now we focus on solving for one column of $\mathbf{G}_{\mathrm{MMSE}}$. Interchanging the expectation and derivative,

$$\frac{\mathrm{d}}{\mathrm{d}\mathbf{g}_k^c}\mathbb{E}\left[\sum_{m=1}^{M} |\mathbf{x}_m - \mathbf{g}_m^*\mathbf{y}|^2\right] = \mathbb{E}\left[\frac{\mathrm{d}}{\mathrm{d}\mathbf{g}_k^c}\sum_{m=1}^{M} |\mathbf{x}_m - \mathbf{g}_m^*\mathbf{y}|^2\right] \tag{3.461}$$

$$= \mathbb{E}\left[\frac{\mathrm{d}}{\mathrm{d}\mathbf{g}_k^c} |\mathbf{x}_k - \mathbf{g}_k^*\mathbf{y}|^2\right] \tag{3.462}$$

$$= \mathbb{E}\left[\mathbf{y}\left(\mathbf{y}^*\mathbf{g}_k - \mathbf{x}_k^*\right)\right]. \tag{3.463}$$

Now taking the expectation and setting the result to zero leads to the MMSE orthogonality equation

$$\mathbf{C_{yy}}\mathbf{g}_k = [\mathbf{C_{yx}}]_{:,k}. \tag{3.464}$$

Solving gives

$$\mathbf{g}_{k,\mathrm{MMSE}} = \mathbf{C}_{\mathbf{yy}}^{-1}[\mathbf{C}_{\mathbf{yx}}]_{:,k}. \tag{3.465}$$

Now reassembling the columns of $\mathbf{G}$ and combining the results gives the key result

$$\mathbf{G}_{\mathrm{MMSE}} = \mathbf{C}_{\mathbf{yy}}^{-1}\mathbf{C}_{\mathbf{yx}}. \tag{3.466}$$

Based on (3.466), the MMSE estimate of $\mathbf{x}$ is then

$$\mathbf{x}_{\mathrm{MMSE}} = \mathbf{G}_{\mathrm{MMSE}}^{*}\mathbf{y} \tag{3.467}$$

$$= \mathbf{C}_{\mathbf{yx}}^{*}\mathbf{C}_{\mathbf{yy}}^{-1}\mathbf{y}. \tag{3.468}$$

The effect of $\mathbf{C}_{\mathbf{yy}}^{-1}$ is to decorrelate $\mathbf{y}$, while $\mathbf{C}_{\mathbf{yx}}^{*} = \mathbf{C}_{\mathbf{xy}}$ is to exploit the joint correlation to extract information about $\mathbf{x}$ from $\mathbf{y}$.

The performance of the MMSE estimator is characterized by its MSE:

$$\sum_{m=1}^{M} \mathbb{E}\left[|\mathbf{x}_m - \mathbf{g}_{m,\mathrm{MMSE}}^{*}\mathbf{y}|^2\right] = \sum_{m=1}^{M} \mathbb{E}\left[|\mathbf{x}_m\mathbf{x}_m^{*} - \mathbf{g}_{m,\mathrm{MMSE}}^{*}\mathbf{y}\mathbf{x}_m^{*}|^2\right] \tag{3.469}$$

$$= \sum_{m=1}^{M} [\mathbf{C}_{\mathbf{xx}}]_{m,m} - \mathbf{g}_{m,\mathrm{MMSE}}^{*}[\mathbf{C}_{\mathbf{yx}}]_{:,m} \tag{3.470}$$

$$= \sum_{m=1}^{M} [\mathbf{C}_{\mathbf{xx}}]_{m,m} - [\mathbf{C}_{\mathbf{yx}}]_{:,m}^{*}\mathbf{C}_{\mathbf{yy}}^{-1}[\mathbf{C}_{\mathbf{yx}}]_{:,m} \tag{3.471}$$

$$= \mathrm{tr}[\mathbf{C}_{\mathbf{xx}}] - \mathrm{tr}\left[\mathbf{C}_{\mathbf{yx}}^{*}\mathbf{C}_{\mathbf{yy}}^{-1}\mathbf{C}_{\mathbf{yx}}\right] \tag{3.472}$$

where the final simplification comes from the $\mathrm{tr}[\mathbf{A}]$ operation that sums the elements of $\mathbf{A}$.

Now consider the model in (3.452) where $\mathbf{A}$ is known, $\mathbf{x}$ is zero mean with covariance $\mathbf{C}_{\mathbf{xx}}$, $\mathbf{v}$ is zero mean with covariance $\mathbf{C}_{\mathbf{vv}}$, and $\mathbf{x}$ and $\mathbf{v}$ are uncorrelated. Then the covariance of $\mathbf{y}$ is

$$\mathbf{C}_{\mathbf{yy}} = \mathbb{E}\left[\mathbf{y}\mathbf{y}^{*}\right] \tag{3.473}$$

$$= \mathbb{E}[\mathbf{A}\mathbf{x}\mathbf{x}^{*}\mathbf{A}^{*}] + \mathbb{E}[\mathbf{v}\mathbf{v}^{*}] \tag{3.474}$$

$$= \mathbf{A}\mathbf{C}_{\mathbf{xx}}\mathbf{A}^{*} + \mathbf{C}_{\mathbf{vv}} \tag{3.475}$$

and the cross-covariance is

$$\mathbf{C}_{\mathbf{yx}} = \mathbb{E}\left[\mathbf{y}\mathbf{x}^{*}\right] \tag{3.476}$$

$$= \mathbb{E}[\mathbf{A}\mathbf{x}\mathbf{x}^{*} + \mathbf{v}\mathbf{x}^{*}] \tag{3.477}$$

$$= \mathbf{A}\mathbf{C}_{\mathbf{xx}}. \tag{3.478}$$

Putting everything together in (3.468):

$$\mathbf{x}_{\mathrm{MMSE}} = \mathbf{C}_{\mathbf{xx}}\mathbf{A}^{*}(\mathbf{A}\mathbf{C}_{\mathbf{xx}}\mathbf{A}^{*} + \mathbf{C}_{\mathbf{vv}})^{-1}\mathbf{y}. \tag{3.479}$$

In the special case where $\mathbf{C_{vv}} = \sigma^2\mathbf{I}$ (the entries of $\mathbf{v}$ are uncorrelated) and $\mathbf{C_{xx}} = \gamma^2\mathbf{I}$ (the entries of $\mathbf{x}$ are uncorrelated), then

$$\mathbf{x}_{\text{MMSE}} = \gamma^2\mathbf{A}^*(\gamma^2\mathbf{AA}^* + \sigma^2\mathbf{I})^{-1}\mathbf{y} \tag{3.480}$$

$$= \mathbf{A}^*\left(\mathbf{AA}^* + \frac{\sigma^2}{\gamma^2}\mathbf{I}\right)^{-1}\mathbf{y}. \tag{3.481}$$

In this case, the MMSE estimate can be viewed as a regularized version of the least squares estimate or maximum likelihood estimate in (3.458). The regularization has the effect of improving the invertibility of $\mathbf{A}$. The ratio of γ^2/σ^2 corresponds to what is usually called the signal-to-noise ratio (SNR) (where $\mathbf{x}$ is the signal and $\mathbf{v}$ is the noise). If the SNR is large, then $\mathbf{A}^*\left(\mathbf{AA}^* + \frac{\sigma^2}{\gamma^2}\mathbf{I}\right)^{-1} \to \mathbf{A}^*(\mathbf{AA}^*)^{-1}$, which is the least squares solution. If the SNR is small, then $\mathbf{A}^*\left(\mathbf{AA}^* + \frac{\sigma^2}{\gamma^2}\mathbf{I}\right)^{-1} \to \mathbf{A}^*\frac{\gamma^2}{\sigma^2}$, which is known as a matched filter.

One major question in the application of MMSE techniques in practical systems is how to obtain the correlation matrices $\mathbf{C_{yy}}$ and $\mathbf{C_{yx}}$. There are two common approaches. The first is to make assumptions about the model that relates $\mathbf{y}$ and $\mathbf{x}$ and the underlying statistics, for example, assuming an AWGN model as in (3.452). Another alternative is to work with random processes and to use ergodicity to estimate the correlation functions. An application of this approach is illustrated in Example 3.47.

Example 3.47 Consider an input-output relationship

$$y[n] = hs[n] + v[n] \tag{3.482}$$

where $s[n]$ is a zero-mean WSS random process with correlation $r_{ss}[n]$, $v[n]$ is a zero-mean WSS random process with correlation $r_{vv}[n]$, and $s[n]$ and $v[n]$ are uncorrelated. Assume that the scalar h is known perfectly.

Suppose that we want to find the estimate of $s[n]$ obtained from $gy[n]$ such that the mean squared error is minimized:

$$\mathbb{E}|e[n]|^2 = \mathbb{E}|s[n] - g\,y[n]|^2. \tag{3.483}$$

- Find an equation for g. First expand the absolute value, then take the derivative with respect to g^* and set the result equal to zero. You can assume you can interchange expectation and differentiation. This gives the orthogonality equation.

  Answer: Since $s[n]$ and $v[n]$ are zero-mean random processes, that is, $m_s = 0$ and $m_v = 0$, the mean of $y[n]$ is $m_y = hm_s + m_v = 0$. Substituting the expression of $y[n]$ into the expression of $\mathbb{E}[|e[n]|^2]$ and expanding,

$$\mathbb{E}\left[|e[n]|^2\right] = \mathbb{E}\left[|s[n] - gy[n]|^2\right] \tag{3.484}$$

$$= \mathbb{E}\left[(s[n]s^*[n] - 2s[n]y^*[n]g^* + gy[n]y^*[n]g^*)\right]. \tag{3.485}$$

Taking the derivative with respect to g^* and setting it to zero, we obtain the following equation:

$$\mathbb{E}\left[(s[n] - gy[n])y^*[n]\right] = 0. \tag{3.486}$$

Using correlations and setting equal to zero,

$$r_{sy}[0] = gr_{yy}[0], \tag{3.487}$$

which is the orthogonality condition.

• Simplify to get an equation for g. This is the MMSE estimator.

Answer: Simplifying the orthogonality condition gives the MMSE estimator:

$$g_{\text{MMSE}} = r_{yy}^{-1}[0]r_{sy}[0] \tag{3.488}$$

$$= (hh^*r_{ss}[0] + r_{vv}[0])^{-1}r_{ss}[0]h^*. \tag{3.489}$$

• Find an equation for the mean squared error (substitute your estimator in and compute the expectation).

Answer: The mean squared error corresponding to the MMSE estimator is given by

$$\mathbb{E}\left[|e[n]|^2\right] = (1 - g_{\text{MMSE}}h)r_{ss}[0] \tag{3.490}$$

$$= \left(1 - (hh^*r_{ss}[0] + r_{vv}[0])^{-1}r_{ss}[0]h^*h\right)r_{ss}[0]. \tag{3.491}$$

• Suppose that you know $r_{ss}[n]$ and you can estimate $r_{yy}[n]$ from the received data. Show how to find $r_{vv}[n]$ from $r_{ss}[n]$ and $r_{yy}[n]$.

Answer: Since $s[n]$ and $v[n]$ are uncorrelated and $m_s = m_v = m_y = 0$, we have

$$r_{yy}[n] = hh^*r_{ss}[n] + r_{vv}[n], \tag{3.492}$$

thus

$$r_{vv}[n] = r_{yy}[n] - hh^*r_{ss}[n]. \tag{3.493}$$

• Suppose that you estimate $r_{yy}[n]$ through sample averaging of N samples exploiting the ergodicity of the process. Rewrite the equation for g using this functional form.

Answer: Assume the samples $\{y[n]\}_{n=0}^{N-1}$ correspond to the input $\{s[n]\}_{n=0}^{N-1}$, which is also known. An estimate of the autocorrelation of the input is

$$r_{ss}[0] = \frac{\sum_{n=0}^{N-1} s[n]s^*[n]}{N}. \tag{3.494}$$

An estimate of the autocorrelation of the output is

$$r_{yy}[0] = \frac{\sum_{n=0}^{N-1} y[n]y^*[n]}{N}. \tag{3.495}$$

Therefore, the MMSE estimator corresponding to the samples is given by

$$\widehat{g}_{\text{MMSE}} = r_{yy}^{-1}[0]r_{ss}[0]h^* \tag{3.496}$$

$$\Rightarrow \frac{\sum_{n=0}^{N-1} s[n]s^*[n]}{\sum_{n=0}^{N-1} y[n]y^*[n]}h^* \tag{3.497}$$

$$\Rightarrow \frac{\sum_{n=0}^{N-1} s[n]y^*[n]}{\sum_{n=0}^{N-1} y[n]y^*[n]}, \tag{3.498}$$

where (3.497) follows from replacing $r_{ss}[0]$ and $r_y[00]$ with their sample estimates and (3.498) follows from $r_{sy}[0] = r_{ss}[0]h^*$ and replacing with iB sample average.

- Now consider the least squares solution where you know $\{s[n]\}_{n=0}^{N-1}$ and $\{y[n]\}_{n=0}^{N-1}$. Write the least squares solution g_{LS}. Explain how the least squares and the linear MMSE are related in this case.

  Answer: The least squares solution is given by

  $$\widehat{g}_{\text{LS}} = \frac{\sum_{n=0}^{N-1} s[n]y^*[n]}{\sum_{n=0}^{N-1} y[n]y^*[n]}. \tag{3.499}$$

Note that $\widehat{g}_{\text{LS}} = \widehat{g}_{\text{MMSE}}$. The explanation is that exploiting ergodicity in the linear MMSE—that is, taking no advantage of knowledge of the statistics of the random variables—does not give better results than the LS.

Example 3.48   Consider an input-output relationship

$$y[k] = H[k]s[k] + v[k] \tag{3.500}$$

where $y[k]$ is the observed signal, $H[k]$ is a known scaling factor, $s[k]$ is an IID sequence with zero mean and variance 1, and $v[k]$ is additive white Gaussian noise with distribution $\mathcal{N}(0, \sigma^2)$. Suppose that $s[k]$ and $v[k]$ are independent.

- Determine the least squares estimate of $s[k]$.

  Answer: This is the usual least squares problem with 1×1 matrices. Thus $\widehat{s}_{\text{LS}}[k] = (H^*[k]H[k])^{-1}H^*[k]y[k] = H[k]^{-1}y[k]$, which follows because we can divide through and cancel $H^*[k]$ assuming it is nonzero.

- Determine the linear MMSE estimate of $s[k]$. In other words, solve the problem

  $$\widehat{s}_{\text{MMSE}}[k] = \widehat{G}_{\text{MMSE}}^*[k]y[k] \tag{3.501}$$

  where

  $$\widehat{G}_{\text{MMSE}}[k] = \arg\min_{g} \mathbb{E}[|g^*y[k] - s[k]|^2]. \tag{3.502}$$

Answer: The formulated problem is just a variation of (3.459). The solution is

$$\widehat{G}_{\text{MMSE}}[k] = C_{yy}^{-1} C_{ys} \tag{3.503}$$
$$= (|H[k]|^2 + \sigma^2)^{-1} H[k] \tag{3.504}$$
$$= \frac{H[k]}{|H[k]|^2 + \sigma^2}. \tag{3.505}$$

Then

$$\widehat{s}_{\text{MMSE}}[k] = \frac{H^*[k]}{|H[k]|^2 + \sigma^2} y[k]. \tag{3.506}$$

3.6 Summary

- Convolution in the time domain corresponds to multiplication in the frequency domain for the CTFT and DTFT. For the DFT, that convolution is periodic.

- According to the Nyquist sampling theorem, a continuous-time signal can be completely represented from its periodically spaced samples if the sampling period is less than one-half the maximum frequency in the signal. Aliasing results if the sample period is not small enough.

- There are many ways to measure the bandwidth of a signal, beyond just absolute bandwidth. The bandwidth of a WSS random process is defined based on the Fourier transform of its covariance function.

- Continuous-time LTI systems that process bandlimited signals can be replaced by a continuous-to-discrete converter, a discrete-time LTI system, and a discrete-to-continuous converter.

- WSS random processes are fully characterized by their mean and correlation function. Sample averages are used to compute the mean and correlation function of an ergodic WSS random process.

- Passband signals are associated with an equivalent complex baseband signal and a complex sinusoid that depends on the carrier f_c. Upconversion creates a passband signal from a complex baseband signal. Downconversion creates a complex baseband signal from a passband signal.

- When an LTI system operates on a passband signal, only the part of the system in the bandwidth of the passband signal is important. An equivalent discrete-time input-output relationship can be written based on complex baseband inputs, outputs, and an equivalent channel. Communication of passband signals through LTI systems is therefore fundamentally related to discrete-time signal processing.

- Upsampling and downsampling are used to change the rate of discrete-time signals. Filtering, upsampling, and downsampling can be interchanged with suitable modifications to the filters.

- The linear least squares estimator is also the maximum likelihood estimator for linear models with AWGN.

Problems

1. There are many occasions to consider the CTFT and the DTFT in wireless communication systems. You should be able to solve the following problems using the appropriate tables and transform properties. They can all be solved without computing any integrals or sums. Use $x(t)$, $x(f)$, $x(e^{j2\pi f})$, and $x[n]$ as appropriate when you are not given the explicit function. Use t for the time variable and n for the sequence index. For example, the CTFT of $x(t - \tau)$ is $x(f)e^{-j2\pi f \tau}$.

(a) CTFT of $x(t)e^{j2\pi f_0 t}$

(b) CTFT of $x(t)y(t)$

(c) CTFT of $\int_{\tau=0}^{T} h(\tau)x(t - \tau)d\tau$

(d) CTFT of $e^{j2\pi f_0 t}\mathrm{rect}\left(\frac{t}{T}\right)$, where f_0 and T are constant

(e) Inverse CTFT of $x(f) + x(-f)$. How does this simplify if $x(t)$ is real?

(f) Inverse CTFT of $x(\alpha f)$, where $\alpha > 1$

(g) DTFT of $x[n]e^{j2\pi f_0 n}$

(h) DTFT of $x[n]y[n]$

(i) DTFT of $\sum_{m=0}^{M} h[m]x[n - m]$

(j) Inverse DTFT of $x(e^{j2\pi f})\cos(2\pi T_0 f)$

(k) Inverse CTFT of $x(f)e^{j2\pi f t_0}$

(l) CTFT of $x(t) * y(t)$

(m) CTFT of $x(\alpha t)$, where $\alpha > 1$

(n) CTFT of $x(t)\cos(2\pi f_c t)$

(o) DTFT of $x(n)\cos(2\pi f_1 n)\cos(2\pi f_2 n)$, where f_1 and f_2 are constant

(p) CTFT of $\sin(2\pi f_1 t)x(t - t_1)$, where f_1 and t_1 are constant

(q) DTFT of $e^{j2\pi f_1 n}(x[n] + ax[n - 1])$, where f_1 and a are constant

(r) CTFT of $\cos(2\pi f_1 t)x(2t) + \sin(2\pi f_1 t)x(2t + 1)$, where f_1 is constant

(s) Inverse CTFT of $\frac{1}{4}x(f - f_1) + \frac{1}{2}x(f) + \frac{1}{4}x(f + f_1)$, where f_1 is constant

(t) Inverse DTFT of $\mathsf{x}(e^{j2\pi(f-f_1)}) - \mathsf{x}^*(e^{j2\pi f})e^{j2\pi fn_1}$, where f_1 and n_1 are constant

(u) CTFT of $e^{j2\pi f_1 t}x(t - t_1)$, where f_1 and t_1 are constant

(v) CTFT of $\cos(2\pi f_1 t)x(2t) + \sin(2\pi f_1 t)x(2t + 1)$, where f_1 is constant

(w) DTFT of $\cos(2\pi f_1 n)x[n - 1]$, where f_1 is constant

(x) Inverse CTFT of $\mathsf{x}(f - f_1) + \mathsf{x}^*(f)e^{-j2\pi f t_1}$, where f_1 and t_1 are constant

(y) Inverse DTFT of $\mathsf{x}(e^{j2\pi f})\cos(2\pi f n_0)$, where n_0 is constant

2. Let $x[n]$ be a discrete-time signal with the discrete-time Fourier transform (DTFT), $\mathsf{x}(e^{j2\pi f})$. We define another signal as follows:

$$y[n] = \frac{(e^{j\pi n}x[n]) + x[n]}{2}. \tag{3.507}$$

(a) Determine the DTFT $Y(e^{j2\pi f})$ of $y[n]$ in terms of $\mathsf{x}(e^{j2\pi f})$.

(b) Show that the DTFT of $y[2n]$ is equal to $\mathsf{y}(e^{j\pi f})$.

3. Evaluate the Fourier transform of the damped sinusoid wave $g(t) = e^{-t}\sin(2\pi f_c t)u(t)$ where $u(t)$ is the unit step function.

4. Verify the following properties of the CTFT and the DTFT.

(a) Time shifting:

$$x(t - \tau) \leftrightarrow e^{-j2\pi f\tau}\mathsf{x}(f) \tag{3.508}$$

$$x[n - k] \leftrightarrow e^{-j2\pi fk}\mathsf{x}(e^{j2\pi f}) \tag{3.509}$$

(b) Frequency shifting:

$$e^{-j2\pi ft}x(t) \leftrightarrow \mathsf{x}(f - f_0) \tag{3.510}$$

$$e^{-j2\pi fn}x[n] \leftrightarrow \mathsf{x}(e^{j2\pi(f-f_0)}) \tag{3.511}$$

5. Let $x(t)$ be a complex signal with bandwidth W and Fourier transform $\mathsf{x}(f)$.

(a) Show that the Fourier transform of $x^*(t)$ is $\mathsf{x}^*(-f)$.

(b) Show that the Fourier transform of real $x(t) = \frac{1}{2}(\mathsf{x}(f) + \mathsf{x}^*(-f))$.

(c) Find the Fourier transform of $x(t) * x(t)$.

(d) Find the Fourier transform of $x(t) * x^*(t)$.

(e) Find the Fourier transform of $x(t) * x^*(-t)$.

6. Sketch the following sequence.

$$x[n] = \begin{cases} 1 & : |n| \leq 4, n \text{ even} \\ 0 & : \text{otherwise.} \end{cases} \tag{3.512}$$

and its corresponding DTFT $\mathsf{x}(e^{j2\pi f})$. Show your work. *Hint:* You may need to use a combination of Fourier transform properties.

7. Let $x[n]$ be a discrete-time signal with a Fourier transform that is zero for $1/8 \leq |f| \leq 1/2$. Prove that

$$x[n] = \sum_{k=-\infty}^{\infty} x[4k] \left(\frac{\sin(\frac{\pi}{4}(n-4k))}{\frac{\pi}{4}(n-4k)} \right). \tag{3.513}$$

8. The output $y(t)$ of a causal LTI system is related to the input $x(t)$ by the following:

$$\frac{dy(t)}{dt} + 11y(t) = \int_{-\infty}^{\infty} x(\tau)z(t-\tau)d\tau - x(t), \tag{3.514}$$

where $z(t) = e^{-2t}u(t) + 5\delta(t)$.

(a) Find the frequency response $\mathsf{h}(f) = \mathsf{y}(f)/\mathsf{x}(f)$ of this system.

(b) Find the impulse response of this system.

9. The even part of a real sequence $x[n]$ is defined by

$$x_e[n] = \frac{x[n] + x[-n]}{2}. \tag{3.515}$$

Suppose that $x[n]$ is a real finite-length sequence defined such that $x[n] = 0$ for $n < 0$ and $n \geq N$. Let $X[k]$ denote the N-point DFT of $x[n]$.

(a) Is $\text{Re}[X[k]]$ the DFT of $x_e[n]$?

(b) What is the inverse DFT of $\text{Re}[X[k]]$ in terms of $x[n]$?

10. Let $x[n]$ be a length N sequence with $X[k]$ denoting its N-point DFT. We represent the DFT operation as $X[k] = \mathcal{F}\{x[n]\}$. Determine the sequence $y[n]$ obtained by applying the DFT operation six times to $x[n]$, that is,

$$y[n] = \mathcal{F}\{\mathcal{F}\{\mathcal{F}\{\mathcal{F}\{\mathcal{F}\{\mathcal{F}\{x[n]\}\}\}\}\}\}. \tag{3.516}$$

11. Show that the circular convolution is commutative.

12. Let $g[n]$ and $h[n]$ be two finite-length sequences of length 7 each. If $y_L[n]$ and $y_C[n]$ denote the linear and 7-point circular convolutions of $g[n]$ and $h[n]$, respectively, express $y_C[n]$ in terms of $y_L[n]$.

13. **The Fast Fourier Transform (FFT)** Suppose that a computer program is available for computing the DFT:

$$X[k] = \sum_{n=0}^{N-1} x[n]e^{-j(2\pi/N)kn}, \quad k = 0, 1, \cdots, N-1; \tag{3.517}$$

that is, the input to the program is the sequence $x[n]$ and the output is the DFT $X[k]$. Show how the input and/or output sequences may be rearranged such that the program can also be used to compute the inverse DFT:

$$x[n] = \frac{1}{N} \sum_{k=0}^{N-1} X[k] e^{\mathrm{j}(2\pi/N)kn}, \quad n = 0, 1, \cdots, N-1; \qquad (3.518)$$

that is, the input to the program should be $X[k]$ or a sequence simply related to $X[k]$, and the output should be either $x[n]$ or a sequence simply related to $x[n]$. There are a couple possible approaches.

(a) Compute the inverse DFT using the modulo property of $X[k]$ and $x[n]$.

(b) Compute the inverse DFT using the complex conjugate of $X[k]$.

14. Given the N-point DFT $X[k] = \sum_{n=0}^{N-1} x[n] W_N^{kn}$, $W_N^{kn} = e^{-\mathrm{j}(2\pi/N)kn}$, answer the following questions:

(a) Break it into four $N/4$-point DFTs in preparation for a radix-4 decimation-in-frequency FFT algorithm. Simplify the expressions as much as you can.

(b) Compare the complexities of direct DFT computation and the suggested decimation-in-frequency FFT algorithm in (a).

15. **Computer**   This problem explores the impact of zero padding on resolution.

(a) Construct the 128-point sequence $x[n]$ to be a cosine wave of 0.125 cycles per sample plus 0.5 times a sine wave of 0.25 cycles per sample. Use the *fft* function in MATLAB to compute the DFT of $x[n]$. Plot the magnitude of the DFT.

(b) Pad 896 zeros at the end of the 128-point sequence and use the *fft* function to compute the 1024-point DFT of the zero-padded signal. Does the 1024-point DFT look the same as the 128-point DFT of part (a)? If not, how is it different?

(c) Now find the 1024-point DFT of a pulse sequence with width 128 ($x[n] = 1$ for $n = 0, 1, \cdots, 127$ and 0 otherwise). How would this DFT explain the differences between (a) and (b)?

16. Consider the continuous-time signal

$$x(t) = \cos(10\pi t) + \sin(3\pi t + \pi/2). \qquad (3.519)$$

Let T be the sampling period and reconstruction period. Let $x_s(t)$ be $x(t)s(t)$ where $s(t) = \sum_k \delta(t - kT)$. We uniformly sample $x(t)$ to form $x[n] = x(nT)$.

(a) What is the largest value of T that we can use to sample the signal and still achieve perfect reconstruction?

(b) Suppose we sample with $T = 1/30$. Please illustrate $\mathsf{x}(f)$, $\mathsf{x}_s(f)$, and $\mathsf{x}(e^{j2\pi f})$ in the interval $-1/2T$ to $1/2T$. Do not forget to label correctly.

(c) What is $x[n]$ for $T = 1/30$?

(d) Suppose we sample with $T = 1/10$. Please illustrate $\mathsf{x}(f)$, $\mathsf{x}_s(f)$, and $\mathsf{x}(e^{j2\pi f})$ in the interval $-1/2T$ to $1/2T$. Do not forget to label correctly.

(e) What is $x[n]$ for $T = 1/10$?

17. Suppose that we wish to create a signal

$$x(t) = 4\cos(2\pi10^3 t) + 3\cos(4\pi10^3 t). \qquad (3.520)$$

(a) What is the Nyquist frequency of $x(t)$?

(b) What is the Nyquist rate of $x(t)$?

(c) Suppose that you generate $x(t)$ using a discrete-to-continuous converter operating at three times the Nyquist rate. What function $x[n]$ do you need to input into the discrete-to-continuous converter to generate $x(t)$?

18. Compute the Nyquist frequency and the Nyquist rate of the following signals:

(a) $x(t) = \cos(2\pi10^2 t)$

(b) $x(t) = \frac{\sin(100\pi t)}{100\pi t}\cos(2\pi10^2 t)$

(c) $x(t) = \frac{\sin(100\pi t)}{100\pi t}\cos(2\pi10^2 t) - \left(\frac{\sin(100\pi t)}{100\pi t}\right)^2 \sin(2\pi10^2 t)$

19. Suppose that we wish to create a signal

$$x(t) = \cos(2\pi10^3 t)\frac{\sin(100\pi t)}{100\pi t}. \qquad (3.521)$$

(a) Compute the Fourier transform of $x(t)$.

(b) Sketch the magnitude of $\mathsf{x}(f)$. What is interesting about the spectrum?

(c) What is the Nyquist frequency of $x(t)$ denoted f_N?

(d) What is the Nyquist rate of $x(t)$?

(e) Suppose that $x(t)$ is sampled with sampling rate $3f_N/4$. Sketch the spectrum of $\mathsf{x}(e^{j2\pi f})$.

(f) Suppose that $x(t)$ is sampled with sampling rate $3f_N$. Sketch the spectrum of $\mathsf{x}(e^{j2\pi f})$.

(g) Suppose that we want to generate $x(t)$ using a discrete-to-continuous converter operating at two times the Nyquist rate. What function $x[n]$ do you need to input into the discrete-to-continuous converter to generate $x(t)$?

20. **Sampling and Quantization** A *discrete-time* signal is generated from a *continuous-time* signal through sampling. In a digital communication system, this discrete-time signal is then quantized in order to generate a *digital* signal, which is then transmitted over a link or channel. If the analog signal is sampled at rate f_s and passed through a quantizer with Q levels, the digital signal at the output of the quantizer has rate $f_s \log_2(Q)$ bits/sec. Now consider the following problem:

A digital communication link carries binary-encoded words representing samples of an input signal

$$x(t) = 3\cos(600\pi t) + 2\cos(1800\pi t). \qquad (3.522)$$

The link is operated at $10,000$ bits/sec, and each input sample is quantized into 1024 levels.

(a) What is the sampling frequency that was used in this system?

(b) What is the Nyquist rate?

21. Let $x(t)$ be a real-valued continuous-time signal with maximum frequency 40Hz. Also, let $y(t) = x(t - 1/160)$.

- If $x[n] = x(n/80)$, is it possible to recover $x(t)$ from $x[n]$? Why or why not?
- If $y[n] = y(n/80)$, is it possible to recover $y(t)$ from $y[n]$? Why or why not?
- Is it possible to find $y[n]$ from $x[n]$ without performing any upsampling or downsampling operations by using an LTI system with frequency response $h(f)$? Find $h(f)$.

22. Consider the systems shown in Figure 3.36. Assume that $h_1(e^{j2\pi f})$ is fixed and known. Determine $h_2(e^{j2\pi f})$, which is the transfer function of an LTI system, such that $y_1[n] = y[n]$ if the inputs to $h_1(e^{j2\pi f})$ and $h_2(e^{j2\pi f})$ are the same.

23. Consider Figure 3.37 where $x[n] = x(nT)$ and $y[n] = y[6n]$.

(a) Assume that $x(t)$ has a Fourier transform such that $x(f) = 0$ for $|f| > 300$. What value of T is necessary such that $x(e^{j2\pi f}) = 0$ for $1/4 < |f| < 1/2$?

(b) How should T be chosen such that $y(t) = x(t)$?

24. Consider the following continuous-time signal:

$$x(t) = \sum_{k=-5}^{5} \cos(0.6k\pi t). \qquad (3.523)$$

Let $x_s(t)$ be $x(t)\mathrm{III}_T(t)$, where $\mathrm{III}_T(t) = \sum_k \delta(t - kT)$.

(a) What is the largest value of T that we can use to sample the signal and still achieve perfect reconstruction?

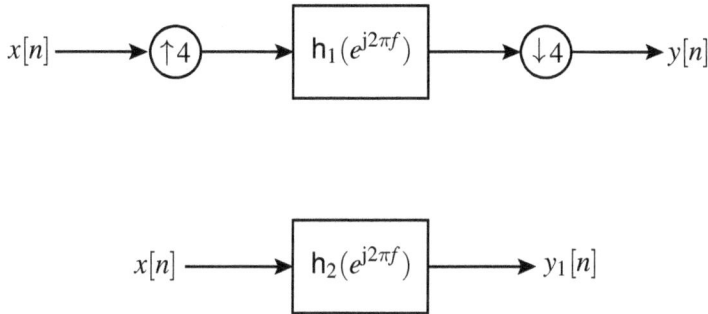

Figure 3.36 Two equivalent LTI systems

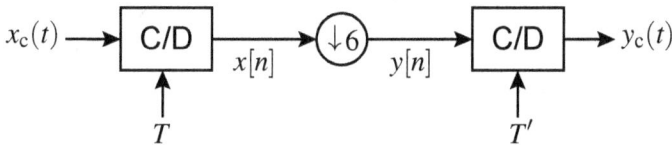

Figure 3.37 Interconnected continuous-to-discrete and discrete-to-continuous converters

    (b) Suppose we sample with $T = 1/2$. Please illustrate $|\mathsf{x}_s(f)|$ in the period $-1/2T$ to $1/2T$. Do not forget to label correctly.

    (c) Suppose we sample with $T = 1/10$. Please illustrate $|\mathsf{x}_s(j\Omega)|$ in the period $-\pi/T$ to π/T. Do not forget to label correctly.

25. Consider a random discrete-time signal $x[n] = s[n] + e[n]$, where both $s[n]$ and $e[n]$ are independent zero-mean WSS random processes. Suppose $s[n]$ has autocorrelation function $R_{ss}[n]$ and that $e[n]$ is white noise with variance σ^2. Determine the mean and autocorrelation function of $x[n]$.

26. Let $x[n]$ and $y[n]$ be two independent zero-mean WSS random signals with autocorrelations $R_{xx}[n]$ and $R_{yy}[n]$, respectively. Consider the random signal $v[n]$ obtained by a linear combination of $x[n]$ and $y[n]$, that is, $v[n] = ax[n] + by[n]$, where a and b are constants. Express the autocorrelation and cross-correlations, $R_{vv}[n]$, $R_{vx}[n]$, and $R_{vy}[n]$, in terms of $R_{xx}[n]$ and $R_{yy}[n]$.

27. Let $x[n]$ and $y[n]$ be two independent zero-mean complex WSS random signals with autocorrelations $R_{xx}[n]$ and $R_{yy}[n]$, respectively. Determine the mean and autocorrelation functions of the following functions:

    (a) $x[n] + y[n]$

    (b) $x[n] - y[n]$

    (c) $x[n]y[n]$

    (d) $x[n] + y^*[n]$

28. Suppose that $s[n]$ is a complex WSS random process with mean m_s and covariance $C_{ss}[k]$. Find the mean and covariance for $x[n]$ defined as

$$x[n] = \sum_{\ell=0}^{L} h[\ell]s[n - \ell]. \tag{3.524}$$

29. Let $x[n]$ be a random process generated by

$$x[n] = \alpha x[n - 1] + w[n] \tag{3.525}$$

where $n \geq 0$, $x[-1] = 0$, and $w[n]$ is an IID $\mathcal{N}(0, \sigma_w^2)$ process. Clearly $x[n]$ is zero mean.

   (a) Find $r_{xx}[n, n + k]$.

   (b) If $|\alpha| < 1$, show that asymptotically the process is WSS; that is, for $k \geq 0$, show that $\lim_{n \to \infty} r_{xx}[n, n + k]$ is only a function of k.

30. Let $x[n]$ be an IID discrete-time, discrete-valued random process such that $x[n] = 1$ with probability p and $x[n] = -1$ with probability $1 - p$.

   (a) Write a general equation for computing the mean m_x of a discrete-valued IID random process.

   (b) Write a general equation for computing the covariance $C_{xx}[\ell]$ of a discrete-valued IID random process.

   (c) Determine the mean of m_x of the random process given in this problem.

   (d) Determine the covariance $C_{xx}[\ell]$ of the random process given in this problem.

31. Suppose that $v[n]$ is an IID complex Gaussian random process with zero mean and variance σ_v^2. Let a and b denote constants. Suppose that $w[n]$ is an IID complex Gaussian random process with zero mean and variance σ_w^2. $v[n]$ and $w[n]$ are independent.

   (a) Compute the mean of $y[n] = aw[n] + bw[n - 2] + v[n]$.

   (b) Compute the correlation of $w[n]$.

   (c) Compute the correlation of $v[n]$.

   (d) Compute the correlation of $y[n]$.

   (e) Compute the covariance of $y[n]$.

   (f) Is $y[n]$ wide-sense stationary? Please justify.

   (g) Is $y[n - 4]$ wide-sense stationary? Please justify.

   (h) How can you estimate the correlation of $y[n]$ given samples $y[0], y[1], \ldots, y[99]$?

32. Suppose that $w[n]$ is an IID complex Gaussian random process with zero mean and variance σ_w^2. Let a, b, and c denote constants.

   (a) Compute the correlation of $w[n]$.

   (b) Compute the mean of $y[n] = aw[n] + bw[n-2]$.

   (c) Compute the correlation and covariance of $y[n]$.

   (d) Is $y[n]$ wide-sense stationary? Please justify.

   (e) Is $y[n-4]$ wide-sense stationary? Please justify.

   (f) Let $v[n]$ be a random process generated by $v[n] = cv[n-1]+w[n]$, where $n \geq 0$ and $v[-1] = 0$. Compute the correlation of $v[n]$; that is, compute $r_{vv}[n, n+k]$.

   (g) If $|c| < 1$, prove that $\lim_{n \to \infty} r_{vv}[n, n+k]$ is only a function of k for $k \geq 0$.

33. Toss a fair coin at each n, $-\infty < n < \infty$, and let

$$
w[n] = \begin{cases} +S, & \text{if heads is the outcome, } \mathbb{P}(H) = \frac{1}{2} \\ -S, & \text{if tails is the outcome, } \mathbb{P}(T) = \frac{1}{2} \end{cases} \tag{3.526}
$$

where $S > 0$. $w[n]$ is an IID random process. Now define a new random process $x[n]$, $n \geq 1$ as

$$x[1] = w[1] \tag{3.527}$$
$$x[2] = w[1] + w[2] \tag{3.528}$$
$$\vdots \qquad \vdots$$
$$x[n] = \sum_{i=1}^{n} w[i]. \tag{3.529}$$

Find the mean and covariance function of the process $x[n]$, which is known as a *random walk*. Is this process wide-sense stationary? Random walks are often used to model node location in simulations and analysis of mobile users in wireless system performance.

34. Consider the following received discrete-time random process defined by the equation

$$y[n] = x[n] - \nu[n] \tag{3.530}$$

where $x[n]$ and $\nu[n]$ are independent WSS random processes with means μ_x and μ_ν and autocovariance $C_{xx}[\tau] = (1/2)^{-\tau}$ and $C_{\nu\nu}[\tau] = (-1/4)^{-\tau}$, respectively.

   (a) What is the mean of $y[n]$?

   (b) What is the autocorrelation of $y[n]$?

   (c) What is the autocovariance of $y[n]$?

   (d) What is the power spectrum of $y[n]$?

(e) Assuming that $x[n]$ is the "signal" and $-\nu[n]$ is the "noise," calculate the signal-to-noise ratio. Remember, this is the ratio of the variance of the signal to the variance of the noise.

35. Consider a continuous-time zero-mean WSS random process $x(t)$ with covariance function $C_{xx}(\tau) = e^{-|\tau|}$.

(a) Compute the power spectrum $P_{xx}(f)$ of $x(t)$.

(b) Determine the half-power bandwidth (or 3dB bandwidth) of the signal.

(c) Determine the sampling period T such that you sample the signal at twice the 3dB frequency.

(d) Compute the covariance function of $x[n] = x(nT)$.

(e) Compute the power spectrum $P_{xx}(e^{j2\pi f})$ of the sampled signal.

(f) Is information lost in going from continuous to discrete time in this case?

36. Consider again the same setup as problem 35, but now we use fractional containment to define bandwidth.

(a) Determine the power spectral density $P_x(f)$ of $x(t)$.

(b) Compute the 3dB bandwidth of $x(t)$.

(c) Compute the fractional power containment bandwidth with $\alpha = 0.9$, that is, the bandwidth that contains 90% of the signal energy.

(d) Find the sampling period T such that you sample $x(t)$ at twice the 3dB frequency.

(e) Determine the covariance function of $x[n] = x(nT)$.

(f) Compute the power spectral density $P_x(e^{j2\pi f})$ of $x[n]$.

37. **Computer** Consider a discrete-time IID complex Gaussian WSS random process with zero mean and unit 1, given by $w[n]$. Let $x[n] = w[n] + 0.5w[n-1] + 0.25w[n-2]$. Generate a realization of length 256 of $x[n]$ in your favorite computer program.

(a) Plot $\widehat{m}_x$ from (3.195) for $N = 5, 10, 20, 50, 75, 100, 256$ and explain the results.

(b) Plot $\widehat{R}_{xx}[k]$ from (3.196) for $k = 0, 1, 2$ for the same values of N. Compare with the true value of $R_{xx}[k]$ that you compute analytically.

38. Consider a continuous-time system with input random process $\mathsf{x}(t)$ and output process $\mathsf{y}(t)$:

$$y(t) = \frac{1}{3} \int_{-7}^{7} \mathsf{x}(t-s)\mathrm{d}s. \qquad (3.531)$$

Assume that $x(t)$ is wide-sense stationary with power spectrum $P_x(f) = 8$ for $-\infty < f < \infty$.

(a) Find the power spectrum of the output $P_y(f)$.

(b) Find the correlation of the output $R_{yy}(t)$.

39. Consider a continuous-time WSS random process $x(t)$ with correlation function $C_{xx}(\tau) = e^{-|t|}$.

(a) Compute the power spectrum $P_x(f)$ of $x(t)$.

(b) Determine the half-power bandwidth (or 3dB bandwidth) of the passband signal.

40. Another important Fourier transform relationship is the *Wiener-Khinchin theorem*, which states that the power spectral density of a WSS random process is equal to the Fourier transform of the autocorrelation function. Consider a time-truncated function $x(t, T)$:

$$x(t, T) = \begin{cases} x(t) & : |t| \le \frac{T}{2} \\ 0 & : \text{otherwise.} \end{cases} \tag{3.532}$$

The energy spectral density of $x(t, T)$ is given as

$$S_{xx}(f, T) = \mathbb{E}[|X(f)|^2], \tag{3.533}$$

and the power spectral density is given as

$$S_{xx}(f) = \lim_{T \to \infty} \frac{S_{xx}(f, T)}{T}. \tag{3.534}$$

For this problem, prove the Wiener-Khinchin theorem. In other words, show that

$$S_{xx}(f) = \mathcal{F}\{R_{xx}(\tau)\}, \tag{3.535}$$

where $\mathcal{F}$ is the Fourier transform operator and

$$R_{xx}(\tau) = \mathbb{E}[x(t)x^*(t - \tau)] \tag{3.536}$$

is the autocorrelation of a function $x(t)$.

41. Suppose that X is a complex normal random variable with distribution $\mathcal{N}_{\mathbb{C}}(m, \sigma^2)$.

(a) What is the mean of X?

(b) What is the variance of X?

(c) What is the mean of $\text{Re}[X]$?

(d) What is the value of $\mathbb{E}\left[\mathrm{Re}[X]\mathrm{Im}[X]\right]$?

42. Consider the following passband signal:

$$x_{\mathrm{p}}(t) = e^{-at}u(t)\cos(2 \times 10^9 \pi t), \qquad (3.537)$$

where $a = 1000$ and

$$u(t) = \begin{cases} 1 & \text{if } t \geq 0 \\ 0 & \text{if } t < 0. \end{cases} \qquad (3.538)$$

(a) Determine and plot $|\mathsf{x}_{\mathrm{p}}(f)|$, where $\mathsf{x}_{\mathrm{p}}(f)$ is the CTFT of $x_{\mathrm{p}}(t)$. Do not forget to label correctly.

(b) Find the complex baseband equivalent signal $x(t)$ of $x_{\mathrm{p}}(t)$. Plot $|\mathsf{x}(f)|$, where $\mathsf{x}(f)$ is the CTFT of $x(t)$. Do not forget to label correctly.

(c) Determine the absolute bandwidth of the passband signal.

(d) Determine the 3dB bandwidth of the passband signal.

(e) Determine the 3dB bandwidth of the baseband signal.

(f) Determine the Nyquist rate of the baseband signal, assuming we define the bandwidth by the 3dB bandwidth.

(g) Draw a block diagram of the entire system to create $x_{\mathrm{p}}(t)$, including D/C and upconversion.

43. Consider a wireless communication system with the carrier frequency of $f_{\mathrm{c}} = 2\mathrm{GHz}$ and the absolute bandwidth of $W = 1\mathrm{MHz}$. The propagation channel consists of a sum of attenuated reflections:

$$h_{\mathrm{c}}(t) = \sum_{n=0}^{99}(0.9)^n \delta(t - n10^{-7}). \qquad (3.539)$$

(a) Determine the channel magnitude in the frequency domain, that is, $|\mathsf{h}_{\mathrm{c}}(f)|$ where $\mathsf{h}_{\mathrm{c}}(f)$ is the continuous-time Fourier transform of $h_{\mathrm{c}}(t)$. Do not forget to label correctly.

(b) Find the passband channel $h_{\mathrm{p}}(t)$ by applying an ideal bandpass filter $p(t)$ with the center frequency at f_{c} and the absolute bandwidth W to the channel $h_{\mathrm{c}}(t)$. Sketch $|\mathsf{h}_{\mathrm{p}}(f)|$, where $\mathsf{h}_{\mathrm{p}}(f)$ is the continuous-time Fourier transform of $h_{\mathrm{p}}(t)$. Do not forget to label correctly.

(c) Find the complex baseband equivalent channel $h(t)$. Sketch $|\mathsf{h}(f)|$, where $\mathsf{h}(f)$ is the continuous-time Fourier transform of $h(t)$. Do not forget to label correctly.

(d) Find the pseudo-baseband equivalent channel $h_{\mathrm{pb}}(t)$.

(e) Find an equation for the discrete-time complex baseband equivalent channel $h[n]$, assuming sampling at the Nyquist rate.

44. Consider the following passband sinc^2 pulse signal:

$$x_\mathrm{p}(t) = 2\mathrm{sinc}^2(2 \times 10^7 t) \cos\left(4.8 \times 10^9 \pi t\right),\qquad (3.540)$$

where $\mathrm{sinc}(a) = \frac{\sin(\pi a)}{\pi a}$.

(a) Determine and plot $|x_\mathrm{p}(f)|$, where $x_\mathrm{p}(f)$ is the continuous-time Fourier transform of $x_\mathrm{p}(t)$. Do not forget to label correctly.

(b) Find the complex baseband equivalent signal $x(t)$ of $x_\mathrm{p}(t)$. Plot $|x(f)|$, where $x(f)$ is the continuous-time Fourier transform of $x(t)$. Do not forget to label correctly.

(c) Determine the absolute bandwidth of the passband signal.

(d) Determine the half-power bandwidth (or 3dB bandwidth) of the passband signal.

45. Consider the following passband sinc^2 pulse signal:

$$x_\mathrm{p}(t) = \mathrm{sinc}^2(3 \times 10^7 t) \cos\left(4 \times 10^9 \pi t\right),\qquad (3.541)$$

where $\mathrm{sinc}(a) = \frac{\sin(\pi a)}{\pi a}$.

(a) Determine and plot $|x_\mathrm{p}(f)|$, where $x_\mathrm{p}(f)$ is the continuous-time Fourier transform of $x_\mathrm{p}(t)$. Do not forget to label correctly.

(b) Find the complex baseband equivalent signal $x(t)$ of $x_\mathrm{p}(t)$. Plot $|x(f)|$, where $x(f)$ is the continuous-time Fourier transform of $x(t)$. Do not forget to label correctly.

(c) Determine the absolute bandwidth of the passband signal.

(d) Determine the absolute bandwidth of the baseband signal.

(e) Determine the Nyquist rate of the baseband signal.

(f) Suppose we create $x(t)$ using a discrete-time sequence $x[n]$ with T chosen corresponding to twice the Nyquist rate. Determine $x[n]$.

(g) Draw a block diagram of the entire system to create $x_\mathrm{p}(t)$, including D/C and upconversion.

46. Let $s(t)$ be a bandlimited signal such that $s(f) = 0$ for $|f| \le B \le f_\mathrm{c}$Hz. Also, let $\widehat{s}(t)$ be the Hilbert transform of $s(t)$, where $\widehat{s}(t)$ is defined as

$$\widehat{s}(t) = \frac{1}{\pi} \int_{-\infty}^{\infty} \frac{s(\tau)}{t - \tau}\,\mathrm{d}\tau.\qquad (3.542)$$

If $x(t) = s(t)\cos(2\pi f_\mathrm{c}t) \pm \widehat{s}(t)\sin(2\pi f_\mathrm{c}t)$, prove that $x(t)$ is a single-sideband signal.

47. Consider the following passband signal:

$$y_\mathrm{p}(t) = \sqrt{2}\, \mathrm{sinc}(3t)\cos(200\pi t). \tag{3.543}$$

(a) What is the carrier frequency of $y_\mathrm{p}(t)$?

(b) Determine and plot $|\mathsf{y}_\mathrm{p}(f)|$, where $\mathsf{y}_\mathrm{p}(f)$ is the continuous-time Fourier transform of $y_\mathrm{p}(t)$. Do not forget to label correctly.

(c) What is the absolute passband bandwidth of $y_\mathrm{p}(t)$?

(d) Find the complex baseband equivalent $y(t)$ of $y_\mathrm{p}(t)$.

(e) What is the absolute bandwidth of $y(t)$?

(f) Draw a block diagram for downconversion for this system.

48. Consider a wireless communication system with the carrier frequency of $f_\mathrm{c} = 1900\mathrm{MHz}$ and the absolute bandwidth of $W = 500\mathrm{kHz}$. The two-path channel model is

$$h(t) = 0.5\, \delta(t - 2 \times 10^{-9}) - 0.25\, \delta(t - 5 \times 10^{-9}). \tag{3.544}$$

(a) Determine and sketch the channel magnitude in both time and frequency domains, that is, $|h(t)|$ and $|\mathsf{h}(f)|$, where $\mathsf{h}(f)$ is the continuous-time Fourier transform of $h(t)$. Do not forget to label correctly.

(b) Find the passband channel $h_\mathrm{p}(t)$ by applying an ideal bandpass filter $p(t)$ with the center frequency at f_c and the absolute bandwidth W to the channel $h(t)$. Sketch $|\mathsf{h}_\mathrm{p}(f)|$, where $\mathsf{h}_\mathrm{p}(f)$ is the continuous-time Fourier transform of $h_\mathrm{p}(t)$. Do not forget to label correctly.

(c) Find the complex baseband equivalent channel $h(t)$. Sketch $|\mathsf{h}(f)|$, where $\mathsf{h}(f)$ is the continuous-time Fourier transform of $h(t)$. Do not forget to label correctly.

(d) Find the pseudo-baseband equivalent channel $h_\mathrm{pb}(t)$.

(e) Find an equation for the discrete-time complex baseband equivalent channel $h[n]$, assuming sampling at the Nyquist rate. Do not forget about the appropriate normalization factors.

49. Consider a wireless communication system with the following channel model:

$$h(t) = \begin{cases} e^{-at}, & t \geq 0 \\ 0, & t < 0. \end{cases} \tag{3.545}$$

(a) Determine the continuous-time frequency response of the channel and sketch its magnitude.

(b) Suppose a baseband signal with bandwidth of B is input into a system with impulse response $h(t)$. Determine the discrete-time equivalent system.

50. Consider a wireless communication system with carrier frequency of $f_c = 2000\text{MHz}$ and absolute bandwidth of $W = 1\text{MHz}$. The channel model is exponential (common in practice):

$$h_c(t) = \exp(-(t - 100 \times 10^{-9}))u(t - 100 \times 10^{-9}). \qquad (3.546)$$

(a) Determine and sketch the channel magnitude in both time and frequency domains, that is, $|h_c(t)|$ and $|h_c(f)|$, where $h_c(f)$ is the continuous-time Fourier transform of $h_c(t)$. Do not forget to label correctly.

(b) Find the passband channel $h_p(t)$ by applying an ideal bandpass filter $p(t)$ with the center frequency at f_c and the absolute bandwidth W to the channel $h(t)$. Sketch $|h_p(f)|$, where $h_p(f)$ is the continuous-time Fourier transform of $h_p(t)$. Do not forget to label correctly. The convolution can be done; you might consider using Mathematica if you have trouble.

(c) Find the complex baseband equivalent channel $h(t)$. Sketch $|h(f)|$, where $h(f)$ is the continuous-time Fourier transform of $h(t)$. Do not forget to label correctly.

(d) Find an equation for the discrete-time complex baseband equivalent channel $h[n]$, assuming sampling at the Nyquist rate.

51. Consider a wireless communication system with the carrier frequency of $f_c = 1700\text{MHz}$ and the absolute bandwidth of $W = 20\text{MHz}$. The channel response is given by

$$h_c(t) = \sum_{k=0}^{\infty} \alpha^k \delta(t - k \times 10^{-6}) \qquad (3.547)$$

where $0 < \alpha < 1$.

(a) Determine and sketch the channel magnitude in both time and frequency domains, that is, $|h_c(t)|$ and $|h_c(f)|$, where $h_c(f)$ is the continuous-time Fourier transform of $h_c(t)$. Do not forget to label correctly.

(b) Find the passband channel $h_p(t)$ by applying an ideal bandpass filter $p(t)$ with the center frequency at f_c and the absolute bandwidth W to the channel $h(t)$. Sketch $|h_p(f)|$, where $h_p(f)$ is the continuous-time Fourier transform of $h_p(t)$. Do not forget to label correctly. The convolution can be done; you might consider using Mathematica if you have trouble.

(c) Find the complex baseband equivalent channel $h(t)$. Sketch $|h(f)|$, where $h(f)$ is the continuous-time Fourier transform of $h(t)$. Do not forget to label correctly.

(d) Find the pseudo-baseband equivalent channel $h_{pb}(t)$.

(e) Find an equation for the discrete-time complex baseband equivalent channel $h[n]$, assuming sampling at the Nyquist rate.

52. Consider a wireless communication system with carrier frequency of $f_c = 2\text{GHz}$ and an absolute bandwidth of 10MHz. Suppose that the sampled complex baseband signal to be sent is

$$x[n] = \exp(j\pi n) + \exp(j3\pi n/10). \qquad (3.548)$$

(a) Suppose that $x[n]$ is fed into a discrete-to-continuous converter operating at five times the Nyquist rate. What is the complex baseband equivalent signal $x(t)$, which is the signal output from this converter?

(b) Suppose that $x(t)$ is modulated by the RF to produce the passband signal $x_p(t)$. Draw a block diagram that describes the modulation operations.

(c) Write an equation for the passband signal $x_p(t)$.

(d) What is the bandwidth of the resulting passband signal $x_p(t)$?

53. **Passband-to-Baseband Conversion**   Consider the following passband signal:

$$y_p(t) = \sqrt{2}\ \text{sinc}(2t)\cos(200\pi t). \qquad (3.549)$$

(a) What is the carrier frequency of $y_p(t)$?

(b) What is the absolute passband bandwidth of $y_p(t)$?

(c) Find the complex baseband equivalent $y(t)$ of $y_p(t)$.

(d) What is the absolute bandwidth of $y(t)$?

54. Consider the continuous-time signal

$$x(t) = \cos(10\pi t) + \sin(3\pi t + \pi/4). \qquad (3.550)$$

Let T be the sampling period and reconstruction period. Let $x_s(t)$ be $x(t)W_t(E)$ when $W_t(E) = \sum_k \delta(t - kT)$. We uniformly sample $x(t)$ to form $x[n] = x(nT)$. We filter $x[n]$ by $h[n] = \delta[n] - \frac{1}{2}\delta[n-1]$ to produce $y[n]$. We produce $y(t)$ using an ideal discrete-to-continuous converter.

(a) What is the largest value of T that we can use to sample the signal and still achieve perfect reconstruction?

(b) Suppose we sample with $T = 1/30$. Please illustrate $|x_s(f)|$ in the period $-1/2T$ to $1/2T$. Do not forget to label correctly.

(c) Again suppose that we sample with $T = 1/30$. Please determine $y[n]$ and illustrate $|y(f)|$ in the period $-1/2T$ to $1/2T$. Do not forget to label correctly.

(d) Suppose we sample with $T = 1/10$. Please illustrate $|x_s(f)|$ in the period $-1/2T$ to $1/2T$. Do not forget to label correctly.

(e) Again suppose that we sample with $T = 1/10$. Please determine $y[n]$ and illustrate $|y(f)|$ in the period $-1/2T$ to $1/2T$. Do not forget to label correctly.

55. Suppose that you have a continuous function $f(x)$ that you want to approximate with a polynomial $\alpha + \beta x + \gamma x^2$. Find the values of α, β, and γ such that the squared error

$$\int_0^1 (f(x) - \alpha - \beta x - \gamma x^2)^2 dx \qquad (3.551)$$

is minimized. You can exchange integration and differentiation operations as needed. Assume that x is real.

56. Suppose you model a desired signal $y[n]$ as

$$y[n] = a_0 + a_1/n \qquad (3.552)$$

and let $e[n] = y[n] - a_0 + a_1 n$. Suppose you are given $y[n]$ for $n = 1, 2, \ldots, N$. Find an expression for the coefficients a_0 and a_1 such that the $\sum_{n=1}^N |e[n]|^2$ is minimized.

57. Suppose you model a desired signal $x[n]$ using a pth-order polynomial as

$$y[n] = a_0 + na_1 + n^2 a_2 + \ldots + n^p a_p. \qquad (3.553)$$

Suppose you are given $y[n]$ for $n = 0, 1, 2, \ldots, N$. Find an expression for the coefficients $\{a_k\}_{k=0}^p$ such that $\sum_{n=0}^N |e[n]|^2$ is minimized when $e[n] = y[n] - (a_0 + na_1 + n^2 a_2 + \ldots + n^p a_p)$.

58. Consider a system that provides an output $y[n]$ for an input $x[n]$ at time n, where the input-output relationship is believed to have the following model:

$$y[n] = \beta_0 x[n] + \beta_1 x^3[n] \qquad (3.554)$$

where β_0 and β_1 are the complex coefficients to be estimated. A variation of this type of memoryless input-output relationship (for real signals) is a good model for a power amplifier.

   Let $e[n] = y[n] - \beta_0 x[n] - \beta_1 x^3[n]$. Suppose that we have N ($N \geq 3$) observations $y[0], y[1], \ldots, y[N-1]$ corresponding to N known training symbols $x[0], x[1], \ldots, x[N-1]$. Our goal is to use the least squares estimation method to find the coefficients β_0 and β_1 based on the known training symbols and their corresponding observations such that $\sum_{n=0}^{N-1} |e[n]|^2$ is minimized.

   (a) Assuming that $\beta_1 = 0$, find the least squares estimation of β_0 by expanding the sum, differentiating, and solving.

   (b) Assuming that $\beta_0 = 0$, find the least squares estimation of β_1 by expanding the sum, differentiating, and solving.

   (c) Find the least squares estimation of β_0 and β_1 by expanding the sum, differentiating, and solving.

(d) Formulate and solve the least squares problem for finding β_0 and β_1 in matrix form.

59. Consider a system that provides an output $y[n]$ for an input $x[n]$ at time n, where the input-output relationship is believed to have the following model:

$$y[n] = \beta_0 x[n] + \beta_1 x[n] x^*[n], \tag{3.555}$$

where β_0 and β_1 are the (possibly complex) coefficients to be estimated. Let $e[n] = y[n] - \beta_0 x[n] - \beta_1 x[n] x^*[n]$. Suppose that we have N ($N \geq 3$) observations $y[0], y[1], \ldots, y[N-1]$ corresponding to N known training symbols $x[0], x[1], \ldots, x[N-1]$. Our goal is to use the least squares estimation method to find the coefficients β_0 and β_1 based on the known training symbols and their corresponding observations such that $\sum_{n=0}^{N-1} |e[n]|^2$ is minimized.

(a) Assuming that $\beta_1 = 0$, find the least squares estimation of β_0 by expanding the sum, differentiating, and solving.

(b) Assuming that $\beta_0 = 0$, find the least squares estimation of β_1 by expanding the sum, differentiating, and solving.

(c) Find the least squares estimation of β_0 and β_1 by expanding the sum, differentiating, and solving.

(d) Formulate and solve the least squares problem for finding β_0 and β_1 in matrix form. Note that because of the matrix dimensions, you can find an expression for the inverse.

60. Consider a continuous function $f(x) = xe^x$ for $0 \leq x \leq 1$ that we would like to model with a simple polynomial $g(x) = \alpha + \beta x$. Our goal is to find the values of α and β that minimize the squared error between $f(x)$ and $g(x)$. *Hint:* You might want to recall that $\int_0^1 f(x)\mathrm{d}x = 1$ and $\int_0^1 xf(x)\mathrm{d}x = e^{-2}$.

(a) Write an expression for the squared error function. *Hint:* You need an integral.

(b) By taking derivatives with respect to α and β, formulate a least squares problem in matrix form.

(c) Solve for the optimal α and β.

(d) Evaluate the mean squared error using the α and β that you find.

61. Suppose that $w[n]$ is an IID random process with zero mean and unit variance. Let $x[n] = w[n] + 0.5w[n-1]$. Suppose that $y[n] = x[n] + v[n]$ where $v[n]$ is an IID random process with zero mean and variance σ^2.

(a) First consider a scalar MMSE problem. Find the value of g such that

$$\mathbb{E}\left[|g^* y[n] - x[n]|^2\right] \tag{3.556}$$

is minimized.

(b) Determine an expression for the resulting mean square error.

(c) Now suppose that we have an MMSE filter with taps $\{g[k]\}_{k=0}^{K-1}$. Find the values of g such that

$$\mathbb{E}\left[\left|\sum_{k=0}^{K-1} g^*[k]y[n-k] - x[n]\right|^2\right] \tag{3.557}$$

is minimized.

(d) Determine an expression for the mean square error.

Chapter 4

Digital Modulation and Demodulation

The foundation of modern wireless communication systems is digital communication. The mapping of digital information onto waveforms at the transmitter is known as digital modulation. The extraction of the transmitted digital information from the noisy received signal is known as digital demodulation. General background on the principles of digital communication was provided in Chapter 2, to introduce important components of an entire digital communication system and relevant terminology. This chapter introduces a special class of digital modulation techniques that we call complex pulse-amplitude modulation and describes important mathematical concepts that underlie this type of modulation. There are two components of complex pulse-amplitude modulation that can be designed: the symbol constellation and the pulse-shaping filter. First we provide background on symbol mapping and constellations, including a review of several common constellations. We assume that the constellations are normalized to have zero mean and unit energy. After defining the bandwidth of the complex pulse-amplitude modulated signal, we introduce the AWGN channel, where the only impairment is additive Gaussian noise. The generalization to include frequency-selective channels is found in Chapter 5. Then we present results on designing the pulse shape for a communication channel where there is only AWGN. We describe how to implement pulse shaping using multirate digital signal processing. Finally, we examine the basic receiver operations. We derive the maximum likelihood symbol detector and then we characterize its performance through bounds on the probability of symbol error.

4.1 Transmitter for Complex Pulse-Amplitude Modulation

There are many different types of digital communication signals. In this book we consider the general class of modulation techniques that can be generated according to the abstract block diagram in Figure 4.1. We call this type of modulation complex pulse-amplitude modulation, since complex symbols amplitude modulate successive pulses. This diagram

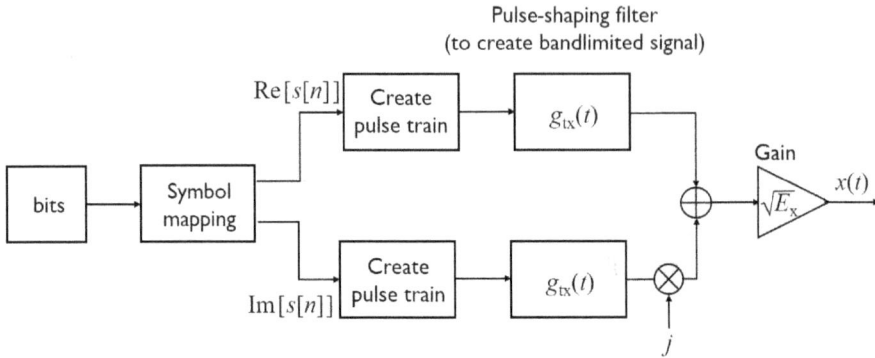

Figure 4.1 Abstract block diagram for generating a complex pulse-amplitude modulated waveform

is meant to reflect the mathematical process of creating a baseband complex bandlimited waveform and does not exactly translate into a hardware implementation. A more realistic implementation of the filtering is provided in Figure 4.10.

Now we explain each block in Figure 4.1, starting from the left and working toward the right.

The source for digital modulation is a sequence of bits $\{b[n]\}$ that the transmitter intends to send to the receiver. The bit sequence could have been generated from an error control code, as described in Chapter 2, but the presence of coding is not exploited in this or subsequent chapters. The physical layer often works with a finite sequence or block of bits $\{b[n]\}_{n=0}^{N-1}$. In this section we consider the usual case where the receiver does not know the bits being sent by the transmitter, which is the entire reason for performing communication in the first place! In Chapter 5, though, we relax this assumption and exploit known information in the form of training signals to estimate and mitigate different channel impairments to further assist in the demodulation of an unknown bit sequence.

The sequence of bits $\{b[n]\}$ is processed by the symbol-mapping block to produce a sequence of symbols $\{s[n]\}$. With complex pulse-amplitude modulation, each value of $s[n]$ is a complex number that comes from a finite set of symbols called the *constellation* and is written as

$$\mathcal{C} = \{c_0, ..., c_{M-2}, c_{M-1}\}. \tag{4.1}$$

The entries of the constellation are different (possibly complex) values. The size of the constellation, or cardinality, is denoted as $|\mathcal{C}| = M$, where M is the number of symbols in the constellation. For practical implementation $M = 2^b$, where b is the number of bits per symbol so that a group of b input bits can be mapped to one symbol. In more sophisticated digital communication systems, the symbols may be vectors or matrices, for example, as discussed in Chapter 6.

The next two operations in Figure 4.1 generate a waveform from the symbol sequence $\{s[n]\}$. The first functional block performs the symbol-to-pulse-train operation to produce

$\sum_n \text{Re}[s[n]]\delta(t-nT)$ and $\sum_n \text{Im}[s[n]]\delta(t-nT)$. Note that the output consists of the real and imaginary parts of the symbols $s[n]$ multiplying an ideal Dirac delta $\delta(t-nT)$. The spacing between symbols is T, the symbol period.

The pulse train sequence is passed to the pulse-shaping filter with impulse response $g_{tx}(t)$. The output is the complex baseband waveform

$$g_{tx}(t) * \sum_n s[n]\delta(t-nT) = \sum_n s[n]g_{tx}(t-nT). \tag{4.2}$$

The right-hand side of (4.2) is reminiscent of the Nyquist reconstruction formula with $g_{tx}(t)$ acting as a reconstruction filter. Unlike perfect reconstruction, though, the pulse-shaping filter $g_{tx}(t)$ is not in general a sinc function as in (3.82). The choice of transmit pulse shaping impacts the bandwidth of the transmit signal and other properties. In applications considered in this chapter, $g_{tx}(t)$ is specially chosen to ensure that the transmitted waveform is bandlimited. Note that $g_{tx}(t)$ is assumed to be real in this book, which is the most common scenario in commercial wireless systems.

The output of the pulse-shaping filter is scaled by $\sqrt{E_x}$ to produce the complex baseband signal

$$x(t) = \sqrt{E_x} \sum_{n=-\infty}^{\infty} s[n]g_{tx}(t-nT). \tag{4.3}$$

The scaling factor E_x is used to model the effect of amplifying $x(t)$ to add power. This chapter makes several normalization assumptions so that E_x/T represents the transmit power of the signal. This book primarily considers digital communication signals where the complex envelope of the signal is generated according to (4.3).

We refer to $x(t)$ as a complex pulse-amplitude signal because complex symbols modulate a succession of pulses given by the function $g_{tx}(t)$. Effectively, symbol $s[n]$ rides the pulse $g_{tx}(t-nT)$. The symbol rate R_s corresponding to $x(t)$ in (4.3) is $1/T$. The units are symbols per second. Typically in wireless systems this is measured in kilosymbols per second or megasymbols per second, but emerging commercial systems like millimeter-wave systems measure data rates in gigasymbols per second [268]. Be careful to note that $1/T$ is not necessarily the bandwidth of $x(t)$, which depends on $g_{tx}(t)$ as will become clear in Section 4.3. The bit rate is b/T (recall that there are 2^b symbols in the constellation) and is measured in bits per second. The *bit rate* is a measure of the number of bits per second the baseband waveform $x(t)$ carries.

Example 4.1 In this example, we provide a visualization of $x(t)$ in (4.3). Suppose that pulse-amplitude modulation is used with $M = 4$, which is called 4-PAM. This is a real constellation that is not widely used in wireless communication but is convenient for illustration purposes. To simplify the plot, set $E_x = 1$ and $T = 1$. Suppose that the bit-to-symbol mapping is given in Table 4.1. This is an example of Gray coding where neighboring symbols have only 1-bit difference.

- Assuming the bit sequence is $\{b[n]\} = \{0010011100\}$, starting from $n = 0$ to $n = 7$, determine the sequence of symbols.

Table 4.1 A Bit-to-Symbol Mapping for 4-PAM

Input	Output
00	$-3/\sqrt{5}$
01	$-1/\sqrt{5}$
10	$3/\sqrt{5}$
11	$1/\sqrt{5}$

Answer: The first pair of bits gives symbol $s[0] = -3/\sqrt{5}$. The rest of the pairs give $s[1] = 3/\sqrt{5}$, $s[2] = -1/\sqrt{5}$, $s[3] = 1/\sqrt{5}$, and $s[4] = -1/\sqrt{5}$.

- Determine a simple expression for $x(t)$, assuming that $s[n] = 0$ for $n < 0$ or $n > 4$.

  Answer: From (4.3), remembering that $E_x = 1$ and $T = 1$,

$$x(t) = s[0]g_{tx}(t) + s[1]g_{tx}(t-1) + s[2]g_{tx}(t-2)$$
$$+ s[3]g_{tx}(t-3) + s[4]g_{tx}(t-4). \tag{4.4}$$

- Plot $x(t)$ for $g_{tx}(t) = \text{rect}(t)$ and $g_{tx} = \text{sinc}(t)$.

  Answer: The plots are shown in Figure 4.2 along with plots of the pulse shapes.

In Example 4.1, we provided a visualization of $x(t)$ for 4-PAM modulation for two different choices of pulse-shaping functions. In subsequent sections we discuss issues related to complex pulse-amplitude modulation with more care, including the choice of constellation, pulse shape, and appropriate receiver design.

4.2 Symbol Mapping and Constellations

This section provides some background on common constellations and establishes some procedures for working with constellations. In particular, the mean of a constellation is determined and removed if present. Then the energy of the zero constellation is computed and the constellation is scaled so it has unit energy. Throughout the remainder of this book, unless otherwise specified, the constellations have zero mean and unit energy.

4.2.1 Common Constellations

In this section we review common constellations used with the complex pulse-amplitude modulation framework. A constellation is defined by two quantities, the alphabet or set of symbols $\mathcal{C}$ and the bit-to-symbol mapping. Often a constellation plot is used to illustrate both the complex values in $\mathcal{C}$ and the corresponding binary mapping. To help in implementation, it is often convenient to order the constellation set according to its binary bit-to-symbol mapping. For example, if $M = 4$, then the symbol for the binary number 01 would be the second entry of $\mathcal{C}$.

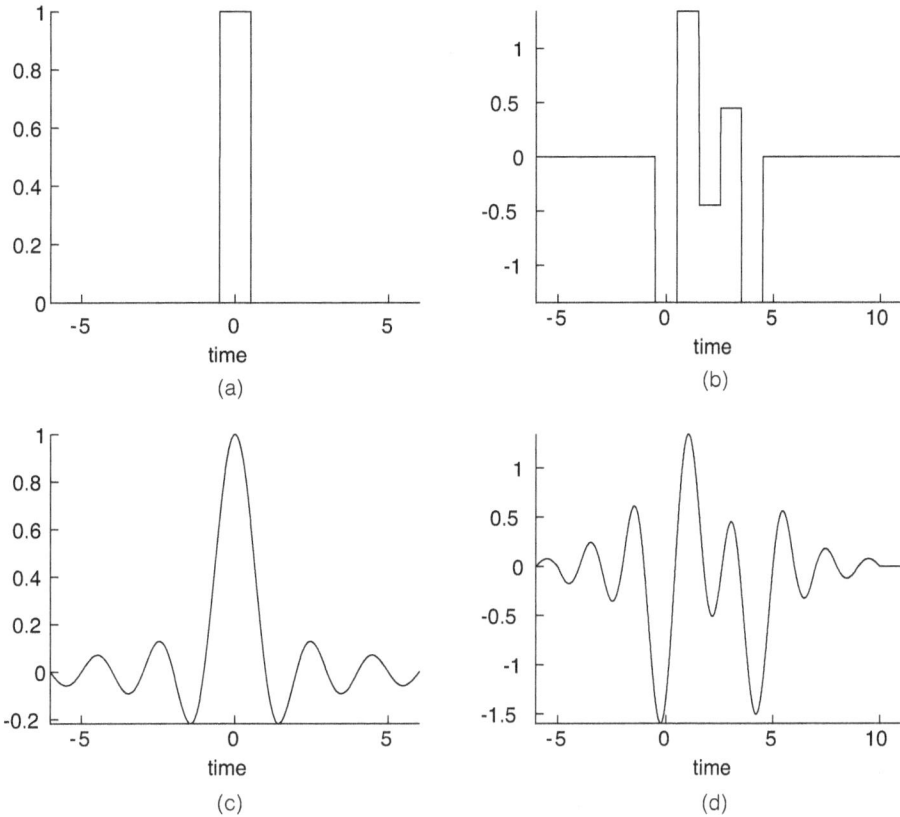

Figure 4.2 (a) A rectangular transmit pulse shape. (b) An encoded sequence of 4-PAM symbols according to Example 4.1 with a rectangular pulse shape. (c) A sinc transmit pulse shape. (d) An encoded sequence of 4-PAM symbols according to Example 4.1 with a sinc pulse shape. The symbols in both cases ride the pulse shape. It is harder to visualize in the case of the sinc function because of the ripples of the sinc. In practice, functions like the sinc are preferred because they are bandlimited.

Several different constellations are used in commercial wireless systems and are summarized here. The normalization of the constellations and their order are designed for pedagogical purposes and may vary in different textbooks or standards. Later in this section, we explain the preferred constellation normalization used in this book.

- Binary phase-shift keying (BPSK) has

$$\mathcal{C} = \{+1, -1\}. \tag{4.5}$$

This is arguably the simplest constellation in use. The bit labeling is typically $s[n] = (-1)^{b[n]}$; thus $b[n] = 0 \rightarrow s[n] = 1$ and $b[n] = 1 \rightarrow s[n] = -1$.

- M-pulse-amplitude modulation (M-PAM) is a generalization of BPSK to any M that is a power of 2. The M-PAM constellation is typically written as

$$\mathcal{C} = \left\{ -\frac{(M-1)}{2}, \dots, -\frac{1}{2}, \frac{1}{2}, \dots, \frac{(M-1)}{2} \right\}. \qquad (4.6)$$

Like BPSK, M-PAM is a real constellation. M-PAM is not common in wireless systems since for $M > 2$ the most energy-efficient constellations have complex values.

- Quadrature phase-shift keying (QPSK) is a complex generalization of BPSK with

$$\mathcal{C} = \{1+\mathrm{j}, -1+\mathrm{j}, -1-\mathrm{j}, 1-\mathrm{j}\}. \qquad (4.7)$$

Essentially QPSK uses BPSK for the real and BPSK for the imaginary component. QPSK is also known as 4-QAM and is used in several commercial wireless systems like IEEE 802.11g and IEEE 802.16.

- M-phase-shift keying (PSK) is a constellation constructed by taking equally spaced points on the complex unit circle

$$\mathcal{C} = \left\{ e^{\frac{\mathrm{j}2\pi k}{M}} \right\}_{k=0}^{M-1}. \qquad (4.8)$$

While not strictly required, in practice M is chosen to be a power of 2. The constellation symbols are the Mth roots of unity. 4-PSK is a different and alternative generalization of BPSK with

$$\mathcal{C} = \{1, \mathrm{j}, -1 - \mathrm{j}\}. \qquad (4.9)$$

The most common PSK constellation is 8-PSK, which fills the gap between 4-QAM and 16-QAM and provides a constellation with 3 bits per symbol. 8-PSK is used in the EDGE standard, an extension of the GSM mobile cellular standard [120].

- M-QAM is a generalization of QPSK to arbitrary M, which is a power of 4. M-QAM is formed as a Cartesian product of two $M/2$-PAM constellations, with symbols of the form

$$\mathcal{C} = \left\{ \cdots, \frac{k}{2} + \mathrm{j}\frac{\ell}{2}, \cdots \right\} \qquad (4.10)$$

where $k, \ell \in \{-(M/2 - 1), -(M/2 - 2), \dots, -1, 1, \dots, (M/2 - 1)\}$. In commercial wireless systems, 4-QAM, 16-QAM, and 64-QAM are common, for example, IEEE 802.11a/g/n, IEEE 802.16, and 3GPP LTE. IEEE 802.11ac adds support for 256-QAM. Microwave links used for backhaul can support much higher values of M [106].

In a wireless system, the performance in terms of coded bit error rate is determined in part by the exact mapping of bits to symbols. There are no unique bit-to-symbol mappings in practice, though the bit labelings in Figure 4.3 are typical. The specific

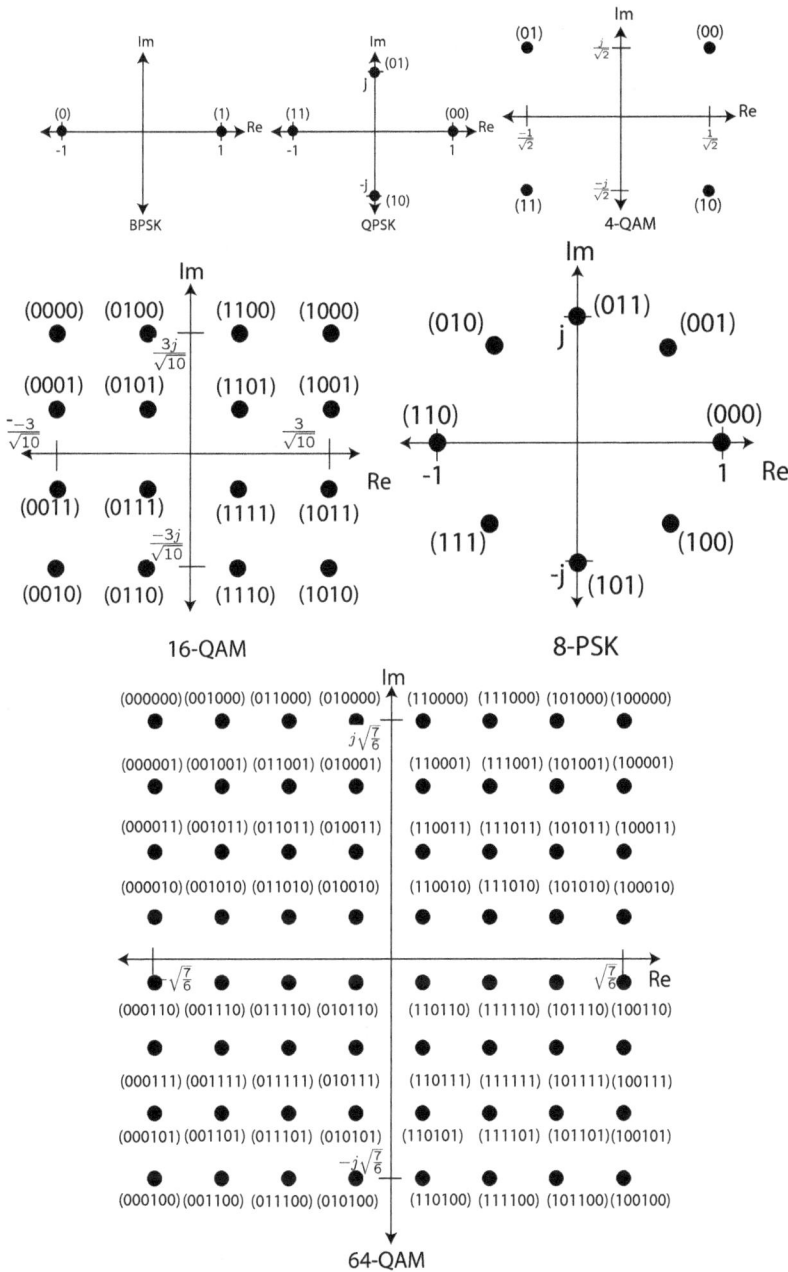

Figure 4.3 Several different constellations with generally accepted bit-to-symbol mappings based on Gray labeling to minimize the number of bit differences in adjacent symbols

labeling in this case is known as Gray coding. Essentially the labeling is done such that the closest symbols differ by only a single bit. It turns out that the closest symbols are the most likely errors; thus this results in a single-bit error for the most typical bit errors. Set-partitioning (SP) labeling is another paradigm for constellation labeling used in trellis-coded modulation. The constellation is partitioned into mutually exclusive subsets of symbols called cosets, and the labels are assigned to maximize the separation between the nearest neighbors in the same coset [334, 335]. Other labelings are associated with more advanced types of error control codes like bit-interleaved coded modulations. These labelings may be determined through optimization of coded bit error rate performance [60].

4.2.2   Symbol Mean

In this section, we compute the mean of a constellation and show how to transform it so that it has zero mean.

Any given finite set of complex numbers $\mathcal{C}$ can be used as a constellation. For analysis and implementation in wireless systems, it is desirable for the symbol mean

$$\mu_s = \mathbb{E}_s\left[s[n]\right] \tag{4.11}$$

to be zero, for example, $\mu_s = 0$. To compute the mean (and the constellation energy), it is usually assumed that the symbols in the constellation are equally likely; thus $P_r[s[n] = c] = 1/M$ for any $c \in \mathcal{C}$, and further that $s[n]$ is an IID random process. Equally likely symbols are reasonable if there is any form of encryption or scrambling prior to transmission, as is usually the case. Under this assumption, the expectation in (4.11) may be computed as

$$\mu_s = \frac{1}{M} \sum_{m=0}^{M-1} c_m. \tag{4.12}$$

Therefore, if the mean of the constellation is defined as the right-hand side of (4.12), then the symbols will also have zero mean. Having zero-mean constellations also ensures that the complex baseband signal $x(t)$ is also zero mean. A nonzero mean corresponds to a DC component, which makes analog baseband implementation more challenging.

Given an arbitrary $\mathcal{C}$, it is desirable to compute an equivalent constellation that has zero mean. It is possible to construct a zero-mean constellation $\mathcal{C}_0$ from $\mathcal{C}$ by simply removing the mean from all the constellation points. To do this, first compute the mean of the constellation:

$$\mu_s = \frac{1}{M} \sum_{m=0}^{M-1} c_m. \tag{4.13}$$

Then construct a new zero-mean constellation by subtracting the mean:

$$\mathcal{C}_0 = \{c_0 - \mu_s, c_1 - \mu_s, \ldots, c_{M-1} - \mu_s\}. \tag{4.14}$$

For the remainder of this book (except in the problems or where otherwise stated) we assume that the constellation has zero mean.

4.2.3 Symbol Energy

Now we define the symbol energy and show how to normalize the constellation so that the symbol energy is unity.

The symbol energy is defined for zero-mean constellations as

$$E_{\mathrm{s}} = \mathbb{E}_s \left[|s[n]|^2 \right]. \tag{4.15}$$

To calculate E_{s} we also use the equally likely and IID assumption on $s[n]$; thus

$$E_{\mathrm{s}} = \mathbb{E}_s \left[|s[n]|^2 \right] \tag{4.16}$$

$$= \frac{1}{M} \sum_{m=0}^{M-1} |c_m|^2. \tag{4.17}$$

In this book, we adjust the constellation so that $E_{\mathrm{s}} = 1$. The reason is that the symbol energy E_{s} is related to the amount of transmit power contained in the signal $x(t)$, as is explained in Section 4.3. If E_{s} is large, then the transmitted signal will have a large amplitude (equivalently the transmit power will also be large). In our mathematical framework, we reserve the term E_{x} to control the gain in the transmitted signal; thus we adjust the constellation so that $E_{\mathrm{s}} = 1$. Different textbooks make different assumptions about the normalization of the constellations. Most practical implementations use constellations that are normalized, as proposed in this book, because gain in the transmitted signal is applied in the analog front end of the radio and not by the baseband signal procesing.

To find an equivalent constellation $\mathcal{C}$ with unit energy given an arbitrary zero-mean constellation, it suffices to scale all the constellation points by some α such that the resulting constellation has $E_{\mathrm{s}} = 1$. All the constellation points are scaled by the same value to preserve the distance properties of the constellation.

Let the scaling factor α be selected to force the new constellation to have $E_{\mathrm{s}} = 1$. Given an arbitrary zero-mean constellation $\mathcal{C} = \{c_0, c_1, \ldots, c_{M-1}\}$, the scaled constellation is

$$\mathcal{C}_\alpha = \{\alpha c_0, \alpha c_1, \ldots, \alpha c_{M-1}\}. \tag{4.18}$$

Now substituting (4.18) into (4.17) and setting $E_{\mathrm{s}} = 1$:

$$\alpha = \sqrt{\frac{1}{\frac{1}{M} \sum_{m=0}^{M-1} |c_m|^2}}. \tag{4.19}$$

The scaled constellation $\mathcal{C}_\alpha$ with the optimum value of α in (4.19) is the normalized constellation. We conclude this section with some example calculations of normalized constellations in Example 4.2 and Example 4.3.

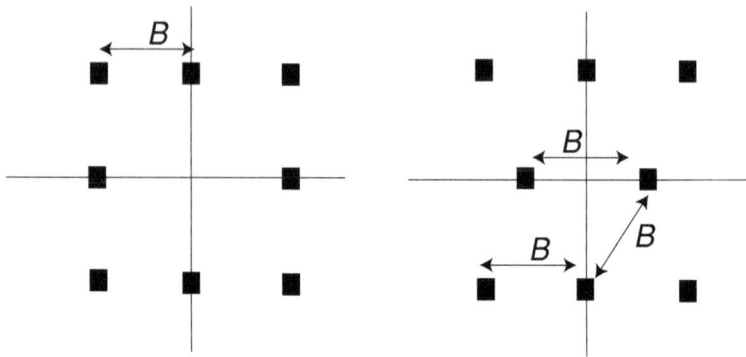

Figure 4.4 Two signal constellations

Example 4.2 Figure 4.4 contains two different 8-point signal constellations. The minimum distance between adjacent points is B. Assuming that the signal points are equally likely in each constellation, find E_s for each constellation and suggest a suitable choice of α to make the constellation have unit energy.

Answer: For the constellation on the left, the average transmitted power is

$$E_s = \frac{1}{8}(B^2 + B^2 + B^2 + B^2) + \frac{1}{8}(2B^2 + 2B^2 + 2B^2 + 2B^2) \tag{4.20}$$

$$= \frac{1}{8}(4B^2) + \frac{1}{8}(8B^2) \tag{4.21}$$

$$= \frac{B^2}{2} + B^2 \tag{4.22}$$

$$= \frac{3B^2}{2}. \tag{4.23}$$

From (4.19), a suitable normalization is

$$\alpha = \sqrt{\frac{2}{3B^2}}. \tag{4.24}$$

For the constellation on the right, the average transmitted power is

$$E_s = \frac{1}{8}\left(\frac{B^2}{4} + \frac{B^2}{4}\right) + \frac{1}{8}\left(\frac{3B^2}{4} + \frac{3B^2}{4}\right) + \frac{1}{8}\left(\frac{7B^2}{4} + \frac{7B^2}{4} + \frac{7B^2}{4} + \frac{7B^2}{4}\right) \tag{4.25}$$

$$= \frac{1}{8}\left(\frac{B^2}{2}\right) + \frac{1}{8}\left(\frac{3B^2}{2}\right) + \frac{1}{8}(7B^2) \tag{4.26}$$

$$= \frac{B^2}{16} + \frac{3B^2}{16} + \frac{7B^2}{8} \tag{4.27}$$

$$= \frac{9B^2}{8}. \tag{4.28}$$

From (4.19), a suitable normalization is

$$\alpha = \sqrt{\frac{8}{9B^2}}. \tag{4.29}$$

Example 4.3   Consider a 4-point constellation $\mathcal{C} = \{j, (1-j)/\sqrt{2}, -j, (-1-j)/\sqrt{2}\}$.

- Compute the mean of the constellation.

  Answer: The mean of the constellation is

  $$\mu_s = \frac{1}{4}\left(j - j + \frac{1}{\sqrt{2}}(-1-j) + \frac{1}{\sqrt{2}}(1-j)\right) \tag{4.30}$$

  $$= -j\frac{1}{2\sqrt{2}}. \tag{4.31}$$

- Compute the zero-mean constellation.

  Answer: The zero-mean constellation is

  $$\mathcal{C}_0 = \left\{\left(1 + \frac{1}{2\sqrt{2}}\right)j, (1-j)/\sqrt{2}, (-1 + \frac{1}{2\sqrt{2}})j, (-1-j)/\sqrt{2}\right\}. \tag{4.32}$$

- Compute the normalized zero-mean constellation.

  Answer: Computing (4.19),

  $$\alpha = \sqrt{\frac{4}{\left(1 + \frac{1}{2\sqrt{2}}\right)^2 + 1 + \left(1 - \frac{1}{2\sqrt{2}}\right)^2 + 1}} \tag{4.33}$$

  $$= \frac{4}{\sqrt{17}}, \tag{4.34}$$

  then the normalized zero-mean constellation is

  $$\mathcal{C}_0 = \left\{\frac{4}{\sqrt{17}}j\left(1 + \frac{1}{2\sqrt{2}}\right), \frac{4}{\sqrt{34}}(1-j), \frac{4}{\sqrt{17}}\left(-1 + \frac{1}{2\sqrt{2}}\right)j, \frac{4}{\sqrt{34}}(-1-j)\right\}. \tag{4.35}$$

For the rest of this book, unless otherwise specified, we assume that the constellation has been normalized so that the symbol energy is 1.

4.3   Computing the Bandwidth and Power of $x(t)$

The signal $x(t)$ is the complex baseband signal that is passed to the analog front end for upconversion and transmission on the wireless channel. As such it is useful to define several quantities of interest related to $x(t)$. Along the way, we also establish our main normalization assumption about $g_{\text{tx}}(t)$.

To compute the bandwidth of a signal $x(t)$, the conventional approach would be to compute the Fourier spectrum and make a bandwidth determination from that spectrum. As $s[n]$ is usually modeled as an IID random process, this direction calculation is not meaningful. As discussed in Chapter 3, the bandwidth of a random process is usually defined based on its power spectral density, for WSS random processes. While $x(t)$ is in fact a random process, it is actually not WSS. In fact it is what is known as a cyclostationary wide-sense stationary random process [121]. The subtleties of this process are beyond the scope of this book; for our purposes it suffices that we can define an operational notion of power spectral density for such processes as

$$P_{xx}(f) = \frac{E_{\text{x}}}{T} E_{\text{s}} \left| G_{\text{tx}}(f) \right|^2. \tag{4.36}$$

Under the unit symbol energy assumption, $\bar{F}_s = 1$, thus

$$P_{xx}(f) = \frac{E_{\text{x}}}{T} \left| G_{\text{tx}}(f) \right|^2, \tag{4.37}$$

and it is apparent that $\left| G_{\text{tx}}(f) \right|^2$ determines the bandwidth of $x(t)$.

To calculate the power associated with $x(t)$, it is useful to make some additional assumptions about $g_{\text{tx}}(t)$. In this book, unless otherwise stated, we always assume that the pulse shape is normalized to have unit energy. This means that

$$\int_{-\infty}^{\infty} \left| G_{\text{tx}}(f) \right|^2 \mathrm{d}f = 1 \tag{4.38}$$

or equivalently by Parseval's theorem

$$\int_{-\infty}^{\infty} \left| g_{\text{tx}}(t) \right|^2 \mathrm{d}t = 1. \tag{4.39}$$

We assume this for the same reason that we assume the constellation has unit energy. Specifically, in our mathematical formulation we want to use E_{x} exclusively to control the transmit power. Under this assumption the transmit power is

$$\int_{-\infty}^{\infty} P_{xx}(f) \mathrm{d}f = \frac{E_{\text{x}}}{T}, \tag{4.40}$$

where E_{x} plays the role of the signal energy and energy per second is the appropriate unit of power.

Example 4.4   Determine the normalized version of the pulse-shaping filter

$$g_{\text{tx,non}}(t) = \text{sinc}^2(t/T), \tag{4.41}$$

the power spectral density of the corresponding $x(t)$, and the absolute bandwidth.

Answer: Compute the normalized pulse shape by finding its energy:

$$\int |g_{\text{tx,non}}(t)|^2 \text{d}t = \int |G_{\text{tx,non}}(f)|^2 \text{d}f \tag{4.42}$$

$$= T^2 \int_{-\frac{1}{T}}^{\frac{1}{T}} (1 - |f|T)^2 \text{d}f \tag{4.43}$$

$$= \frac{2}{3}T. \tag{4.44}$$

Therefore, the normalized pulse shape is

$$g_{\text{tx}}(t) = \sqrt{\frac{3}{2T}} \text{sinc}^2\left(\frac{t}{T}\right). \tag{4.45}$$

The power spectral density of $x(t)$ with the normalized pulse shape comes from (4.37). Assuming the constellation has been properly normalized,

$$P_{xx}(f) = \frac{E_{\text{x}}}{T} \left| \sqrt{\frac{3}{2T}} T\Lambda(fT) \right|^2 \tag{4.46}$$

$$= \frac{3}{2} E_{\text{x}} \Lambda^2(fT). \tag{4.47}$$

The absolute bandwidth of $P_{xx}(f)$ is just the absolute bandwidth of $|G_{\text{tx}}(f)|^2$. Since $G_{\text{tx}}(f) = \sqrt{\frac{3}{2T}} T\Lambda(fT)$, from Table 3.2 and using the time-scaling theorem from Table 3.1, the absolute bandwidth is $1/T$.

4.4   Communication in the AWGN Channel

In this section, we explore communication in the presence of the most basic impairment: AWGN. We start with an introduction to AWGN. Then we propose a pulse shape design that achieves good performance in AWGN channels. Important concepts like Nyquist pulse shapes and the matched filter are a by-product of the discussion. Then, assuming an optimum pulse shape, we derive the maximum likelihood symbol detector for AWGN. We conclude the section with an analysis of a bound on the probability of symbol error, which reveals the importance of the SNR.

4.4.1   Introduction to the AWGN Channel

Beyond bandwidth and power, other characteristics of the pulse-shaping filter $g_{tx}(t)$ play a role in determining how well the receiver can detect the transmitted symbols. The design of the pulse-shaping filter depends on the impairments that are created by the propagation channel and also the electronics at the receiver. The most basic impairment—which is a steppingstone for dealing with more complicated impairments—is AWGN. The transmit pulse-shaping filter is commonly chosen assuming AWGN is the main impairment, even in more complicated systems. In this section, we introduce the AWGN channel model and the associated transmitter and receiver.

Thermal noise is present in every communication receiver. It is created from the variation of electrons due to temperature and occurs in the different analog receiver components. Because the received voltages can be very small in a wireless system (because of distance-dependent path loss as discussed in Section 2.3.3), thermal noise has a measurable impact on the received signal. There are other kinds of noise in wireless systems, but thermal noise is the most common impairment. The design of the RF front end has a great impact on the severity of thermal noise through the effective noise temperature influenced by the choice of components [271]. Further discussion, though, is beyond the scope of this book.

The AWGN communication channel is a mathematical model for the impairment due to thermal noise. Because we deal with the complex baseband equivalent system, we consider AWGN as applied at baseband. A simple model is illustrated in Figure 4.5. Mathematically

$$y(t) = x(t) + v(t), \tag{4.48}$$

where $x(t)$ is the transmitted complex baseband signal, $v(t)$ is the AWGN, and $y(t)$ is the observed complex baseband signal.

The assumption that $v(t)$ is AWGN has the following implications:

- The noise is additive. Other types of noise, like multiplicative noise and phase noise, are also possible but lead to other types of additive impairments.

- The noise is IID (this is where the term *white* comes from, since the power spectral density of an IID signal is flat in the frequency domain) and is therefore also a WSS

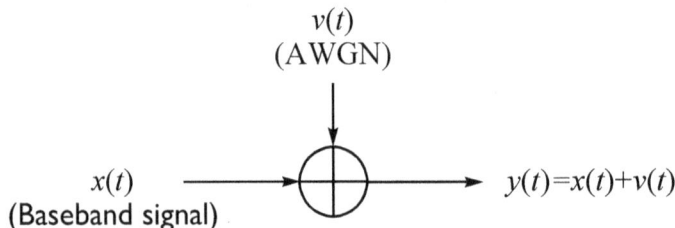

Figure 4.5 The additive white Gaussian noise communication channel

random process. The autocorrelation function is $R_{vv}(\tau) = \sigma^2 \delta(\tau)$, and the power spectral density is $P_{vv}(f) = \sigma^2$.

- The first-order distribution of $v(t)$ is $\mathcal{N}_c(0, \sigma^2)$.

- The total variance is $\sigma^2 = N_o$. The noise spectral density is $N_o = kT_e$, where k is Boltzmann's constant $k = 1.38 \times 10^{-23} J/K$ and the effective noise temperature of the device is T_e in kelvins. Assume $T_e = 290$K in the absence of other information. The effective noise temperature is a function of the ambient temperature, the type of antennas, as well as the material properties of the analog front end. Sometimes N_o is expressed in decibels. For example, if $T_e = 290$K, then $N_o = -228$dB/Hz.

The signal-to-noise ratio (SNR) is the critical measure of performance for the AWGN communication channel. We define it as

$$\text{SNR} = \frac{E_x}{N_o}. \tag{4.49}$$

Many different performance measures in AWGN channels are a function of the SNR, including the probability of symbol error and the channel capacity [28]. In other types of noise or receiver impairments, different measures of performance may be more suitable. The SNR, though, is a widely used parameter in wireless systems.

4.4.2 Receiver for Complex Pulse-Amplitude Modulation in AWGN

The receiver in a digital communication system is tasked with taking an observation of $y(t)$ and making its best guess about the corresponding $\{s[n]\}$ that were transmitted. The theory behind this "best guess" comes from detection theory [340, 339, 341] and depends on the assumed channel model that describes the transformation of $x(t)$ to $y(t)$. Detection theory is related to hypothesis testing in statistics [176]; in this case the receiver tests different hypotheses about the possible values of the symbols that were sent.

In this book, we focus on the receiver structure for complex pulse-amplitude modulated signals $x(t)$ as described in (4.3). The derivation (from first principles) of this optimum receiver structure for the AWGN channel for arbitrary modulations is beyond the scope of this book. To simplify the presentation, yet capture the salient details of the receiver function in an AWGN system, we start with the observation that the optimal receiver has the form in Figure 4.6.

Now we summarize the key receiver blocks in Figure 4.6. The filter $g_{\text{rx}}(t)$ is known as the receiver pulse-shaping filter. It performs a bandlimiting operation among other functions, as will become clearer in Section 4.4.3. The continuous-to-discrete converter samples the received signal at the symbol rate of $1/T$. The detection block produces a good guess about the $s[n]$ that might have been transmitted based on the sampled output of the C/D. The inverse symbol mapping determines the bit sequence corresponding to the detected symbol $\widehat{s}[n]$ output from the detection block. The blocks in the receiver in Figure 4.6 work together to reverse the corresponding operations at the transmitter.

Tx

bits → [Symbol mapping] → [Create pulse train] → [$g_{\mathrm{tx}}(t)$] → [▷ $\sqrt{E_{\mathrm{x}}}$] → $x(t)$

Rx

$y(t)$ → [$g_{\mathrm{rx}}(t)$] → [C/D] → $y[n]$ → [Detection] → [Inverse symbol mapping] → $\widehat{\text{bits}}$

↑
T

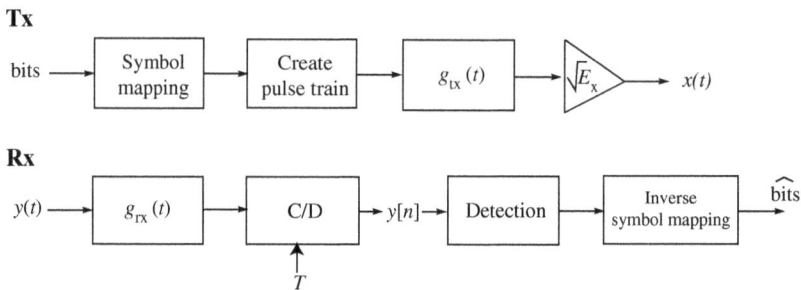

Figure 4.6 Transmitter and receiver for an AWGN communication channel

There are three important design decisions in specifying the operation of the receiver:

1. Determine the best receiver filter $g_{\mathrm{rx}}(t)$ and how it depends on the transmitted signal. It turns out that a matched filter is optimum, as described further in Section 4.4.3.

2. Specify what happens in the detection block using mathematical principles. We show in Section 4.4.4 that the detector should implement a minimum Euclidean distance detector, where the (scaled) constellation symbol that is closest to the observed sample is output from the detector.

3. Determine a more realizable structure for the pulse-shaping and filtering operations in Figure 4.6. In Section 4.4.5, we use multirate signal processing concepts from Chapter 3 to implement both the transmit pulse shaping and the receive filtering in discrete time.

The first two design decisions are addressed in this section, and the solution to the third is found in Section 4.5.

4.4.3 Pulse Shape Design for the AWGN Channel

In this section, we devise designs for the transmit pulse shape $g_{\mathrm{tx}}(t)$ and the receive filter $g_{\mathrm{rx}}(t)$. The derivation is based on the idea of maximizing the signal-to-interference-plus-noise ratio (SINR), assuming a memoryless detector. Other types of signal processing, like equalization followed by detection, are explored in Chapter 5.

The derivation starts with the received signal after AWGN:

$$y(t) = \sqrt{E_{\mathrm{x}}} g_{\mathrm{rx}}(t) * g_{\mathrm{tx}}(t) * \sum_m s[m]\delta(t - mT) + g_{\mathrm{rx}}(t) * v(t). \qquad (4.50)$$

Let $g(t) = g_{\mathrm{rx}}(t) * g_{\mathrm{tx}}(t)$ denote the combined transmit and receive pulse-shaping filters. After sampling at the symbol rate,

$$y[n] = \sqrt{E_{\mathrm{x}}} \sum_m s[m]g((n - m)T) + \widetilde{v}[n] \qquad (4.51)$$

where $\widetilde{v}[n] = \int v(\tau)g_{\mathrm{rx}}(nT - \tau)\mathrm{d}\tau$ is the filtered noise. The detector extracts $s[n]$ from $y[n]$; thus we rewrite (4.51) as

$$y[n] = \sqrt{E_{\mathrm{x}}}s[n] + \sqrt{E_{\mathrm{x}}}\sum_{m \neq n} s[m]g((n - m)T) + \widetilde{v}[n]. \tag{4.52}$$

The first term is the desired signal, the second term is intersymbol interference, and the third term is sampled noise. As such, it makes sense to select $g_{\mathrm{tx}}(t)$ and $g_{\mathrm{rx}}(t)$ to permit the best detection performance.

We propose to design the transmit pulse shape and received filter to maximize the SINR. The signal energy is

$$\mathbb{E}\left[\left|\sqrt{E_{\mathrm{x}}}s[n]g(0)\right|^2\right] = E_{\mathrm{x}}|g(0)|^2. \tag{4.53}$$

The only way to increase the signal energy is by increasing E_{x} (the transmit power) or increasing $g(0)$. It will be clear shortly why increasing $g(0)$ is not an option.

The ISI energy is

$$\mathbb{E}\left[\left|\sqrt{E_{\mathrm{x}}}\sum_{m,m \neq n} s[m]g((n - m)T)\right|^2\right] = E_{\mathrm{x}}\sum_{m \neq 0} |g(mT)|^2. \tag{4.54}$$

The ISI can be decreased by reducing E_{x}, but this also reduces the signal power. The only other way to reduce ISI is to make $|g(mT)|$ as small as possible for $m \neq 0$. There are special choices of $g(t)$ that can completely eliminate ISI.

The noise energy is

$$\mathbb{E}\left[\left|g_{\mathrm{rx}}(t) \star v(t)|_{nT}\right|^2\right] = N_{\mathrm{o}}\int |G_{\mathrm{rx}}(f)|^2\mathrm{d}f. \tag{4.55}$$

The larger the bandwidth of $g_{\mathrm{rx}}(t)$, the higher the noise density. Reducing the bandwidth also reduces noise, as long as the bandwidth is at least as large as the signal bandwidth. Otherwise the transmitted signal may be filtered out.

Treating the ISI as an additional source of Gaussian noise, the SINR becomes a relevant performance metric for AWGN channels. Computing the ratio of (4.53) with the sum of (4.54) and (4.55),

$$\mathrm{SINR} = \frac{E_{\mathrm{x}}|g(0)|^2}{N_{\mathrm{o}}\int |G_{\mathrm{rx}}(f)|^2\mathrm{d}f + E_{\mathrm{x}}\sum_{m \neq n} |g(mT)|^2}. \tag{4.56}$$

Now we proceed to find a good choice of $g_{\mathrm{tx}}(t)$ and $g_{\mathrm{rx}}(t)$ to maximize the SINR.

Our approach to maximizing the SINR is to find an upper bound that can be achieved with equality for the right pulse shape design. First, examine the average received signal power term $E_{\mathrm{x}}|g(0)|^2$. From the definition of $g(t)$, notice that

$$g(0) = \int g_{\mathrm{rx}}^*(-t)g_{\mathrm{tx}}(t)\mathrm{d}t. \tag{4.57}$$

To upper bound $g(t)$, we apply the Cauchy-Schwarz inequality. For two complex integrable functions $a(t)$ and $b(t)$ with finite energy, the Cauchy-Schwarz inequality says that

$$\left(\int_{-\infty}^{\infty} a^*(t)b(t)\mathrm{d}t \right)^2 \leq \int_{-\infty}^{\infty} |a(t)|^2 \mathrm{d}t \int_{-\infty}^{\infty} |b(t)|^2 \mathrm{d}t \qquad (4.58)$$

with equality if and only if $b(t) = \alpha a(t)$ where α is a (possibly complex) constant. Applying Cauchy-Schwarz to (4.57),

$$|g(0)|^2 = \left| \int g_{\mathrm{rx}}(-t)g_{\mathrm{tx}}(t)\mathrm{d}t \right|^2 \qquad (4.59)$$

$$\leq \int |g_{\mathrm{tx}}(t)|^2 \, \mathrm{d}t \int |g_{\mathrm{rx}}(t)|^2 \, \mathrm{d}t \qquad (4.60)$$

$$= \int |g_{\mathrm{rx}}(t)|^2 \, \mathrm{d}t \qquad (4.61)$$

where the last step follows because we have assumed that $g_{\mathrm{tx}}(t)$ has unit energy. Since $g_{\mathrm{rx}}(t)$ shows up in every term of the numerator and denominator of the SINR, we can without loss of generality also assume that $g_{\mathrm{rx}}(t)$ has unit energy. Consequently, take $g_{\mathrm{rx}}(t) = g_{\mathrm{tx}}^*(-t)$. Such a choice of $g_{\mathrm{rx}}(t)$ is known as a *matched filter* and leads to $g(0) = 1$. Effectively, the matched filter correlates the received signal with the transmit pulse shape to achieve the largest signal energy.

Now we further upper bound the SINR by minimizing the denominator of (4.56). Since $g_{\mathrm{rx}}(t)$ is determined from $g_{\mathrm{tx}}(t)$, the noise power at this point is constant and cannot be further minimized. Instead we focus on the ISI term. Because the terms in the summation are nonnegative, the summation is minimized if all the terms are zero, that is, if $E_x \sum_{m \neq 0} |g(mT)|^2 = 0$. This is possible only if $g(mT) = 0$ for $m \neq 0$. Combined with the fact that $g(0) = 1$ from our assumed normalization of the receive filter $g_{\mathrm{rx}}(t)$, the ISI is zero if it is possible to find $g_{\mathrm{tx}}(t)$ such that

$$g(nT) = \delta[n]. \qquad (4.62)$$

Notice that (4.62) places requirements on the samples of $g(t)$ but does not otherwise constrain the waveform. Assuming such pulse shapes are realized, optimum SINR becomes simply $\mathrm{SNR} = E_x/N_o$.

To discover the implications of (4.62), let us go into the frequency domain. Treating $g_{\mathrm{d}}[n] = g(nT)$ as a discrete-time sequence and taking the Fourier transform of both sides gives

$$G_{\mathrm{d}}(e^{\mathrm{j}2\pi f}) = 1. \qquad (4.63)$$

But recall from (3.83) that the CTFT of a continuous-time signal $G(f)$ and the DTFT of the sampled signal $G_{\mathrm{d}}(e^{\mathrm{j}2\pi f})$ are related through

$$G_{\mathrm{d}}(e^{\mathrm{j}2\pi f}) = \sum_{k=-\infty}^{\infty} G(fT + k). \qquad (4.64)$$

Combining (4.63) and (4.64) gives

$$\sum_{k=-\infty}^{\infty} G(fT + k) = 1. \tag{4.65}$$

Effectively, the aliased sampled pulse shape should be a constant.

Functions that satisfy (4.62) or (4.65) are known as Nyquist pulse shapes. They are special. Sampling at symbol rate T, the sampled function $g(nT)$ is zero for $n \neq 0$. Note that this is true only if the function is sampled at the exact correct place and does not imply that the function is otherwise zero.

Example 4.5 Consider a rectangular pulse shape

$$g_{\text{tx}}(t) = \sqrt{\frac{2}{T}} \text{rect}(t/T - 1/2). \tag{4.66}$$

- Find the matched filter $g_{\text{rx}}(t)$.

  Answer:

$$g_{\text{rx}}(t) = g_{\text{tx}}(-t) \tag{4.67}$$

$$= \sqrt{\frac{2}{T}} \text{rect}(-t/T - 1/2). \tag{4.68}$$

- Find the combined filter $g(t) = \int g_{\text{rx}}(\tau) g_{\text{tx}}(t - \tau) \mathrm{d}\tau$.

  Answer:

$$g(t) = \int g_{\text{tx}}(\tau) g_{\text{rx}}(t - \tau) \mathrm{d}\tau \tag{4.69}$$

$$= \Lambda\left(\frac{t}{T}\right) \tag{4.70}$$

  where Λ is the triangle pulse in Table 3.2.

- Is $g(t)$ a Nyquist pulse shape?

  Answer: Yes, since $g(nT) = \Lambda(n) = \delta[n]$. It is not a particularly good pulse shape choice, though, because $g_{\text{tx}}(t)$ is not bandlimited.

Perhaps the best-known example of a Nyquist pulse shape is the sinc:

$$g_{\text{sinc}}(t) = \text{sinc}(t/T). \tag{4.71}$$

With the sinc function, the baseband bandwidth of the pulse shape is $1/2T$, and there is no overlap in the terms in $G\left(f + \frac{k}{T}\right)$ in (4.65). Other choices of $g(t)$ have larger bandwidth and the aliases add up so that equality is maintained in (4.65).

The sinc pulse-shaping filter has a number of implementation challenges. Ideal implementations of $g(t)$ do not exist in analog in practice. Digital implementations require truncating the pulse shape, which is a problem since it decays in time with $1/t$, requiring a lot of memory. Furthermore, the sinc function is sensitive to sampling errors (not sampling at exactly the right point). For these reasons it is of interest to consider pulse shapes that have excess bandwidth, which means the baseband bandwidth of $g(t)$ is greater than $1/2T$, or equivalently the passband bandwidth is greater than $1/T$.

The most common Nyquist pulse, besides the sinc pulse, is the raised cosine. The raised cosine pulse shape has Fourier spectrum

$$G_{\rm rc}(f) = \begin{cases} T, & 0 \le |f| \le (1-\alpha)/2T \\ \frac{T}{2}\left[1 + \cos\left(\frac{\pi T}{\alpha}\left(|f| - \frac{1-\alpha}{2T}\right)\right)\right], & \frac{1-\alpha}{2T} \le |f| \le \frac{1+\alpha}{2T} \\ 0, & |f| > \frac{1+\alpha}{2T} \end{cases} \tag{4.72}$$

and transform

$$g_{\rm rc}(t) = \text{sinc}(t/T)\frac{\cos(\pi\alpha t)}{1 - 4\alpha^2 t^2/T^2}. \tag{4.73}$$

The parameter α is the rolloff factor, $0 \le \alpha \le 1$. Sometimes β is used instead of α. Often the rolloff is expressed as a percentage of excess bandwidth. In other words, 50% excess bandwidth would correspond to $\alpha = 0.5$.

Example 4.6 Show that for any value of α, the raised cosine spectrum $G_{\rm rc}(f)$ satisfies

$$\int_{-\infty}^{\infty} G_{\rm rc}(f)\mathrm{d}f = 1. \tag{4.74}$$

Answer: Let $g_{\rm rc}(t)$ be the inverse Fourier transform of $G_{\rm rc}(f)$. At the sampling instants $t = nT$, we have

$$g_{\rm rc}(nT) = \int_{-\infty}^{\infty} G_{\rm rc}(f)e^{\mathrm{j}2\pi fnT}\mathrm{d}f. \tag{4.75}$$

We divide the integral in (4.75) into several integrals covering length $1/T$ intervals, which yields

$$g_{\rm rc}(nT) = \sum_{m=-\infty}^{\infty} \int_{(2m-1)/(2T)}^{(2m+1)/(2T)} G_{\rm rc}(f)e^{\mathrm{j}2\pi fnT}\mathrm{d}f \tag{4.76}$$

$$= \sum_{m=-\infty}^{\infty} \int_{-1/(2T)}^{1/(2T)} G_{\rm rc}(f + m/T)e^{\mathrm{j}2\pi fnT}\mathrm{d}f \tag{4.77}$$

$$= \int_{-1/(2T)}^{1/(2T)} \left\{\sum_{m=-\infty}^{\infty} G_{\rm rc}(f + m/T)\right\}e^{\mathrm{j}2\pi fnT}\mathrm{d}f \tag{4.78}$$

$$= \int_{-1/(2T)}^{1/(2T)} Te^{\mathrm{j}2\pi fnT}\mathrm{d}f \tag{4.79}$$

where (4.79) follows from the fact that the raised cosine function satisfies the Nyquist criterion. Thus, by setting n = 0, we have that

$$g_{\mathrm{rc}}(0) = \int_{-\infty}^{\infty} G_{\mathrm{rc}}(f)\mathrm{d}f \tag{4.80}$$

$$= \int_{-1/(2T)}^{1/(2T)} T\mathrm{d}f \tag{4.81}$$

$$= 1. \tag{4.82}$$

Example 4.7   Using the results from Example 4.6, consider now the behavior of the raised cosine pulse shape when either oversampled or undersampled.

- Is it still a Nyquist pulse shape if sampled at $1/(2T)$?

  Answer: For all f note that

$$\sum_k G_{\mathrm{rc}}\left(f + \frac{k}{2T}\right) = \sum_h G_{\mathrm{rc}}\left(f + \frac{2h}{2T}\right) + \sum_h G\left(f + \frac{2h+1}{2T}\right) \tag{4.83}$$

$$= \underbrace{\sum_h G_{\mathrm{rc}}\left(f + \frac{h}{T}\right)}_{A} + \underbrace{\sum_h G_{\mathrm{rc}}\left(f + \frac{h}{T} + \frac{1}{2T}\right)}_{B} \tag{4.84}$$

$$= T + T \tag{4.85}$$

$$= 2T \tag{4.86}$$

where the first term is T from Example 4.6 and the second term is T since shifting a flat spectrum maintains a flat spectrum.

- What about sampling at $2/T$?

  Answer: The bandwidth of $G_{\mathrm{rc}}(f)$ is $\frac{1+\alpha}{2T}$, which is at most $1/T$. Sampling with $2/T$ means that there is no overlap in the functions in $\sum_k G_{\mathrm{rc}}\left(f + \frac{2k}{T}\right)$. Because there is no overlap, the spectrum cannot be flat since the spectrum of $G_{\mathrm{rc}}(f)$ is not flat.

The absolute bandwidth of the raised cosine pulse shape at baseband is $(1 + \alpha)/2T$ and at passband is $(1 + \alpha)/T$. The Nyquist rate of the baseband digital communication signal $x(t)$ with a raised cosine pulse shape is $(1 + \alpha)/T$. Note that $1/T$ is the rate at which $x(t)$ is sampled. Thus there is aliasing in $x(t)$ since Nyquist is not satisfied. The zero-ISI condition, though, ensures that aliasing does not lead to intersymbol interference in the received signal.

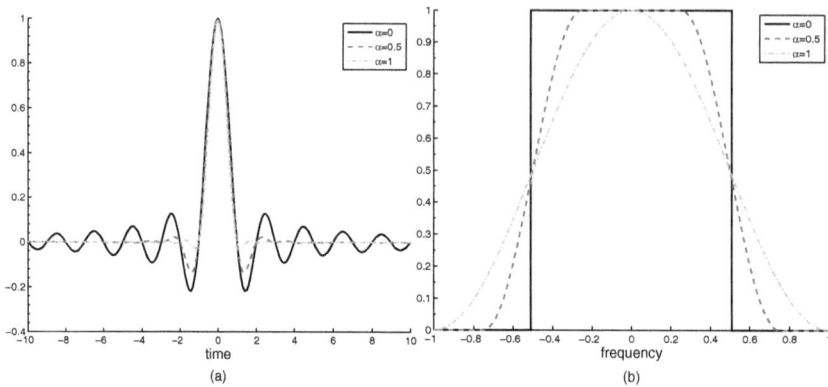

Figure 4.7 (a) The raised cosine pulse shape for various choices of α in the time domain. (b) The raised cosine pulse shape for various choices of α in the frequency domain. A value of $T = 1$ is assumed.

The raised cosine is plotted in Figure 4.7. For larger values of α the spectrum rolloff is smoother and the time-domain decay of the pulse is faster. This makes implementation easier—at the expense of excess bandwidth.

In practice we do not usually use the raised cosine pulse directly. The reason is that the raised cosine pulse shape is $g(t)$ whereas we use $g_{\mathrm{tx}}(t)$ at the transmitter and $g_{\mathrm{rx}}(t)$ at the receiver. Recall that $g(t) = \int g_{\mathrm{rx}}(\tau)g_{\mathrm{tx}}(t - \tau)\mathrm{d}\tau$ and that $g_{\mathrm{rx}}(t) = g_{\mathrm{tx}}(-t)$. In the frequency domain this means that $G(f) = G_{\mathrm{tx}}(f)G_{\mathrm{tx}}^*(f)$. Consequently, we choose $g_{\mathrm{tx}}(t)$ to be a "square root" of the raised cosine. Such a pulse shape is known as a square root raised cosine or a root raised cosine and is written

$$g_{\mathrm{sqrc}}(t) = \frac{4\alpha}{\pi\sqrt{T}}\frac{\cos\left((1+\alpha)\pi t/T\right) + \frac{\sin((1-\alpha)\pi t/T)}{4\alpha t/T}}{1 - (4\alpha t/T)^2}. \tag{4.87}$$

Note that this pulse shape is normalized. Note also that $g_{\mathrm{sqrc}}(t)$ is even; thus if $g_{\mathrm{tx}}(t) = g_{\mathrm{sqrc}}(t)$, then $g_{\mathrm{rx}}(t) = g_{\mathrm{tx}}(-t) = g_{\mathrm{tx}}(t)$ and the transmit pulse shape and the receive pulse shape are identical. The square root raised cosine is its own matched filter!

Root raised cosines are a common transmit pulse for complex pulse-amplitude modulation. We now summarize key relationships between the bandwidth, the symbol rate, and the excess bandwidth with a square root raised transmit pulse shape. Equivalently when $g(t)$ is a raised cosine:

- The symbol rate is $R = 1/T$ where T is the symbol period.

- The absolute bandwidth of the complex pulse-amplitude modulated signal with root raised cosine pulse shaping at baseband is $(1 + \alpha)/2T$.

- The absolute bandwidth of a complex pulse-amplitude modulated signal with root raised cosine pulse shaping modulated at some carrier frequency is $(1+\alpha)/T$. This is twice the bandwidth at baseband per the difference in bandwidth definitions between baseband and passband signals.

- The Nyquist frequency for a root raised cosine pulse-shaped complex pulse-amplitude modulated signal at baseband is $(1 + \alpha)/2T$.

- The Nyquist rate for a root raised cosine pulse-shaped complex pulse-amplitude modulated signal at baseband is $(1 + \alpha)/T$.

A side benefit of the square root raised cosine filter is that the sampled noise becomes uncorrelated. Specifically, consider the discrete-time sampled noise in (4.51). The noise is still zero mean since $\mathbb{E}_v[\tilde{v}[n]] = 0$. The autocovariance of the noise is

$$C_{vv}[k] = \mathbb{E}_v\left[\tilde{v}[n]\tilde{v}^*[n + k]\right] \tag{4.88}$$

$$= \mathbb{E}_v\left[\int_{\tau_1}\int_{\tau_2} v(\tau_1)v^*(\tau_2)g_{\text{sqrc}}(nT - \tau_1)g_{\text{sqrc}}((n - k)T - \tau_2)\mathrm{d}\tau_1\mathrm{d}\tau_2\right] \tag{4.89}$$

$$= \sigma_v^2 \int_{\tau} g_{\text{sqrc}}(nT_s - \tau)g_{\text{sqrc}}((n - k)T - \tau)\mathrm{d}\tau \tag{4.90}$$

$$= \sigma_v^2 \int_{\tau} g_{\text{sqrc}}(\tau)g_{\text{sqrc}}(-kT - \tau)\mathrm{d}\tau \tag{4.91}$$

where we have used the IID property $\mathbb{E}_v[v(\tau_1)v(\tau_2)] = \delta(\tau_2 - \tau_1)$ to simplify (4.89) to (4.90). Now recognizing that $g_{\text{sqrc}}(t) = g_{\text{sqrc}}(-t)$, it follows that

$$C_{vv}[k] = \sigma_v^2 \int_{\tau} g_{\text{sqrc}}(\tau)g_{\text{sqrc}}(kT + \tau)\mathrm{d}\tau. \tag{4.92}$$

Notice, though, from the construction of the square root filter that

$$g_{\text{rc}}(t) = \int_{\tau} g_{\text{sqrc}}(t - \tau)g_{\text{sqrc}}(\tau)\mathrm{d}\tau \tag{4.93}$$

$$= \int_{\tau} g_{\text{sqrc}}(t + \tau)g_{\text{sqrc}}(\tau)\mathrm{d}\tau \tag{4.94}$$

because of even symmetry. Since the raised cosine is a Nyquist pulse shape, it follows that

$$C_{vv}[k] = \sigma_v^2 g_{\text{rc}}(kT) \tag{4.95}$$

$$= \sigma_v^2 \delta[k] \tag{4.96}$$

where $\sigma_v^2 = N_{\text{o}}$ for complex AWGN. Therefore, the square root raised cosine preserves the IID property of the noise (the whiteness). This property is true for any appropriately chosen real square root of a Nyquist pulse. Detection with noise that is not white, which is correlated, is more complicated and is not discussed in this book. In practice, correlated noise is dealt with at the receiver by using a whitening filter.

4.4.4   Symbol Detection in the AWGN Channel

In this section, we address the problem of inferring the transmitted symbols given noisy observations at the receiver. This process is known as symbol detection.

With a suitable combination of Nyquist pulse shaping and matched filtering (using the square root raised cosine, for example), the received signal after sampling can be written as

$$y[n] = \sqrt{E_x} \sum_m s[m]g((n-m)T) + g_{rx}(t) * v(t)|_{nT} \qquad (4.97)$$

$$y[n] = \sqrt{E_x}s[n] + v[n] \qquad (4.98)$$

where $v[n]$ is IID complex Gaussian noise with $\mathcal{N}_c(0, N_o)$.

The detection problem is to answer the following question:

Based on $y[n]$, what is my best guess as to the value of $s[n]$?

This is an example of symbol detection. Other kinds of detection, such as bit detection or sequence detection, are used when detection is combined with joint forward error correction decoding.

A symbol detector is an algorithm that given the observation $y[n]$ produces the best $\widehat{s}[n] \in \mathcal{C}$ according to some criterion. In this book, we consider the maximum likelihood (ML) detector

$$\widehat{s}[n] = \arg\max_{s \in \mathcal{C}} f_{y|s}(y[n]|s[n] = s) \qquad (4.99)$$

where $f_{y|s}(\cdot)$ is the conditional PDF of $y[n]$ given $s[n]$ and is known as the likelihood function.

To implement ML detection we need to find an algorithm for solving the equation in (4.99). To make implementation easier, it often pays to simplify the formula. For the AWGN channel, the conditional distribution of $y[n]$ given $s[n] = s$ is Gaussian with mean $\sqrt{E_x}s$ and variance σ_v^2. Therefore,

$$f_{y|s}(y[n]|s[n] = s) = f_v(y[n] - \sqrt{E_x}s) \qquad (4.100)$$

$$= \frac{1}{\pi\sigma_v^2} e^{\frac{|y[n] - \sqrt{E_x}s|^2}{\sigma_v^2}}. \qquad (4.101)$$

The ML detector solves the optimization

$$\arg\max_{s \in \mathcal{C}} f_{y|s}(y[n]|s[n] = s) = \arg\min_{s \in \mathcal{C}} \frac{1}{\pi\sigma_v^2} e^{-\frac{|y[n] - \sqrt{E_x}s|^2}{\sigma_v^2}}. \qquad (4.102)$$

Note that the objective is to find the symbol $s \in \mathcal{C}$ that minimizes the conditional likelihood, not the value of the conditional likelihood at the minimum value. This allows us to search for the *minimizer* and not the minimum and in turn simplify the optimization even further. We can neglect the scaling factor since it does not change minimizer s but only the value of the minimum:

$$\arg\max_{s \in \mathcal{C}} f_{y|s}(y[n]|s[n] = s) = \arg\min_{s \in \mathcal{C}} e^{-\frac{|y[n] - \sqrt{E_x}s|^2}{\sigma_v^2}}. \qquad (4.103)$$

The $\ln(\cdot)$ function is a monotonically increasing function. Therefore, the minimizer of $\ln(f(x))$ is the same as the minimizer of $f(x)$. Consequently,

$$\arg\max_{s \in \mathcal{C}} \ln\left(f_{y|s}(y[n]|s[n] = s)\right) = \arg\max_{s \in \mathcal{C}} -|y[n] - \sqrt{E_x}s|^2. \qquad (4.104)$$

Instead of minimizing the negative of a function we maximize the function:

$$\arg\max_{s \in \mathcal{C}} f_{y|s}(y[n]|s[n] = s) = \arg\min_{s \in \mathcal{C}} |y[n] - \sqrt{E_x}s|^2. \qquad (4.105)$$

This gives the desired result for ML detection in an AWGN channel.

The final expression in (4.105) provides a simple form for the optimal ML detector: given an observation $y[n]$, determine the transmitted symbol $s \in \mathcal{C}$, scaled by E_x, that is closest to $y[n]$ in terms of the squared error or Euclidean distance. An example calculation of (4.105) is illustrated in Example 4.8.

Example 4.8   Consider a digital communication system that employs BPSK. Suppose that $\sqrt{E_x} = 2$. Symbol sequences $s[0] = 1$ and $s[1] = -1$ are transmitted through an AWGN channel and the received signals are $r[0] = -0.3 + 0.1\mathrm{j}$ and $r[1] = -0.1 - 0.4\mathrm{j}$. Assuming ML detection, what will be the detected sequence $\widehat{s}[n]$ at the receiver?

Answer: The detection problem involves solving (4.105). For this purpose, we make the following calculations to find $\widehat{s}[0]$ first by hypothesizing that $s[0] = -1$,

$$|r[0] - 2|^2 = |-0.3 + 0.1\mathrm{j} - 2|^2 \qquad (4.106)$$
$$= 4.42, \qquad (4.107)$$

and then by hypothesizing that $s[0] = 1$:

$$|r[0] + 2|^2 = |-0.3 + 0.1\mathrm{j} + 2|^2 \qquad (4.108)$$
$$= 3.62. \qquad (4.109)$$

Since $3.62 < 4.42$, it follows that $\widehat{s}[0] = -1$ is the most likely symbol. Repeating the calculations to find $\widehat{s}[1]$,

$$|r[1] - 2|^2 = |-0.1 - 0.4\mathrm{j} - 2|^2 \qquad (4.110)$$
$$= 4.57, \qquad (4.111)$$

and then by hypothesizing that $s[1] = 1$:

$$|r[1] + 2|^2 = |-0.1 - 0.4\mathrm{j} + 2|^2 \qquad (4.112)$$
$$= 3.77. \qquad (4.113)$$

Since $3.77 < 4.57$, it follows that $\widehat{s}[1] = -1$ is the most likely symbol. During the detection process, the detector makes an error in detecting $s[0]$ since $\widehat{s}[0] \neq s[0]$.

In general, an ML detector for a linear modulation scheme like the one we have considered solves a detection problem by computing the squared distance between the scaled constellation point and the observation $y[n]$. For some constellations, the decision can be further simplified by exploiting symmetry and structure in the constellation, as shown in Example 4.9.

Example 4.9 Show how to simplify the detector for BPSK by exploiting the fact that for $s \in \mathcal{C}$, s is real and $|s|^2 = 1$.

Answer: Starting with the argument of the minimization in (4.105) and expanding:

$$\arg \min_{s \in \mathcal{C}} |y[n] - \sqrt{E_x}s|^2 = \arg \min_{s \in \mathcal{C}} |y[n]|^2 + |E_x s|^2 - 2\mathrm{Re}\left[y^*[n]\sqrt{E_x}s\right] \quad (4.114)$$

$$= \arg \min_{s \in \mathcal{C}} |y[n]|^2 + |E_x|^2 - 2\mathrm{Re}\left[y^*[n]\sqrt{E_x}s\right]. \quad (4.115)$$

$$= \arg \max_{s \in \mathcal{C}} \mathrm{Re}\left[y^*[n]s\right] \quad (4.116)$$

$$= \arg \max_{s \in \mathcal{C}} s\mathrm{Re}\left[y^*[n]\right]. \quad (4.117)$$

The first step follows by expanding the square. The second step follows from $|s| = 1$. The third step neglects constants that do not affect the argument minimization and removes the negative sine. The fourth step recognizes that s is real and can be pulled out of the real operation. The final detector simply computes $\mathrm{Re}[y[n]]$ and $-\mathrm{Re}[y[n]]$ and chooses the largest value. Similar simplifications are possible for M-QAM, M-PAM, and M-PSK signals.

A concept that helps both in the analysis of the detector and in simplifying the detection process is the Voronoi region. Consider symbol $s_\ell \in \mathcal{C}$. The set of all possible observed y that will be detected as s_ℓ is known as the Voronoi region or Voronoi cell for s_ℓ. Mathematically it is written as

$$\mathcal{V}_{s_\ell} = \left\{ y : \left| y - \sqrt{E_x}s_\ell \right|^2 < \left| y - \sqrt{E_x}s_k \right|^2, \quad s_\ell, s_k \in C, \quad k \neq \ell \right\}. \quad (4.118)$$

The union of all such cells is the entire complex space.

Example 4.10 Compute and plot the Voronoi regions for 4-QAM.

Answer: We compute the Voronoi region by determining the points where $\left| y - \sqrt{E_x}s_k \right|^2 - \left| y - \sqrt{E_x}s_\ell \right|^2 > 0$. Expanding the squares along the lines of (4.115), exploiting the fact that $|s_k|^2 = |s_\ell|^2 = 1$, and canceling terms:

$$\left| y - \sqrt{E_x}s_k \right|^2 - \left| y - \sqrt{E_x}s_\ell \right|^2 = 2\mathrm{Re}\left[y^*\sqrt{E_x}s_\ell\right] - 2\mathrm{Re}\left[y^*\sqrt{E_x}s_k\right]. \quad (4.119)$$

The factor of $2\sqrt{E_x}$ can be canceled to leave

$$\mathcal{V}_{s_\ell} = \left\{ y : \mathrm{Re}\left[y^*s_\ell\right] > \mathrm{Re}\left[y^*s_k\right] \right\}. \quad (4.120)$$

The regions can be calculated by recognizing that all the symbols have the form $(\pm 1 \pm j)\sqrt{2}$ for a normalized QPSK constellation. The $\sqrt{2}$ can also be canceled and the values of ± 1

and $\pm j$ used to further simplify $\mathrm{Re}\,[y^* s_\ell] = \mathrm{Re}[y]\mathrm{Re}[s] + \mathrm{Im}[y]\mathrm{Im}[s]$. Consider the point $(1+j)/\sqrt{2}$. For it to be larger than $(1-j)/\sqrt{2}$, $(-1-j)/\sqrt{2}$, and $(-1+j)/\sqrt{2}$ it must be true that

$$\mathrm{Re}[y] + \mathrm{Im}[y] > \mathrm{Re}[y] - \mathrm{Im}[y] \tag{4.121}$$
$$\mathrm{Re}[y] + \mathrm{Im}[y] > -\mathrm{Re}[y] - \mathrm{Im}[y] \tag{4.122}$$
$$\mathrm{Re}[y] + \mathrm{Im}[y] > -\mathrm{Re}[y] + \mathrm{Im}[y]. \tag{4.123}$$

Combining

$$\mathrm{Im}[y] > 0 \tag{4.124}$$
$$\mathrm{Re}[y] + \mathrm{Im}[y] > 0 \tag{4.125}$$
$$\mathrm{Re}[y] > 0 \tag{4.126}$$

where the second equality is redundant with the other two gives the following simplified Voronoi regions for 4-QAM:

$$\mathcal{V}_{(1+j)/\sqrt{2}} = \{y : \mathrm{Re}[y] > 0 \ \text{ and } \ \mathrm{Im}[y] > 0\} \tag{4.127}$$
$$\mathcal{V}_{(1-j)/\sqrt{2}} = \{y : \mathrm{Re}[y] > 0 \ \text{ and } \ \mathrm{Im}[y] < 0\} \tag{4.128}$$
$$\mathcal{V}_{(-1+j)/\sqrt{2}} = \{y : \mathrm{Re}[y] < 0 \ \text{ and } \ \mathrm{Im}[y] > 0\} \tag{4.129}$$
$$\mathcal{V}_{(-1-j)/\sqrt{2}} = \{y : \mathrm{Re}[y] < 0 \ \text{ and } \ \mathrm{Im}[y] < 0\}. \tag{4.130}$$

The Voronoi regions for the 4-QAM constellation are illustrated in Figure 4.8. The four Voronoi regions are the four quadrants. Based on this result, a simplified ML detector for 4-QAM can simply compute the sign of $\mathrm{Re}[y[n]]$ and $\mathrm{Im}[y[n]]$, which determines the quadrant and thus the corresponding closest symbol.

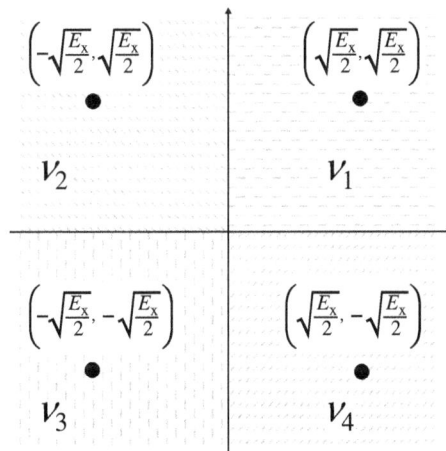

Figure 4.8 Voronoi regions for the 4-QAM constellation are the quadrants of the complex plane. Any point $y[n]$ that falls in a Voronoi region is mapped to the corresponding symbol that generates that region through the ML detection process.

4.4.5   Probability of Symbol Error Analysis

The performance of a detector is measured by the probability that it makes an error. In Section 4.4.4, we derived the maximum likelihood symbol detector. Consequently, in this section we derive the natural measure of performance for this detector: the probability of symbol error, also known as the symbol error rate. The field of communication theory is especially concerned with computing the probability of error for different detectors.

The probability of error for a detector depends on the probabilistic assumptions made about the channel. Different kinds of noise distributions, for example, lead to different detectors and thus different probabilities of symbol error expressions. In this section, we derive the maximum likelihood detector for the AWGN channel: $P_e(\frac{E_x}{N_o})$. It turns out that for the AWGN channel, the probability of error is a function only of the SNR, which simplifies as E_x/N_o; consequently the notation reflects that fact.

The probability of symbol error is the expected value of the conditional probability of symbol error $P_{e|s_m}(\frac{E_x}{N_o})$ where $P_{e|s_m}(\frac{E_x}{N_o})$ is the probability that the detector makes an error given $s_m \in \mathcal{C}$ was transmitted. Assuming the symbols are equally likely,

$$P_e\left(\frac{E_x}{N_o}\right) = \frac{1}{M} \sum_{m=0}^{M-1} P_{e|s_m}\left(\frac{E_x}{N_o}\right). \tag{4.131}$$

The conditional probability of symbol error is

$$P_{e|s_m}\left(\frac{E_x}{N_o}\right) = \mathbb{P}\left[s_m \text{ is detected incorrectly} | s_m \text{ was transmitted}\right] \tag{4.132}$$

$$= \sum_{\substack{\ell=0 \\ \ell \neq m}}^{M-1} \mathbb{P}\left[s_m \text{ is decoded as } s_\ell | s_m \text{ was transmitted}, m \neq \ell\right]. \tag{4.133}$$

Computing the probability or error term in (4.133) requires an integration of the conditional probability distribution function $f_{y|s}(x)$ over the corresponding Voronoi regions.

The exact calculation of (4.133) is possible for some constellations, though it can be tedious [219]. As a consequence, we focus on calculating the union bound, which is a function of the pairwise error probability. Let $\mathbb{P}[s_m \rightarrow s_\ell]$ denote the probability that s_m is decoded as s_ℓ, assuming that the constellation consists of two symbols $\{s_m, s_\ell\}$. This can be written as

$$\mathbb{P}[s_m \rightarrow s_\ell] = \mathbb{P}\left[\left|y - \sqrt{E_x}s_m\right|^2 > \left|y - \sqrt{E_x}s_\ell\right|^2\right]. \tag{4.134}$$

For complex AWGN, it can be shown that [73]

$$\mathbb{P}[s_m \rightarrow s_\ell] = Q\left(\sqrt{\frac{E_x}{N_o} \frac{|s_m - s_\ell|^2}{2}}\right) \tag{4.135}$$

where

$$Q(x) = \frac{1}{\sqrt{2\pi}} \int_x^\infty e^{-t^2/2} dt \tag{4.136}$$

is the Gaussian Q-function. While $Q(x)$ is usually tabulated or computed numerically, the Chernoff bound is sometimes useful for providing some intuition on the behavior of $Q(x)$:

$$Q(x) \leq \frac{1}{2}e^{-x^2/4}. \tag{4.137}$$

Notice that $Q(x)$ decreases exponentially as a function of x. This is the reason that the Q-function is usually plotted on a log-log scale.

The pairwise error probability provides an upper bound on the conditional error probability,

$$\mathbb{P}\left[s_m \text{ is decoded as } s_\ell | s_m \text{ was transmitted}, m \neq l\right] \leq \mathbb{P}\left[s_m \to s_\ell\right], \tag{4.138}$$

because the Voronoi regions for a two-symbol constellation are the same size as or larger than the Voronoi regions for a larger constellation. Consequently, this is a pessimistic evaluation of the probability that s_m is decoded as s_ℓ; thus the reason for the upper bound.

Substituting (4.138) into (4.133),

$$P_{e|s_m}\left(\frac{E_x}{N_o}\right) \leq \sum_{\substack{\ell=0 \\ \ell \neq m}}^{M-1} \mathbb{P}\left[s_m \to s_\ell\right], \tag{4.139}$$

and then using (4.131),

$$P_e\left(\frac{E_x}{N_o}\right) \leq \frac{1}{M} \sum_{m=0}^{M-1} \sum_{\ell=0, \ell \neq m}^{M-1} Q\left(\sqrt{\frac{E_x}{N_o} \frac{|s_m - s_\ell|^2}{2}}\right). \tag{4.140}$$

Now define the minimum distance of the constellation as

$$d_{\min}^2 = \min_{s_\ell \in \mathcal{C}, s_m \in \mathcal{C}, s_m \neq s_\ell} |s_\ell - s_m|^2. \tag{4.141}$$

The minimum distance is the quantity that characterizes the quality of a constellation and must be computed with the normalized constellation. The minimum distance provides a lower bound

$$d_{\min}^2 \leq |s_\ell - s_m|^2 \tag{4.142}$$

for any distinct pair of symbols s_ℓ and s_m. Since the Q-function monotonoically decreases as a function of its argument, a lower bound on the argument can be used to derive an upper bound on the probability of error:

$$P_e\left(\frac{E_x}{N_o}\right) \leq \frac{1}{M} \sum_{m=0}^{M-1} (M-1)Q\left(\sqrt{\frac{E_x}{N_o} \frac{d_{\min}^2}{2}}\right) \tag{4.143}$$

$$= (M-1)Q\left(\sqrt{\frac{E_x}{N_o} \frac{d_{\min}^2}{2}}\right). \tag{4.144}$$

The final expression in (4.144) is the union bound on the probability of symbol error for the constellation $\mathcal{C}$.

Example 4.11   Compute the union upper bound for M-QAM given that for the normalized constellation

$$d_{\min}^2 = \frac{6}{M-1}. \tag{4.145}$$

Answer: Substituting into (4.144):

$$P_e^{\text{QAM}}\left(\frac{E_x}{N_o}\right) \le (M-1)Q\left(\sqrt{\frac{E_x}{N_o}\frac{3}{M-1}}\right). \tag{4.146}$$

Typical values of the probability of symbol error are in the range of 10^{-1} to 10^{-4}. To represent the values correctly, it is common to plot the probability of symbol error on a log scale. The values of SNR that are of interest range from around 1 to 1000. To capture this range, it is common for SNR to be calculated and plotted in its decibel form where SNR dB $= 10\log_{10}(\text{SNR})$. Since double the SNR $10\log_{10}(2\text{SNR}) \approx 3\text{dB} + 10\log_{10}(\text{SNR})$, doubling E_x leads to a 3dB increase in the SNR. The probability of error for M-QAM is examined in Example 4.12 and plotted in Figure 4.9.

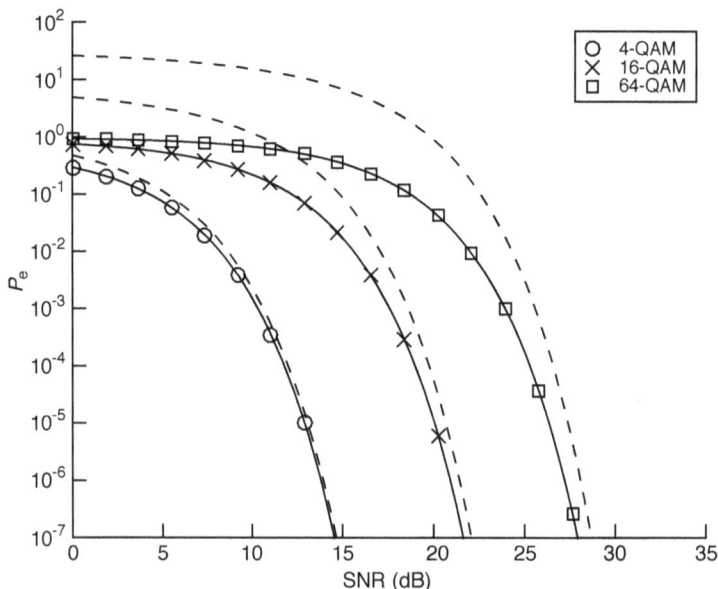

Figure 4.9 Probability of symbol error for M-QAM for different choices of M. Solid lines correspond to the exact probability of symbol error from (4.147). Dashed lines correspond to the union bound computed using.(4.144)

Example 4.12 Plot the union upper bound on the probability of symbol error and compare with the exact solution from [73] and given by

$$P_e^{\text{QAM}} \left(\frac{E_x}{N_o} \right) = 4 \left(1 - \frac{1}{\sqrt{M}} \right) Q \left(\sqrt{\frac{E_x}{N_o} \frac{3}{M-1}} \right)$$

$$- 4 \left(1 - \frac{1}{\sqrt{M}} \right)^2 \left[Q \left(\sqrt{\frac{E_x}{N_o} \frac{3}{M-1}} \right) \right]^2 \qquad (4.147)$$

for $M = 4$, $M = 16$, and $M = 64$. Plot using SNR in decibels and plot using a semilogy axis.

Answer: The plots are found in Figure 4.9. The union bound provides a reasonable approximation of the probability of symbol error at high SNR but is quite loose at low SNR. The reason is that nearest-neighbor error events dominate at high SNR but not at low SNR. The behavior of the exact and upper bounds is consistent at high SNR. Comparing the different modulation orders, we see that for low values of $P_e \left(\frac{E_x}{N_o} \right)$, for example, 10^{-6}, there is approximately a 6dB SNR gap between the curves. This means that to achieve the same symbol error rate performance, 16-QAM needs four times more power since $10 \log_{10} 6 \approx 6\text{dB}$. For a fixed SNR, the symbol error rate is larger for larger values of M. The reason is that the constellation has more points; thus the minimum distance between points is smaller.

The final derivations are a function of SNR $= E_x/N_o$, the size of the constellation, and the minimum distance of the constellation. For a given constellation, to reduce the probability of error and increase system performance, either E_x must be increased or N_o must be reduced. The average signal power E_x can be increased by using more transmit power, though not without bound. The maximum power is constrained due to health concerns in most wireless systems. Because of path loss and fading in the channel, as discussed in Chapter 5, the signal attenuates with distance. As a result, the SNR measured at the receiver becomes the appropriate performance measure, accounting for losses in the channel. This can be increased by reducing the distance between the transmitter and the receiver (this will be clearer after the discussion of path loss). The effective noise power density is $N_o = kT_e$ where T_e is the effective noise temperature of the device. The noise temperature can be reduced to some extent by changing the RF design, providing better cooling, or using higher-quality components [271].

For a fixed SNR, the constellation can be changed to improve the probability of symbol error. For example, 16-QAM has a higher $d_{\min}$ than 16-PSK because it achieves a better packing of points. Because of the constellation normalization, there are limited gains from going to more elaborate constellations than QAM. Alternatively, the number of points in the constellation may be reduced. Then the remaining points can be spaced farther apart. This is evident when looking at $d_{\min}$ for M-QAM in (4.145). Unfortunately, reducing the number of points also reduces the number of bits per symbol. In systems that support link adaptation, the constellation size is varied adaptively to achieve a particular target probability of symbol error, for example, 10^{-2}, as the SNR changes. The application usually dictates a target probability of error. This is usually the probability of bit error after error control decoding, but it can be translated into an effective probability of bit or symbol error before decoding.

For many systems, especially with more elaborate impairments, it is difficult to calculate the exact probability of symbol error. An alternative is to use Monte Carlo simulation. While it seems ad hoc, this approach is commonly used to evaluate the true error performance of complex systems. A Monte Carlo approach for estimating the probability of symbol error directly from the discrete-time input-output relationship is as follows.[1] First, choose a number of iterations N. For $n = 0, 1, \ldots, N - 1$, generate a realization of a symbol $s[n]$ by choosing a symbol uniformly from the constellation $\mathcal{C}$. Generate a realization of complex noise $v[n]$ from a Gaussian distribution. Most numerical software packages have Gaussian random number generators for $\mathcal{N}(0, 1)$. You can generate complex noise with variance N_o as $v[n] = \sqrt{N_\mathrm{o}/2}(x + \mathrm{j}y)$ where x and y are generated from $\mathcal{N}(0, 1)$. Next, form $y[n] = \sqrt{E_\mathrm{x}}s[n] + v[n]$ and pass $y[n]$ to an ML detector. If the output $\widehat{s}[n]$ is different from $s[n]$, count it as an error. The probability of symbol error is then estimated as

$$\widehat{P}_e\left(\frac{E_\mathrm{x}}{N_\mathrm{o}}\right) = \frac{\#\text{ errors}}{N}. \tag{4.148}$$

To get a good estimate, a rule of thumb is that N must be chosen such that at least ten errors are observed. It is often useful to compare the final result with the upper bound in (4.144) as a sanity check, to make sure that the estimated probability of error is below the theoretical upper bound.

4.5 Digital Implementation of Pulse Shaping

In this section, we explain a more practical implementation of transmit pulse shaping and receive filtering. The system in Figure 4.6 implements transmit pulse shaping and receive filtering somewhat idealistically in analog. This structure does not map to a realizable implementation for several reasons. First, there is a mix of analog and digital implementations without an explicit discrete-to-continuous conversion at the transmitter. Second, the structure does not allow for the possibility of performing transmit and receive pulse shaping entirely in the digital domain. Pulse shaping for digital can be an advantage since it is easier to implement than analog pulse shaping and is flexible since it can be combined with other receiver processing. In this section, we explain how to implement pulse shaping at the transmitter and matched filtering at the receiver using multirate signal processing concepts from Section 3.4.

4.5.1 Transmit Pulse Shaping

In this section we develop an approach for generating $x(t)$ in discrete time, using the fact that when the pulse shaping is bandlimited, it follows that $x(t)$ is bandlimted.

---

1. To simulate more complex receiver impairments, as in Chapter 5, a baseband simulation with the continuous-time waveforms may be required.

We need to be careful to remember that $1/T$ is the symbol rate and is not necessarily the Nyquist rate. For example, for a square root raised cosine pulse shape, the Nyquist rate is $(1 + \alpha)/T$, which is greater than $1/T$ when there is excess bandwidth, that is, $\alpha > 0$. For this reason, we need multirate signal processing from Section 3.4 to change between the symbol rate and the sample rate.

We focus on generating the transmit signal prior to scaling by $\sqrt{E_x}$ as given by

$$\widetilde{x}(t) = \sum_{n=-\infty}^{\infty} s[n]g_{tx}(t - nT). \tag{4.149}$$

This is reasonable because the scaling by $\sqrt{E_x}$ is usually applied by the analog front end.

Since $x(t)$ is bandlimited by virtue of the bandlimited pulse shape $g_{tx}(t)$, there exists a sampling period T_x such that $1/T_x$ is greater than twice the maximum frequency of $g_{tx}(t)$. According to the reconstruction equation in the Nyquist sampling theorem,

$$\widetilde{x}(t) = \sum_{n=-\infty}^{\infty} \widetilde{x}[n]\mathrm{sinc}((t - nT_x)/T_x) \tag{4.150}$$

where

$$\widetilde{x}[n] = \widetilde{x}(nT_x) \tag{4.151}$$

$$= \sum_{m=-\infty}^{\infty} s[m]g_{tx}(nT_x - mT). \tag{4.152}$$

It remains now to create the convolution sum in discrete time.

Now suppose that $T_x = T/M_{tx}$ for some positive integer M_{tx}, where M_{tx} is the oversampling factor. If the available sampling rate of the discrete-to-continuous converter does not satisfy this property, the results of this section can be followed by the sample rate conversion in Chapter 3. Let

$$g_{tx}[n] = g_{tx}(nT_x) \tag{4.153}$$

be the oversampled transmit pulse-shaping filter. Then

$$\widetilde{x}[n] = \sum_{m=-\infty}^{\infty} s[m]g_{tx}[n - M_{tx}m]. \tag{4.154}$$

Recognizing the structure in (4.154), using (3.395),

$$\widetilde{x}[n] = \left(\sum_{m=-\infty}^{\infty} s[m]\delta[n - M_{tx}m] \right) * g_{tx}[n] \tag{4.155}$$

where the first term corresponds to $s[n]$ upsampled by M_{tx}. Recall from Chapter 3 that upsampling by M_{tx} corresponds to inserting $M_{tx} - 1$ zeros after every $s[n]$. With upsampling, a discrete-time implementation of the transmitter takes on the simple form as

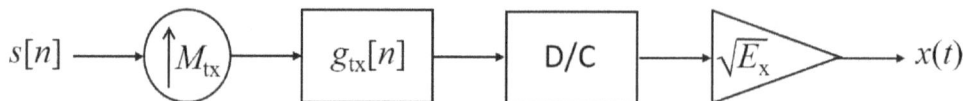

Figure 4.10 An implementation of transmit pulse shaping with a combination of upsampling and filtering

illustrated in Figure 4.10; recall that $\uparrow M_{\text{tx}}$ is the block diagram notation for upsampling. The symbol sequence is upsampled by M_{tx} and filtered with the oversampled transmit pulse shape prior to being sent to the discrete-to-continuous converter. This allows pulse shaping to be implemented completely in digital and makes the generation of $\tilde{x}(t)$ practical.

The complexity of (4.155) can be reduced through application of multirate identities from Section 3.4.4. Complexity is a concern because the convolution after upsampling does not exploit the many zeros after the upsampling. A more efficient approach is to use a filter bank. Define the ℓth polyphase component of $\tilde{x}[n]$ as

$$\tilde{x}^{(\ell)}[n] = \tilde{x}[nM_{\text{tx}} + \ell] \tag{4.156}$$

$$= \sum_{m=-\infty}^{\infty} s[m]g_{\text{tx}}[nM_{\text{tx}} + \ell - mM_{\text{tx}}]. \tag{4.157}$$

In a similar way, define the polyphase components of the transmit pulse-shaping filter $g_{\text{tx}}^{(\ell)}[n] = g_{\text{tx}}[nM_{\text{tx}} + \ell]$. Using this definition, it is possible to build the ℓth subsequence as a convolution between the symbol stream $\{s[n]\}$ and the subsampled transmit filters:

$$\tilde{x}^{(\ell)}[n] = \sum_{m=-\infty}^{\infty} s[m]g_{\text{tx}}^{(\ell)}[n - m]. \tag{4.158}$$

To reconstruct $\tilde{x}[n]$, let $\bar{x}^{(\ell)}[n]$ be $\tilde{x}^{(\ell)}[n]$ upsampled by M_{tx} using the notation in (3.395). Then

$$\tilde{x}[n] = \sum_{\ell=0}^{M_{\text{tx}}-1} \bar{x}^{(\ell)}[n - \ell]. \tag{4.159}$$

Substituting (4.157) and recognizing that $\bar{x}[n] = \sum_{k=-\infty}^{\infty} \tilde{x}[k]\delta[n - kM_{\text{tx}}]$, then

$$\tilde{x}[n] = \sum_{\ell=0}^{M_{\text{tx}}-1} \sum_{k=-\infty}^{\infty} \left(\sum_{m=-\infty}^{\infty} s[m]g_{\text{tx}}^{(\ell)}[k - m] \right) \delta[n - kM_{\text{tx}} - \ell]. \tag{4.160}$$

In words, the symbol stream $s[n]$ is first convolved by each polyphase filter $g_{\text{tx}}^{(\ell)}[n]$. The length of $g_{\text{tx}}^{(\ell)}[n]$ is approximately M_{tx} times smaller than $g_{\text{tx}}^{(\ell)}[n]$, thus resulting in a complexity reduction. Then the different filtered sequences are upsampled and delayed to create the desired output $\tilde{x}[n]$. This filter bank is illustrated in Figure 4.11.

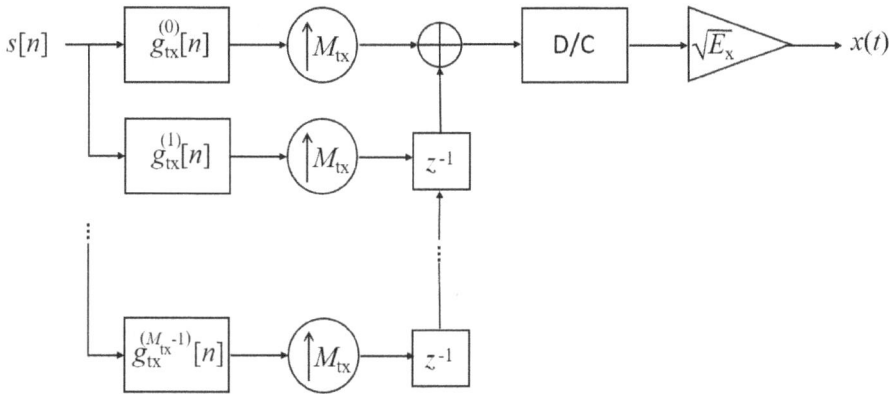

Figure 4.11 A lower-complexity implementation of pulse shaping. We use the notation z^{-1} to denote a delay of one sample.

4.5.2 Receiver Matched Filtering

In this section, we develop an approach for performing matched filtering in discrete time. Unlike the transmitter, the receiver signal processing in Figure 4.6 can be performed using realistic hardware. A main disadvantage, though, is that it requires an analog implementation of the matched filter. Usually this involves solving a filter design problem to approximate the response of $g_{rx}(t)$ with available hardware components. Digital implementation of matched filtering leads to a simpler analog design and a more flexible receiver implementation. For example, with a digital implementation only a lowpass filter is required prior to sampling. Analog lowpass filters are widely available. It also makes the receiver operation more flexible, making it easier to adapt to waveforms with different bandwidths. Finally, it makes other functions like symbol synchronization easier to implement.

Our approach is based on oversampling the received signal. Let us denote the received signal prior to matched filtering as $r(t)$. We assume that $r(t)$ has already been filtered so that it can be treated as a bandlimited complex baseband signal. Let $T_r = T/M_{rx}$ for some positive integer M_{rx} such that $1/T_r$ is greater than the Nyquist rate of the signal. Let $r[n] = r(nT_r)$. From the results in Chapter 3, since $r(t)$ is bandlimited, it is possible to replace the continuous-time matched filter in Figure 4.6 with a continuous-to-discrete converter, a discrete-time filter, and a discrete-time continuous converter as shown in Figure 4.12.

As the pulse shape is already bandlimited, and thus does not need to be lowpass filtered, the discrete-time filter is just the scaled oversampled filter

$$g_{rx}[n] = T_r g_{rx}(nT_r). \tag{4.161}$$

The scaling by T_r can be neglected since it does not impact the SNR.

The combination of discrete-to-continuous conversion with T_r followed by continuous-to-discrete conversion $T_r M_{rx}$ can be further simplified. In fact, this is just downsampling

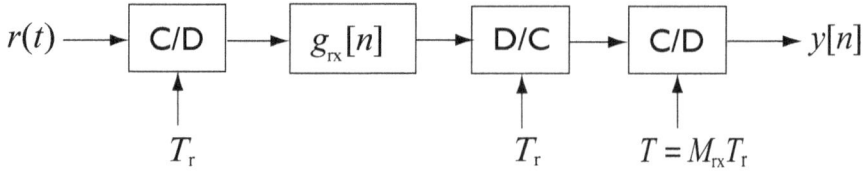

Figure 4.12 An implementation of receive matched filtering in continuous time using discrete-time processing

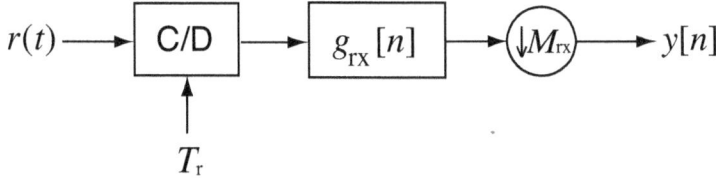

Figure 4.13 An implementation of receive matched filtering in continuous time using discrete-time processing after simplifying

by M_{rx}. This leads to a simplified system as illustrated in Figure 4.13. Mathematically, the output can be written as

$$y[n] = \sum_{k=-\infty}^{\infty} g_{\mathrm{rx}}[k]r[nM_{\mathrm{rx}} - k]. \tag{4.162}$$

The system in Figure 4.13 convolves the sampled received signal by $g_{\mathrm{rx}}[n]$, then downsamples by M_{rx} to produce $y[n]$.

The computational complexity of Figure 4.13 can be further reduced by recognizing that many computations are performed to obtain samples that are subsequently discarded by the downsampling operation. This is not inefficient. As an alternative, consider the application of some multirate signal processing identities. First note that

$$y[n] = \sum_{k=-\infty}^{\infty} r[k]g_{\mathrm{rx}}[nM_{\mathrm{rx}} - k] \tag{4.163}$$

$$= \sum_{p=-\infty}^{\infty} \sum_{m=0}^{M-1} r[pM_{\mathrm{rx}} + m]g_{\mathrm{rx}}[nM_{\mathrm{rx}} - pM_{\mathrm{rx}} - m] \tag{4.164}$$

$$= \sum_{m=0}^{M_{\mathrm{rx}}-1} \sum_{p=-\infty}^{\infty} r^{(m)}[p]g_{\mathrm{rx}}[M_{\mathrm{rx}}(n - p) - m] \tag{4.165}$$

$$= \sum_{m=0}^{M_{\mathrm{rx}}-1} \sum_{p=-\infty}^{\infty} r^{(m)}[p]g_{\mathrm{rx}}[M_{\mathrm{rx}}(n - p - 1) + M_{\mathrm{rx}} - m] \tag{4.166}$$

$$= \sum_{m=0}^{M_{\mathrm{rx}}-1} \sum_{p=-\infty}^{\infty} r^{(m)}[p]g_{\mathrm{rx}}^{(M_{\mathrm{rx}}-m)}[n - p - 1] \tag{4.167}$$

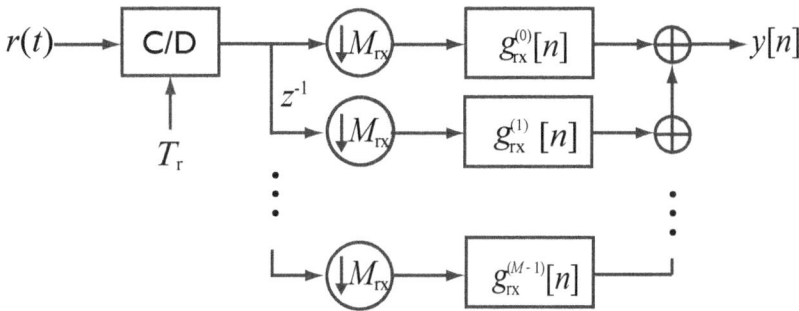

Figure 4.14 Exchange of downsampling and filtering operations

where $g_{\mathrm{rx}}^{(m)}[n]$ is the mth polyphase component of $g_{\mathrm{rx}}[n]$ (with $g^{(M_{\mathrm{rx}})}[n] = g_{\mathrm{rx}}^{(0)}[n]$) and $r^{(m)}[n]$ is the mth polyphase component of $r[n]$. An implementation of this alternative structure is presented in Figure 4.14. While it does not look neat, counting the number of additions and multiplications reveals that it is a far more efficient implementation.

4.6   Summary

- The complex pulse-amplitude modulator involves bit-to-symbol mapping, pulse shaping, and gain. Normalizations of the constellation and transmit pulse shape are required so that the power of the transmitted signal comes only from the added gain.

- The optimal receiver for the AWGN channel involves filtering and symbol rate sampling. To maximize the received SINR, the receive filter should be matched to the transmit pulse shape. The combined transmit pulse shape and matched filter should be a Nyquist pulse shape, to satisfy the zero-ISI condition.

- Maximum likelihood detection of symbols in an AWGN channel involves finding the scaled symbol in the constellation that is closest to the observed sample. The detection is applied separately on each symbol thanks to the Nyquist pulse shape. Sometimes the detection algorithm can be simplified by recognizing symmetries and structure in the constellation.

- The performance of a maximum likelihood symbol detector is evaluated through the probability of symbol error, which is a function of the SNR and the constellation. Since it is hard to compute directly, it is often useful to compute the union upper bound, which depends on the SNR, the minimum distance of the constellation, and the number of points. The union bound is loose except at high SNR.

- Pulse shaping can be performed in discrete time at the transmitter, through a combination of upsampling and filtering with the oversampled pulse shape. The derivation is based on the Nyquist reconstruction formula. Complexity can be reduced by using a filter-bank structure that performs filtering prior to upsampling.

- Matched filtering can be performed in discrete time at the receiver through a combination of filtering and downsampling. The derivation uses results on discrete-time processing of continuous-time bandlimited signals. Complexity can be further reduced using a filter-bank structure, where the filtering happens after the downsampling.

Problems

1. Mr. Q. A. Monson has come up with a new $M = 8$-point constellation, illustrated in Figure 4.15.

   (a) Compute E_s for this constellation, assuming all symbols are equally likely.

   (b) Compute the scaling factor that makes the constellation in Figure 4.15 have $E_s = 1$.

   (c) Plot the constellation diagram for the normalized constellation.

   (d) Determine a bit-to-symbol mapping and add it to your constellation plot. Argue why your mapping makes sense.

2. Consider a general complex pulse-amplitude modulation scheme with transmitted signal $x(t) = \sum_n s[n]g(t - nT_s)$ where the pulse shape $g(t) = e^{-t^2/4}$. Let the constellation be $\mathcal{C} = \{2, -2, 2j, -2j, 0\}$.

   (a) What is the appropriate matched filter to apply to this signal?

   (b) What is the average energy of the constellation?

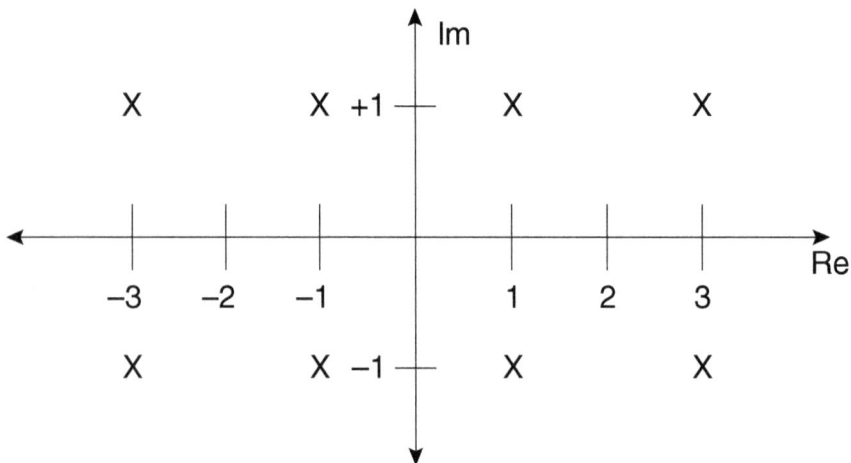

Figure 4.15 An 8-point constellation

   (c) Suppose that you generate complex Gaussian noise by generating the real and imaginary components separately with distribution $\mathcal{N}(0, 1)$. How much do you need to scale the transmit constellation to achieve a signal-to-noise ratio of 10dB?

3. Consider a digital communication system that uses Gaussian pulses $g_{\text{tx}}(t) = \alpha \exp(-\pi a^2 t^2)$ and $g_{\text{tx}}(t) = g_{\text{rx}}(t)$. Let T be the symbol interval. Let $g(t) = g_{\text{tx}}(t) * g_{\text{rx}}(t)$.

   (a) Find the value of α such that the pulse shape satisfies the unit norm property. Use the normalized pulse shape for the rest of the problem.

   (b) Prove that Gaussian pulse shapes do not satisfy the zero-ISI criterion.

   (c) Suppose that a is chosen such that $g(T) = 0.001$. Find a.

   (d) Given that $G(f)$ is the Fourier transform of $g(t)$, define the bandwidth W of $g(t)$ such that $G(W)/G(0) = 0.001$. Find the value of W for the value of a given previously. How does this value compare to that obtained from a raised cosine filter with 100% rolloff?

4. Let

$$g(t) = 3T \left(\frac{\sin(3\pi t/T)}{3\pi t/T} \right) \cos(3\pi t/T) \tag{4.168}$$

and

$$g[n] = g_\beta(nT) \tag{4.169}$$

where $g_\beta(t)$ is the normalized version of $g(t)$ with unit energy.

   (a) Compute the normalized pulse shape $g_\beta(t)$.

   (b) Compute $g[n]$.

   (c) Compute $G_\beta(e^{j2\pi f})$.

   (d) Is $g[n]$ a Nyquist pulse shape? Please justify your answer.

5. Consider a QAM communication system that employs raised cosine pulse shaping at the transmitter and a carrier frequency of 2GHz. In this problem you will answer various questions about the relationships between symbol rates and bandwidths.

   (a) Suppose that the symbol rate is 1 megasymbol per second and that the raised cosine has 50% excess bandwidth. What is the baseband absolute bandwidth?

   (b) Suppose that the symbol rate is 1 megasymbol per second and that the raised cosine has 50% excess bandwidth. What is the passband absolute bandwidth?

(c) Suppose that the absolute bandwidth of a signal is 1MHz at baseband. What is the best symbol rate that can be achieved with a raised cosine pulse? *Hint:* You can choose any raised cosine pulse.

(d) Suppose that you have a raised cosine pulse shape with absolute bandwidth of 4MHz at baseband and excess bandwidth of 50%. What is the symbol rate?

(e) Suppose that the absolute bandwidth of your modulated signal is 3MHz at baseband with an excess bandwidth of 25%. What is the Nyquist frequency? What is the Nyquist rate?

6. Consider a variation of problem 5 with a QAM communication system that employs raised cosine pulse shaping at the transmitter and a carrier frequency of 3GHz. In this problem you will answer various questions about the relationships between symbol rates and bandwidths.

(a) Suppose that the symbol rate is 1 megasymbol per second and that the raised cosine has 50% excess bandwidth. What is the baseband absolute bandwidth?

(b) Suppose that the symbol rate is 1 megasymbol per second and that the raised cosine has 50% excess bandwidth. What is the passband absolute bandwidth?

(c) Suppose that the absolute bandwidth of a signal is 1MHz at baseband. What is the maximum symbol rate that can be achieved with a raised cosine pulse?

(d) Suppose that you have a raised cosine pulse shape with absolute bandwidth of 2MHz at baseband and excesss bandwidth of 50%. What is the symbol rate?

(e) Suppose that the absolute bandwidth of your modulated signal is 3MHz at baseband with an excess bandwidth of 25%. What is the Nyquist rate? What is the sampling rate at the receiver?

7. Consider a QAM system with symbol rate $1/T$ and transmit pulse-shaping filter $g_{tx}(t)$. Suppose that the D/C operates at a rate of L/MT where L and M are coprime integers, $L > M$, and L/MT is greater than the Nyquist rate of $g_{tx}(t)$.

(a) Determine a straightforward implementation of transmit pulse shaping using the theory of multirate signal processing. Find the coefficients of the filters and provide a block diagram of the transmitter.

(b) Using the upsampling and downsampling identities, interchange the upsampling and downsampling operations to create an alternative implementation. Find the coefficients of the filters and provide a block diagram of the transmitter.

(c) Which approach is more computationally efficient?

8. Consider a complex pulse-amplitude modulated digital communication system. Suppose that a square root raised cosine is used for the transmit pulse shape

and received matched filter with 50% excess bandwidth and a symbol rate of 1 megasymbol per second.

(a) Suppose that you employ the structure in Figure 4.10 to implement the pulse shaping at the transmitter. Determine the smallest value of $M_{\text{tx}} \in \mathbb{Z}^+$ such that $T_{\text{x}} = T/M_{\text{tx}}$ will still result in perfect reconstruction.

(b) Determine the coefficients of the finite impulse response filter $\{g_{\text{tx}}[k]\}_{k=-K}^{K}$ such that $g_{\text{tx}}[k]$ has 99% of the energy of $g_{\text{tx}}(t)$. Essentially choose K, then find $g_{\text{tx}}[k]$. This is an example of creating a practical approximation of an infinite-length filter.

(c) Determine the complexity per symbol required for the filter computations, assuming the filter is implemented directly in the time domain.

(d) Now suppose that you employ a filter bank to implement pulse shaping at the transmitter as in Figure 4.11. Determine the values of $g_{\text{tx}}^{(m)}[k]$ given the M_{tx} and $g_{\text{tx}}[k]$ you determined previously.

(e) Determine the complexity per symbol required for the FIR (finite impulse response) filter computations, assuming the filters are implemented directly in the time domain.

(f) Explain the advantage of the filter-bank approach.

9. Prove that the cascade of a discrete-to-continuous converter operating at a rate of f_{s} followed by a continuous-to-discrete converter operating at a rate of $f_{\text{s}/M}$ is equivalent to downsampling by M as shown in Figure 4.16.

10. Prove that the zero-ISI property is maintained in the architecture in Figure 4.13 assuming that $g_{\text{rx}}(t)$ is matched to $g_{\text{tx}}(t)$, the composite filter $g(t) = g_{\text{tx}}(t) \star g_{\text{rx}}(t)$ is a Nyquist pulse shape, and $T_{\text{r}} = T/M_{\text{rx}}$. In other words, show that $y[n] = \sqrt{E_{\text{x}}} s[n]$ in the absence of noise.

11. Consider the problem of designing an approximate linear interpolator. Suppose that $x[n]$ are samples from a bandlimited continuous-time function, sampled at

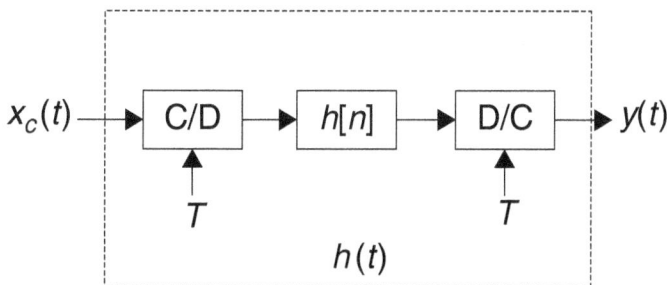

Figure 4.16 See problem 9

the Nyquist rate. Now suppose that we want to obtain samples $y[n]$ that were taken with twice the frequency. In other words, if the sampling period is T and $x[n] = x(nT)$, then the interpolated function is $y[n] = x(nT/2)$.

(a) First consider optimum interpolation.

    i. Draw the block diagram for the ideal interpolator. The input is $x[n]$ and the output is $y[n]$. *Hint:* It involves an upsampling by 2 followed by a filter and uses some multirate signal processing from Chapter 3.

    ii. Write an equation for $y[2n]$ and $y[2n + 1]$. Simplify to the extent possible. *Hint:* $y[2n]$ is very simple.

    iii. Draw a block diagram that shows how to implement the ideal interpolator using filtering before the upsampling. Again, this uses multirate signal processing results from Chapter 3.

(b) The computation of $y[2n + 1]$ is computationally expensive. Suppose that we want to implement a simple averaging instead. Suppose that we average the two adjacent samples to perform the interpolation. Let the coefficients of the filter be $\{h[0], h[1]\}$, and let

$$\widehat{y}[2n + 1] = h[0]y[2n - 2] + h[1]y[2n + 2] \tag{4.170}$$

be our low-complexity estimate of $y[2n+1]$. Suppose that you have observations $\{y[n]\}_{n=0}^{K-1}$ enough to compute N errors. To be precise, assuming K is odd, then we need $K = 2N + 3$ to compute N errors. Let $e[n] = y[2n + 1] - (h[0]y[2n - 2] + h[1]y[2n + 2])$ be the error. The goal in the remainder of this problem is to use least squares methods to estimate $\{h[0], h[1]\}$ such that $\sum_{n=0}^{N-1} |e[n]|^2$ is minimized.

    i. Assuming that $h[1] = 0$, find the least squares estimate of $h[0]$ by expanding the sum, differentiating, and solving.

    ii. Assuming that $h[0] = 0$, find the least squares estimate of $h[1]$ by expanding the sum, differentiating, and solving.

    iii. Find the least squares estimate of $h[0]$ and $h[1]$ by expanding the sum, differentiating, and solving.

    iv. Formulate and solve the least squares problem for finding $h[0]$ and $h[1]$ in matrix form. Note that because of the matrix dimensions, you can find an expression for the inverse.

    v. Now rewrite the solution as a function of estimates of the autocorrelation function $R_{yy}[k]$ and interpret your answer.

12. Suppose that we can model the received signal using the following equation,

$$y = x + v, \tag{4.171}$$

where x is generated uniformly from a 16-QAM constellation (that has been normalized to have $E_s = 1$) and v is distributed as $\mathcal{N}_c(0, 0.025)$. Note that x and v are independent random variables.

(a) What are the mean and variance of x?

(b) What are the mean and variance of $\text{Re}[v]$?

(c) What are the mean and variance of v?

(d) What are the mean and variance of y?

(e) We define the SNR as $\mathbb{E}[|x|^2] / \mathbb{E}[|v|^2]$. What is the SNR in this case?

13. Consider an AWGN equivalent system model where

$$y = \sqrt{E_s}s + v \tag{4.172}$$

where s is a symbol from constellation $\mathcal{C} = \{s_1, s_2\}$ and v is AWGN with $\mathcal{N}(0, N_o)$.

(a) Derive the probability of symbol error. Essentially, you need to derive $\mathbb{P}[s_1 \to s_2]$, not use the result given in the book. Of course, you can check your result with the book.

(b) Suppose that $\mathcal{C} = \{1+j, -1-j\}$. Derive the zero-mean normalized constellation.

(c) Using the normalized constellation, plot the probability of symbol error as a function of SNR. You should have SNR in decibels and use a semilogy plot. Also plot using the Chernoff upper bound on the $Q(\cdot)$ function.

14. Consider the maximum likelihood detector for AWGN channels. Suppose there is a scalar channel α, potentially complex, that operates on the transmitted symbols $x[n]$. This is known as a flat-fading channel. In this case the signal at the receiver after sampling is

$$y[n] = \alpha x[n] + v[n] \tag{4.173}$$

where $x[n]$ is a symbol chosen from constellation $\mathcal{C}$ and $v[n]$ is AWGN with variance N_o. Find the maximum likelihood detector given that you know α (which can be a random variable). *Hint*: Solve for $\arg\max_{s \in \mathcal{C}} f_{Y|X}(y[n]|x[n], \alpha)$.

15. Recall the derivation of the maximum likelihood detector for AWGN channels. Now suppose that $v[n]$ is an IID Bernoulli random variable that takes value 0 with

probability p and value 1 with probability $1 - p$. Consider the channel model given in (4.174) known as the binary symmetric channel, which uses addition modulo 2 (i.e., like an XOR function). Suppose the received signal is

$$y = (s + v) \bmod 2, \tag{4.174}$$

where s is a symbol chosen from a constellation $\mathcal{C} = 0, 1$. Find the maximum likelihood detector. Since the noise is a discrete random variable, we formulate the maximum likelihood problem as

$$\arg \max_{S \in \mathcal{C}} P_{y|s}[y|s = S] \tag{4.175}$$

where $P_{y|s}[y|s = S]$ is the conditional probability mass function. *Hint:* (a) Solve for the likelihood function $P_{y|s}[y|s = S]$. (b) Find the detection rule for 0 and 1 that maximizes the likelihood; that is, choose 0 if $P_{y|s}[y|s = 0] \geq P_{y|s}[y|s = 1]$.

16. This problem considers an abstraction of a visible light communication system.

    (a) Consider the BPSK modulation format with constellation $\{1, -1\}$. Because we need to keep the light on while transmitting, we want to create a *nonzero-mean* constellation. Create a modified constellation such that the mean is 5.

    (b) Let s_1 denote the scaled 1 symbol and let s_2 denote the scaled -1. Suppose that the noise is signal dependent. In particular, suppose that if s_1 is sent,

$$y = s_1 + v_1 \tag{4.176}$$

    where v_1 is $N(0, \sigma^2)$, and if s_2 is sent,

$$y = s_2 + v_2 \tag{4.177}$$

    where v_2 is $N(0, 2\sigma^2)$.

    Determine the maximum likelihood detection rule. Specifically, come up with an algorithm or formula that takes the observation y and produces as an output either s_1 or s_2.

17. Consider a digital communication system that sends QAM data over a voice-band telephone link with a rate of 1200 symbols/second. Assume that the data is corrupted by AWGN. Find the SNR that is required, using both the union bound and the exact error expressions, to obtain a symbol error probability of 10^{-4} if the bit rate is:

    (a) 2400 bits/sec

    (b) 4800 bits/sec

    (c) 9600 bits/sec

18. Consider a modulation scheme with $M = 3$ and the following constellation:

$$\mathcal{C} = \{1 + j, 1 + 4j, 1 + 7j\}. \tag{4.178}$$

(a) Plot the constellation.

(b) Find and plot the equivalent zero-mean constellation.

(c) Find and plot the equivalent unit energy constellation.

(d) What is the normalized constellation's minimum distance?

(e) Find the union bound on the probability of symbol error as a function of SNR.

19. Suppose that we use the following constellation in a general pulse-amplitude modulation system: $\mathcal{C} = \{0, 2, 2j, -2\}$.

(a) Determine the mean of the constellation. Construct the zero-mean constellation.

(b) Determine the appropriate scaling factor to make the constellation unit energy. Construct the zero-mean normalized constellation.

(c) What is the minimum distance of the zero-mean unit norm constellation?

(d) Plot this constellation and sketch the Voronoi regions.

(e) What is the union bound on the probability of error for this constellation assuming an AWGN channel?

20. Consider a modulation scheme with $M = 3$ and complex constellation

$$\mathcal{C} = \{1, \cos(\pi 7/12) + j\sin(\pi 7/12), \cos(\pi 11/12) + j\sin(\pi 11/12)\}. \qquad (4.179)$$

(a) Normalize the constellation and plot it. Use the normalized constellation for all subsequent parts of this problem.

(b) Find the minimum distance of the constellation.

(c) Find the union bound on the probability of symbol error as a function of the SNR in an AWGN channel. You may assume AWGN with unit variance (i.e., SNR $= E_x/1$).

(d) Plot the union bound on a semilogy graph. Make sure that SNR is in decibels. You will need to figure out how to compute the Q-function (you can use the *erfc* function in part) and how to reverse this process.

21. Consider a modulation scheme with $M = 8$ and the complex constellation

$$\mathcal{C} = \{0, 4, -4j, 4 - 4j, 2 + 2j, 2 - 2j, -2 - 2j, -2 + 2j\} \qquad (4.180)$$

and pulse-shaping filter

$$g_{tx,non}(t) = \text{sinc}^2(t/T) \qquad (4.181)$$

where T is the symbol period.

(a) Determine the mean of the constellation, assuming that each symbol is equally likely. In other words, compute $\mu = \mathbb{E}_s[s[n]]$, where $s[n]$ is the complex symbol randomly chosen from the constellation.

(b) Find the zero-mean constellation, obtained by removing the mean from $\mathcal{C}$.

(c) Determine the average energy of the zero-mean constellation E_s.

(d) We need to ensure that the constellation is zero mean. Form a new constellation by removing the mean, which we refer to as the zero-mean constellation. Plot the resulting constellation on a real/imaginary plot. Do not forget to label properly.

(e) Compute the scaling factor that makes the zero-mean constellation have unit average power. Plot the resulting constellation, which is in fact the normalized constellation with zero mean and unit average power. Do not forget to label properly.

(f) Compute the minimum distance of the normalized constellation, $d_{\min}$.

(g) Compute the normalized pulse-shaping filter $g_{tx}(t)$ from $g_{tx,non}(t)$.

(h) Let the transmitted signal be

$$x(t) = \sqrt{E_x} \sum_n s[n] g_{tx}(t - nT) \qquad (4.182)$$

where $s[n]$ is an element of the zero-mean unit energy constellation. Determine the power spectral density of $x(t)$.

(i) Determine the absolute bandwidth of $x(t)$.

(j) Determine the union bound on the probability of symbol error as a function of the SNR, that is, the ratio E_x/N_o. Plot the union bound on a semilogy graph. Make sure that the SNR is in decibels. Do not forget to label properly.

22. Consider the transmission of the superposition of two independent messages x_1 and x_2 over an AWGN channel, where the received signal after sampling is given by

$$y = x_1 + x_2 + v, \qquad (4.183)$$

where x_1 is chosen from the ternary constellation $\mathcal{C}_1 = \{0, 1, j\}$, with equal probability, x_2 is drawn independently of x_1 from the ternary constellation $\mathcal{C}_2 = \{c_1, c_2, c_3\}$, also with equal probability, and v is zero-mean AWGN with variance σ^2.

(a) Choose the constellation points $\{c_1, c_2, c_3\}$ of $\mathcal{C}_2$ such that $\mathcal{C}1 \cap \mathcal{C}2 = \emptyset$ and the constellation resulting from $x = x_1 + x_2$ is zero mean.

(b) Find the normalized constellation of x; that is, find the constellation with unit average power.

(c) Find the minimum distance of the normalized constellation.

(d) Find the maximum likelihood (ML) detector for detecting x given the observation y.

(e) Draw the constellation of x and the Voronoi regions for the ML detector in part (d).

23. Consider a modulation scheme with $M = 4$ and complex constellation

$$\mathcal{C} = \{0, 2, 2j, 1 + j\}. \tag{4.184}$$

(a) We need to ensure that the constellation is zero mean. Determine the mean of the constellation assuming that each symbol is equally likely. In other words, compute $\mu = \mathbb{E}[x[n]]$ where $x[n]$ is randomly chosen from the constellation. Form a new constellation by removing the mean. Plot the resulting constellation on a real/imaginary plot.

(b) Given the modified constellation, with the mean removed, find the normalized constellation—that is, find the constellation with unit average power—and plot the resulting constellation.

(c) Find the minimum distance of the normalized constellation.

(d) Find the union bound on the probability of symbol error as a function of the SNR.

(e) Compute the SNR such that your bound on the probability of symbol error is 10^{-5}. You will need to figure out how to compute the Q-function (you can use the *erfc* function in part) and how to reverse this process. Explain how you got your answer.

24. Consider a noise-free system transmitting QPSK symbols operating at $\frac{E_x}{N_o} = 10\text{dB}$. The input to the symbol detector $y[n]$ is given by

$$y[n] = \sqrt{E_x} \sum_m s[m]g((n-m)T - \tau_d) + v[n], \tag{4.185}$$

where $g(t) = g_{tx}(t) * h(t) * g_{rx}(t)$ is the raised cosine pulse shape with 50% excess bandwidth, and where $0 < \tau_d < T$ is an unknown symbol delay. Also, let

$$y_0[n] = \sqrt{E_x} \sum_m s[m]g((n-m)T) + v[n] \tag{4.186}$$

denote the ideal synchronized version.

(a) Find $\mathbb{E}[|y[n] - y_0[n]|^2]$ as a function of τ_d. Plot for $0 \le \tau_d \le T$ and interpret the results.

(b) Treating the intersymbol interference due to τ_0 as noise, using the exact QPSK probability of symbol error expression, plot the probability of error for $\tau = 0, 0.25T, 0.5T, 0.75T$. The x-axis of your plot should be the SNR in decibels and the y-axis should be the probability of symbol error on a semilogy curve.

25. **Computer**   Implement an ML detector for 4-QAM, 16-QAM, and 64-QAM. Use the procedure outlined in Section 4.4.5 where you generate $y[n] = \sqrt{E_x}s[n] + v[n]$ and apply the detector to the result. Use the Monte Carlo estimation approach in (4.148) to estimate the error rate. Plot the theoretical curves in (4.147) and verify that they line up when enough symbols are used in the estimator.

26. **Computer**   Create a program that generates an M-QAM waveform.

    (a) Create a function to generate a block of N_{bits}, each of which is equally likely.

    (b) Create a function that implements the bit-to-symbol mapping using the labelings in Figure 4.3. Your code should support $M = 4$, $M = 16$, and $M = 64$. Be sure to use the normalized constellation.

    (c) Create a function to generate the oversampled square root raised cosine pulse shape. The amount of oversampling M_{tx} should be a parameter. You should use a finite impulse response approximation with suitably chosen length.

    (d) Create a function to generate the scaled sampled waveform $x[n] = x(nT/M_{\text{tx}})$ using the functions you have created.

    (e) Demonstrate that your function works by plotting the output for an input bit sequence 00011110101101 with 4-QAM and a rolloff of $\alpha = 0.5$. For your plot, choose $\sqrt{E_x} = 5$. Plot the real and imaginary outputs.

Chapter 5

Dealing with Impairments

Communication in wireless channels is complicated, with additional impairments beyond AWGN and more elaborate signal processing to deal with those impairments. This chapter develops more complete models for wireless communication links, including symbol timing offset, frame timing offset, frequency offset, flat fading, and frequency-selective fading. It also reviews propagation channel modeling, including large-scale fading, small-scale fading, channel selectivity, and typical channel models.

We begin the chapter by examining the case of frequency-flat wireless channels, including topics such as the frequency-flat channel model, symbol synchronization, frame synchronization, channel estimation, equalization, and carrier frequency offset correction. Then we consider the ramifications of communication in considerably more challenging frequency-selective channels. We revisit each key impairment and present algorithms for estimating unknown parameters and removing their effects. To remove the effects of the channel, we consider several types of equalizers, including a least squares equalizer determined from a channel estimate, an equalizer directly determined from the unknown training, single-carrier frequency-domain equalization (SC-FDE), and orthogonal frequency-division multiplexing (OFDM). Since equalization requires an estimate of the channel, we also develop algorithms for channel estimation in both the time and frequency domains. Finally, we develop techniques for carrier frequency offset correction and frame synchronization. The key idea is to use a specially designed transmit signal to facilitate frequency offset estimation and frame synchronization prior to other functions like channel estimation and equalization. The approach of this chapter is to consider specific algorithmic solutions to these impairments rather than deriving optimal solutions.

We conclude the chapter with an introduction to propagation channel models. Such models are used in the design, analysis, and simulation of communication systems. We begin by describing how to decompose a wireless channel model into two submodels: one based on large-scale variations and one on small-scale variations. We then introduce large-scale models for path loss, including the log-distance and LOS/NLOS channel models. Next, we describe the selectivity of a small-scale fading channel, explaining how to determine if it is frequency selective and how quickly it varies over time. Finally, we present several small-scale fading models for both flat and frequency-selective channels, including some analysis of the effects of fading on the average probability of symbol error.

5.1 Frequency-Flat Wireless Channels

In this section, we develop the frequency-flat AWGN communication model, using a single-path channel and the notion of the complex baseband equivalent. Then we introduce several impairments and explain how to correct them. Symbol synchronization corrects for not sampling at the correct point, which is also known as symbol timing offset. Frame synchronization finds a known reference point in the data—for example, the location of a training sequence—to overcome the problem of frame timing offset. Channel estimation is used to estimate the unknown flat-fading complex channel coefficient. With this estimate, equalization is used to remove the effects of the channel. Carrier frequency offset synchronization corrects for differences in the carrier frequencies between the transmitter and the receiver. This chapter provides a foundation for dealing with impairments in the more complicated case of frequency-selective channels.

5.1.1 Discrete-Time Model for Frequency-Flat Fading

The wireless communication channel, including all impairments, is not well modeled simply by AWGN. A more complete model also includes the effects of the propagation channel and filtering in the analog front end. In this section, we consider a single-path channel with impulse response

$$h_c(t) = \alpha \delta(t - \tau_d). \tag{5.1}$$

Based on the derivations in Section 3.3.3 and Section 3.3.4, this channel has a complex baseband equivalent given by

$$h(t) = B\alpha e^{-j2\pi f_c \tau_d} \operatorname{sinc}(B(t - \tau_d)) \tag{5.2}$$

and pseudo-baseband equivalent channel

$$h_{pb}(t) = \alpha e^{-j2\pi f_c \tau_d} \delta(t - \tau_d). \tag{5.3}$$

See Example 3.37 and Example 3.39 to review this calculation. The $B\operatorname{sinc}(t)$ term is present because the complex baseband equivalent is a baseband bandlimited signal. In the frequency domain

$$h(f) = \operatorname{rect}(f/B)\alpha e^{-j2\pi f_c \tau_d} e^{-j2\pi \tau_d f} \tag{5.4}$$

we observe that $|H(f)|$ is a constant for $f \in [-B/2, B/2]$. This channel is said to be frequency flat because it is constant over the bandwidth of interest to the signal. Channels that consist of multipaths can be approximated as frequency flat if the signal bandwidth is much less than the coherence bandwidth, which is discussed further in Section 5.8.

Now we incorporate this channel into our received signal model. The channel in (5.2) convolves with the transmitted signal prior to the noise being added at the receiver.

As a result, the complex baseband received signal prior to matched filtering and sampling is

$$r(t) = \sqrt{E_\mathrm{x}} h(t) * g_\mathrm{tx}(t) * \sum_{m=-\infty}^{\infty} s[m]\delta(t - mT) + v(t) \tag{5.5}$$

$$= \sqrt{E_\mathrm{x}} \alpha e^{-\mathrm{j}2\pi f_\mathrm{c}\tau_\mathrm{d}} \sum_{m=-\infty}^{\infty} s[m] g_\mathrm{tx}(t - mT - \tau_\mathrm{d}) + v(t). \tag{5.6}$$

To ensure that $r(t)$ is bandlimited, henceforth we suppose that the noise has been lowpass filtered to have a bandwidth of $B/2$. For simplicity of notation, and to be consistent with the discrete-time representation, we let $h = \alpha e^{-\mathrm{j}2\pi f_\mathrm{c}\tau_\mathrm{d}} \sqrt{E_\mathrm{x}}$ and write

$$r(t) = h \sum_{m=-\infty}^{\infty} s[m] g_\mathrm{tx}(t - mT - \tau_\mathrm{d}) + v(t). \tag{5.7}$$

The factor of $\sqrt{E_\mathrm{x}}$ is included in h since only the combined scaling is important from the perspective of receiver design. Matched filtering and sampling at the symbol rate give the received signal

$$y[n] = h \sum_{m=-\infty}^{\infty} s[m] g((n - m)T - \tau_\mathrm{d}) + v[n] \tag{5.8}$$

where $g(t)$ is a Nyquist pulse shape. Compared with the AWGN received signal model $y[n] = s[n] + v[n]$, there are several sources of distortion, which must be recognized and corrected.

One impairment is caused by symbol timing error. Suppose that τ_d is a fraction of a symbol period, that is, $\tau_\mathrm{d} \in [0, T)$. This models the effect of sample timing error, which happens when the receiver does not sample at precisely the right point in time. Under this assumption

$$y[n] = \underbrace{hs[n]g(\tau_\mathrm{d})}_{\text{desired}} + \underbrace{h \sum_{m \neq n} s[m]g((n-m)T - \tau_\mathrm{d})}_{\text{ISI}} + \underbrace{v[n]}_{\text{noise}}. \tag{5.9}$$

Intersymbol interference is created when the Nyquist pulse shape is not sampled exactly at nT, since $g(nT + \tau_\mathrm{d})$ is not generally equal to $\delta[n]$. Correcting for this fractional delay requires *symbol synchronization, equalization,* or a more complicated detector.

A second impairment occurs with larger delays. For illustration purposes, suppose that $\tau_\mathrm{d} = dT$ for some integer d. This is the case where symbol timing has been corrected but an unknown propagation delay, which is a multiple of the symbol period, remains. Under this assumption

$$y[n] = h \sum_{m=-\infty}^{\infty} s[m] g((n - m)T - \tau_\mathrm{d}) + v[n] \tag{5.10}$$

$$= h \sum_{m=-\infty}^{\infty} s[m] g((n - m)T - dT) + v[n] \tag{5.11}$$

$$= hs[n - d] + v[n]. \tag{5.12}$$

Essentially integer offsets create a mismatch between the indices of the transmitted and received symbols. Frame synchronization is required to correct this frame timing error impairment.

Finally, suppose that the unknown delay τ_d has been completely removed so that $\tau_d = 0$. Then the received signal is

$$y[n] = hs[n] + v[n]. \tag{5.13}$$

Sampling and removing delay leaves distortion due to the attenuation and phase shift in h. Dealing with h requires either a specially designed modulation scheme, like DQPSK (differential quadrature phase-shift keying), or channel estimation and equalization.

It is clear that amplitude, phase, and delay if not compensated can have a drastic impact on system performance. As a consequence, every wireless system is designed with special features to enable these impairments to be estimated or directly removed in the receiver processing. Most of this processing is performed on small segments of data called bursts, packets, or frames. We emphasize this kind of batch processing in this book. Many of the algorithms have adaptive extensions that can be applied to continually estimate impairments.

5.1.2 Symbol Synchronization

The purpose of symbol synchronization, or timing recovery, is to estimate and remove the fractional part of the unknown delay τ_d, which corresponds to the part of the error in $[0, T)$. The theory behind these algorithms is extensive and their history is long [309, 226, 124]. The purpose of this section is to present one of the many algorithm approaches for symbol synchronization in complex pulse-amplitude modulated systems.

Philosophically, there are several approaches to synchronization as illustrated in Figure 5.1. There is a pure analog approach, a combined digital and analog approach where digital processing is used to correct the analog, and a pure digital approach. Following the DSP methodology, this book considers only the case of pure digital symbol synchronization.

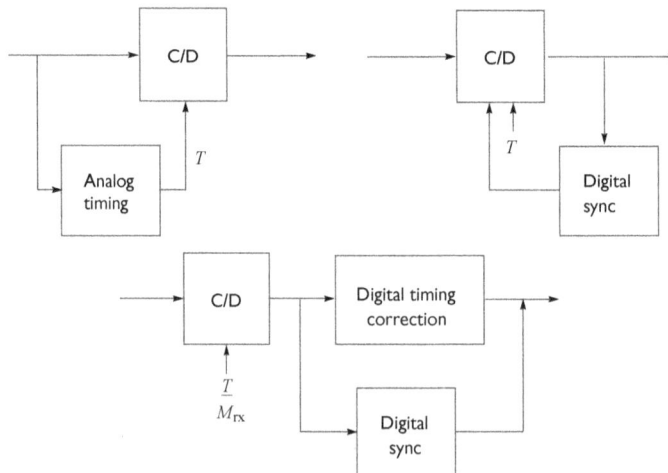

Figure 5.1 Different options for correcting symbol timing. The bottom approach relies only on DSP.

We consider two different strategies for digital symbol synchronization, depending on whether the continuous-to-discrete converter can be run at a low sampling rate or a high sampling rate:

- The oversampling method, illustrated in Figure 5.2, is suitable when the oversampling factor M_{rx} is large and therefore a high rate of oversampling is possible. In this case the synchronization algorithm essentially chooses the best multiple of T/M_{rx} and adds a suitable integer delay $k^\star$ prior to downsampling.

- The resampling method is illustrated in Figure 5.3. In this case a resampler, or interpolator, is used to effectively create an oversampled signal with an effective sample period of T/M_{rx} even if the actual sampling is done with a period less than T/M_{rx} but still satisfying Nyquist. Then, as with the oversampling method, the multiple of T/M_{rx} is estimated and a suitable integer delay $k^\star$ is added prior to the downsampling operation.

A proper theoretical approach for solving for the best of τ_{d} would be to use estimation theory. For example, it is possible to solve for the maximum likelihood estimator. For simplicity, this section considers an approach based on a cost function known as the maximum output energy (MOE) criterion, the same approach as in [169]. There are other approaches based on maximum likelihood estimation such as the early-late gate approach. Lower-complexity implementations of what is described are possible that exploit filter banks; see, for example, [138].

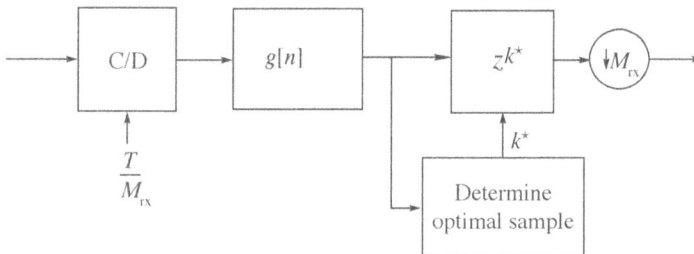

Figure 5.2 Oversampling method suitable when M_{rx} is large

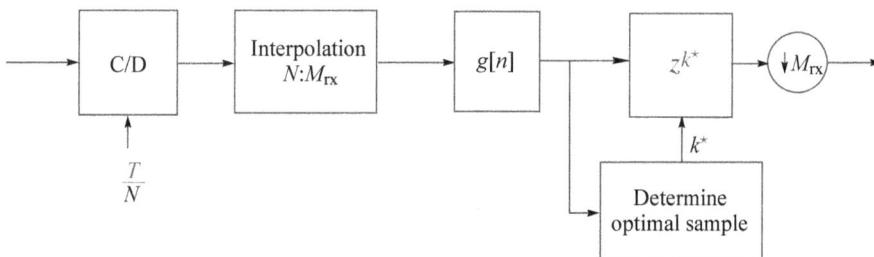

Figure 5.3 Resampling or interpolation method suitable when M_{rx} is small, for example, $N = 2$

First we compute the energy output as a way to justify each approach for timing synchronization. Denote the continuous-time output of the matched filter as

$$y(t) = h \sum_{m=-\infty}^{\infty} s[m]g(t - mT - \tau_\mathrm{d}) + v(t). \tag{5.14}$$

The output energy after sampling by $nT + \tau$ is

$$J_\mathrm{MOE}(\tau) = \mathbb{E}\left[|y(nT + \tau)|^2\right] \tag{5.15}$$

$$= |h|^2 \sum_{m=-\infty}^{\infty} |g(mT + \tau - \tau_\mathrm{d})|^2 + N_\mathrm{o}. \tag{5.16}$$

Suppose that $\tau_\mathrm{d} = dT + \tau_\mathrm{frac}$ and $\tau = \widehat{d}T + \widehat{\tau}_\mathrm{frac}$; then

$$\mathbb{E}\left[|y(nT + \tau)|^2\right] = |h|^2 \sum_{m=-\infty}^{\infty} |g(mT + dT + \tau_\mathrm{frac} - \widehat{d}T - \widehat{\tau}_\mathrm{frac})|^2 + N_\mathrm{o} \tag{5.17}$$

$$= |h|^2 \sum_{m=-\infty}^{\infty} |g(mT + \tau_\mathrm{frac} - \widehat{\tau}_\mathrm{frac})|^2 + N_\mathrm{o} \tag{5.18}$$

with a change of variables. Therefore, while the delay τ can take an arbitrary positive value, only the offset that corresponds to the fractional delay has an impact on the output energy.

The maximum output energy approach to symbol timing attempts to find the τ that maximizes $J_\mathrm{MOE}(\tau)$ where $\tau \in [0, T]$. The maximum output energy solution is

$$\widehat{\tau}_d = \arg\max_{\tau \in [0,T)} J_\mathrm{MOE}(\tau). \tag{5.19}$$

The rationale behind this approach is that

$$\mathbb{E}\left[|y(nT + \tau)|^2\right] \le |h|^2|g(0)|^2 + N_\mathrm{o} \tag{5.20}$$

where the inequality is true for most common pulse-shaping functions (but not for arbitrarily chosen pulse shapes). Thus the unique maximum of $J_\mathrm{MOE}(\tau)$ corresponds to $\tau = \tau_\mathrm{d}$. The values of $J_\mathrm{MOE}(\tau)$ are plotted in Figure 5.4 for the raised cosine pulse shape. A proof of (5.20) for sinc and raised cosine pulse shapes is established in Example 5.1 and Example 5.2.

Example 5.1 Show that the inequality in (5.20) is true for sinc pulse shapes.
  Answer: For convenience denote

$$\mathrm{sinc}\left(\frac{mT + \tau - \tau_\mathrm{d}}{T}\right) = \mathrm{sinc}(m + a) \tag{5.21}$$

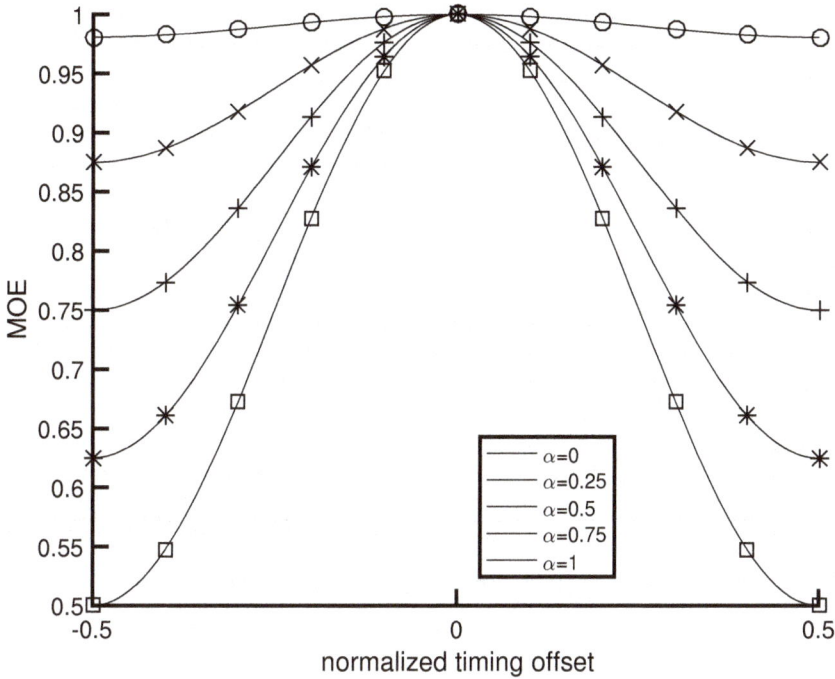

Figure 5.4 Plot of $J_{\mathrm{MOE}}(\tau)$ for $\tau \in [-T/2, T/2]$, in the absence of noise, assuming a raised cosine pulse shape (4.73). The x-axis is normalized to T. As the rolloff value of the raised cosine is increased, the output energy peaks more, indicating that the excess bandwidth provided by choosing larger values of rolloff α translates into better symbol timing using the maximum output energy cost function.

where $a = \frac{\tau - \tau_d}{T}$. Let $g(t) = \mathrm{sinc}(t+a)$ and $g[m] = g(mT) = \mathrm{sinc}(m+a)$, and let $\mathrm{g}(f)$ and $\mathrm{g}(e^{j2\pi f})$ be the CTFT of $g(t)$ and DTFT of $g[m]$ respectively. Then $\mathrm{g}(f) = e^{j2\pi a f}\mathrm{rect}(f)$ and

$$\mathrm{g}(e^{j2\pi f}) = \frac{1}{1}\sum_{n=-\infty}^{\infty} \times\left(\frac{f}{1} - \frac{n}{1}\right) \tag{5.22}$$

$$= \sum_{n=-\infty}^{\infty} e^{j2\pi a(f-n)}\mathrm{rect}(f-n). \tag{5.23}$$

By Parseval's theorem for the DTFT we have

$$\sum_{m=-\infty}^{\infty} |\mathrm{sinc}(m+a)|^2 = \int_{-1/2}^{1/2} |g(e^{j2\pi f})|^2 \mathrm{d}f \tag{5.24}$$

$$= \int_{-1/2}^{1/2} \left|\sum_{n=-\infty}^{\infty} e^{j2\pi a(f-n)}\mathrm{rect}(f-n)\right|^2 \mathrm{d}f \tag{5.25}$$

$$\leq \int_{-1/2}^{1/2} \sum_{n=-\infty}^{\infty} \left| e^{j2\pi a(f-n)} \mathrm{rect}(f-n) \right|^2 \mathrm{d}f \tag{5.26}$$

$$= \int_{-1/2}^{1/2} \sum_{n=-\infty}^{\infty} (\mathrm{rect}(f-n))^2 \mathrm{d}f \tag{5.27}$$

$$= \int_{-1/2}^{1/2} 1 \mathrm{d}f \tag{5.28}$$

$$= 1 \tag{5.29}$$

$$= |g(0)|^2, \tag{5.30}$$

where the inequality follows because $|\sum a_i| \leq \sum |a_i|$. Therefore,

$$\sum_{m=-\infty}^{\infty} |g(mT + \tau - \tau_\mathrm{d})|^2 \leq |g(0)|^2. \tag{5.31}$$

Example 5.2   Show that the MOE inequality in (5.20) holds for raised cosine pulse shapes.

Answer: The raised cosine pulse

$$g_\mathrm{rc}(t) = \mathrm{sinc}(t/T) \frac{\cos(\pi \alpha t)}{1 - 4\alpha^2 t^2/T^2} \tag{5.32}$$

is a modulated version of the sinc pulse shape. Since

$$\left| \frac{\cos(\pi \alpha t)}{1 - 4\alpha^2 t^2/T^2} \right|^2 \leq 1 \tag{5.33}$$

it follows that

$$|g_\mathrm{rc}(t)|^2 \leq |\mathrm{sinc}(t/T)|^2 \tag{5.34}$$

and the result from Example 5.1 can be used.

Now we develop the direct solution to maximizing the output energy, assuming the receiver architecture with oversampling as in Figure 4.10. Let $r[n]$ denote the signal after oversampling or resampling, assuming there are M_rx samples per symbol period. Let the output of the matched receiver filter prior to downsampling at the symbol rate be

$$\widetilde{y}[n] = \sum_{m=-\infty}^{\infty} r[m] g_\mathrm{rx}[n-m]. \tag{5.35}$$

We use this sampled signal to compute a discrete-time version of $J_\mathrm{MOE}(\tau)$ given by

$$J_\mathrm{MOE,d}[k] = \mathbb{E}\left[|\widetilde{y}[nM_\mathrm{rx} + k]|^2 \right] \tag{5.36}$$

where k is the sample offset between $0, 1, \ldots, M_{\mathrm{rx}} - 1$ corresponding to an estimate of the fractional part of the timing offset given by kT/M_{rx}. To develop a practical algorithm, we replace the expectation with a time average over P symbols, thanks to ergodicity, so that

$$J_{\mathrm{MOE,e}}[k] = \frac{1}{P} \sum_{p=0}^{P-1} |r[pM_{\mathrm{rx}} + k]|^2. \tag{5.37}$$

Looking for the maximizer of $J_{\mathrm{MOE,e}}[k]$ over $k = 0, 1, \ldots, M_{\mathrm{rx}} - 1$ gives the optimum sample $k^\star$ and an estimate of the symbol timing offset $k^\star T/M_{\mathrm{rx}}$.

The optimum correction involves advancing the received signal by $k^\star$ samples prior to downsampling. Essentially, the synchronized data is $y[n] = \widetilde{y}[nM_{\mathrm{rx}} + k^\star]$. Equivalently, the signal can be delayed by $k^\star - M_{\mathrm{rx}}$ samples, since subsequent signal processing steps will in any case correct for frame synchronization.

The main parameter to be selected in the symbol timing algorithms covered in this section is the oversampling factor M_{rx}. This decision can be based on the residual ISI created because the symbol timing is quantized. Using the SINR from (4.56), assuming h is known perfectly, matched filtering at the receiver, and a maximum symbol timing offset $T/2M_{\mathrm{rx}}$, then

$$\mathrm{SINR} = \frac{|h|^2 \left| g\left(\frac{T}{2M_{\mathrm{rx}}}\right) \right|^2}{|h|^2 \sum_{m \neq 0} \left| g\left(mT + \frac{T}{2M_{\mathrm{rx}}}\right) \right|^2 + N_{\mathrm{o}}}. \tag{5.38}$$

This value can be used, for example, with the probability of symbol error analysis in Section 4.4.5 to select a value of M_{rx} such that the impact of symbol timing error is acceptable, depending on the SNR and target probability of symbol error. An illustration is provided in Figure 5.5 for the case of 4-QAM. In this case, a value of $M_{\mathrm{rx}} = 8$ leads to less than 1dB of loss at a symbol error rate of 10^{-4}.

5.1.3 Frame Synchronization

The purpose of frame synchronization is to resolve multiple symbol period delays, assuming that symbol synchronization has already been performed. Let d denote the remaining offset where $d = \tau_{\mathrm{d}}/T - k^\star/M_{\mathrm{rx}}$, assuming the symbol timing offset was corrected perfectly. Given that, ISI is eliminated and

$$y[n] = hs[n - d] + v[n]. \tag{5.39}$$

To reconstruct the transmitted bit sequence it is necessary to know "where the symbol stream starts." As with symbol synchronization, there is a great deal of theory surrounding frame synchronization [343, 224, 27]. This section considers one common algorithm for frame synchronization in flat channels that exploits the presence of a known training sequence, inserted during a training phase.

Most wireless systems insert reference signals into the transmitted waveform, which are known by the receiver. Known information is often called a training sequence or a pilot symbol, depending on how the known information is inserted. For example, a training

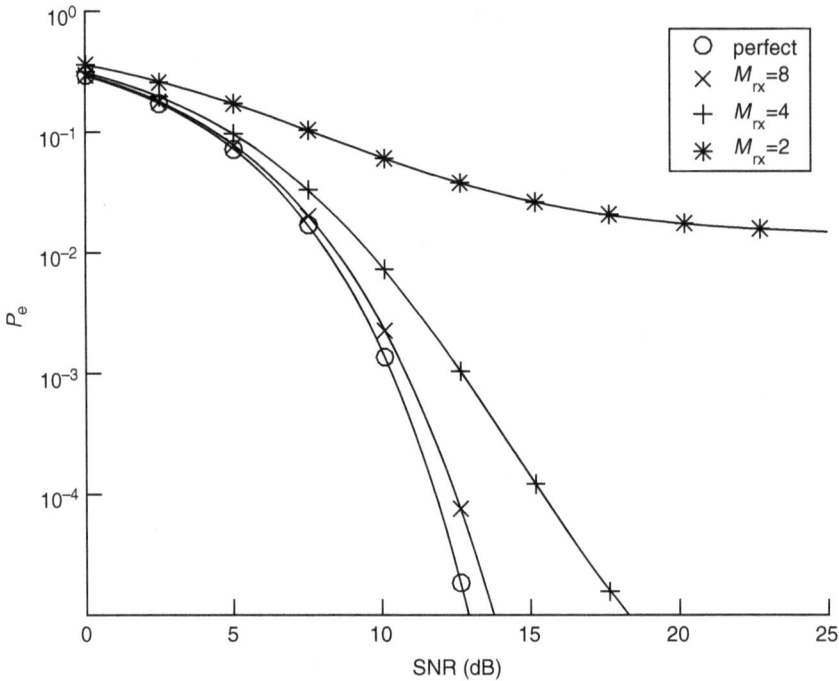

Figure 5.5 Plot of the exact symbol error rate for 4-QAM from (4.147), substituting SINR for SNR, with $|h| = \sqrt{E_x}$ for different values of M_{rx} in (5.38), assuming a raised cosine pulse shape with $\alpha = 0.25$. With enough oversampling, the effects of timing error are small.

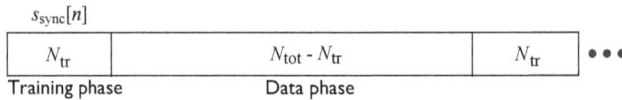

Figure 5.6 A frame structure that consists of periodically inserted training and data

sequence may be inserted at the beginning of a transmission as illustrated in Figure 5.6, or a few pilot symbols may be inserted periodically. Most systems use some combination of the two where long training sequences are inserted periodically and shorter training sequences (or pilot symbols) are inserted more frequently. For the purpose of explanation, it is assumed that the desired frame begins at discrete time $n = 0$. The total frame length is N_{tot}, including a length N_{tr} training phase and an $N_{tot} - N_{tr}$ data phase. Suppose that $\{t[n]\}_{n=0}^{N_{tr}-1}$ is the training sequence known at the receiver.

One approach to performing frame synchronization is to correlate the received signal with the training sequence to compute

$$R[n] = \sum_{k=0}^{N_{tr}-1} t^*[k]y[n+k] \tag{5.40}$$

and then to find

$$\widehat{d} = \arg\max_n |R[n]|. \tag{5.41}$$

The maximization usually occurs by evaluating $R[n]$ over a finite set of possible values. For example, the analog hardware may have a carrier sense feature that can determine when there is a significant signal of interest. Then the digital hardware can start evaluating the correlation and looking for the peak. A threshold can also be used to select the starting point, that is, finding the first value of n such that $|R[n]|$ exceeds a target threshold. An example with frame synchronization is provided in Example 5.3.

Example 5.3   Consider a system as described by (5.39) with $h = 0.5e^{\mathrm{j}\pi/3}\sqrt{E_{\mathrm{x}}}$, $N_{\mathrm{tr}} = 7$, and $N_{\mathrm{tot}} = 21$ with $d = 0$. We assume 4-QAM for the data transmission and that the training consists of the Barker code of length 7 given by $\{t[n]\}_{n=0}^{7} = \{1, 1, 1, -1, -1, 1, -1\}$. The SNR is 5dB. We consider a frame snippet that consists of 14 data symbols, 7 training symbols, 14 data symbols, 7 training symbols, and 14 data symbols. In Figure 5.7, we plot $|R[n]|$ computed from (5.40). There are two peaks that correspond to locations of our training data. The peaks happen at 21 and 42 as expected. If the snippet was delayed, then the peaks would shift accordingly.

A block diagram of a receiver including both symbol synchronization and frame synchronization operations is illustrated in Figure 5.8. The frame synchronization happens

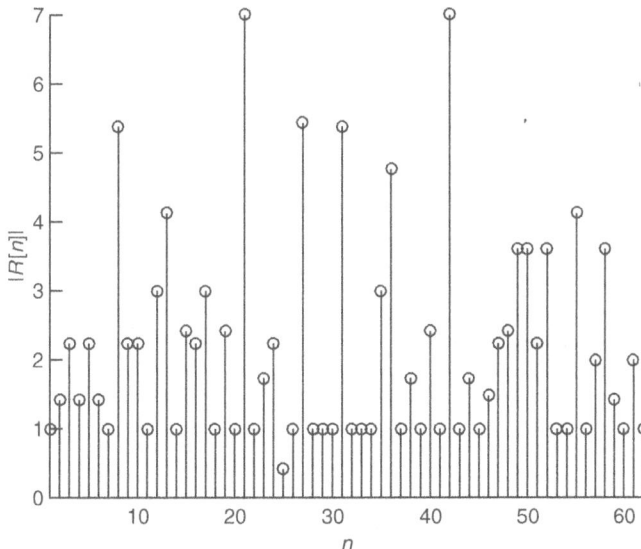

Figure 5.7 The absolute value of the output of a correlator for frame synchronization. The details of the simulation are provided in Example 5.3. Two correlation peaks are seen, corresponding to the location of the two training sequences.

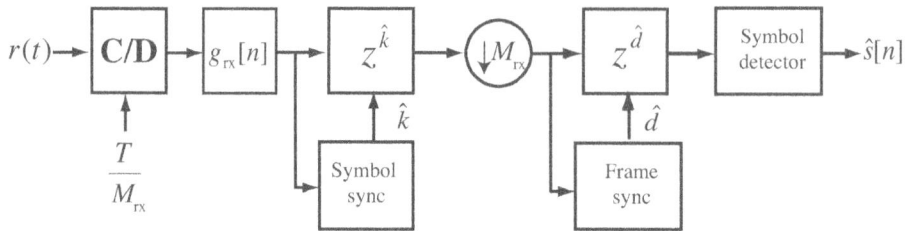

Figure 5.8 Receiver with symbol synchronization based on oversampling and frame synchronization

after the downsampling and prior to the symbol detection. Fixing the frame synchronization requires advancing the signal by $\widehat{d}$ symbols.

The frame synchronization algorithm may find a false peak if the data is the same as
the training sequence. There are several approaches to avoid this problem. First, training
sequences can be selected that have good correlation properties. There are many kinds
of sequences known in the literature that have good autocorrelation properties [116, 250,
292], or periodic autocorrelation properties [70, 267]. Second, a longer training sequence
may be used to reduce the likelihood of a false positive. Third, the training sequence
can come from a different constellation than the data. This was used in Example 5.3
where a BPSK training sequence was used but the data was encoded with 4-QAM.
Fourth, the frame synchronization can average over multiple training periods. Suppose
that the training is inserted every N_{tot} symbols. Averaging over P periods then leads to
an estimate

$$\widehat{d} = \arg \max_n \sum_{p=0}^{P-1} \left| \sum_{k=0}^{N_{\text{tr}}-1} t^*[k]y[n+k+pN_{\text{tot}}] \right|. \tag{5.42}$$

Larger amounts of averaging improve performance at the expense of higher complexity and more storage requirements. Finally, complementary training sequences can be
used. In this case a pair of training sequences $\{t_1[n]\}$ and $\{t_2[n]\}$ are designed such that
$\sum_{k=0}^{N_{\text{tr}}-1} t_1^*[k]y[n+k] + t_2^*[k]y[n+k+N_{\text{tot}}]$ has a sharp correlation peak. Such sequences
are discussed further in Section 5.3.1.

5.1.4 Channel Estimation

Once frame synchronization and symbol synchronization are completed, a good model
for the received signal is

$$y[n] = hs[n] + v[n]. \tag{5.43}$$

The two remaining impairments are the unknown flat channel h and the AWGN $v[n]$.
Because h rotates and scales the constellation, the channel must be estimated and either
incorporated into the detection process or removed via equalization.

The area of channel estimation is rich and the history long [40, 196, 369]. In general,
a channel estimation problem is handled like any other estimation problem. The formal

approach is to derive an optimal estimator under assumptions about the signal and noise. Examples include the least squares estimator, the maximum likelihood estimator, and the MMSE estimator; background on these estimators may be found in Section 3.5.

In this section we emphasize the use of least squares estimation, which is also the ML estimator when used for linear parameter estimation in Gaussian noise. To use least squares, we build a received signal model from (5.43) by exploiting the presence of the known training sequence from $n = 0, 1, \ldots, N_{\text{tr}} - 1$. Stacking the observations in (5.43) into vectors,

$$\underbrace{\begin{bmatrix} y[0] \\ y[1] \\ \vdots \\ y[N_{\text{tr}} - 1] \end{bmatrix}}_{\mathbf{y}} = \underbrace{\begin{bmatrix} t[0] \\ t[1] \\ \vdots \\ t[N_{\text{tr}} - 1] \end{bmatrix}}_{\mathbf{t}} h + \underbrace{\begin{bmatrix} v[0] \\ v[1] \\ \vdots \\ v[N_{\text{tr}} - 1] \end{bmatrix}}_{\mathbf{v}} \tag{5.44}$$

which becomes compactly

$$\mathbf{y} = \mathbf{t}h + \mathbf{v}. \tag{5.45}$$

We already computed the maximum likelihood estimator for a more general version of (5.46) in Section 3.5.3. The solution was the least squares estimate, given by

$$\widehat{h} = (\mathbf{t}^*\mathbf{t})^{-1}\mathbf{t}^*\mathbf{y} \tag{5.46}$$

$$= \frac{\sum_{n=0}^{N_{\text{tr}}-1} t^*[n]y[n]}{\sum_{n=0}^{N_{\text{tr}}-1} t^*[n]t[n]}. \tag{5.47}$$

Essentially, the least squares estimator correlates the observed data with the training data and normalizes the result. The denominator is just the energy in the training sequence, which can be precomputed and stored offline. The numerator is calculated as part of the frame synchronization process. Therefore, frame synchronization and channel estimation can be performed jointly. The receiver, including symbol synchronization, frame synchronization, and channel estimation, is illustrated in Figure 5.9.

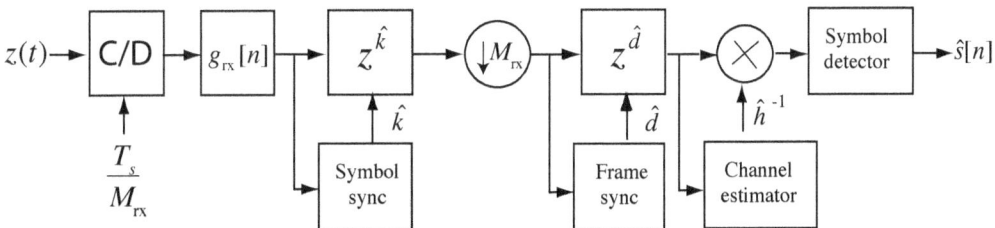

Figure 5.9 Receiver with symbol synchronization based on oversampling, frame synchronization, and channel estimation

Example 5.4   In this example, we evaluate the squared error in the channel estimate. We consider a similar system to the one described in Example 5.3 with a length $N_{\mathrm{tr}} = 7$ Barker code used for a training sequence, and a least squares channel estimator as in (5.46). We perform a Monte Carlo estimate of the channel by generating a noise realization and estimating the channel $\widehat{h}[n]$ for $n = 1, 2, \ldots, 1000$ realizations. In Figure 5.10, we evaluate the estimation error as a function of the SNR, which is defined as $|h|^2/\sigma_v^2$, and is 5dB in this example. We plot the error for one realization of a channel estimation, which is given by $|h-\widehat{h}[0]|^2$, and the mean squared error, which is given by $\frac{1}{1000}\sum_{n=0}^{999}|h-\widehat{h}[n]|^2$. The plot shows how the estimation error, based both on one realization and on average, decreases with SNR.

Example 5.5   In this example, we evaluate the squared error in the channel estimate for different lengths of 4-QAM training sequences, an SNR of 5dB, the channel as in Example 5.3, and a least squares channel estimator as in (5.46). We perform a Monte Carlo estimate of the squared error of one realization and the mean squared error, as described in Example 5.4. We plot the results in Figure 5.11, which show how longer training sequences reduce estimation error. Effectively, longer training increases the effective SNR through the addition of energy through coherent combining in $\sum_{n=0}^{N_{\mathrm{tr}}-1} t^*[n]t[n]$ and more averaging in $\sum_{n=0}^{N_{\mathrm{tr}}-1} t^*[n]v[n]$.

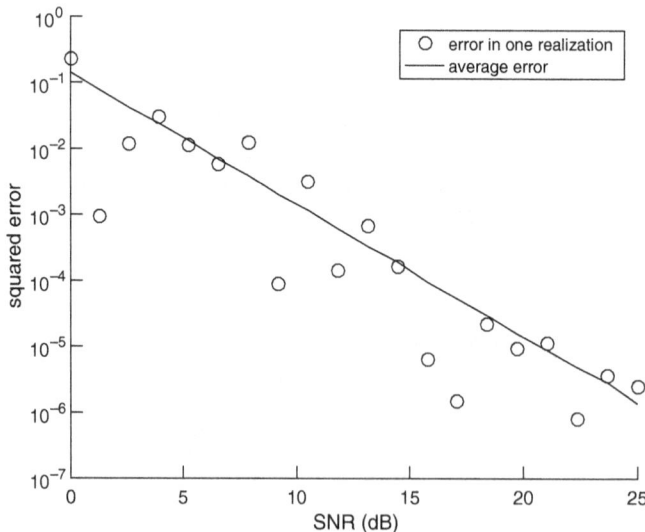

Figure 5.10 Estimation error as a function of SNR for the system in Example 5.4. The error estimate reduces as SNR increases.

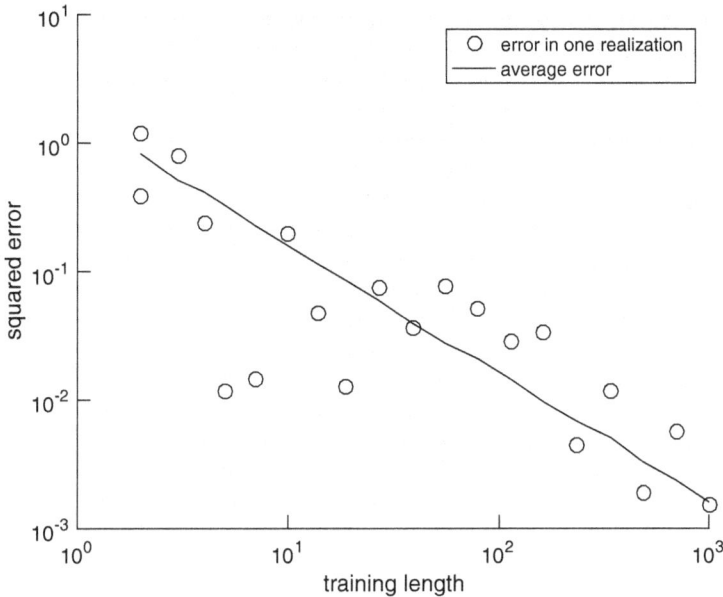

Figure 5.11 Estimation error as a function of the training length for the system in Example 5.5. The error estimate reduces as the training length increases.

5.1.5 Equalization

Assuming that channel estimation has been completed, the next step in the receiver processing is to use the channel estimate to perform symbol detection. There are two reasonable approaches; both involve assuming that $\widehat{h}$ is the true channel estimate. This is a reasonable assumption if the channel estimation error is small enough.

The first approach is to incorporate $\widehat{h}$ into the detection process. Replacing $\sqrt{E_{\mathrm{x}}}$ by h in (4.105), then

$$\widehat{s}[n] = \arg\min_{s \in \mathcal{C}} |y[n] - \widehat{h}s|^2. \tag{5.48}$$

Consequently, the channel estimate becomes an input into the detector. It can be used as in (5.48) to scale the symbols during the computation of the norm, or it can be used to create a new constellation $\bar{\mathcal{C}} = \{\widehat{h}c_1, \widehat{h}c_2, \dots, \widehat{h}c_M\}$ and detection performed using the scaled constellation as

$$\widehat{s}[n] = \arg\min_{s \in \bar{\mathcal{C}}} |y[n] - s|^2. \tag{5.49}$$

The latter approach is useful when N_{tot} is large.

An alternative approach to incorporating the channel into the ML detector is to remove the effects of the channel prior to detection. For $\widehat{h} \neq 0$,

$$\arg\min_{s \in \mathcal{C}} |y[n] - \widehat{h}s|^2 = \arg\min_{s \in \mathcal{C}} |\widehat{h}|^{-1} \left| \frac{y[n]}{\widehat{h}} - s \right|^2 = \arg\min_{s \in \mathcal{C}} \left| \frac{y[n]}{\widehat{h}} - s \right|^2. \tag{5.50}$$

The process of creating the signal $y[n]/\widehat{h}$ is an example of equalization. When equalization is used, the effects of the channel are removed from $y[n]$ and a standard detector can be applied to the result, leveraging constellation symmetries to reduce complexity. The receiver, including symbol synchronization, frame synchronization, channel estimation, and equalization, is illustrated in Figure 5.9.

The probability of symbol error with channel estimation can be computed by treating the estimation error as noise. Let $h = \widehat{h} + \widehat{h}_e$ where $h - \widehat{h} = \widehat{h}_e$ is the estimation error. For a given channel h and estimate $\widehat{h}$ the equalized received signal is

$$\widehat{s}[n] = \frac{1}{\widehat{h}}(\widehat{h} + \widehat{h}_e)s[n] + \frac{1}{\widehat{h}}v[n] \tag{5.51}$$

$$= s[n] + \frac{\widehat{h}_e}{\widehat{h}}s[n] + \frac{1}{\widehat{h}}v[n]. \tag{5.52}$$

It is common to treat the middle interference term as additional noise. Moving the common $|\widehat{h}|^2$ to the numerator,

$$\mathrm{SINR}_{\widehat{h}_e} = \frac{|\widehat{h}|^2}{|\widehat{h}_e|^2 + \sigma_v^2}. \tag{5.53}$$

This can be used as part of a Monte Carlo simulation to determine the impact of channel estimation. Since the receiver does not actually know the estimation error, it is also common to consider a variation of the SINR expression where the variance of the estimate $\mathbb{E}\left[|\widehat{h}_e|^2\right]$ is used in place of the instantaneous value (assuming that the estimator is unbiased so zero mean). Then the SINR becomes

$$\mathrm{SINR} = \frac{|\widehat{h}|^2}{\mathbb{E}\left[|\widehat{h}_e|^2\right] + \sigma_v^2}. \tag{5.54}$$

This expression can be used to study the impact of estimation error on the probability of error, as the mean squared error of the estimate is a commonly computed quantity. A comparison of the probability of symbol error for these approaches is provided in Example 5.6.

Example 5.6   In this example we evaluate the impact of channel estimation error. We consider a system similar to the one described in Example 5.4 with a length $N_{\mathrm{tr}} = 7$ Barker code used for a training sequence, and a least squares channel estimator as in (5.46). We perform a Monte Carlo estimate of the channel by generating a noise realization and estimating the channel $\widehat{h}[n]$ for $n = 1, 2, \ldots, 1000$ realizations. For each realization, we compute the error, insert into the $\mathrm{SINR}_{\widehat{h}_e}$ in (5.53), and use that to compute the exact probability of symbol error from Section 4.4.5. We then average this over 1000 Monte Carlo simulations. We also compare in Figure 5.12 with the probability of error assuming no estimation error with SNR $|h|^2 E_{\mathrm{x}}/N_{\mathrm{o}}$ and with the probability of error computed using the average error from the SINR in (5.54). We see in this example

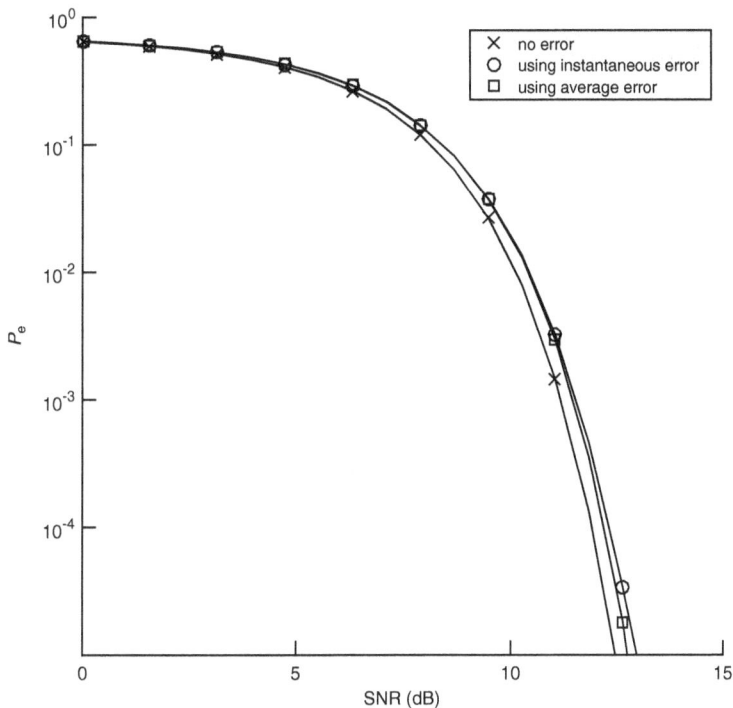

Figure 5.12 The probability of symbol error with 4-QAM and channel estimation using a length $N_{\mathrm{tr}} = 7$ training sequence. The probability of symbol error with no channel estimation error is compared with the average of the probability of symbol error including the instantaneous error, and the probability of symbol error using the average estimation error. There is little loss in using the average estimation error.

that the loss due to estimation error is about 1dB and that there is little difference in the average probability of symbol error and the probability of symbol error computed using the average estimation error.

5.1.6 Carrier Frequency Offset Synchronization

Wireless communication systems use passband communication signals. They can be created at the transmitter by upconverting a complex baseband signal to carrier f_{c} and can be processed at the receiver to produce a complex baseband signal by downconverting from a carrier f_{c}. In Section 3.3, the process of upconversion to carrier f_{c}, downconversion to baseband, and the complex equivalent notation were explained under the important assumption that f_{c} is known perfectly at both the transmitter and the receiver. In practice, f_{c} is generated from a local oscillator. Because of temperature variations and the fact that the carrier frequency is generated by a different local oscillator and transmitter and receiver, in practice f_{c} at the transmitter is not equal to f_{c}' at the receiver as illustrated

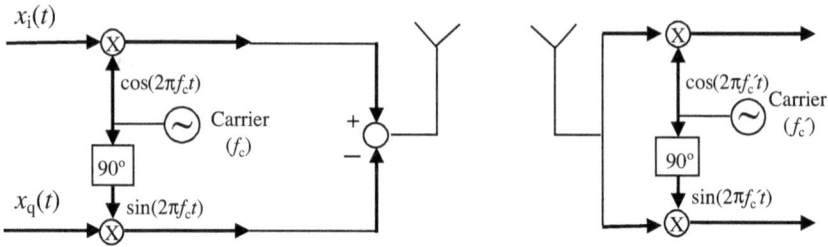

Figure 5.13 Abstract block diagram of a wireless system with different carrier frequencies at the transmitter and the receiver

in Figure 5.13. The difference between the transmitter and the receiver $f_e = f_c' - f_c$ is the carrier frequency offset (or simply frequency offset) and is generally measured in hertz. In device specification sheets, the offset is often measured as $|f_e|/f_c$ and is given in units of parts per million. In this section, we derive the system model for carrier frequency offset and present a simple algorithm for carrier frequency offset estimation and correction.

Let

$$x_p(t) = \text{Re}\left[x(t)e^{j2\pi f_c t}\right] \tag{5.55}$$

be the passband signal generated at the transmitter. Let the observed passband signal at the receiver at carrier f_c be

$$r_p(t) = \int h_c(t - \tau)x_p(\tau)d\tau \tag{5.56}$$

$$= r_i(t)\cos(2\pi f_c t) - r_q(t)\sin(2\pi f_c t) \tag{5.57}$$

$$= \frac{1}{2}\left(r(t)e^{j2\pi f_c t} + r^*(t)e^{-j2\pi f_c t}\right) \tag{5.58}$$

where $r(t) = r_i(t) + jr_q(t)$ is the complex baseband signal that corresponds to $r_p(t)$ and we use the fact that $\text{Re}(x) = \frac{1}{2}(x + x^*)$ for the last step. Now suppose that $r_p(t)$ is downconverted using carrier f_c' to produce a new signal $r'(t)$. Then the extracted complex baseband signal is (ignoring noise)

$$r'(t) = 2B\text{sinc}(tB) * \left(r_p(t)e^{-j2\pi f_c' t}\right). \tag{5.59}$$

Focusing on the last term in (5.59) with $f_e = f_c - f_c'$, then

$$r_p(t)e^{-j2\pi f_c' t} = r_p(t)e^{-j2\pi(f_c - f_e)t} \tag{5.60}$$

$$= e^{j2\pi f_e t}r_p(t)e^{-j2\pi f_c t}. \tag{5.61}$$

Substituting in $r_p(t)$ from (5.58),

$$r_p(t)e^{-j2\pi f_c' t} = \frac{1}{2}r(t)e^{j2\pi f_e t} + r^*(t)e^{j2\pi f_e t}e^{-j4\pi f_c t}. \tag{5.62}$$

Lowpass filtering (assuming, strictly speaking, a bandwidth of $B + |f_e|$) and correcting for the factor of $\frac{1}{2}$ leaves the complex baseband equivalent

$$r'(t) = e^{j2\pi f_e t} r(t). \qquad (5.63)$$

Carrier frequency offset results in a rolling phase shift that happens after the convolution and depends on the difference in carrier f_e. As the phase shift accumulates, failing to synchronize can quickly lead to errors.

Following the methodology of this book, it is of interest to formulate and solve the frequency offset estimation and correction problem purely in discrete time. To that end a discrete-time complex baseband equivalent model is required.

To formulate a discrete-time model, we focus on the sampled signal after matched filtering, still neglecting noise, as

$$y(t) = \int r'(t - \tau) g_{rx}(\tau) d\tau \qquad (5.64)$$

$$= \int e^{j2\pi f_e(t-\tau)} r(t - \tau) g_{rx}(\tau) d\tau \qquad (5.65)$$

$$= e^{j2\pi f_e t} \int r(t - \tau) e^{-j2\pi f_e \tau} g_{rx}(\tau) d\tau. \qquad (5.66)$$

Suppose that the frequency offset f_e is sufficiently small that the variation over the duration of $g_{rx}(t)$ can be assumed to be a constant; then $e^{-j2\pi f_e \tau} g_{rx}(\tau) \approx g_{rx}(\tau)$. This is reasonable since most of the energy in the matched filter $g_{rx}(t)$ is typically concentrated over a few symbol periods. Assuming this holds,

$$y(t) = e^{j2\pi f_e t} \int r(t - \tau) g_{rx}(\tau) d\tau. \qquad (5.67)$$

As a result, it is reasonable to model the effects of frequency offset as if they occurred on the matched filtered signal.

Now we specialize to the case of flat-fading channels. Substituting for $r(t)$, adding noise, and sampling gives

$$y[n] = e^{j2\pi\epsilon n} h \sum_{m=-\infty}^{\infty} s[m] g((n - m)T - \tau_d) + v[n] \qquad (5.68)$$

where $\epsilon = f_e T$ is the normalized frequency offset. Frequency offset introduces a multiplication by the discrete-time complex exponential $e^{j2\pi f_e T n}$.

To visualize the effects of frequency offset, suppose that symbol and frame synchronization has been accomplished and only equalization and channel estimation remain. Then the Nyquist property of the pulse shape can be exploited so that

$$y[n] = e^{j2\pi\epsilon n} h s[n] + v[n]. \qquad (5.69)$$

Notice that the transmit symbol is being rotated by $\exp(j2\pi\epsilon n)$. As n increases, the offset increases, and thus the symbol constellation rotates even further. The impact of this is

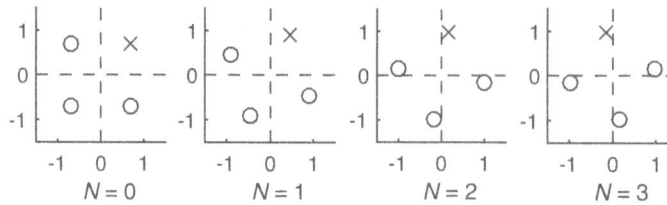

Figure 5.14 The successive rotation of a 4-QAM constellation due to frequency offset. The Voronoi regions for each constellation point are indicated with the dashed lines. To make the effects of the rotation clearer, one of the constellation points is marked with an x.

an increase in the number of symbol errors as the symbols rotate out of their respective Voronoi regions. The effect is illustrated in Example 5.7.

Example 5.7   Consider a system with the frequency offset described by (5.69). Suppose that $\epsilon = 0.05$. In Figure 5.14, we plot a 4-QAM constellation at times $n = 0, 1, 2, 3$. The Voronoi regions corresponding to the unrotated constellation are shown on each plot as well. To make seeing the effect of the rotation easier, one point is marked with an x. Notice how the rotations are cumulative and eventually the constellation points completely leave their Voronoi regions, which results in detection errors even in the absence of noise.

Correcting for frequency offset is simple: just multiply $y[n]$ by $e^{-j2\pi\epsilon n}$. Unfortunately, the offset is unknown at the receiver. The process of correcting for ϵ is known as frequency offset synchronization. The typical method for frequency offset synchronization involves first estimating the offset $\hat{\epsilon}$, then correcting for it by forming the new sequence $\exp(-j2\pi\hat{\epsilon}n)y[n]$ with the phase removed. There are several different methods for correction; most employ a frequency offset estimator followed by a correction phase. Blind offset estimators use some general properties of the received signal to estimate the offset, whereas non-blind estimators use more specific properties of the training sequence.

Now we review two algorithms for correcting the frequency offset in a flat-fading channel. We start by observing that frequency offset does not impact symbol synchronization, since the $e^{j2\pi f_o T n}$ cancels out the maximum output energy maximization caused by the magnitude function (see, for example, (5.15)). As a result, ISI cancels and a good model for the received signal is

$$y[n] = e^{j2\pi\epsilon n} h s[n - d] + v[n] \tag{5.70}$$

where the offset ϵ, the channel h, and the frame offset d are unknowns. In Example 5.8, we present an estimator based on a specific training sequence design. In Example 5.9, we use properties of 4-QAM to develop a blind frequency offset estimator. We compare their performance to highlight differences in the proposed approaches in Figure 5.15. We also explain how to jointly estimate the delay and channel with each approach. These algorithms illustrate some ways that known information or signal structure can be exploited for estimation.

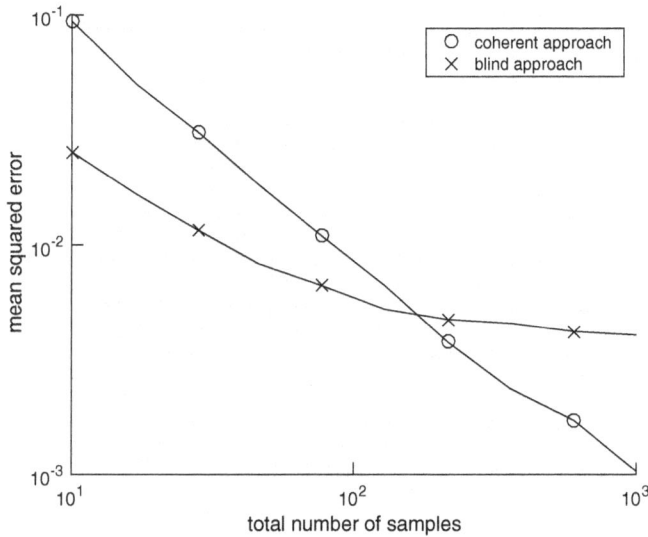

Figure 5.15 The mean squared error performance of two different frequency offset estimators: the training-based approach in Example 5.8 and the blind approach in Example 5.9. Frame synchronization is assumed to have been performed already for the coherent approach. The channel is the same as in Example 5.3 and other examples, and the SNR is 5dB. The true frequency offset is $\epsilon = 0.01$. The estimators are compared assuming the number of samples (N_{tr} for the coherent case and N_{tot} for the blind case). The error in each Monte Carlo simulation is computed from $\text{phase}(e^{\mathrm{j}2\pi(\epsilon-\hat{\epsilon})})$ to avoid any phase wrapping effects.

Example 5.8   In this example, we present a coherent frequency offset estimator that exploits training data. We suppose that frame synchronization has been solved. Let us choose as a training sequence $t[n] = \exp(\mathrm{j}2\pi f_t n)$ for $n = 0, 1, \ldots, N_{\text{tr}} - 1$ where f_t is a fixed frequency. This kind of structure is available from the Frequency Correction Burst in GSM, for example [100]. Then for $n = 0, 1, \ldots, N_{\text{tr}} - 1$,

$$y[n] = e^{\mathrm{j}2\pi\epsilon n} e^{\mathrm{j}2\pi f_t n} h + v[n]. \qquad (5.71)$$

Correcting the known offset introduced by the training data gives

$$e^{-\mathrm{j}2\pi f_t n} y[n] = e^{\mathrm{j}2\pi\epsilon n} h + e^{-\mathrm{j}2\pi f_t n} v[n]. \qquad (5.72)$$

The rotation by $e^{-\mathrm{j}2\pi f_t n}$ does not affect the distribution of the noise. Solving for the unknown frequency in (5.72) is a classic problem in signal processing and single-tone parameter estimation, and there are many approaches [330, 3, 109, 174, 212, 218]. We explain the approach used in [174] based on the model in [330]. Approximate the corrected signal in (5.72) as

$$e^{-\mathrm{j}2\pi f_t n} y[n] \approx |h| e^{\mathrm{j}2\pi\epsilon n + \theta + \nu[n]} \qquad (5.73)$$

where θ is the phase of h and $\nu[n]$ is Gaussian noise. Then, looking at the phase difference between two adjacent samples, a linear system can be written from the approximate model as

$$\text{phase}(e^{\mathrm{j}2\pi f_t n} y^*[n] e^{-\mathrm{j}2\pi f_t(n+1)} y[n+1]) = 2\pi\epsilon + \nu[n+1] - \nu[n]. \quad (5.74)$$

Aggregating the observations in (5.79) from $n = 1, \ldots, N_{\text{tot}} - 1$, we can create a linear estimation problem

$$\mathbf{p} = 2\pi\epsilon\mathbf{1} + \boldsymbol{\nu}, \quad (5.75)$$

where $[\mathbf{p}]_n = \text{phase}(e^{\mathrm{j}2\pi f_t n} y^*[n] e^{-\mathrm{j}2\pi f_t(n+1)} y[n+1])$, $\mathbf{1}$ is an $N_{\text{tr}} - 1 \times 1$ vector, and $\boldsymbol{\nu}$ is an $N_{\text{tr}} - 1 \times 1$ noise vector. The least squares solution is given by

$$\widehat{\epsilon} = \frac{1}{2\pi}(\mathbf{p}^*\mathbf{p})^{-1}\mathbf{1}^*\mathbf{p} \quad (5.76)$$

$$= \frac{1}{2\pi(N_{\text{tr}}-1)} \sum_{n=1}^{N_{\text{tr}}-1} \text{phase}(e^{\mathrm{j}2\pi f_t} y^*[n] y[n+1]), \quad (5.77)$$

which is also the maximum likelihood solution if $\nu[n]$ is assumed to be IID [174].

Frame synchronization and channel estimation can be incorporated into this estimator as follows. Given that the frequency offset estimator in (5.77) solves a least squares problem, there is a corresponding expression for the squared error; see, for example, (3.426). Evaluate this expression for many possible delays and choose the delay that has the lowest squared error. Correct for the offset and delay, then estimate the channel as in Section 5.1.4.

Example 5.9 In this example, we exploit symmetry in the 4-QAM constellation to develop a blind frequency offset estimator, which does not require training data. The main observation is that for 4-QAM, the normalized constellation symbols are points on the unit circle. In particular for 4-QAM, it turns out that $s^4[n] = -1$. Taking the fourth power,

$$y^4[n] = -e^{\mathrm{j}2\pi\epsilon 4n} h^4 + \nu[n], \quad (5.78)$$

where $\nu[n]$ contains noise and products of the signal and noise. The unknown parameter d disappears because $s^4[n-d] = -1$, assuming a continuous stream of symbols. The resulting equation has the form of a complex sinusoid in noise, with unknown frequency, amplitude, and phase similar to (5.72). Then a set of linear equations can be written from

$$\text{phase}(y^4[n+1] y^{*4}[n]) = 8\pi\epsilon + \widetilde{\nu}[n] \quad (5.79)$$

in a similar fashion to (5.75), and then

$$\widehat{\epsilon} = \frac{1}{8\pi(N_{\text{tot}}-1)} \sum_{n=1}^{N_{\text{tot}}-1} \text{phase}(y^4[n+1] y^{*4}[n]). \quad (5.80)$$

Frame synchronization and channel estimation can be incorporated as follows. Use all the data to estimate the carrier frequency offset from (5.80). Then correct for the offset and perform frame synchronization and channel estimation based on training data, as outlined in Section 5.1.3 and Section 5.1.4.

5.2 Equalization of Frequency-Selective Channels

In this and subsequent sections, we generalize the development in Section 5.1 to frequency-selective fading channels. We focus specifically on equalization, assuming that channel estimation, frame synchronization, and frequency offset synchronization have been performed. We solve the channel estimation problem in Section 5.3 and the frame and frequency offset synchronization problems in Section 5.4. First we develop the discrete-time received signal model, including a frequency-selective channel and AWGN. Then we develop three approaches for linear equalization. The first approach is based on constructing an FIR filter that approximately inverts the effective channel. The second approach is to insert a special prefix in the transmitted signal to permit frequency-domain equalization at the receiver, in what is called SC-FDE. The third approach also uses a cyclic prefix but precodes the information in the frequency domain, in what is called OFDM modulation. As the equalization removes intersymbol interference, standard symbol detection follows the equalization operations.

5.2.1 Discrete-Time Model for Frequency-Selective Fading

In this section, we develop a received signal model for general frequency-selective channels, generalizing the results for a single path in Section 5.1.1. Assuming perfect synchronization, the received complex baseband signal after matched filtering but prior to sampling is

$$y(t) = g_{\text{rx}}(t) * h(t) * \sqrt{E_{\text{x}}} \sum_{m=-\infty}^{\infty} s[m] g_{\text{tx}}(t - mT) + g_{\text{rx}}(t) * v(t) \qquad (5.81)$$

$$= \sum_{m=-\infty}^{\infty} s[m] h_{\text{eff}}(t - mT) + g_{\text{rx}}(t) * v(t). \qquad (5.82)$$

Essentially, $y(t)$ takes the form of a complex pulse-amplitude modulated signal but where $g(t)$ is replaced by $h_{\text{eff}}(t) = g_{\text{rx}}(t) * g_{\text{tx}}(t) * \sqrt{E_{\text{x}}} h(t)$. Except in special cases, this new effective pulse is no longer a Nyquist pulse shape.

We now develop a sampled signal model. Let

$$h[n] = h_{\text{eff}}(nT) \qquad (5.83)$$

denote the sampled effective discrete-time channel. This channel combines the complex baseband equivalent model of the propagation channel, the transmit pulse-shaping filter, the receive pulse matched transmit filter, and the scaled transmit energy $\sqrt{E_{\text{x}}}$.

Sampling (5.82) at the symbol rate and with the effective discrete-time channel gives the received signal

$$y[n] = \sum_{\ell=-\infty}^{\infty} h[\ell]s[n-\ell] + v[n]. \qquad (5.84)$$

The main distortion is ISI, since every observation $y[n]$ is a linear combination of all the transmitted symbols through the convolution integral.

Example 5.10   To get some insight into the impact of intersymbol interference, suppose that $h[n] = \sqrt{E_x}\delta[n] + \sqrt{E_x}h_1\delta[n-1]$. Then

$$y[n] = \sqrt{E_x}s[n] + \sqrt{E_x}h_1s[n-1] + v[n]. \qquad (5.85)$$

The nth symbol $s[n]$ is subject to interference from $s[n-1]$, sent in the previous symbol period. Without correcting for this interference, detection performance will be bad. Treating ISI as noise,

$$\text{SINR} = \frac{E_x}{E_x|h_1|^2 + N_o}. \qquad (5.86)$$

For example, for SNR = 10dB = 10 and $|h_1|^2 = 1$, SINR = $10/(10+1) = 0.91$ or -0.4dB. Figure 5.16 shows the SINR as a function of $|h_1|^2$ for SNR = 10dB and SNR = 5dB.

To develop receiver signal processing algorithms, it is reasonable to treat $h[n]$ as causal and FIR. The causal assumption is reasonable because the propagation channel cannot predict the future. Furthermore, frame synchronization algorithms attempt to align, assuming a causal impulse response. The FIR assumption is reasonable because (a) there are no perfectly reflecting environments and (b) the signal energy decays as a function of distance between the transmitter and the receiver. Essentially, every time the signal reflects, some of the energy passes through the reflector or is scattered and thus it loses energy. As the signal propagates, it also loses power as it spreads in the environment. Multipaths that are weak will fall below the noise threshold. With the FIR assumption,

$$y[n] = \sum_{\ell=0}^{L} h[\ell]s[n-\ell] + v[n]. \qquad (5.87)$$

The channel is fully specified by the $L+1$ coefficients $\{h[\ell]\}_{\ell=0}^{L}$. The order of the channel, given by L, determines to a large extent the severity of the ISI. Flat fading corresponds to the special case of $L = 0$. We develop equalizers specifically for the system model in (5.87), assuming that the channel coefficients are perfectly known at the receiver.

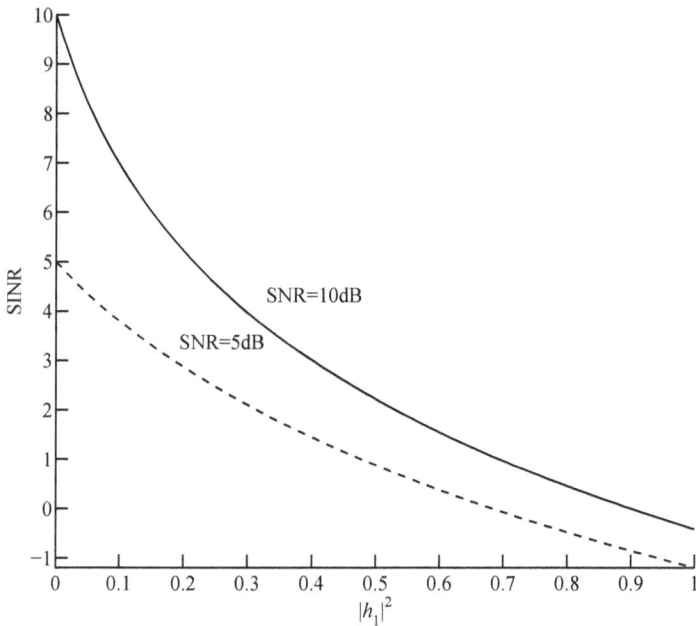

Figure 5.16 The SINR as a function of $|h_1|^2$ for the discrete-time channel $h[n] = \sqrt{E_x}\delta[n] + \sqrt{E_x}h_1\delta[n-1]$ in Example 5.10.

5.2.2 Linear Equalizers in the Time Domain

In this section, we develop an FIR equalizer to remove (approximately) the effects of ISI. We suppose that the channel coefficients are known perfectly; they can be estimated using training data via the method described in Section 5.3.1.

There are many strategies for equalization. One of the best approaches is to apply what is known as maximum likelihood sequence detection [348]. This is a generalization of the AWGN detection rule, which incorporates the memory in the channel due to $L > 0$. Unfortunately, this detector is complex to implement for channels with large L, though note that it is implemented in practical systems for modest values of L. Another approach is decision feedback equalization, where the contributions of detected symbols are subtracted, reducing the extent of the ISI [17, 30]. Combinations of these approaches are also possible.

In this section, we develop an FIR linear equalizer that operates on the time-domain signal. The goal of a linear equalizer is to find a filter that removes the effects of the channel. Let $\{f[\ell]\}_{\ell=0}^{L_f}$ be an FIR equalizer. The length of the equalizer is given by L_f. The equalizer is also associated with an equalizer delay n_d, which is a design parameter. Generally, allowing $n_d > 0$ improves performance. The best equalizers consider several values of n_d and choose the best one.

An ideal equalizer, neglecting noise, would satisfy

$$\sum_{\ell=0}^{L_f} f[\ell]h[n-\ell] = \delta[n-n_d] \tag{5.88}$$

for $n = 0, 1, \ldots, L_f + L$. Unfortunately, there are only $L_f + 1$ unknown parameters, so (5.88) can be satisfied exactly only in trivial cases like flat fading. This is not a surprise, as we know from DSP that the inverse of an FIR filter is an IIR filter. Alternatively, we pursue a least squares solution to (5.88).

We write (5.88) as a linear system and then find the least squares solution. The key idea is to write a set of linear equations and solve for the filter coefficients that ensure that (5.88) minimizes the squared error. First incorporate the channel coefficients into the $L_f + L + 1 \times L_f + 1$ convolution matrix:

$$\mathbf{H} = \begin{bmatrix} h[0] & 0 & \cdots & \cdots & 0 \\ h[1] & h[0] & 0 & \cdots & \vdots \\ \vdots & \ddots & & & h[0] \\ h[L] & & & & h[1] \\ 0 & h[L] & \cdots & & \vdots \\ \vdots & & & & h[L] \end{bmatrix}. \tag{5.89}$$

Then write the equalizer coefficients in a vector

$$\mathbf{f} = \begin{bmatrix} f[0] \\ f[1] \\ \vdots \\ f[L_f] \end{bmatrix} \tag{5.90}$$

and the desired response as the vector $\mathbf{e}_{n_d}$ with zeros everywhere except for a 1 in the $n_d + 1$ position. With these definitions, the desired linear system is

$$\mathbf{H}\mathbf{f}_{n_d} = \mathbf{e}_{n_d}. \tag{5.91}$$

The least squares solution is

$$\mathbf{f}_{\mathrm{LS},n_d} = (\mathbf{H}^*\mathbf{H})^{-1}\mathbf{H}^*\mathbf{e}_{n_d} \tag{5.92}$$

with squared error

$$J[n_d] = \mathbf{e}_{n_d}^*(\mathbf{I} - \mathbf{H}(\mathbf{H}^*\mathbf{H})^{-1}\mathbf{H}^*)\mathbf{e}_{n_d}. \tag{5.93}$$

The squared error can be minimized further by choosing n_d such that $J[n_d]$ is minimized. This is known as optimizing the equalizer delay.

Example 5.11 In this example, we compute the least squares optimal equalizer of length $L_{\mathrm{f}} = 6$ for a channel with impulse response $h[0] = 0.5$, $h[1] = \mathrm{j}/2$, and $h[2] = 0.4 \exp(\mathrm{j} * \pi/5)$. First we construct the convolution matrix

$$
\mathbf{H} =
\begin{bmatrix}
0.5 & 0 & 0 & 0 & 0 & 0 & 0 \\
\mathrm{j}/2 & 0.5 & 0 & 0 & 0 & 0 & 0 \\
0.4 & \mathrm{j}/2 & 0.5 & 0 & 0 & 0 & 0 \\
0 & 0.4 & \mathrm{j}/2 & 0.5 & 0 & 0 & 0 \\
0 & 0 & 0.4 & \mathrm{j}/2 & 0.5 & 0 & 0 \\
0 & 0 & 0 & 0.4 & \mathrm{j}/2 & 0.5 & 0 \\
0 & 0 & 0 & 0 & 0.4 & \mathrm{j}/2 & 0.5 \\
0 & 0 & 0 & 0 & 0 & 0.4 & \mathrm{j}/2 \\
0 & 0 & 0 & 0 & 0 & 0 & 0.4
\end{bmatrix}
\tag{5.94}
$$

and use it to compute (5.93) to determine the optimal equalizer length. As illustrated in Figure 5.17(a), the optimum delay occurs at $n_{\mathrm{d}} = 5$ with $J[5] = 0.0266$. The optimum equalizer derived from (5.92) is

$$
\mathbf{f}_{\mathrm{LS},5} =
\begin{bmatrix}
-0.1051 - \mathrm{j}0.1054 \\
-0.1848 + \mathrm{j}0.1665 \\
0.2100 + \mathrm{j}0.3607 \\
0.6065 - \mathrm{j}0.2521 \\
-0.2146 - \mathrm{j}0.9521 \\
0.4835 + \mathrm{j}0.0926 \\
-0.1907 + \mathrm{j}0.1905
\end{bmatrix}.
\tag{5.95}
$$

The equalized impulse response is plotted in Figure 5.17(b).

The matrix $\mathbf{H}$ has a special kind of structure. Notice that the diagonals are all constant. This is known as a Toeplitz matrix, sometimes called a filtering matrix. It shows up often when writing a convolution in matrix form. The structure in Toeplitz matrices also leads to many efficient algorithms for solving least squares equations, implementing adaptive solutions, and so on [172, 294]. This structure also means that $\mathbf{H}$ is full rank as long as at least one coefficient is nonzero. Therefore, the inverse in (5.92) exists.

The equalizer is applied to the sampled received signal to produce

$$
\widehat{s}[n - n_{\mathrm{d}}] = \sum_{\ell=0}^{L_{\mathrm{f}}} f_{n_{\mathrm{d}}}[\ell] y[n - \ell].
\tag{5.96}
$$

The delay n_{d} is known and can be corrected by advancing the output by the corresponding number of samples.

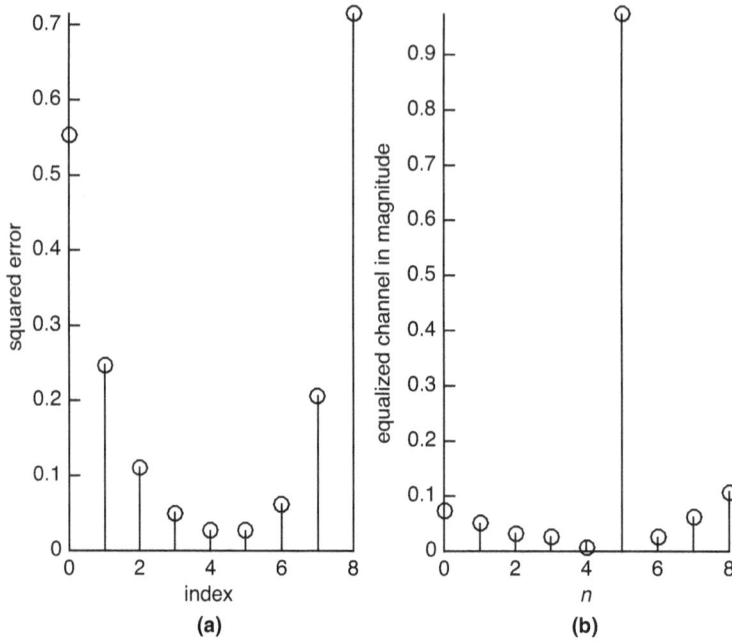

Figure 5.17 The squared error for the equalizer $J[n_{\mathrm{d}}]$ in (a) and the equalized channel $h[n] * f_{n_{\mathrm{d}}}[n]$ in (b), corresponding to the parameters in Example 5.11

An alternative to the least squares equalizer is the LMMSE equalizer. Let us rewrite (5.96) to place the equalizer coefficients into a single vector

$$\widehat{s}[n - n_{\mathrm{d}}] = \mathbf{f}_{n_{\mathrm{d}}}^{\mathrm{T}} \mathbf{y}[n] \tag{5.97}$$

where $\mathbf{y}^{\mathrm{T}}[n] = [y[n], y[n-1], \ldots, y[n - L_{\mathrm{f}}]^{\mathrm{T}}$,

$$\mathbf{y}[n] = \mathbf{H}^{\mathrm{T}} \mathbf{s}[n] + \mathbf{v}[n] \tag{5.98}$$

with $\mathbf{s}^{\mathrm{T}}[n] = [s[n], s[n-1], \ldots, s[n - L]]^{\mathrm{T}}$ and $\mathbf{H}$ as in (5.89). We seek the equalizer that minimizes the mean squared error

$$\mathbb{E}\left[\left| s[n - n_{\mathrm{d}}] - \mathbf{f}_{n_{\mathrm{d}}}^{\mathrm{T}} \mathbf{y}[n] \right|^2 \right]. \tag{5.99}$$

Assume that $s[n]$ is IID with zero mean and unit variance, $v[n]$ is IID with variance σ_v^2, and $s[n]$ and $v[n]$ are independent. As a result,

$$\mathbf{C_{yy}} = \mathbb{E}\left[\mathbf{y}[n] \mathbf{y}^*[n] \right] \tag{5.100}$$

$$= \mathbf{H}^{\mathrm{T}} \mathbf{H}^{\mathrm{c}} + \sigma_v^2 \mathbf{I} \tag{5.101}$$

and

$$\mathbf{C_{y}}_s = \mathbb{E}\left[\mathbf{y}[n]s^*[n - n_\mathrm{d}]\right] \tag{5.102}$$

$$= \mathbf{H}^\mathrm{T}\mathbf{e}_{n_\mathrm{d}}. \tag{5.103}$$

Then we can apply the results from Section 3.5.4 to derive

$$\mathbf{f}_{n_\mathrm{d},\mathrm{MMSE}} = \mathbf{C}_{\mathbf{yy}}^{-c}\mathbf{C}_{\mathbf{y}s}^{c} \tag{5.104}$$

$$= \left(\mathbf{H}^*\mathbf{H} + \sigma_v^2\mathbf{I}\right)^{-1}\mathbf{H}^*\mathbf{e}_{n_\mathrm{d}} \tag{5.105}$$

where the conjugate results from the use of conjugate transpose in the formulation of the equalizer in (3.459). The LMMSE equalizer gives an inverse that is regularized by the noise power σ_v^2, which is N_o if the only impairment is AWGN. This should improve performance at lower values of the SNR where equalization creates the most noise enhancement.

The asymptotic equalizer properties are interesting. As $\sigma_v^2 \to 0$, $\mathbf{f}_{n_\mathrm{d},\mathrm{MMSE}} \to \mathbf{f}_{\mathrm{LS},n_\mathrm{d}}$. In the absence of noise, the MMSE solution becomes the LS solution. As $\sigma_v^2 \to \infty$, $\mathbf{f}_{n_\mathrm{d},\mathrm{MMSE}} \to \frac{1}{\sigma_v^2}\mathbf{H}^*\mathbf{e}_{n_\mathrm{d}}$. This can be seen as a spatially matched filter.

A block diagram of linear equalization and channel estimation is illustrated in Figure 5.18. Note that the optimization over delay and correction for delay are included in the equalization computation, though they could be separated in additional blocks. Symbol synchronization is included in the diagram, despite the fact that linear equalization can correct for symbol timing errors through equalization. The reason is that SNR performance can nonetheless be improved by taking the best sample, especially when the pulse shape has excess bandwidth. An alternative is to implement a fractionally spaced equalizer, which operates on the signal prior to downsampling [128]. This is explored further in the problems at the end of the chapter.

A final comment on complexity. The length of the equalizer L_f is a design parameter that depends on L. The parameter L is the extent of the multipath in the channel and is determined by the bandwidth of the signal as well as the maximum delay spread derived from propagation channel measurements. The equalizer is an approximate FIR inverse of an FIR filter. As a consequence, performance will improve if L_f is large, assuming perfect channel knowledge. The complexity required per symbol, however, also grows with L_f. Thus there is a trade-off between choosing large L_f to have better equalizer performance and smaller L_f to have more efficient receiver implementation. A rule of thumb is to take L_f to be at least $4L$.

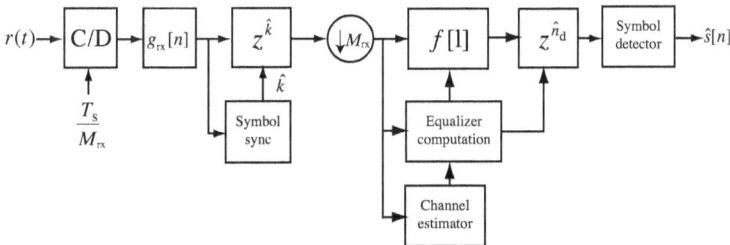

Figure 5.18 Receiver with channel estimation and linear equalization

5.2.3 Linear Equalization in the Frequency Domain with SC-FDE

Both the direct and the indirect equalizers require a convolution on the received signal to remove the effects of the channel. In practice this can be done with a direct implementation using the overlap-and-add or overlap-and-save methods for efficiently computing convolutions in the frequency domain. An alternative to FIR in the time domain is to perform equalization completely in the frequency domain. This has the advantage of allowing an ideal inverse of the channel to be computed. Application of frequency-domain equalization, though, requires additional mathematical structure in the transmitted waveform as we now explain.

In this section, we describe a technique known as SC-FDE [102]. At the transmitter, SC-FDE divides the symbols into blocks and adds redundancy in the form of a cyclic prefix. The receiver can exploit this extra information to permit equalization using the DFT. The result is an equalization strategy that is capable of perfectly equalizing the channel, in the absence of noise. SC-FDE is supported in IEEE 802.11ad, and a variation is used in the uplink of 3GPP LTE.

We now explain the motivation for working with the DFT and the key ideas behind the cyclic prefix. First, we explain why direct application of the DTFT is not feasible. Consider the received signal with intersymbol interference but without noise in (5.87). In the frequency domain

$$\mathsf{y}\left(e^{j2\pi f}\right) = \mathsf{h}\left(e^{j2\pi f}\right)\mathsf{s}\left(e^{j2\pi f}\right). \tag{5.106}$$

The ideal zero-forcing equalizer is

$$\mathsf{f}\left(e^{j2\pi f}\right) = \frac{1}{\mathsf{h}\left(e^{j2\pi f}\right)}. \tag{5.107}$$

Unfortunately, it is not possible to directly implement the ideal zero-forcing equalizer in the DTFT frequency domain. First of all, the equalizer does not exist at f for which $H\left(e^{j2\pi f}\right)$ is zero. This can be solved by using a pseudo-inverse rather than an inverse equalizer. Several more important problems occur as a by-product of the use of the DTFT. It is often not possible to compute the ideal DTFT in practice. For example, the whole $\{s[n]\} \to S\left(e^{j2\pi f}\right)$ is required, but typically only a few samples of $s[n]$ are available. Even when $\{s[n]\}$ is available, the DTFT may not even exist since the sum may not converge. Furthermore, it is not possible to observe over a long interval since $h[\ell]$ is time invariant only over a short window.

A solution to this problem is to use a specially designed $\{s[n]\}$ and leverage the principles of the discrete Fourier transform. A comprehensive discussion of the DFT and its properties is provided in Chapter 3. To review, recall that the DFT is a basis expansion for finite-length signals:

$$\text{Analysis} \quad \mathsf{x}[k] = \sum_{n=0}^{N-1} x[n]e^{-j\frac{2\pi}{N}kn} \quad k = 0, 1, ..., N-1 \tag{5.108}$$

$$\text{Synthesis} \quad x[n] = \frac{1}{N}\sum_{k=0}^{N-1} \mathsf{x}[k]e^{j\frac{2\pi}{N}kn} \quad n = 0, 1, ..., N-1. \tag{5.109}$$

The DFT can be computed efficiently with the FFT for N as a power of 2 and certain other special cases. Practical implementations of the DFT use the FFT. The key properties of the DFT are summarized in Table 3.5.

A distinguishing feature of the DFT is that multiplication in frequency corresponds to *circular* convolution in time. Consider two sequences $\{x_1[n]\}_{n=0}^{N-1}$ and $\{x_2[n]\}_{n=0}^{N-1}$. Let $\mathcal{F}_N(\cdot)$ denote the DFT operation on a length N signal, and let $\mathcal{F}_N^{-1}(\cdot)$ denote its inverse. With $\mathsf{x}_1[k] = \mathcal{F}_N(x_1[n])$ and $\mathsf{x}_2[k] = \mathcal{F}_N[x_2[n]]$, then

$$\mathcal{F}_N^{-1}(\mathsf{x}_1[k]\mathsf{x}_2[k]) = \sum_{m=0}^{N-1} x_1[m]x_2[((n-m))_N]. \tag{5.110}$$

Unfortunately, linear convolution, not circular convolution, is a good model for the effects of wireless propagation per (5.87).

It is possible to mimic the effects of circular convolution by modifying the transmitted signal with a suitably chosen guard interval. The most common choice is what is called a cyclic prefix, illustrated in Figure 5.19. To understand the need for a cyclic prefix, consider the circular convolution between a block of symbols of length N where $N > L$: $\{s[n]\}_{n=0}^{N-1}$ and the channel $\{h[\ell]\}_{\ell=0}^{L}$, which has been zero padded to have length N, that is, $h[n] = 0$ for $n \in [L+1, N-1]$. The output of the circular convolution is

$$y[n] = \sum_{\ell=0}^{N-1} h[\ell]s\left[((n-\ell))_N\right] \tag{5.111}$$

$$= \sum_{\ell=0}^{L} h[\ell]s\left[((n-\ell))_N\right] \tag{5.112}$$

$$= \begin{cases} \sum_{\ell=0}^{n} h[\ell]s[n-\ell] + \sum_{\ell=N+1}^{L} h[\ell]s[n+n-\ell] & 0 \le n < L \\ \sum_{\ell=0}^{L} h[\ell]s[n-\ell] & n \ge L. \end{cases} \tag{5.113}$$

The portion for $n \ge L$ looks like a linear convolution; the circular wrap-around occurs only for the first L samples.

Now we modify the transmitted sequence to obtain a circular convolution from the linear convolution introduced by the channel. Let $L_c \ge L$ be the length of the cyclic prefix. Form the signal $\{w[n]\}_{n=0}^{N+L_c-1}$ where the cyclic prefix is

$$w[n] = s[n+N-L_c] \quad n = 0, 1, \dots, L_c - 1 \tag{5.114}$$

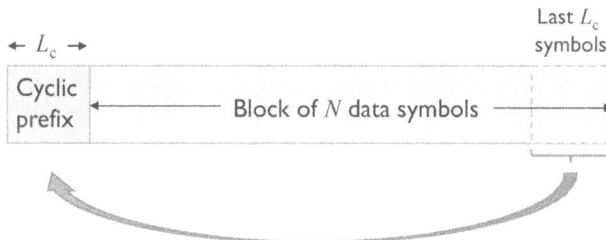

Figure 5.19 The cyclic prefix

and the data is

$$w[n] = s[n - L_c] \quad n = L_c, L_c + 1, \ldots, L_c + N - 1. \tag{5.115}$$

After convolution with an $L + 1$ tap channel, and neglecting noise for the derivation,

$$y[n] = \sum_{\ell=0}^{L} h[\ell] w[n - \ell]. \tag{5.116}$$

Now we neglect the first L_c terms of the convolution, which is called *discarding the cyclic prefix*, to form the new signal:

$$\bar{y}[n] = y[n + L_c] \quad n = 0, 1, \ldots, N - 1 \tag{5.117}$$

$$= \sum_{\ell=0}^{L} h[\ell] w[n + L_c - \ell]. \tag{5.118}$$

To see the effect, it is useful to evaluate $\bar{y}[n]$ for a few values of n:

$$\bar{y}[0] = \sum_{\ell=0}^{L} h[\ell] w[0 + L_c - \ell] \tag{5.119}$$

$$= h[0]w[L_c] + h[1]w[L_c - 1] + \cdots + h[\ell]w[L_c - \ell] \tag{5.120}$$

$$= h[0]s[0] + h[1]s[N - 1] + \cdots + h[\ell]s[N - L - 1] \tag{5.121}$$

$$= \sum_{\ell=0}^{L} h[\ell] s[((0 - \ell))_N] \tag{5.122}$$

$$\bar{y}[L - 1] = \sum_{\ell=0}^{L} h[\ell] w[L - 1 + L_c - \ell] \tag{5.123}$$

$$= h[0]w[L - 1 + L_c] + h[1]w[L - 1 + L_c - 1]$$
$$+ \cdots + h[\ell]w[L - 1 + L_c - \ell] \tag{5.124}$$

$$= h[0]s[L - 1] + h[1]s[L - 2] + \cdots + h[L - 1]s[0] + h[\ell]s[N - 1] \tag{5.125}$$

$$= \sum_{\ell=0}^{L} h[\ell] s[((L - 1 - \ell))_N] \tag{5.126}$$

$$\bar{y}[L] = \sum_{\ell=0}^{L} h[\ell] w[L + L_c - \ell] \tag{5.127}$$

$$= h[0]w[L + L_c] + h[1]w[L + L_c - 1] + \cdots + h[\ell]w[L + L_c - \ell] \tag{5.128}$$

$$= h[0]s[\ell] + h[1]s[\ell] + \cdots + h[\ell]s[0] \tag{5.129}$$

$$= \sum_{\ell=0}^{L} h[\ell] s[((L - \ell))_N] \tag{5.130}$$

$$= \sum_{\ell=0}^{L} h[\ell] s[L - \ell]. \tag{5.131}$$

For values of $n < L$, the cyclic prefix replicates the effects of the linear convolution observed in (5.113). For values of $N \geq L$, the circular convolution becomes a linear convolution, also as expected from (5.113) because $L < N$. In general, the truncated sequence satisfies

$$\bar{y}[n] = \sum_{\ell=0}^{L} h[\ell]s[((n-\ell))_N].$$

(5.132)

Therefore, thanks to the cyclic prefix, it is possible to implement frequency-domain equalization simply by computing $\mathsf{y}[k] = \mathcal{F}_N(\bar{y}[n])$, $\mathsf{s}[k] = \mathcal{F}_N(s[n])$, and then

$$\widehat{s}[n] = \mathcal{F}_N^{-1}\left(\frac{\mathsf{y}[k]}{\mathsf{h}[k]}\right)$$

(5.133)

$$= \mathcal{F}_N^{-1}\left(\frac{\mathcal{F}_N(\bar{y}[n])}{\mathcal{F}_N(h[n])}\right).$$

(5.134)

This is the key idea behind the SC-FDE equalizer.

The cyclic prefix also acts as a guard interval, separating the contributions of different blocks of data. To see this, note that $\{\bar{y}[n]\}_{n=0}^{N-1}$ depends only on symbols $\{s[n]\}_{n=0}^{N-1}$ and not symbols sent in previous blocks like $s[n]$ for $n < 0$ or $n > N$. Zero padding can alternatively be used as a guard interval, as explored in Example 5.12.

Example 5.12   Zero padding is an alternative to the cyclic prefix [236]. With zero padding, L_c zero values replace the cyclic prefix where

$$w[n] = 0 \quad n = 0, 1, \dots, L_c - 1.$$

(5.135)

The data is encoded as in (5.115) and $L \leq L_c$. We neglect noise for this problem.

- Show how zero padding enables successive decoding of $s[n]$ from $y[n]$. *Hint:* Start with $n = 0$ and show that $s[0]$ can be derived from $y[L_c]$. Let $\widehat{s}[0]$ denote the detected symbol. Then show how, by subtracting off $\widehat{s}[0]$, you can detect $\widehat{s}[1]$. Then assume it is true for a given n and show that it works for $n + 1$.

  Answer: To decode $s[0]$, we look at the expression of $y[L_c]$. After expanding and simplifying, because of the zeros, we obtain $y[L_c] = h[0]w[L_c] = h[0]s[0]$. Because $h[0]$ is known, we can decode $s[0]$ as

$$\widehat{s}[0] = \frac{y[L_c]}{h[0]}.$$

(5.136)

  To decode $s[1]$, we use $y[L_c + 1]$ and the previously detected value of $\widehat{s}[0]$. Because $y[L_c+1] = h[0]w[L_c+1] + h[1]w[L_c] = h[0]s[1] + h[1]\widehat{s}[0]$, and assuming the detected symbol is correct,

$$\widehat{s}[1] = \frac{y[L_c + 1] - h[1]\widehat{s}[0]}{h[0]}.$$

(5.137)

Finally, assume that $\{\widehat{s}[n]\}_{n=0}^{k-1}$ has been decoded; then

$$y[L_c + k] = \sum_{\ell=0}^{L} h[\ell]w[L_c + k - \ell] \tag{5.138}$$

$$= h[0]s[k] + \sum_{\ell=1}^{L} h[\ell]\widehat{s}[k - \ell]. \tag{5.139}$$

We thus can decode $s[k]$ as follows:

$$\widehat{s}[k] = \frac{y[L_c + k] - \sum_{\ell=1}^{L} h[\ell]\widehat{s}[k - \ell]}{h[0]}. \tag{5.140}$$

The main drawback of this approach is that it suffers from error propagation: a symbol error in $\widehat{s}[k]$ affects the detection of all symbols after k.

- Consider the following alternative to discarding the cyclic prefix:

$$\widetilde{y}[n] = y[n + L_c] + y[n + N + L_c] \quad \text{for} \quad n = 0, 1, \ldots, L - 1 \tag{5.141}$$
$$\widetilde{y}[n] = y[n + L_c] \quad \text{for} \quad n = L, L + 1, \ldots, N - 1. \tag{5.142}$$

Show how this structure also permits frequency domain equalization (neglect noise for the derivation).

Answer: For $n = 0, 1, \ldots, L - 1$:

$$\widetilde{y}[n] = y[n + L_c] + y[n + N + L_c] \tag{5.143}$$

$$= \sum_{\ell=0}^{n} h[\ell]w[n + L_c - \ell] + \sum_{\ell=n+1}^{L} h[\ell]\cancel{w[n + L_c - \ell]}$$

$$+ \sum_{\ell=0}^{n} h[\ell]\cancel{w[n + N + L_c - \ell]} + \sum_{\ell=n+1}^{L} h[\ell]w[n + N + L_c - \ell] \tag{5.144}$$

$$= \sum_{\ell=0}^{n} h[\ell]s[n - \ell] + \sum_{\ell=n+1}^{L} h[\ell]s[n + N - \ell] \tag{5.145}$$

$$= \sum_{\ell=0}^{L} h[\ell]s[((n - \ell))_N] \tag{5.146}$$

where the cancellation is due to the cyclic prefix. For $n = L, L - 1, \ldots, N - 1$:

$$\widetilde{y}[n] = \sum_{\ell=0}^{L} h[\ell]w[n + L_c - \ell] \tag{5.147}$$

$$= \sum_{\ell=0}^{L} h[\ell]s[n - \ell] \tag{5.148}$$

$$= \sum_{\ell=0}^{L} h[\ell]s[((n - \ell))_N]. \tag{5.149}$$

Therefore, $\widetilde{y}[n]$ is the same as (5.132), and the same equalization as SC-FDE can be applied, with a little performance penalty due to the double addition of noise. Zero padding has been used in UWB (ultra-wideband) [197] to make multiband OFDM easier to implement, and in millimeter-wave SC-FDE systems [82], where it allows for reconfiguration of the RF parameters without signal distortion.

Noise has an adverse effect on the performance of the SC-FDE equalizer. Now we examine what happens to the effective noise variance after equalization. In the presence of noise, and with perfect channel state information,

$$\widehat{s}[n] = s[n] + \mathcal{F}_N^{-1}\left(\mathsf{v}[k]/\mathsf{h}[k]\right). \tag{5.150}$$

The second term is augmented during the equalization process in what is called noise enhancement. Let $\widetilde{v}[n]$ be the enhanced noise component. It remains zero mean. To compute its variance, we expand the enhanced noise as

$$\widetilde{v}[n] = \frac{1}{N}\sum_{n=0}^{N-1}\frac{\sum_{m=0}^{N-1}v[m]e^{-\mathrm{j}\frac{2\pi}{N}km}}{\mathsf{h}[k]}e^{\mathrm{j}\frac{2\pi}{N}kn} \tag{5.151}$$

and then compute

$$\mathbb{E}[\widetilde{v}[n]\widetilde{v}^*[n]] = \mathbb{E}\left[\frac{1}{N}\sum_{k_1=0}^{N-1}\frac{\sum_{m_1=0}^{N-1}v[m_1]e^{-\mathrm{j}\frac{2\pi}{N}k_1m_1}}{\mathsf{h}[k_1]}e^{\mathrm{j}\frac{2\pi}{N}k_1n_1}\right.$$

$$\left.\times\frac{1}{N}\sum_{k_2=0}^{N-1}\frac{\sum_{m_2=0}^{N-1}v^*[m_2]e^{\mathrm{j}\frac{2\pi}{N}k_2m_2}}{\mathsf{h}^*[k_2]}e^{-\mathrm{j}\frac{2\pi}{N}k_2n_2}\right] \tag{5.152}$$

$$= \frac{\sigma_v^2}{N^2}\sum_{k_1=0}^{N-1}\sum_{k_2=0}^{N-1}\sum_{m_1=0}^{N-1}\sum_{m_2=0}^{N-1}\frac{\delta[m_1-m_2]}{\mathsf{h}[k_1]\mathsf{h}^2[k_2]}e^{-\mathrm{j}\frac{2\pi}{N}k_1m_1}e^{\mathrm{j}\frac{2\pi}{N}k_2m_2}e^{\mathrm{j}\frac{2\pi}{N}k_1n_1}e^{-\mathrm{j}\frac{2\pi}{N}k_2n_2} \tag{5.153}$$

$$= \frac{\sigma_v^2}{N^2}\sum_{k_1=0}^{N-1}\sum_{k_2=0}^{N-1}\sum_{m=0}^{N-1}\frac{1}{\mathsf{h}[k_1]\mathsf{h}^*[k_2]}e^{-\mathrm{j}\frac{2\pi}{N}(k_1-k_2)m}e^{\mathrm{j}\frac{2\pi}{N}k_1n_1}e^{-\mathrm{j}\frac{2\pi}{N}k_2n_2} \tag{5.154}$$

$$= \frac{\sigma_v^2}{N}\sum_{k=0}^{N-1}\frac{1}{|\mathsf{h}[k]|^2}e^{\mathrm{j}\frac{2\pi}{N}kn}e^{-\mathrm{j}\frac{2\pi}{N}kn} \tag{5.155}$$

$$= \frac{\sigma_v^2}{N}\sum_{k=0}^{N-1}\frac{1}{|\mathsf{h}[k]|^2}, \tag{5.156}$$

where the results follow from the IID property of $v[n]$ and repeated applications of the orthogonality of discrete-time complex exponentials. The main conclusion is that we can model (5.150) as

$$\widehat{s}[n] = s[n] + \nu[n] \tag{5.157}$$

where $\nu[n]$ is AWGN with variance given in (5.156), which is the geometric mean of the inverse of the channel in the frequency domain. We contrast this result with that obtained using OFDM in the next section.

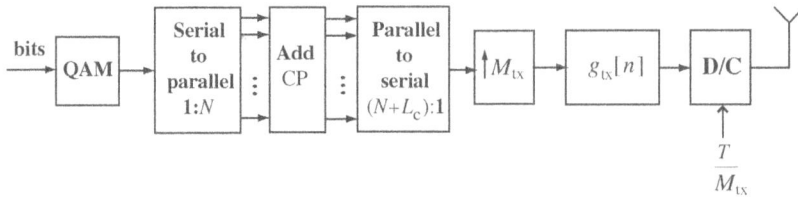

Figure 5.20 QAM transmitter for a single-carrier frequency-domain equalizer, with CP denoting cyclic prefix

An implementation of a QAM system with frequency-domain equalization is illustrated in Figure 5.20. The operation of collecting the input bits into groups of N symbols is given by the serial-to-parallel converter. A cyclic prefix takes the N input symbols, copies the last L_c symbols to the beginning, and outputs $N + L_c$ symbols. The resulting symbols are then converted from parallel to serial, followed by upsampling and the usual matched filtering. The serial-to-parallel and parallel-to-serial blocks can be implemented from a DSP perspective using simple filter banks.

Compared with linear equalization, SC-FDE has several advantages. The channel inverse can be done perfectly since the inverse is exact in the DFT domain (assuming, of course, that none of the $h[k]$ are zero), and the time-domain equalizer is an approximate inverse. SC-FDE also works regardless of the value of L as long as $L_c \geq L$. The equalizer complexity is fixed and is determined by the complexity of the FFT operation, proportional to $N \log_2 N$. The complexity of the time-domain equalizer is a function of K and generally grows with L (assuming that K grows with L), unless it is itself implemented in the frequency domain. As a general rule of thumb it becomes more efficient to equalize in the frequency domain for L around 5.

The main parameters to select in SC-FDE are N and L_c. To minimize complexity it makes sense to take N to be small. The amount of overhead, though, is $L_c/(N + L_c)$. Consequently, taking N to be large reduces the system overhead incurred by redundancy in the cyclic prefix. Too large an N, however, may mean that the channel varies over the N symbols, violating the LTI assumption. In general, L_c is selected to be large enough that $L < L_c$ for most channel realizations, throughout the use cases of the wireless system. For example, for a personal area network application, indoor channel measurements may be used to establish the power delay profile and other channel statistics from which a maximum value of L_c can be derived.

5.2.4   Linear Equalization in the Frequency Domain with OFDM

The SC-FDE receiver illustrated in Figure 5.21 performs a DFT on a portion of the received signal, equalizes with the DFT of the channel, and takes the IDFT (inverse discrete Fourier transform) to form the equalized sequence $\hat{s}[n]$. This offloads the primary equalization operations to the receiver. In some cases, however, it is of interest to have a more balanced load between transmitter and receiver. A solution is to shift the IDFT to the transmitter. This results in a framework known as multicarrier modulation or OFDM modulation [63, 72].

Several wireless standards have adopted OFDM modulation, including wireless LAN standards like IEEE 802.11a/b/n/ac/ad [279, 290], fourth-generation cellular systems like 3GPP LTE [299, 16, 93, 253], digital audio broadcasting (DAB) [333], and digital video broadcasting (DVB) [275, 104].

In this section, we describe the key operations of OFDM at the transmitter as illustrated in Figure 5.22 and the receiver as illustrated in Figure 5.23. We present OFDM from the perspective of having already derived SC-FDE, though historically OFDM was developed several decades prior to SC-FDE. We conclude with a discussion of OFDM versus SC-FDE versus linear equalization techniques.

The key idea of OFDM, from the perspective of SC-FDE, is to insert the IDFT after the first serial-to-parallel converter in Figure 5.22. Given $\{s[n]\}_{n=0}^{N-1}$ and a cyclic prefix of length L_{c}, the transmitter produces the sequence

$$w[n] = \frac{1}{N} \sum_{m=0}^{N-1} s[m] e^{j2\pi \frac{m(n-L_{\mathrm{c}})}{N}} \quad n = 0, ..., N + L_{\mathrm{c}} - 1, \qquad (5.158)$$

which is passed to the transmit pulse-shaping filter. The signal $w[n]$ satisfies $w[n] = w[n+N]$ for $n = 0, 1, \ldots, L_{\mathrm{c}} - 1$; therefore, it has a cyclic prefix. The samples $\{w[L_{\mathrm{c}}],$

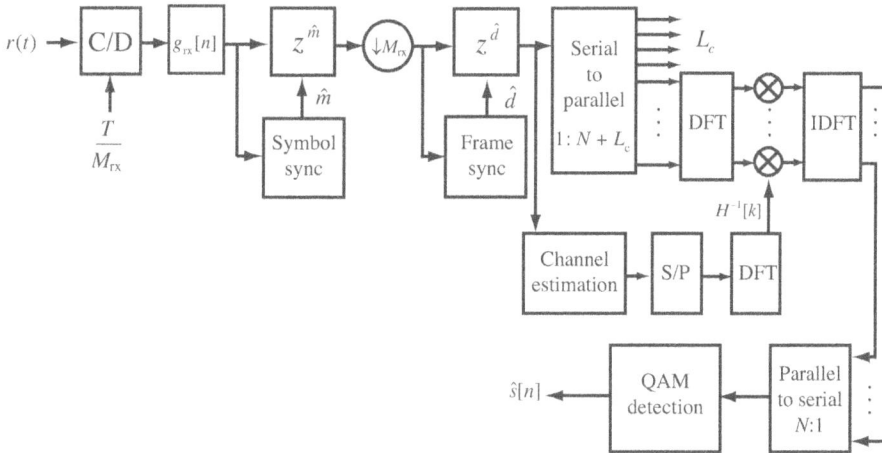

Figure 5.21 Receiver with cyclic prefix removal and a frequency-domain equalizer

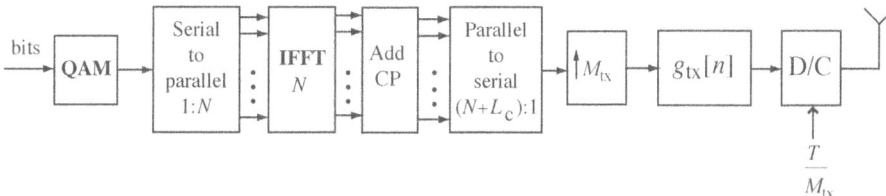

Figure 5.22 Block diagram of an OFDM transmitter. Often the upsampling and digital pulse shaping are omitted when rectangular pulse shapes are used. The inverse DFT is implemented using an inverse fast Fourier transform (IFFT).

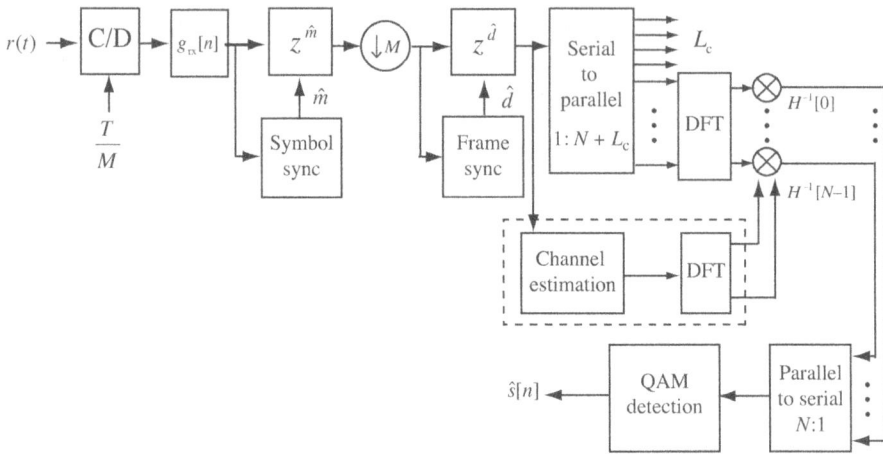

Figure 5.23 Block diagram of an OFDM receiver. Normally the matched filtering and symbol synchronization functions are omitted.

$w[L_c + 1], \ldots, w[N + L_c - 1]\}$ correspond to the IDFT of $\{s[n]\}_{n=0}^{N-1}$. Unlike the SC-FDE case, the transmitted symbols can be considered to originate in the frequency domain. We do not use the frequency-domain notation for $s[n]$ for consistency with the signal model.

We present the transmitter structure in Figure 5.22 to make the connection with SC-FDE clear. In OFDM, though, it is common to use rectangular pulse shaping where

$$g_{tx}(t) = \frac{1}{NT}\text{rect}\left(\frac{1}{NT}\right). \tag{5.159}$$

This function is not bandlimited, so it cannot strictly speaking be implemented using the upsampling followed by digital pulse shaping as shown in Figure 5.22. Instead, it is common to simply use the "stair-step" response of the digital-to-analog converter to perform the pulse shaping. This results in a signal at the output of the DAC that takes the form

$$w(t) = \frac{1}{N}\sum_{m=-N/2}^{N/2-1} s[m]e^{j2\pi\frac{m(t-L_cT)}{TN}} \quad \text{for} \quad t \in [0, (N + L_c)T]. \tag{5.160}$$

This interpretation shows how symbol $s[m]$ rides on continuous-time carrier $\exp(j2\pi t m/NT)$ at baseband with a carrier of frequency $1/NT$, which is also called the subcarrier bandwidth. This is one reason that OFDM is also called multicarrier modulation. We write the summation as shown in (5.160) to make it clear that frequencies around $\ldots, N-3, N-2, N-1, 0, 1, 2, \ldots$ are low frequencies whereas those around $N/2$ are high frequencies. Subcarrier $n = 0$ is known as the DC subcarrier, which is often assigned a zero symbol to avoid DC offset issues. The subcarriers near $N/2$ are often assigned zero symbols to facilitate spectral shaping.

Now we show how OFDM works. Consider the received signal as in (5.116). Discard the first L_c samples to form

$$\bar{y}[n] = y[n + L_c] \quad n = 0, 1, \ldots, N - 1. \tag{5.161}$$

Inserting (5.158) for $w[n]$ and interchanging summations gives

$$\bar{y}[n] = \sum_{\ell=0}^{L} h[\ell] w[n + L_c - \ell] \tag{5.162}$$

$$= \frac{1}{N} \sum_{\ell=0}^{L} h[\ell] \sum_{m=0}^{N-1} s[m] e^{j2\pi \frac{m(n+L_c-L_c-l)}{N}} \tag{5.163}$$

$$= \frac{1}{N} \sum_{\ell=0}^{L} h[\ell] \sum_{m=0}^{N-1} s[m] e^{j2\pi \frac{mn}{N}} e^{-j2\pi \frac{m\ell}{N}} \tag{5.164}$$

$$= \frac{1}{N} \sum_{m=0}^{N-1} \underbrace{\sum_{\ell=0}^{L} h[\ell] e^{-j2\pi \frac{m\ell}{N}}}_{\mathsf{h}[m]} s[m] e^{j2\pi \frac{mn}{N}} \tag{5.165}$$

$$= \mathcal{F}_N^{-1} \left(\mathsf{h}[m] s[m] \right). \tag{5.166}$$

Therefore, taking the DFT gives

$$\mathsf{y}[n] = \mathsf{h}[n] s[n] \tag{5.167}$$

for $n = 0, 1, \ldots, N-1$, and equalization simply involves multiplying by $\mathsf{h}^{-1}[n]$. Low SNR performance could be improved by multiplying by the LMMSE equalizer $(|\mathsf{h}[n]|^2 + \sigma_v^2)^{-1}\mathsf{h}^*[n]$ instead of by $\mathsf{h}^{-1}[n]$. In terms of time-domain quantities,

$$\hat{s}[n] = \frac{\mathcal{F}_N(\bar{y}[n])}{\mathcal{F}_N(h[n])}. \tag{5.168}$$

The effective channel experienced by $s[n]$ is $\mathsf{h}[n]$, which is a flat-fading channel. OFDM effectively converts a problem of equalizing a frequency-selective channel into that of equalizing a set of parallel flat-fading channels. Equalization thus simplifies a great deal versus time-domain linear equalization.

The terminology in OFDM systems is slightly different from that in single-carrier systems. Usually in OFDM, the collection of samples including the cyclic prefix $\{w[n]\}_{n=0}^{N-1}$ is called an *OFDM symbol*. The constituent symbols $\{s[n]\}_{n=0}^{N-1}$ are called subsymbols. The OFDM symbol period is $(N + L_c)T$, and T is called the sample period. The guard interval, or cyclic prefix duration, is $L_c T$. The subcarrier spacing is $1/(NT)$ and is the spacing between adjacent subcarriers as measured on a spectrum analyzer. The passband bandwidth is $1/T$, assuming the use of a sinc pulse-shaping filter (which is not common; a rectangular pulse shape is used along with zeroing certain subcarriers).

There are many trade-offs associated with selecting different parameters. Making N large while L_c is fixed reduces the fraction of overhead $N/(N+L_c)$ due to the cyclic prefix. A larger N, though, means a longer block length and shorter subcarrier spacing, increasing the impact of time variation in the channel, Doppler, and residual carrier frequency offset. Complexity also increases with larger values as the complexity of processing per subcarrier grows with $\log_2 N$.

Example 5.13   Consider an OFDM system where the OFDM symbol period is $3.2\mu s$, the cyclic prefix has length $L_c = 64$, and the number of subcarriers is $N = 256$. Find the sample period, the passband bandwidth (assuming that a sinc pulse-shaping filter is used), the subcarrier spacing, and the guard interval.

Answer: The sample period T satisfies the relation $T(256 + 64) = 3.2\mu s$, so the sample period is $T = 10$ns. Then, the passband bandwidth is $1/T = 100$MHz. Also, the subcarrier spacing is $1/(NT) = 390.625$kHz. Finally, the guard interval is $L_c T = 640$ns.

Noise impacts OFDM differently than SC-FDE. Now we examine what happens to the effective noise variance after equalization. In the presence of noise, and with perfect channel state information,

$$\widehat{s}[n] = s[n] + \frac{\mathsf{v}[n]}{\mathsf{h}[n]} \tag{5.169}$$

where $\mathsf{v}[n] = \mathcal{F}_N(v[n])$. Because linear combinations of Gaussian random variables are Gaussian, and the DFT is an orthogonal transformation, $\mathsf{v}[n]$ remains Gaussian with zero mean and variance σ_v^2. Therefore, $\mathsf{v}[n]/\mathsf{h}[n]$ is AWGN with zero mean and variance $\sigma_v^2/|\mathsf{h}[n]|^2$. Unlike SC-FDE, the noise enhancement varies with different subcarriers. When $\mathsf{h}[n]$ is small for a particular value of n (close to a null in the spectrum), substantial noise enhancement is created. With SC-FDE there is also noise enhancement, but each detected symbol sees the same effective noise variance as in (5.156). With coding and interleaving, the error rate differences between SC-FDE and OFDM are marginal, unless other impairments like low-resolution DACs and ADCs or nonlinearities are included in the comparison (see, for example, [291, 102, 300, 192, 277]) when SC-FDE has a slight edge.

OFDM is in general more sensitive to RF impairments compared with SC-FDE and standard complex pulse-amplitude modulated signals. It is sensitive to nonlinearities because the ratio between the peak of the OFDM signal and its average value (called the peak-to-average power ratio) is higher in an OFDM system compared with a standard complex pulse-amplitude modulated signal. The reason is that the IDFT operation at the transmitter makes the signal more likely to have all peaks. Signal processing techniques can be used to mitigate some of the differences [166]. OFDM signals are also more sensitive to phase noise [264], gain and phase imbalance [254], and carrier frequency offset [75].

The OFDM waveform offers additional degrees of flexibility not found in SC-FDE. For example, the information rate can be adjusted to current channel conditions based on the frequency selectivity of the channel by changing the modulation and coding on different subcarriers. Spectral shaping is possible by zeroing certain symbols, as already described. Different users can even be allocated to subcarriers or groups of subcarriers in what is called orthogonal frequency-division multiple access (OFDMA). Many systems like IEEE 802.11a/g/n/ac use OFDM exclusively for transmission and reception. 3GPP LTE Advanced uses OFDM on the downlink and a variation of SC-FDE on the uplink where power backoff is more critical. IEEE 802.11ad supports SC-FDE as a mandatory mode and OFDM in a higher-rate optional mode. Despite their differences, both

OFDM and SC-FDE maintain an important advantage of time-domain linear equalization: the equalizer complexity does not scale with L, as long as the cyclic prefix is long enough. Going forward, OFDM and SC-FDE are likely to continue to see wide commercial deployment.

5.3 Estimating Frequency-Selective Channels

When developing algorithms for equalization, we assumed that the channel state information $\{h[\ell]\}_{\ell=0}^{L}$ was known perfectly at the receiver. This is commonly known as genie-aided channel state information and is useful for developing analysis of systems. In practice, the receiver needs to estimate the channel coefficients as engineering an all-knowing genie has proved impossible. In this section, we describe one method for estimating the channel in the time domain, and another for estimating the channel in the frequency domain. We also describe an approach for direct channel equalization in the time domain, where the coefficients of the equalizer are estimated from the training data instead of first estimating the channel and then computing the inverse. All the proposed methods make use of least squares.

5.3.1 Least Squares Channel Estimation in the Time Domain

In this section, we formulate the channel estimation problem in the time domain, making use of a known training sequence. The idea is to write a linear system of equations where the channel convolves only the known training data. From this set of equations, the least squares solution follows directly.

Suppose as in the frame synchronization case that $\{t[n]\}_{n=0}^{N_{\text{tr}}-1}$ is a known training sequence and that $s[n] = t[n]$ for $n = 0, 1, \ldots, N_{\text{tr}} - 1$. Consider the received signal in (5.87). We have to write $y[n]$ only in terms of $t[n]$. The first few samples depend on symbols sent prior to the training data. For example,

$$y[0] = \sum_{\ell=0}^{L} h[\ell]s[n-\ell] + v[n] \tag{5.170}$$

$$= h[0]t[0] + \underbrace{h[1]s[-1] + h[2]s[-2] + \ldots + h[L]s[-L]}_{\text{includes unknown symbols}} + v[n]. \tag{5.171}$$

Since prior symbols are unknown (they could belong to a previous packet or a message sent to another user, or they could even be zero, for example), they should not be included in the formulation. Taking $n \in [L, N_{\text{tr}} - 1]$, though, gives

$$y[n] = \sum_{\ell=0}^{L} h[\ell]t[n-\ell] + v[n], \tag{5.172}$$

which is only a function of the unknown training data. We use these samples to form our channel estimator.

Collecting all the known samples together,

$$
\begin{bmatrix} y[L] \\ y[L+1] \\ \vdots \\ y[N_{\mathrm{tr}}-1] \end{bmatrix}
$$

$$
= \begin{bmatrix} t[L] & t[L-1] & \cdots & t[0] \\ t[L+1] & t[L] & \cdots & t[1] \\ \vdots & \ddots & \ddots & \vdots \\ t[N_{\mathrm{tr}}-1] & t[N_{\mathrm{t}}-2] & \cdots & t[N_{\mathrm{tr}}-1-L] \end{bmatrix} \begin{bmatrix} h[0] \\ h[1] \\ \vdots \\ h[L] \end{bmatrix} + \begin{bmatrix} v[L] \\ v[L+1] \\ \vdots \\ v[N_{\mathrm{tr}}-1] \end{bmatrix}, \qquad (5.173)
$$

which is simply written in matrix form as

$$
\mathbf{y} = \mathbf{T}\mathbf{h} + \mathbf{v}. \qquad (5.174)
$$

As a result, this problem takes the form of a linear parameter estimation problem, as reviewed in Section 3.5.

The least squares channel estimate, which is also the maximum likelihood estimate since the noise is AWGN, is given by

$$
\widehat{\mathbf{h}}_{\mathrm{LS}} = (\mathbf{T}^*\mathbf{T})^{-1}\mathbf{T}^*\mathbf{y}, \qquad (5.175)
$$

assuming that the Toeplitz training matrix $\mathbf{T}$ is invertible. Notice that the product $(\mathbf{T}^*\mathbf{T})^{-1}\mathbf{T}^*$ can be computed offline ahead of time, so the actual complexity is simply a matrix multiplication.

To be invertible, the training matrix must be square or tall, which requires that

$$
N_{\mathrm{tr}} - L \geq L + 1 \to N_{\mathrm{tr}} \geq 2L + 1. \qquad (5.176)
$$

Generally, choosing N_{tr} much larger than $L + 1$ (the length of the channel) gives better performance. The full-rank condition can be guaranteed by ensuring that the training sequence is persistently exciting. Basically this means that it looks random enough. A typical design objective is to find a sequence such that $\mathbf{T}^*\mathbf{T}$ is (nearly) a scaled identity matrix $\mathbf{I}$. This ensures that the noise remains nearly IID among other properties. Random training sequences with sufficient length usually perform well whereas $t[n] = 1$ will fail. Training sequences with good correlation properties generally satisfy this requirement, because the entries of $\mathbf{T}^*\mathbf{T}$ are different (partial) correlations of $t[n]$.

Special training sequences can be selected from those known to have good correlation properties. One such design uses a cyclically prefixed training sequence. Let $\{p[n]\}_{n=0}^{N_{\mathrm{p}}-1}$ be a sequence with good periodic correlation properties. This means that the periodic correlation $R_p[k] = \sum_{n=0}^{N_{\mathrm{p}}-1} p[n]p^*[((n+k))_{N_{\mathrm{p}}}]$ satisfies the property that $|R_p[k]|$ is small or zero for $k > 0$. Construct the training sequence $\{t[n]\}_{n=0}^{N_{\mathrm{tr}}-1}$ by prefixing $\{p[n]\}_{n=0}^{N_{\mathrm{p}}-1}$

with $\{p[n]\}_{n=N_\mathrm{p}-L-1}^{N_\mathrm{p}-1}$, which gives an $N_\mathrm{tr} = N_\mathrm{p} + L$ training sequence. With this construction

$$\mathbf{T} = \begin{bmatrix} p[0] & p[N_\mathrm{p}-1] & \cdots & p[N_\mathrm{p}-L] \\ p[1] & p[0] & \cdots & p[N_\mathrm{p}-L+1] \\ \vdots & \ddots & \ddots & \vdots \\ p[N_\mathrm{p}-1] & p[N_\mathrm{p}-2] & \cdots & p[N_\mathrm{p}-L-1] \end{bmatrix}. \tag{5.177}$$

Then $[\mathbf{T}^*\mathbf{T}]_{k,\ell} = R_p[k - \ell]$ contains lags of the periodic correlation function. Therefore, sequences with good periodic autocorrelation properties should work well when cyclically prefixed.

In the following examples, we present designs of $\mathbf{T}$ based on sequences with perfect periodic autocorrelation. This results in $\mathbf{T}^*\mathbf{T} = N_\mathrm{p}\mathbf{I}$ and a simplified least squares estimate of $\widehat{\mathbf{h}}_\mathrm{LS} = \mathbf{T}^*\mathbf{y}(N_\mathrm{p})^{-1}$. We consider the Zadoff-Chu sequences [70, 117] in Example 5.14 and Frank sequences [118] in Example 5.15. These designs can be further generalized to families of sequences with good cross-correlations as well; see, for example, the Popovic sequences in [267].

Example 5.14 Zadoff-Chu sequences are length N_p sequences with perfect periodic autocorrelation [70, 117], which means that $R_p[k] = 0$ for $k \neq 0$. They have the form

$$p[n] = e^{\mathrm{j}\frac{M\pi n^2}{N_\mathrm{p}}} \quad \text{for } N_\mathrm{p} \text{ even} \tag{5.178}$$

$$p[n] = e^{\mathrm{j}\frac{M\pi (n)(n+1)}{N_\mathrm{p}}} \quad \text{for } N_\mathrm{p} \text{ odd} \tag{5.179}$$

where M is an integer that is relatively coprime with N_p, which could be as small as 1. Chu sequences are drawn from the N_p-PSK constellation and have length N_p.

Find a Chu sequence of length $N_\mathrm{p} = 16$.

Answer: We need to find an M that is coprime with N_p. Since 16 has divisors of 2, 4, and 8, we can select any other number. For example, choosing $M = 3$ gives

$$p[n] = e^{\mathrm{j}\frac{3\pi n^2}{16}} \tag{5.180}$$

which gives the sequence

$$\{p[n]\}_{n=0}^{15} = \left\{ 1, e^{\mathrm{j}\frac{3\pi}{16}}, e^{\mathrm{j}\frac{3\pi}{4}}, e^{-\mathrm{j}\frac{5\pi}{16}}, -1, e^{\mathrm{j}\frac{11\pi}{16}}, e^{\mathrm{j}\frac{3\pi}{4}}, e^{-\mathrm{j}\frac{3\pi}{16}}, 1, \ldots \right.$$
$$\left. e^{-\frac{3\pi}{16}}, e^{\mathrm{j}\frac{3\pi}{4}}, e^{\mathrm{j}\frac{11\pi}{16}}, -1, e^{-\mathrm{j}\frac{5\pi}{16}}, e^{\mathrm{j}\frac{3\pi}{4}}, e^{\mathrm{j}\frac{3\pi}{16}} \right\}. \tag{5.181}$$

With this choice of training, it follows that $\mathbf{T}^*\mathbf{T} = N_\mathrm{p}\mathbf{I}$.

Example 5.15 The Frank sequence [118] gives another construction of sequences with perfect periodic correlation, which uses a smaller alphabet compared to the Zadoff-Chu sequences. For $n = mq + k$,

$$p[mq + k] = e^{\mathrm{j}2\pi \frac{rkm}{q}} \quad 0 \le k, m < q \tag{5.182}$$

where r must be coprime with q, q is any positive integer, and $n \in [0, q^2 - 1]$. The length of the Frank sequence is then $N_p = q^2$ and comes from a q-PSK alphabet.

Find a length 16 Frank sequence.

Answer: Since the length of the Frank sequence is $N_p = q^2$, we must take $q = 4$. The only viable choice of r is 3. With this choice

$$p[m4 + k] = e^{j2\pi \frac{3km}{4}} \tag{5.183}$$

for $m \in [0, 3]$ and $k \in [0, 3]$. This gives the sequence

$$\{p[n]\}_{n=0}^{15} = \{1, 1, 1, 1, 1, -j, -1, j, 1, -1, 1, -1, 1, j, -1, -j\}. \tag{5.184}$$

The Frank sequence also satisfies **T** but does so with a smaller constellation size, in this case 4-PSK. Note that a 4-QAM signal can be obtained by rotating each entry by $\exp(j\pi/4)$.

A variation of the cyclic prefixed training sequence design uses both a prefix and a postfix. This approach was adopted in the GSM cellular standard. In this case, a sequence with good periodic correlation properties $\{p[n]\}_{n=0}^{N_p-1}$ is prefixed by L samples $\{p[n]\}_{n=N_p-L-1}^{N_p-1}$ and postfixed by the samples $\{p[n]\}_{n=0}^{L-1}$. This results in a sequence with length $N_{tr} + 2L$. The motivation for this design is that

$$R_{pt}[m] = \sum_{n=0}^{N_p-1} p[n]t^*[n+m] = R_p[m] \quad \text{for } m \in [-L+1, L-1]. \tag{5.185}$$

With perfect periodic correlation, the cross-correlation between $p[n]$ and $t[n]$ then has the form of

$$\{u, \ldots, u, 0, \ldots, 0, N_p, 0, \ldots, 0, u, \ldots, u\}, \tag{5.186}$$

which looks like a discrete-time delta function plus some additional cross-correlation terms (represented by u; these are the values of $R_{pt}[k]$ for $k \geq L$). As a result, correlation with $p[n]$, equivalently convolving with $p^*[(N_p - n + 1)]$, directly produces an estimate of the channel for certain correlation lags. For example, assuming that $s[n] = t[n]$ for $n = 0, 1, \ldots, N_{tr} - 1$,

$$y[n] * p^*[(N_p - n + 1)] = N_p h[n] + p^*[N_p - n + 1] * v[n] \tag{5.187}$$

for $n = N_p, N_p + 1, \ldots, N_p + L$. In this way, it is possible to produce an estimate of the channel just by correlating with the sequence $\{p[n]\}$ that is used to compose the training sequence $\{t[n]\}$. This has the additional advantage of also helping with frame synchronization, especially when the prefix size is chosen to be larger than L, and there are some extra zero values output from the correlation.

Another training design is based on Golay complementary sequences [129]. This approach is used in IEEE 802.ad [162]. Let $\{a[n]\}_{n=0}^{N_g-1}$ and $\{b[n]\}_{n=0}^{N_g-1}$ be length N_g Golay complementary sequences. Such sequences satisfy a special periodic correlation property

$$\sum_{n=0}^{N_g-1} a[n]a^* \left[((n+k))_{N_g} \right] + b[n]b^* \left[((n+k))_{N_g} \right] = 2N_g \delta[((k))_{N_g}]. \tag{5.188}$$

Furthermore, they also satisfy the aperiodic correlation property

$$\sum_{n=0}^{N_g-k-1} a[n]a^*[n+k] + b[n]b^*[n+k] = 2N_g\delta[k]. \tag{5.189}$$

It turns out that the family of Golay complementary sequences is much more flexible than sequences with perfect periodic correlation like the Frank and Zadoff-Chu sequences. In particular, it is possible to construct such pairs of sequences from BPSK constellations without going to higher-order constellations. Furthermore, such sequence pairs exist for many choices of lengths, including for any N_g as a power of 2. There are generalizations to higher-order constellations. Overall, the Golay complementary approach allows a flexible sequence design with some additional complexity advantages.

Now construct a training sequence by taking $\{a[n]\}_{n=0}^{N_g-1}$, with a cyclic prefix and postfix each of length L, and naturally $N_g > L$. Then append another sequence constructed from $\{b[n]\}_{n=0}^{N_g-1}$ with a cyclic prefix and postfix as well. This gives a length $N_{tr} = 2N_g + 4L$ training sequence. Assuming that $s[n] = t[n]$ for $n = 0, 1, \ldots, N_{tr} - 1$,

$$\sum_{k=0}^{N_g-1} y[k+L]a^* \left[((k+n))_{N_g} \right] + \sum_{k=0}^{N_g-1} y[k+3L+N_g]b^* \left[((k+n))_{N_g} \right]$$
$$= 2N_g h[n] + \text{noise} \tag{5.190}$$

for $n = 0, 1, \ldots, N_g - 1$. As a result, channel estimation can be performed directly by correlating the received data with the complementary Golay pair and adding the results. In IEEE 802.11ad, the Golay complementary sequences are repeated several times, to allow averaging over multiple periods, while also facilitating frame synchronization. They can also be used for rapid packet identification; for example, SC-FDE and OFDM packets use different sign patterns on the complementary Golay sequences, as discussed in [268].

An example construction of binary Golay complementary sequences is provided in Example 5.16. We provide an algorithm for constructing sequences and then use it to construct an example pair of sequences and a corresponding training sequence.

Example 5.16 There are several constructions of Golay complementary sequences [129]. Let $a_m[n]$ and $b_m[n]$ denote a Golay complementary pair of length 2^m. Then binary

Golay sequences can be constructed recursively using the following algorithm:

$$a_0[n] = \delta[n] \tag{5.191}$$
$$b_0[n] = \delta[n] \tag{5.192}$$
$$a_{m+1}[n] = a_m[n] \qquad n \in [0, 2^m - 1] \tag{5.193}$$
$$a_{m+1}[n] = b_m[n - 2^m] \qquad n \in [2^m, 2^{m+1} - 1] \tag{5.194}$$
$$b_{m+1}[n] = a_m[n] \qquad n \in [0, 2^m - 1] \tag{5.195}$$
$$b_{m+1}[n] = -b_m[n - 2^m] \qquad n \in [2^m, 2^{m+1} - 1] \tag{5.196}$$

Find a length $N_g = 8$ Golay complementary sequence, and use it to construct a training sequence assuming $L = 2$.

Answer: Applying the recursive algorithm gives

$$\{a_1[n]\}_{n=0}^1 = \{1, 1\} \tag{5.197}$$
$$\{b_1[n]\}_{n=0}^1 = \{1, -1\} \tag{5.198}$$
$$\{a_2[n]\}_{n=0}^3 = \{1, 1, 1, -1\} \tag{5.199}$$
$$\{b_2[n]\}_{n=0}^3 = \{1, 1, -1, 1\} \tag{5.200}$$
$$\{a_3[n]\}_{n=0}^7 = \{1, 1, 1, -1, 1, 1, -1, 1\} \tag{5.201}$$
$$\{b_3[n]\}_{n=0}^7 = \{1, 1, 1, -1, -1, -1, 1, -1\}. \tag{5.202}$$

The resulting training sequence, composed of Golay complementary sequences with length $L = 2$ cyclic prefix and postfix, is

$$\{t[n]\}_{n=0}^{19} = \{-1, 1, 1, 1, 1, -1, 1, 1, -1, 1, 1, 1, 1, 1, -1, 1, 1, 1, -1, -1, -1, 1, -1, 1, 1\}. \tag{5.203}$$

Example 5.17 IEEE 802.15.3c 60GHz is a standard for high-data-rate WPAN. In this example, we review the preamble for what is called the high-rate mode of the SC-PHY. Each preamble has a slightly different structure to make them easy to distinguish. The frame structure including the preamble is shown in Figure 5.24. Let $\mathbf{a}_{128}$ and $\mathbf{b}_{128}$ denote

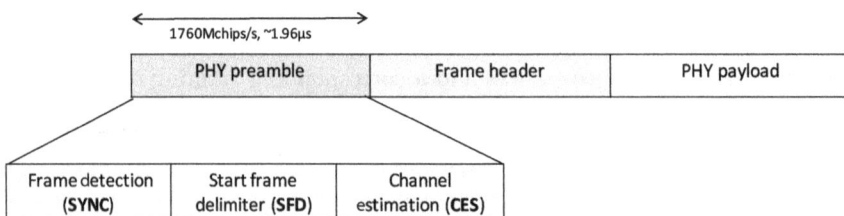

Figure 5.24 Frame structure defined in the IEEE 802.15.3c standard

vectors that contain length 128 complementary Golay sequences (in terms of BPSK symbols), and $\mathbf{a}_{256}$ and $\mathbf{b}_{256}$ vectors that contain length 256 complementary Golay sequences. From the construction in Example 5.16, $\mathbf{a}_{256} = [\mathbf{a}_{128}; \mathbf{b}_{128}]$ and $\mathbf{b}_{256} = [\mathbf{a}_{128}; -\mathbf{b}_{128}]$ where we use ; to denote vertical stacking as in MATLAB. The specific Golay sequences used in IEEE 802.15.3c are found in [159]. The preamble consists of three components:

1. The frame detection (SYNC) field, which contains 14 repetitions of $\mathbf{a}_{128}$ and is used for frame detection

2. The start frame delimiter (SFD), which contains $[\mathbf{a}_{128}; \mathbf{a}_{128}; -\mathbf{a}_{128}; \mathbf{a}_{128}]$ and is used for frame and symbol synchronization

3. The channel estimation sequence (CES), which contains $[\mathbf{b}_{128}; \mathbf{b}_{256}; \mathbf{a}_{256}; \mathbf{b}_{256}; \mathbf{a}_{256}]$ and is used for channel estimation

- Why does the CES start with $\mathbf{b}_{128}$?

  Answer: Since $\mathbf{a}_{256} = [\mathbf{a}_{128}; \mathbf{b}_{128}]$, $\mathbf{b}_{128}$ acts as a cyclic prefix.

- What is the maximum channel order supported?

  Answer: Given that $\mathbf{b}_{128}$ is a cyclic prefix, then $L_c = 128$ is the maximum channel order.

- Give a simple channel estimator based on the CES (assuming that frame synchronization and frequency offset correction have already been performed).

  Answer: Based on (5.190),

  $$\widehat{h}[\ell] = \sum_{n=0}^{255} y[n+128]a_{256}^*[((\ell+n))_{256}] + y[n+384]b_{256}^*[((\ell+n))_{256}]$$

  $$+ y[n+640]a_{256}^*[((\ell+n))_{256}] + y[n+896]b_{256}^*[((\ell+n))_{256}]. \qquad (5.204)$$

  for $\ell = 0, 1, \ldots, L_c$.

In this section, we presented least squares channel estimation based on a training sequence. We also presented several training sequence designs and showed how they can be used to simplify the estimation. This estimator was constructed assuming that the training data contains symbols. This approach is valid for standard pulse-amplitude modulated systems and SC-FDE systems. Next, we present several approaches for channel estimation that are specifically designed for OFDM modulation.

5.3.2   Least Squares Channel Estimation in the Frequency Domain

Channel estimation is required for frequency-domain equalization in OFDM systems. In this section, we describe several approaches for channel estimation in OFDM, based on either the time-domain samples or the frequency-domain symbols.

First, we describe the time-domain approach. We suppose that $N_{\text{tr}} = N$, so the training occupies exactly one complete OFDM symbol. It is straightforward to generalize to multiple OFDM symbols. Suppose that the training $\{t[n]\}_{n=0}^{N_{\text{tr}}-1}$ is input to the IDFT, and a length L_c cyclic prefix is appended, to create $\{w[n]\}_{n=0}^{N+L_c-1}$. In the time-domain approach, the IDFT output samples $\{w[n]\}_{n=0}^{N+L_c-1}$ are used to estimate the time-domain channel coefficients, by building $\mathbf{T}$ in (5.173) from $w[n]$ and then estimating the channel from (5.175). The frequency-domain channel coefficients are then obtained from $\widehat{H}[k] = \sum_{\ell=0}^{L_c} \widehat{h}[\ell] \exp(-j2\pi k l/N)$. The main disadvantage of this approach is that the nice properties of the training sequence are typically lost in the IDFT transformation.

Now we describe an approach where training is interleaved with unknown data in what are called pilots. Essentially the presence of pilots means that a subset of the symbols on a given OFDM symbol are known. With enough pilots, we show that it is possible to solve a least squares channel estimation problem using only the data that appears after the IDFT operation [338].

Suppose that training data is sent on a subset of all the carriers in an OFDM symbol; using all the symbols is a special case. Let $\mathcal{P} = \{p_1, p_2, \ldots, p_P\}$ denote the indices of the subcarriers that contain known training data. The approach we take is to write the observations as a function of the unknown channel coefficients. First observe that by writing the frequency-domain channel in terms of the time-domain coefficients,

$$\mathsf{y}[p_1] = \mathsf{h}[p_1]t[p_1] + \mathsf{v}[p_1] \tag{5.205}$$

$$= \begin{bmatrix} 1 & e^{-j2\pi \frac{p_1}{N}} & \cdots & e^{-j2\pi \frac{p_1 L}{N}} \end{bmatrix} \begin{bmatrix} h[0] \\ h[1] \\ \vdots \\ h[L] \end{bmatrix} t[p_1] + \mathsf{v}[p_1]. \tag{5.206}$$

Collecting the data from the different pilots together:

$$\underbrace{\begin{bmatrix} \mathsf{y}[p_1] \\ \mathsf{y}[p_2] \\ \vdots \\ \mathsf{y}[p_P] \end{bmatrix}}_{\mathsf{y}} = \underbrace{\begin{bmatrix} t[p_1] & 0 & \cdots & \\ 0 & t[p_2] & 0 & \cdots \\ & & \ddots & \\ & & & t[p_P] \end{bmatrix}}_{\mathbf{P}}$$

$$\times \underbrace{\begin{bmatrix} 1 & e^{-j2\pi \frac{p_1}{N}} & \cdots & e^{-j2\pi \frac{p_1 L}{N}} \\ 1 & e^{-j2\pi \frac{p_2}{N}} & \cdots & e^{-j2\pi \frac{p_2 L}{N}} \\ \vdots & \vdots & \ddots & \vdots \\ 1 & e^{-j2\pi \frac{p_M}{N}} & \cdots & e^{-j2\pi \frac{p_P L}{N}} \end{bmatrix}}_{\mathbf{E}} \underbrace{\begin{bmatrix} h[0] \\ h[1] \\ \vdots \\ h[L] \end{bmatrix}}_{\mathbf{h}} + \begin{bmatrix} \mathsf{v}[p_1] \\ \mathsf{v}[p_2] \\ \vdots \\ \mathsf{v}[p_M] \end{bmatrix}. \tag{5.207}$$

This can now be written compactly as

$$\mathbf{y} = \mathbf{PEh} + \mathbf{v}. \tag{5.208}$$

The matrix $\mathbf{P}$ has the training pilots on its diagonal. It is invertible for any choice of nonzero pilot symbols. The matrix $\mathbf{E}$ is constructed from samples of DFT vectors. It is tall and full rank as long as the number of pilots $P \geq L + 1$. Therefore, with enough pilots, we can compute the least squares solution as

$$\widehat{\mathbf{h}}_{\mathrm{LS}} = (\mathbf{E}^* \mathbf{P}^* \mathbf{P} \mathbf{E})^{-1} \mathbf{E}^* \mathbf{P}^* \mathbf{y}. \qquad (5.209)$$

This approach allows least squares channel estimation to be applied based only on training at known pilot locations. The choice of pilot sequences with pilot estimation is not as important as in the time-domain approach; the selection of pilot locations is more important as this influences the rank properties of $\mathbf{E}$. The approach can be extended to multiple OFDM symbols, possibly at different locations for improved performance.

An alternative approach for channel estimation in the frequency domain is to interpolate the frequency-domain channel estimates instead of solving a least squares problem [154]. This approach makes sense when training is sent on the same subcarrier across OFDM symbols, to allow some additional averaging over time.

Consider a comb-type pilot arrangement where P pilots are evenly spaced and $N_c = N/P$. This uniform spacing is optimum under certain assumptions for LMMSE channel estimation in OFDM systems [240] and for least squares estimators if $\mathbf{P}^* \mathbf{P}$ is a scaled identity matrix. Let $\widehat{\mathbf{h}}_p[cN_c]$ be the channel estimated at the pilot locations for $c = 0, 1, \ldots, P-1$ with $\widehat{\mathbf{h}}_p[0] = \widehat{\mathbf{h}}_p[N]$ and $\widehat{\mathbf{h}}_p[-N_c] = \widehat{\mathbf{h}}_p[N - N_c]$ to account for wrapping effects. This estimate is obtained by averaging over multiple observations of $\mathsf{y}[p_k]$ in multiple OFDM symbols. Using second-order interpolation, the missing channel coefficients are estimated as [77]

$$\widehat{\mathsf{h}}[cN_c + \ell] = \frac{\ell(\ell - 1)}{2N} \widehat{\mathsf{h}}_p[(c-1)N_c]$$
$$- \left(\frac{\ell}{N} - 1 \right) \left(\frac{\ell}{N} + 1 \right) \widehat{\mathsf{h}}_p[cN_c] + \frac{\ell(\ell - 1)}{2N} \widehat{\mathsf{h}}_p[(c+1)N_c]. \qquad (5.210)$$

In this way, the frequency-domain channel coefficients may be estimated directly from the pilots without solving a higher-dimensional least squares problem. With a similar total number of pilots, the performance of the time-domain and interpolation algorithms can be similar [154].

There are other approaches for channel estimation that may further improve performance. For example, if some assumptions are made about the second-order statistics of the channel, and these statistics are known at the receiver, then LMMSE methods may be used. Several decibels of performance improvement, in terms of uncoded symbol error rate, may be obtained through MMSE [338]. Pilots may also be distributed over multiple OFDM symbols, on different subcarriers in different OFDM symbols. This allows two-dimensional interpolation, which is valuable in channels that change over time [196]. The 3GPP LTE cellular standard includes pilots distributed across subcarriers and across time.

Example 5.18 In this example, we explore the preamble structure used in IEEE 802.11a, shown in Figure 5.25, and explain how it is used for channel estimation [160,

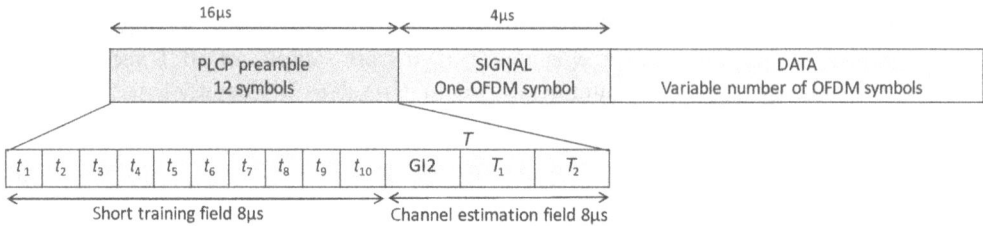

Figure 5.25 Frame structure defined in the IEEE 802.11a standard. In the time domain, the waveforms t_1 through t_{10} are identical (ten repetitions of short training) and T_1 and T_2 are the same (two repetitions of long training). The Physical Layer Convergence Procedure (PLCP) is used to denote the extra training symbols inserted to aid synchronization and training.

Chapter 17]. A similar structure is used in IEEE 802.11g, with some backward compatibility support for 802.11b, and in IEEE 802.11n, with some additions to facilitate multiple-antenna transmission. The preamble in IEEE 802.11a consists of a short training field (STF) and a channel estimation field (CEF). Each field has the duration of two OFDM symbols ($8\mu s$) but is specially constructed. In this problem, we focus on channel estimation based on the CEF, assuming that frame synchronization and carrier frequency offset correction have been performed.

The CEF of IEEE 802.11a consists of two repeated OFDM symbols, each containing training data, preceded by an extra-long cyclic prefix. Consider a training sequence of length $N = 64$ with values

$$\{t[n]\}_{n=0}^{15} = \{0, 1, -1, -1, 1, 1, -1, 1, -1, 1, -1, -1, -1, -1, -1, 1\} \tag{5.211}$$

$$\{t[n]\}_{n=16}^{26} = \{1, -1, -1, 1, -1, 1, -1, 1, 1, 1, 1, 1\} \tag{5.212}$$

$$\{t[n]\}_{n=38}^{49} = \{1, -1, -1, 1, 1, -1, 1, -1, 1, 1, 1, 1\} \tag{5.213}$$

$$\{t[n]\}_{n=50}^{63} = \{1, 1, -1, -1, 1, 1, -1, 1, -1, 1, 1, 1, 1\} \tag{5.214}$$

where all other subcarriers are zero. IEEE 802.11a uses zero subcarriers to help with spectral shaping; a maximum of only 52 subcarriers is used. The $n = 0$ subcarrier corresponds to DC and is also zero. The CEF is constructed as

$$w[n] = \frac{1}{64} \sum_{k=0}^{63} t[k] e^{\mathrm{j}\frac{2\pi(kn-32)}{64}} \tag{5.215}$$

for $n = 0, 1, \ldots, 159$. Essentially the CEF has one extended cyclic prefix of length $L_c = 32$ (extended because it is longer than the normal $L_c = 16$ prefix used in IEEE 802.11a) followed by the repetition of two IDFTs of the training data. The repetition is also useful for frequency offset estimation and frame synchronization described in Section 5.4.4. Time-domain channel estimation can proceed directly using (5.215). For estimation from the frequency domain, we define two slightly different truncated received signals $\bar{y}_1[n] = y[n + 32]$ and $\bar{y}_2[n] = y[n + 96]$ for $n = 0, 1, \ldots, 63$. Taking the DFT gives $\{\mathsf{y}_1[k]\}_{k=0}^{64}$ and $\{\mathsf{y}_2[k]\}_{k=0}^{63}$. The parts of the received signal corresponding to the nonzero training $\{y_1[k]\}_{k=1}^{26}$, $\{y_1[k]\}_{k=38}^{63}$, $\{y_2[k]\}_{k=1}^{26}$, and $\{y_2[k]\}_{k=38}^{63}$ can then be used for channel estimation by using a linear equation of the form

$$\begin{bmatrix} \mathbf{y}_1 \\ \mathbf{y}_2 \end{bmatrix} = \begin{bmatrix} \mathbf{PE} \\ \mathbf{PE} \end{bmatrix} \mathbf{h} + \begin{bmatrix} \mathbf{v}_1 \\ \mathbf{v}_2 \end{bmatrix} \tag{5.216}$$

and finding the least squares solution.

5.3.3 Direct Least Squares Equalizer

In the previous sections, we developed channel estimators based on training data. These channel estimates can be used to devise equalizer coefficients. At each step in the process, both channel estimation and equalizer computation create a source of error since they involve solving least squares problems. As an alternative, we briefly review an approach where the equalizer coefficients are estimated directly from the training data. This approach avoids the errors that occur in two steps, giving somewhat more noise robustness, and does not require explicit channel estimation.

To formulate this problem, again consider the received signal in (5.87), and the signal after equalization in (5.96), which gives an expression for $\hat{s}[n - n_{\mathrm{d}}]$. We now formulate a least squares problem based on the presence of the known training data $s[n] = t[n]$ for $n = 0, 1, \ldots, N_{\mathrm{tr}}$. Given the location of this data, and suitably time shifting the observed sequence, gives

$$t[n] = \sum_{\ell=0}^{L_{\mathrm{f}}} f_{n_{\mathrm{d}}}[\ell] y[n + n_{\mathrm{d}} - \ell] \tag{5.217}$$

for $n = 0, 1, \ldots, N_{\mathrm{tr}}$. Building a linear equation:

$$\underbrace{\begin{bmatrix} t[0] \\ t[1] \\ \vdots \\ t[N_{\mathrm{tr}} - 1] \end{bmatrix}}_{\mathbf{t}} = \underbrace{\begin{bmatrix} y[n_{\mathrm{d}}] & \cdots & y[n_{\mathrm{d}} - L_{\mathrm{f}}] \\ y[n_{\mathrm{d}} + 1] & \ddots & \vdots \\ \vdots & & \vdots \\ y[n_{\mathrm{d}} + N_{\mathrm{tr}} - 1] & \cdots & y[n_{\mathrm{d}} + N_{\mathrm{tr}} - L_{\mathrm{f}}] \end{bmatrix}}_{\mathbf{Y}_{n_{\mathrm{d}}}} \underbrace{\begin{bmatrix} f_{n_{\mathrm{d}}}[0] \\ f_{n_{\mathrm{d}}}[1] \\ \vdots \\ f_{n_{\mathrm{d}}}[L_{\mathrm{f}}] \end{bmatrix}}_{\mathbf{f}_{n_{\mathrm{d}}}}. \tag{5.218}$$

In matrix form, the objective is to solve the following linear system for the unknown equalizer coefficients:

$$\mathbf{t} = \mathbf{Y}_{n_{\mathrm{d}}} \mathbf{f}_{n_{\mathrm{d}}}. \tag{5.219}$$

If $L_{\mathrm{f}} + 1 > N_{\mathrm{tr}}$, then $\mathbf{Y}_{n_{\mathrm{d}}}$ is a fat matrix, and the system of equations is undetermined. This means that there is an infinite number of exact solutions. This, however, tends to lead to overfitting, where the equalizer is perfectly matched to the observations in $\mathbf{Y}_{n_{\mathrm{d}}}$. A more robust approach is to take $L_{\mathrm{f}} \leq N_{\mathrm{tr}} - 1$, turning $\mathbf{Y}_{n_{\mathrm{d}}}$ into a tall matrix. It is reasonable to assume that $\mathbf{Y}_{n_{\mathrm{d}}}\mathbf{x}$ is full rank because of the presence of additive noise in $y[n]$. Under this assumption, the least squares solution is

$$\widehat{\mathbf{f}}_{n_{\mathrm{d}}} = \left(\mathbf{Y}_{n_{\mathrm{d}}}^* \mathbf{Y}_{n_{\mathrm{d}}} \right)^{-1} \mathbf{Y}_{n_{\mathrm{d}}}^* \mathbf{t}. \tag{5.220}$$

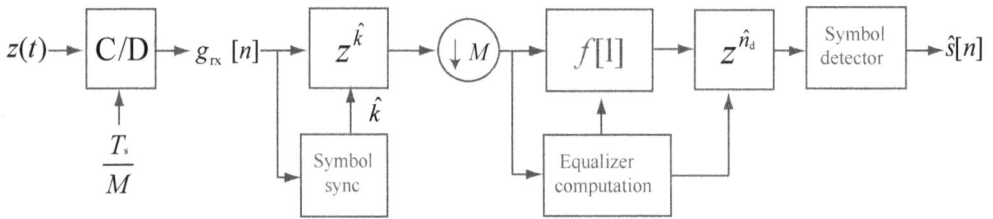

Figure 5.26 A complex pulse-amplitude modulation receiver with direct equalizer estimation and linear equalization

The squared error is measured as $J_f[n_d] = \mathbf{t}^* \mathbf{t} - \mathbf{t}^* \mathbf{Y}_{n_d} \left(\mathbf{Y}_{n_d}^* \mathbf{Y}_{n_d} \right)^{-1} \mathbf{Y}_{n_d}^* \mathbf{t}$. The squared error can be minimized further by choosing n_d such that $J_f[n_d]$ is minimized.

A block diagram including direct equalization is illustrated in Figure 5.26. The block diagram assumes that the equalization computation block optimizes over the delay and outputs the corresponding filter and required delay. Direct equalization avoids the separate channel estimation and equalizer computation blocks.

The direct and indirect equalization techniques have different design trade-offs. With the direct approach, the length of the training determines the maximum length of the equalizer. This is a major difference between the direct and indirect methods. With the indirect method, an equalizer of any order L_f can be designed. The direct method, however, avoids the error propagation where the estimated channel is used to compute the estimated equalizer. Note that with a small amount of training the indirect method may perform better since a larger L_f can be chosen, whereas a direct method may be more efficient when N_{tr} is larger. The direct method also has some robustness as it will work even when there is model mismatch, that is, when the LTI model is not quite valid or there is interference. Direct equalization can be done perfectly in multichannel systems as discussed further in Chapter 6.

5.4 Carrier Frequency Offset Correction in Frequency-Selective Channels

In this section, we describe the carrier frequency offset impairment for frequency-selective channels. We present a discrete-time received signal model that includes frequency offset. Then we examine several carrier frequency offset estimation algorithms. We also remark how each facilitates frame synchronization.

5.4.1 Model for Frequency Offset in Frequency-Selective Channels

In Section 5.1.6, we introduced the carrier frequency offset problem. In brief, carrier frequency offset occurs when the carrier used for upconversion and the carrier used for downconversion are different. Even a small difference can create a significant distortion in the received signal.

We have all the ingredients to develop a signal model for frequency offset in frequency-selective channels. Our starting point is to recall (5.67), which essentially says that carrier frequency offset multiplies the matched filtered received signal by $e^{\mathrm{j}2\pi f_e t}$. Sampling at the symbol rate, and using our FIR model for the received signal in (5.87), we obtain

$$y[n] = e^{\mathrm{j}2\pi\epsilon n} \sum_{\ell=0}^{L} h[\ell]s[n-\ell] + v[n]. \tag{5.221}$$

It is possible to further generalize (5.221) to include frame synchronization by including a delay of d. Correction of carrier frequency offset therefore amounts to estimating ϵ and then derotating the received signals $e^{-\mathrm{j}2\pi\epsilon n}y[n]$.

In the frequency-flat case, the frequency offset creates a successive rotation of each symbol by $\exp(\mathrm{j}2\pi\epsilon n)$. In the frequency-selective case, this rotation is applied to the convolutive mixture between the symbols and the channel. As a result, the carrier frequency offset distorts the data that would otherwise be used for channel estimation and equalization. Even if training data is available, it does not lead directly to a simple estimator. A direct extension of what we have studied leads to a joint carrier frequency offset and channel estimation problem, which has high complexity. In this section, we review several lower-complexity methods for frequency offset estimation, which rely on special signal structure to implement.

5.4.2 Revisiting Single-Frequency Estimation

For our first estimator, we revisit the estimator proposed in Example 5.8 and show how it can also be used in frequency-selective channels with minimal changes. This type of sinusoidal training was used in the GSM system through a special input to the GMSK (Gaussian minimum shift keying) modulator.

Let us choose as a training sequence $t[n] = \exp(\mathrm{j}2\pi f_t n)$ for $n = 0, 1, \ldots, N_{\mathrm{tr}} - 1$ where f_t is a discrete-time design frequency. Consider the received signal (5.87) with $s[n] = t[n]$ for $n = 0, 1, \ldots, N_{\mathrm{tr}} - 1$. Discarding samples that do not depend on the training, we have the received signal

$$y[n] = e^{\mathrm{j}2\pi\epsilon n} \sum_{\ell=0}^{L} h[\ell]t[n-\ell] + v[n] \tag{5.222}$$

for $n = L, L+1, \ldots, N_{\mathrm{tr}} - 1$. Substituting for the training signal,

$$y[n] = e^{\mathrm{j}2\pi\epsilon n} \sum_{\ell=0}^{L} h[\ell]e^{\mathrm{j}2\pi f_t(n-\ell)} + v[n] \tag{5.223}$$

$$= e^{\mathrm{j}2\pi\epsilon n}e^{\mathrm{j}2\pi f_t n} \sum_{\ell=0}^{L} h[\ell]e^{-\mathrm{j}2\pi f_t \ell} + v[n] \tag{5.224}$$

$$= e^{\mathrm{j}2\pi\epsilon n}e^{\mathrm{j}2\pi f_t n}\mathsf{h}\left(e^{\mathrm{j}2\pi f_t}\right) + v[n] \tag{5.225}$$

where $\mathsf{h}\left(e^{\mathrm{j}2\pi f_t}\right)$ is the DTFT of $h[n]$. Derotating by the known frequency of the training signal,

$$e^{-\mathrm{j}2\pi f_t n}y[n] = e^{\mathrm{j}2\pi\epsilon n}\mathsf{h}\left(e^{\mathrm{j}2\pi f_t}\right) + e^{-\mathrm{j}2\pi f_t}v[n]. \tag{5.226}$$

This has the form of (5.72) from Example 5.8. As a result, the estimators for frequency-flat frequency offset estimation using sinusoidal training can also be applied to the frequency-selective case.

The main signal processing trick in deriving this estimator was to recognize that signals of the form $\exp(\mathrm{j}2\pi f_t n)$ are eigenfunctions of discrete-time LTI systems. Because the unknown channel is FIR with order L, and the training $N_{\mathrm{tr}} > L$, we can discard some samples to remove the edge effects. The frequency f_t could be selected of the form k/N_{tr} to take advantage, for example, of subsequent frequency-domain processing in the modem.

The main advantage of single-frequency estimation is that any number of good algorithms can be used for estimation; see, for example, [330, 3, 109, 174, 212, 218] among others. The disadvantage of this approach is that performance is limited by $\mathsf{h}\left(e^{\mathrm{j}2\pi f_t}\right)$. Since the channel is frequency selective by assumption and has L zeros because it is FIR of order L, it could happen that the chosen frequency of f_t is close to a zero of the channel. Then the SNR for the estimator would be poor. The sinusoidal training also does not lead to a sharp correlation peak that can be used for frame detection. Furthermore, if used to construct $\mathbf{T}$ in (5.175), it would in fact not have full rank, and thus it could not be used for channel estimation. As a result, we now review other approaches for carrier frequency offset estimation.

5.4.3 Frequency Offset Estimation and Frame Synchronization Using Periodic Training for Single-Carrier Systems

There are several different algorithms for frequency offset estimation using different properties of the transmitted signal such as periodicity, constant modulus, and so on. One of the most elegant methods was proposed by Moose [231] and has since been studied extensively. This method relies on a special periodic training sequence that permits joint carrier frequency offset estimation and frame synchronization. The training sequences can also be used for channel estimation. Periodic training has found application in IEEE 802.11a/g/n/ac systems and others. In this section, we consider the application of periodic training to a single-carrier system, though note that Moose's original application was to OFDM. We deal with the case of OFDM, including another algorithm called the Schmidl-Cox approach [296], in the next section.

Now we explain the key idea behind the Moose algorithm from the perspective of using just two repeated training sequences. Other extensions are possible with multiple repetitions. Consider the framing structure illustrated in Figure 5.27. In this frame, a pair of training sequences are sent together, followed by a data sequence.

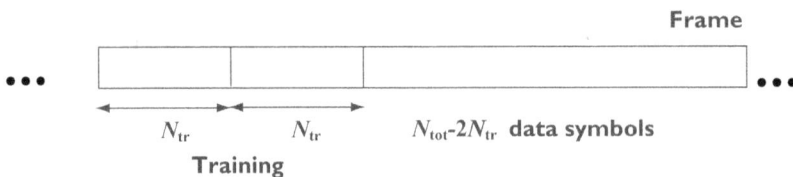

Figure 5.27 The framing used in the Moose estimator. A pair of training sequences are followed by data.

The Moose algorithm exploits the periodicity in the training sequence. Let the training sequence start at $n = 0$. Then

$$s[n] = s[n + N_{tr}] = t[n] \tag{5.227}$$

for $n = 0, 1, \ldots, N_{tr} - 1$. Note that symbols $s[n]$ for $n < 0$ and $n \geq 2N_{tr}$ are unknown (they are either zero or correspond to the unknown portions of the data).

Now we explain the trick associated with periodicity in the training data. For $n \in [L, N_{tr} - 1]$,

$$y[n] = e^{j2\pi\epsilon n} \sum_{\ell=0}^{L} h[\ell]s[n - \ell] + v[n] \tag{5.228}$$

$$y[n + N_{tr}] = e^{j2\pi\epsilon(n+N_{tr})} \sum_{\ell=0}^{L} h[\ell]s[n + N_{tr} - l] + v[n + N_{tr}]. \tag{5.229}$$

Using the fact that $s[n + N_{tr}] = s[n] = t[n]$ for $n = 0, 1, \ldots, N_{tr} - 1$:

$$y[n + N_{tr}] = e^{j2\pi\epsilon N_{tr}} e^{j2\pi\epsilon n} \sum_{\ell=0}^{L} h[\ell]t[n - \ell] + v[n + N_{tr}] \tag{5.230}$$

$$\approx e^{j2\pi\epsilon N_{tr}} y[n]. \tag{5.231}$$

To see the significance of this result, remember that the channel coefficients $\{h[\ell]\}_{\ell=0}^{L}$ are *unknown*! The beauty of (5.231) is that it is a function of known observations and only the unknown frequency offset. Essentially, the periodicity establishes a relationship between different parts of the received signal $y[n]$. This is an amazing observation, since it does not require any assumptions about the channel.

There are several possible approaches for solving (5.231). Note that this is slightly different from the single-frequency estimator because of the presence of $y[n]$ on the right-hand side of (5.231). The direct least squares approach is not possible since the unknown parameter is present in the exponent of the exponential. A solution is to consider a relaxed problem

$$y[n + N_{tr}] = ay[n], \tag{5.232}$$

solving for $\hat{a}_{LS}$ and then finding ϵ from the phase. This problem is similar to flat-fading channel estimation in Section 5.1.4. Using (5.47), we can write

$$\hat{\epsilon}_{LS} = \frac{1}{2\pi N_{tr}} \text{phase} \left(\frac{\sum_{n=L}^{N_{tr}-1} y^*[n]y[n + N_{tr}]}{\sum_{n=L}^{N_{tr}-1} y^*[n]y[n]} \right) \tag{5.233}$$

$$= \frac{1}{2\pi N_{tr}} \text{phase} \left(\sum_{n=L}^{N_{tr}-1} y^*[n]y[n + N_{tr}] \right) \tag{5.234}$$

where we can neglect the denominator since it does not contribute to the phase estimate. Despite the relaxation in this derivation, it turns out that this is also the maximum likelihood estimator [231].

There is an important limitation in the Moose algorithm. Because of the periodicity of discrete-time exponentials, the estimate of ϵ is accurate only for $|\epsilon N_{\mathrm{tr}}| \leq \frac{1}{2}$ or equivalently

$$|\epsilon| \leq \frac{1}{2N_{\mathrm{tr}}}, \tag{5.235}$$

which in terms of the actual frequency offset means

$$|f_{\mathrm{e}}| \leq \frac{1}{2TN_{\mathrm{tr}}}. \tag{5.236}$$

This reveals an interesting trade-off in the estimator performance. Choosing larger N_{tr} improves the estimate since there is more noise averaging, but it reduces the range of offsets that can be corrected. A way of solving this problem is to use multiple repetitions of a short training sequence. IEEE 802.11a and related standards use a combination of both repeated short training sequences and long training sequences.

Example 5.19   Compute the maximum allowable offset for a $1Ms/s$-QAM signal, with $f_c = 2\mathrm{GHz}$ and $N_t = 10$.
    Answer:

$$\max |f_{\mathrm{e}}| = \frac{1}{2TN_{\mathrm{tr}}} \tag{5.237}$$

$$= \frac{1}{2}10^6 \frac{1}{10} \tag{5.238}$$

$$= \frac{1}{2}10^5 = 50\mathrm{kHz}. \tag{5.239}$$

In terms of parts per million, we need an oscillator that can generate a 2GHz carrier with an accuracy of $50e3/2e9 = 25ppm$.

Example 5.20   Consider a wireless system where each data frame is preceded by two training blocks, each consisting of $N_{\mathrm{tr}} = 12$ training symbols. Let the symbol period be $T = 4\mu s$. What is the maximum frequency offset that can be corrected using training?
    Answer: The maximum frequency offset that can be corrected using training is $\epsilon = 1/(2N_{\mathrm{tr}}) \approx 0.0417$ or $f_e = 1/(2TN_{\mathrm{tr}}) \approx 10.417\mathrm{kHz}$.

The Moose algorithm also provides a nice way of performing frame synchronization. The observation is that the correlation peak should occur when the pair of training sequences is encountered at the receiver. Essentially, this involves solving for the offset d such that

$$\widehat{d} = \arg_\Delta \max \frac{\left|\sum_{n=L}^{N_{\mathrm{tr}}-1} y[n+\Delta+N_{\mathrm{tr}}]y^*[n+\Delta]\right|^2}{\left(\sum_{n=L}^{N_{\mathrm{tr}}-1} |y[n+\Delta]|^2\right)^2} \tag{5.240}$$

where the search is performed over a reasonable set of possible values of d. For example, the analog RF may perform carrier sensing, activating the ADCs only when a sufficiently strong signal is present.

The denominator normalization in the frame synchronization is required to avoid false positives when there is no signal present. An alternative solution, which is somewhat more robust in this case, involves normalizing by both observed vectors as in

$$\widehat{d} = \arg_\Delta \max \frac{\left|\sum_{n=L}^{N_{\mathrm{tr}}-1} y[n+\Delta+N_{\mathrm{tr}}]y^*[n+\Delta]\right|^2}{\left(\sum_{n=L}^{N_{\mathrm{tr}}-1} |y[n+\Delta]|^2\right)\left(\sum_{n=L}^{N_{\mathrm{tr}}-1} |y[n+\Delta+N_{\mathrm{tr}}]|^2\right)}. \tag{5.241}$$

No matter which algorithm is used, both result in the same thing: a joint solution to the frame synchronization and frequency offset estimation and correction problem in an intersymbol interference channel.

Channel estimation is also facilitated using the Moose algorithm. Once the frequency offset is estimated and corrected, and the frame is synchronized, the pair of training sequences can be combined for channel estimation. As a result, periodic training provides a flexible approach for solving key receiver functions in frequency-selective channels.

An illustration of the complete receiver can be found in Figure 5.28. The frequency offset estimation and correction are performed prior to channel estimation and equalization but after the downsampling operation. Better performance could be achieved by operating before the downsampling, replacing the symbol synchronizer, but this is usually practical only for smaller values of M_{rx}.

We conclude this section with a detailed example that describes how to use the preamble structure in IEEE 802.15.3c single-carrier mode and in IEEE 802.11a.

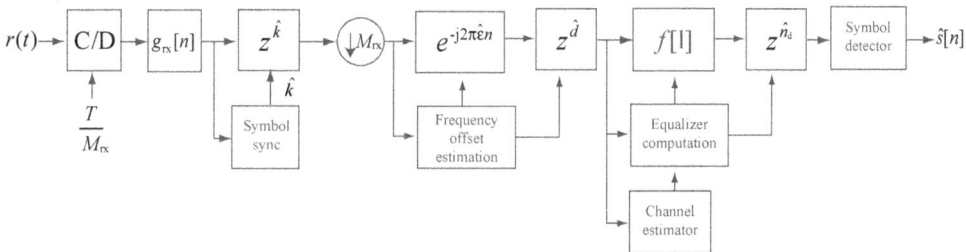

Figure 5.28 A single-carrier receiver with frequency offset estimation and correction, frame synchronization, channel estimation, and linear equalization. The linear equalization could be replaced by an SC-FDE receiver structure with no other changes required.

Example 5.21   Consider the IEEE 802.15.3c preamble structure as described in Example 5.17 for the high-rate mode of the SC-PHY. In this example, we describe how the SYNC field is used for carrier frequency offset estimation and frame synchronization. Note that the standard just describes the preamble itself; it does not describe how the receiver signal processing should be applied to the preamble. A good description of digital synchronization algorithms for IEEE 802.15.3c is found in [204].

The SYNC field can be used for frame synchronization and carrier frequency offset estimation. The SYNC field contains 14 repetitions of $\mathbf{a}_{128}$. Because the maximum supported channel length is $L_c = 128$, though, the first repetition acts as a cyclic prefix. Therefore, we can use the following frame detection algorithm:

$$\widehat{d} = \arg\max_{\Delta} \sum_{p=1}^{13} \frac{\left|\sum_{n=0}^{127} y[n + \Delta + 128p + 128]y^*[n + \Delta + 128p]\right|^2}{\left(\sum_{n=0}^{127} |y[n + \Delta + 128 + 128p]|^2\right)^2} \tag{5.242}$$

where we have put the packet energy in the denominator to make the algorithm more robust when averaging over multiple repetitions. The carrier frequency offset can then be estimated from either averaging the offset with each period and combining or taking the phase of the average as

$$\widehat{\epsilon} = \frac{1}{2\pi 128} \sum_{p=1}^{13} \text{phase}\left(\sum_{n=0}^{127} y[n + \widehat{d} + 128p + 128]y^*[n + \widehat{d} + 128p]\right). \tag{5.243}$$

The maximum absolute offset correctable is $1/256$.

In packet-based wireless systems, the digital part of the receiver may be in sleep mode to save power. As a result, the analog may make an initial determination that there is a signal of interest, for example, if the signal exceeds a threshold. The SYNC field can be used for this initial determination. In the process, though, some samples may be lost, and not all repetitions will be available for averaging. As a result, it may make sense to average over fewer repetitions in (5.242). It also may make sense to either defer frequency offset correction or make a correction based on this tentative estimate (called a coarse correction), then proceed with further correction and refinement using the SFD field.

Example 5.22   Consider the IEEE 802.11a preamble structure as described in Example 5.18. In this example, we explain the coarse frequency offset using the STF field and the fine frequency offset using the CEF field. We assume $B = 20\text{MHz}$ bandwidth. The STF field (in the time domain) consists of ten repetitions (created by zeroing subcarriers as described in Section 5.4.4). The CEF field has two repetitions. Sampled at $T = 1/B$, each repetition of the STF contains 16 samples, whereas the CEF has two repetitions of 64 samples with a length 32 cyclic prefix.

The STF field is used for functions such as packet detection, automatic gain control (AGC), and coarse synchronization. Because the RF may be powering up, and the

AGC adjusting, not all of the ten repetitions are typically used for synchronization. For example, suppose that three repetitions are used with the first acting as a cyclic prefix. Then we use the following frame detection algorithm:

$$\widehat{d}_{\text{coarse}} = \arg\max_{\Delta} \frac{\left|\sum_{n=0}^{15} y[n+\Delta+16]y^*[n+\Delta]\right|^2}{\left(\sum_{n=0}^{15} |y[n+\Delta+16]|^2\right)^2}. \tag{5.244}$$

The coarse carrier frequency offset estimate is then

$$\widehat{\epsilon}_{\text{coarse}} = \frac{1}{2\pi 16}\text{phase}\left(\sum_{n=0}^{15} y[n+\widehat{d}_{\text{coarse}}+16]y^*[n+\widehat{d}_{\text{coarse}}]\right). \tag{5.245}$$

The maximum absolute frequency offset correctable is $1/32$.

Next, we correct for the coarse frequency offset to create the new signal $\widetilde{y}[n] = \exp(-j2\pi\widehat{\epsilon}_{\text{coarse}}n)y[n]$. Using this corrected signal, we then search for the long training sequence:

$$\widehat{d}_{\text{fine}} = \arg\max_{\Delta} \frac{\left|\sum_{n=0}^{63} \widetilde{y}[n+\Delta+64]\widetilde{y}^*[n+\Delta]\right|^2}{\left(\sum_{n=0}^{63} |\widetilde{y}[n+\Delta+32]|^2\right)^2}. \tag{5.246}$$

The coarse frame synchronization estimate can be used to reduce the search space of possible offsets. The fine carrier frequency offset estimate is then

$$\widehat{\epsilon}_{\text{fine}} = \frac{1}{2\pi 64}\text{phase}\left(\sum_{n=0}^{63} \widetilde{y}[n+\widehat{d}_{\text{fine}}+64]\widetilde{y}^*[n+\widehat{d}_{\text{fine}}]\right). \tag{5.247}$$

The two-step approach allows an initial frequency offset estimate with a larger correction range, but it is noisier in the first step because of the use of only 16 samples. Then, in the second step, it is possible to get a more accurate estimate by using 64 samples with a smaller range, assuming the first phase corrected for the larger errors.

5.4.4 Frequency Offset Estimation and Frame Synchronization Using Periodic Training for OFDM Systems

OFDM is sensitive to carrier frequency offset. The reason is essentially that carrier frequency offset over the duration of an OFDM symbol creates problems because the DFT is taken over the whole OFDM symbol period. Small variations are compounded by the DFT operation into bigger variations.

Frequency offset correction algorithms for OFDM systems are similar to their time-domain counterparts. The method in Section 5.4.2 can be applied to OFDM by sending a OFDM symbol with just a single subcarrier active. The method in Section 5.4.3 can be applied directly only to multiple repetitions of OFDM systems. In this section, we develop another carrier frequency offset estimator that works on the periodicity created in one OFDM symbol. It also has an advantage over the Moose method in that it allows a larger range of offsets to be corrected, when used with a second specially designed OFDM symbol.

We start by explaining how to create periodicity in a single OFDM symbol. The key insight is to make a portion of the transmitted signal look periodic and then to apply the same technique as before. Suppose that we construct an OFDM symbol where all the odd subcarriers are zero. Then

$$w[n] = \frac{1}{N} \sum_{m=0}^{N-1} s[m] e^{j\frac{2\pi m(n-L_c)}{N}} \tag{5.248}$$

$$= \frac{1}{N} \sum_{m=0}^{\frac{N}{2}-1} s[2m] e^{j\frac{2\pi m(n-L_c)}{N/2}}. \tag{5.249}$$

This looks like an extended discrete-time sinusoid with carrier $2\pi(N/2)$. As a result, for $n \in [L_c, L_c + N/2]$,

$$w[n] = w\left[n + \frac{N}{2}\right], \tag{5.250}$$

which means that the OFDM signal contains a portion that is periodic. As a result, the Moose algorithm can be applied directly to this setting.

Consider the case where frame synchronization has already been performed. The received signal model is

$$y[n] = e^{j2\pi\epsilon n} \sum_{\ell=0}^{L} h[\ell] w[n-\ell] + v[n]. \tag{5.251}$$

Discarding the cyclic prefix,

$$\bar{y}[n] = y[n + L_c]. \tag{5.252}$$

Because of the periodicity,

$$\bar{y}[n + N/2] \approx e^{j2\pi\epsilon N/2} \bar{y}[n]. \tag{5.253}$$

Then the Moose frequency offset estimator is

$$\widehat{\epsilon}_{LS} = \frac{1}{2\pi N_{tr}} \text{phase}\left(\sum_{n=L}^{N/2-1} \bar{y}^*[n]\bar{y}[n+N]\right). \tag{5.254}$$

As before, it is possible to perform OFDM symbol synchronization (or frame synchronization) and frequency offset estimation jointly using this correlation method. The OFDM training symbol can even be used for channel estimation as this is just a special case of the comb-type pilot arrangement described in Section 5.3.2.

The maximum correctable carrier frequency offset is

$$|\epsilon| \leq \frac{1}{2N/2} \tag{5.255}$$

$$= \frac{1}{N}. \tag{5.256}$$

As a result, the Moose algorithm can essentially correct for continuous-time offsets that are less than one subcarrier spacing $1/(NT)$. In an OFDM system, N may be quite large. This would in turn reduce the range of correctable offsets. Thus, it is of interest to develop techniques to enhance the range. One such approach is the Schmidl-Cox algorithm [296].

Consider an OFDM system with two training OFDM symbols. In the first symbol, all the odd subcarriers are zero. The even subcarriers $\{t_1[2n]\}_{n=0}^{N/2-1}$ are 4-QAM training symbols, which are nominally just a pseudo-noise sequence. The presence of zeros ensures that there is periodicity in the first OFDM symbol. Let $\{t_2[n]\}_{n=0}^{N-1}$ be a set of 4-QAM training symbols sent on the second OFDM symbol, without zeros. Further suppose that the even training symbols are selected such that $t_2[2n]t_1^*[2n] = t_3[2n]$ where $\{t_3[n]\}_{n=0}^{N/2-1}$ is another sequence with good periodic correlation properties. Effectively, the training data on even subcarriers of the second OFDM symbol is differentially encoded.

The Schmidl-Cox algorithm works as follows. Let the frequency offset be written as

$$\epsilon = \frac{2q}{N} + \epsilon_{\text{frac}} \tag{5.257}$$

where q is called the integer offset and ϵ_{frac} is the fractional offset. Because

$$e^{j2\pi\epsilon N/2} = e^{j2\pi\left(\frac{2q}{N} + \epsilon_{\text{frac}}\right)\frac{N}{2}} \tag{5.258}$$

$$= e^{j2\pi\left(\frac{qN2}{N2} + \epsilon_{\text{frac}}\right)\frac{N}{2}} \tag{5.259}$$

$$= e^{j2\pi\epsilon_{\text{frac}}\frac{N}{2}}, \tag{5.260}$$

the expression (5.253) is able to account for only fractional frequency offsets. Suppose that the estimator in (5.254) is used to estimate the fractional frequency offset. Further suppose that the estimate is perfect and is then removed by multiplying by $\exp(-j2\pi\epsilon_{\text{frac}}N/2)$, leaving the received signal after correction as

$$y[n] = e^{j2\pi\frac{2q}{N}n} \sum_{\ell=0}^{L} h[\ell]w[n-\ell] + v[n]. \tag{5.261}$$

Discarding the cyclic prefix and taking the DFT gives

$$\mathsf{y}[k] = e^{-j2\pi\frac{2q}{N}L_c}\mathsf{h}[((k-2q))_N]\mathsf{s}[((k-2q))_N] + \widetilde{v}[k] \tag{5.262}$$

for $k = 0, 1, \ldots, N-1$. The received signals corresponding to each training symbol use the index $k = 0, 1, \ldots, N-1$:

$$\mathsf{y}_1[k] = e^{-j2\pi\frac{2q}{N}L_c}\mathsf{h}[((k-2q))_N]t_1[((k-2q))_N] + \widetilde{v}[k] \tag{5.263}$$

$$\mathsf{y}_2[k] = e^{-j2\pi\frac{2q}{N}L_c}\mathsf{h}[((k-2q))_N]t_2[((k-2q))_N] + \widetilde{v}[k+N+L_c]. \tag{5.264}$$

Now we exploit the fact that the training on the even subcarriers was differentially encoded. Consider

$$\mathsf{y}_2[2k]\mathsf{y}_1^*[2k] = |\mathsf{h}[((2k-2q))_N]|^2 t_2[((2k-2q))_N]t_1^*[((2k-2q))_N] + \widetilde{v}'[k] \quad (5.265)$$
$$= |\mathsf{h}[((k-2q))_N]|^2 t_3[((2k-2q))_N] + \widetilde{v}'[k] \quad (5.266)$$

where $\widetilde{v}'[k]$ contains the products with the noise terms. Because $\{t_3[2n]\}_{n=0}^{N/2-1}$ has good periodic correlation properties, a least-squares-type problem can be formulated and used to find the shift with a correlation peak:

$$\widehat{q} = \max_{p=0,1,\dots,N/2-1} \frac{\left|\sum_{k=0}^{N/2-1} \mathsf{y}_2[2k+2p]\mathsf{y}_1^*[2k+2p]t_3^*[2k+2p]\right|^2}{\left(\sum_{k=0}^{N/2-1} |\mathsf{y}_2[2k]|^2\right)^2}. \quad (5.267)$$

Based on this integer offset estimate, the received samples can be corrected by $e^{-j2\pi\frac{2\widehat{q}}{N}n}$. Then the second OFDM training symbol can be used to estimate the channel (possibly combined with the first OFDM symbol as well).

The final frequency offset estimate obtained with the Schmidl-Cox algorithm is

$$\epsilon = \frac{2\widehat{q}}{N} + \widehat{\epsilon}_{\text{frac}}. \quad (5.268)$$

When both fractional and integer offset corrections are applied, a large range of offsets can be corrected. While it seems like very large offsets can be corrected, recall that the system model was derived assuming a small offset. A large offset would shift part of the desired signal outside the baseband lowpass filter, rendering the signal models inaccurate. In practice, this approach should be able to correct for offsets corresponding to several subcarriers, depending on the analog filtering, digital filtering, and extent over oversampling.

Like the Moose method, the Schmidl-Cox approach can also be used for OFDM symbol synchronization, by looking for a correlation peak around the first OFDM symbol as

$$\widehat{d} = \arg\max_{\Delta} \frac{\left|\sum_{n=0}^{N/2-1} \bar{y}[n+\Delta+N/2]\bar{y}^*[n+\Delta]\right|^2}{\left(\sum_{n=0}^{N/2-1} |\bar{y}[n+\Delta+N/2]|^2\right)^2}. \quad (5.269)$$

There is one distinct point of difference, though. With OFDM, there is also a cyclic prefix and often $L_c > L$. In this case, then

$$y[N+L_c+n] \approx e^{j2\pi\epsilon n}y[n] \quad (5.270)$$

for $n = L, L+1, \dots, L_c, L_c+1, \dots, N+L_c-1$. This means that there will be some ambiguity in (5.269), in that there may be several values of d that are close, especially when L_c is much larger than L. This creates a plateau in the symbol synchronization algorithm [296]. Any value within the plateau, however, should give acceptable performance if the estimated channel has order L_c, but if the smaller channel length of L is assumed, then performance may suffer. Other training sequence designs can improve the sharpness of the frame synchronization, specifically with multiple repetitions along with a sign change among the different repetitions [306].

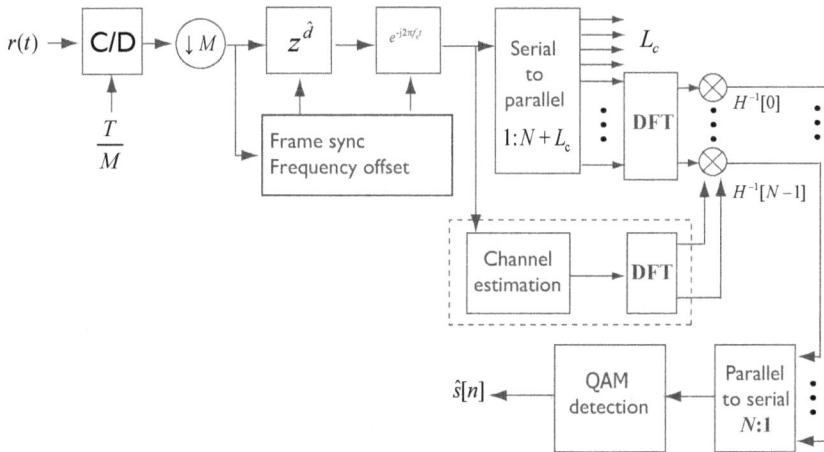

Figure 5.29 OFDM receiver frequency offset estimation and correction, channel estimation, and linear equalization. The matched filtering and symbol sampling are omitted as they are often not performed in OFDM.

A system block diagram with OFDM and frame synchronization, channel estimation, and carrier frequency offset correction is illustrated in Figure 5.29. One frequency offset correction block is shown, but this could be broken up into two pieces corresponding to integer and fine offsets.

Example 5.23   In this example, we show how the STF in IEEE 802.11a is constructed to have periodicity. The STF is generated from a special training sequence with 12 nonzero values constructed as

$$\{t[4k]\}_{k=0}^{15} = \{0, -1-j, -1-j, 1+j, 1+j, 1+j, 1+j,$$
$$0, 0, 0, 0, 1+j, -1-j, 1+j, -1-j, 1+j\} \tag{5.271}$$

for $k = 0, 1, \dots, 15$ and the other undefined values of the training are zero.

The presence of zeros in (5.271) is a result of the need for eliminating the DC subcarrier and spectral shaping, similar to the design of the CEF. Based on this training sequence,

$$w[n] = \frac{1}{64} \sum_{k=0}^{63} t[k] e^{j\frac{2\pi k(n-16)}{64}} \tag{5.272}$$

$$= \frac{1}{64} \sum_{k=0}^{15} t[4k] e^{j\frac{2\pi 4k(n-4)}{64}} \tag{5.273}$$

$$= \frac{1}{64} \sum_{k=0}^{15} t[4k] e^{j\frac{2\pi k(n-4)}{16}} \tag{5.274}$$

for $n = 0, 1, \dots, 159$. The time-domain waveform thus contains ten repetitions of length 16.

5.5   Introduction to Wireless Propagation

Of all the impairments, the propagation channel has the most impact on the design of a wireless receiver. The wireless channel causes the transmitted signal to lose power as it propagates from the transmitter to the receiver. Reflections, diffraction, and scattering create multiple propagation paths between the transmitter and the receiver, each with a different delay. The net result is that wireless propagation leads to a loss of received signal power as well as the presence of multipath, which creates frequency selectivity in the channel. In this section, we provide an introduction to the key mechanisms of propagation. Then we rationalize the need for developing models and explain the distinction between large-scale models (discussed further in Section 5.6) and small-scale models (discussed further in Section 5.7 and Section 5.8). In this section, we briefly describe important factors affecting propagation in wireless channels to explain the need for several different channel models. The interested reader is referred to [270] or [165] for more extensive treatment of wireless propagation.

5.5.1   Mechanisms of Propagation

In a wireless communication system, a transmitted signal can reach the receiver via a number of propagation mechanisms. In this section, we review at a high level these key mechanisms, each potentially associated with a different propagation path. These mechanisms are illustrated in Figure 5.30.

When a signal reaches the receiver from the transmitter in a single path, without suffering any reflections, diffractions, or scattering, this is known as propagation along the *line-of-sight* (LOS) path. An LOS component has the shortest time delay among all the received signals and is usually the strongest signal received. The exact classification of a path being LOS requires that any obstructions be sufficiently far away from the path, which is quantified by the idea of the Fresnel zone [270].

In *non-line-of-sight* (NLOS) propagation, a signal transmitted into a wireless medium reaches the receiver via one or more indirect paths, each having different attenuations

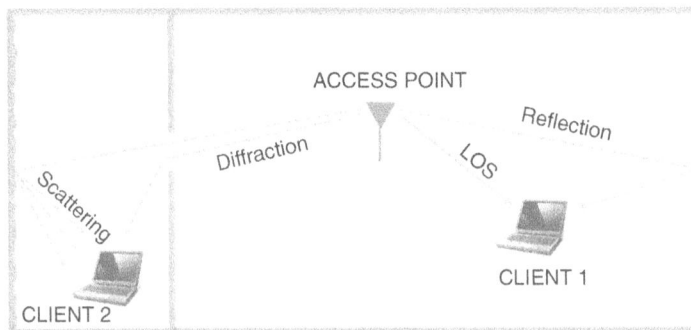

Figure 5.30 The mechanisms of propagation in the context of an indoor wireless local area network. The LOS path is unobstructed between the access point and client 1. That client also receives weaker signals as a result of a reflection off of a wall. The LOS path is obstructed for client 2, who instead receives signals through diffraction in the doorway and also scattering off of a rough wall.

and delays. When a transmitted signal travels through communication paths other than the LOS path to reach the receiver, it is said to have undergone NLOS propagation. NLOS propagation is responsible for coverage behind buildings and other obstructions. The main NLOS propagation mechanisms are reflection, scattering, and diffraction.

Reflection occurs when a wave impinges on an object that is smooth, which means that any protrusions have dimensions much larger than a wavelength. Reflection is accompanied by refraction (transmission of the wave through the object). The strengths of the reflected and refracted waves depend on the type of material. The angles and indices of reflection and refraction are given by Snell's law.

Scattering is what happens when a wave impinges on an object that is rough or has irregularities with dimensions on the order of the wavelength. It is similar to reflection but results in a smearing of the signal around the angle of reflection. This leads to a larger loss of energy as the signal is spread over a wider area. It also results in multiple paths arriving at the receiver from a similar location with slight differences in delay.

Diffraction is the "bending" of waves around sharp corners. Important examples of diffraction include waves bending over the tops of buildings, around street corners, and through doorways. Diffraction is one of the main ways that it is possible to provide cellular coverage in cities and is one reason why lower frequencies, say less than 3GHz, are considered beachfront property in the world of cellular spectrum.

There are other mechanisms of propagation as well, such as tropospheric or ionospheric scattering, but these are not common in land mobile systems. They do have relevance, though, for battlefield networks and for amateur radio enthusiasts.

5.5.2   Propagation Modeling

Propagation has an impact on several aspects of radio link performance. It determines the received signal strength and thus the signal-to-noise ratio, throughput, and probability of error. Propagation also plays a major role in system design and implementation. For example, it determines the length of the discrete-time channel, which determines how much equalization is required and in turn how much training is required. The speed at which the channel varies determines the frequency of equalization and the frequency of training—in other words, how often the channel must be reestimated.

An important component of the study of wireless communication is propagation modeling. A propagation model is a mathematical model (typically stochastic) to characterize either the propagation channel or some function of the propagation channel. Some models try to model the impulse response of the channel, whereas others try to model specific characteristics of the channel like the received power. Propagation models are usually inspired by measurement campaigns. Some models have many parameters and are designed to model specified real-world propagation scenarios. Other models have few parameters and are more amenable for tractable mathematical analysis.

There are many ways to classify propagation models. A common first-order classification is whether they describe large-scale or small-scale phenomena. The term *scale* refers to a wavelength. Large-scale phenomena refer to propagation characteristics over hundreds of wavelengths. Small-scale phenomena refer to propagation characteristics in an area on the order of a wavelength.

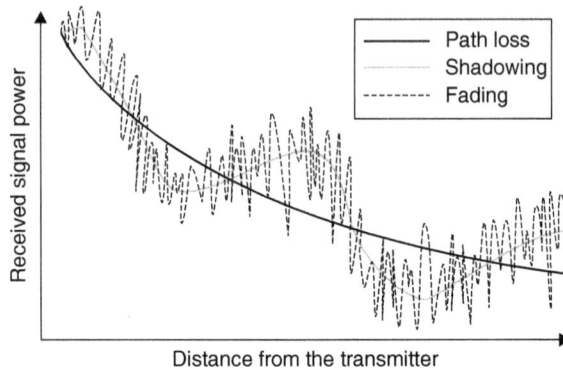

Figure 5.31 Representation of the large-scale (distance-dependent path loss and shadowing) and small-scale (fading) propagation effects

To illustrate different large-scale and small-scale phenomena, consider the average received signal power as a function of distance in Figure 5.31. Three different received signal power realizations are plotted. The first model is the mean (or median) path loss, which characterizes the average signal behavior, where the average is taken over hundreds of wavelengths. In most models, this is an exponential decay with distance. The second model is path loss with shadow fading. Here large obstructions like buildings or foliage are included to provide variability around the mean on a large scale. Shadow fading is often modeled by adding a Gaussian random component (in decibels) parameterized by the standard deviation to the mean path loss. The third model also includes small-scale fading, where the signal level experiences many small fluctuations as a result of the constructive and destructive addition of the multipath components of a transmitted signal. Because the fluctuations happen over a much smaller spatial scale, it is common to develop separate models for large-scale effects (like mean path loss and shadowing) and small-scale effects (like multipath fading).

In essence, large-scale fading models describe the average behavior of the channel in a small area and are used to infer channel behavior over longer distances. Small-scale fading models describe the localized fluctuations in a given area and may be location dependent.

Models for both large-scale and small-scale propagation phenomena are important. Large-scale trends influence system planning, the link budget, and network capacity predictions, and they capture the "typical" loss in received signal strength as a function of distance. Small-scale trends influence physical-layer link design, modulation schemes, and equalization strategies by capturing local constructive and destructive multipath effects. The received signal processing algorithms depend more strongly on small-scale models, but the net performance of those algorithms in a system depends on the large-scale models as well.

Propagation models are widely used for wireless system design, evaluation, and algorithm comparison. Standards bodies (IEEE 802.11, 3GPP, etc.) find them useful for enabling different companies to compare and contrast performance of different candidate schemes. Often these models are chosen to suit certain characteristics like propagation environment (urban, rural, suburban) or receiver speed (fixed, pedestrian, or high speed). The models typically have many possible parameter choices and are used primarily as part of system simulation.

In this section, we focus on developing models for the discrete-time equivalent channel. We decompose the channel taps as a product of a large-scale coefficient and a small-scale coefficient as

$$h[\ell] = \sqrt{G}h_{\mathrm{s}}[\ell] \quad \text{for} \quad \ell = 0, 1, \dots, L. \tag{5.275}$$

The large-scale gain is $G = E_{\mathrm{x}}/P_{\mathrm{rx,lin}}(d)$ where $P_{\mathrm{rx,lin}}(d)$ is the distance-dependent path-loss term in linear (to distinguish it from the more common decibel measure $P_{\mathrm{rx}}(d)$). The small-scale fading coefficient is denoted by $h_{\mathrm{s}}[\ell]$. The path loss is the ratio of the transmit power to the receive power and is often called the path gain for this reason. With complex pulse-amplitude modulation, for example, the transmit power is E_{x}/T, so the received power is $(E_{\mathrm{x}}/T)/P_{\mathrm{rx,lin}}(d)$. In Section 5.6, we describe models for $P_{\mathrm{rx}}(d)$, including free space, log distance, and shadowing. In Section 5.8, we describe models for $\{h_{\mathrm{s}}[\ell]\}_{\ell=0}^{L}$, some that are good for analysis like the IID Rayleigh fading model, and others that are inspired by physical mechanisms of propagation like the clustered model.

5.6 Large-Scale Channel Models

In this section, we review several large-scale models for signal strength path loss as a function of distance, denoted by $P_{\mathrm{r}}(d)$. We review the Friis free-space equation as a starting point, followed by the log-distance model and the LOS/NLOS path-loss model. We conclude with some performance calculations using path-loss models.

5.6.1 Friis Free-Space Model

For many propagation prediction models, the starting point has been the Friis free-space model [165, 270]. This model is most appropriate for situations where there are no obstructions present in the propagation environment, like satellite communication links or in millimeter-wave LOS links. The Friis free-space equation is given by

$$P_{\mathrm{rx,lin}}(d) = \frac{P_{\mathrm{tx,lin}}G_{\mathrm{t,lin}}G_{\mathrm{r,lin}}\lambda^2}{(4\pi)^2 d^2} \tag{5.276}$$

where the key quantities are summarized as follows:

- $P_{\mathrm{tx,lin}}$ is the transmit power in linear. For complex pulse-amplitude modulation $P_{\mathrm{tx,lin}} = E_{\mathrm{x}}/T$.

- d is the distance between the transmitter and the receiver. It should be in the same units as λ, normally meters.

- λ is the wavelength of the carrier, normally in meters.

- $G_{\mathrm{t,lin}}$ and $G_{\mathrm{r,lin}}$ are the transmit and receive antenna gains, normally assumed to be unity unless otherwise given.

A loss factor may also be included in the denominator of (5.276) to account for cable loss, impedance mismatch, and other impairments.

The Friis free-space equation implies that the isotropic path loss (assuming unity antenna gains $G_{r,\text{lin}} = G_{t,\text{lin}} = 1$) increases inversely with the wavelength squared, λ^{-2}. This fact implies that higher carrier frequencies, which have smaller λ, will have higher path loss. For a given physical antenna aperture, however, the maximum gains for a nonisotropic antenna generally scale as $G_{\text{lin}} = 4\pi A_e/\lambda^2$, where the effective aperture A_e is related to the physical size of the antenna and the antenna design [270]. Therefore, if the antenna aperture is fixed (the antenna area stays "the same size"), then path loss can actually be lower at higher carrier frequencies. Compensating for path loss in this manner, however, requires directional transmission with high-dimensional antenna arrays and is most feasible at higher frequencies like millimeter wave [268].

Most path-loss equations are written in decibels because of the small values involved. Converting the Friis free-space equation into decibels by taking the log of both sides and multiplying by 10 gives

$$P_{\text{rx}}(d) = P_{\text{tx}} + G_t + G_r + 20\log_{10}(\lambda) - 20\log_{10}(4\pi) - 20\log_{10}(d) \quad (5.277)$$

where $P_{\text{rx}}(d)$ and P_{tx} are measured in decibels referenced to 1 watt (dBW) or decibels referenced to 1 milliwatt (dBm), and G_t and G_r are in decibels. All other path-loss equations are given directly in terms of decibels.

The path loss is the ratio of $P_{r,\text{lin}}(d) = P_{\text{tx,lin}}/P_{\text{rx,lin}}(d)$, or in decibels $P_{\text{tx}} - P_{\text{rx}}(d)$. Path loss is essentially the inverse of the path gain between the transmitter and the receiver. Path loss is used because the inverse of a small gain becomes a large number. The path loss for the Friis free-space equation in decibels is

$$P_r(d) = 20\log_{10}(d) + 20\log_{10}(4\pi) - G_t - G_r - 20\log_{10}(\lambda). \quad (5.278)$$

Example 5.24 Calculate the free-space path loss at 100m assuming $G_t = G_r = 0$dB and $\lambda = 0.01$m.
Answer:

$$P_r(100) = 20\log_{10}(100) + 20\log_{10}(4\pi) - G_t - G_r - 20\log_{10}(0.01) \quad (5.279)$$
$$= 102\text{dB}. \quad (5.280)$$

To build intuition, it is instructive to explore what happens as one parameter changes while keeping the other parameters fixed. For example, observe the following (keeping in mind the caveats about how antenna gains might also scale):

- If the distance doubles, the path loss increases by 6dB.
- If the wavelength doubles, the received power increases by 6dB.
- If the frequency doubles (inversely proportional to the wavelength), the received power drops by 6dB.

From a system perspective, it also makes sense to fix the maximum value of loss $P_r(d)$ and then to see how d behaves if other parameters are changed. For example, if the wavelength doubles, then the distance can also be doubled while maintaining the same loss. Equivalently, if the frequency doubles, then the distance decreases by half. This effect is also observed in more complicated real-world propagation settings. For example, with the same power and antenna configuration, a Wi-Fi system operating using 2.4GHz has a longer range than one operating at 5.2GHz. For similar reasons, spectrum in cellular systems below 1GHz is considered more valuable than spectrum above 1GHz (though note that spectrum is very expensive in both cases).

The loss predicted by the free-space equation is optimistic in many terrestrial wireless systems. The reason is that in terrestrial systems, ground reflections and other modes of propagation result in measured powers decreasing more aggressively as a function of distance. This means that the actual received signal power, on average, decays faster than predicted by the free-space equation. There are many other path-loss models inspired by analysis that solve this problem, including the two-ray model or ground bounce model [270], as well as models derived from experimental data like the Okumura [246] or Hata [141] models.

5.6.2 Log-Distance Path-Loss Model

The most common extension of the free-space model is the log-distance path-loss model. This classic model is based on extensive channel measurements. The log-distance path-loss model is

$$P_r(d) = \alpha + 10\beta \log_{10}(d) + \eta \tag{5.281}$$

where α and β are the linear parameters and η is a random variable that corresponds to shadowing. The equation is valid for $d \geq d_0$, where d_0 is a reference distance. Often the free-space equation is used for values of $d < d_0$. This results in a two-slope path-loss model, where there may be one distribution of η for $d < d_0$ and another one for $d > d_0$.

The linear model parameters can be chosen to fit measurement data [95, 270]. It is common to select α based on the path loss derived from the free-space equation (5.278) at the reference distance d_0. In this way, the α term incorporates the antenna gain and wavelength-dependent effects [312]. A reasonable choice for $d_0 = 1$m; see, for example, [322]. The path-loss exponent is β. Values of β vary a great deal depending on the environment. Values of $\beta < 2$ can occur in urban areas or in hallways, in what is called the urban canyon effect. Values of β near 2 correspond to free-space conditions, for example, the LOS link in a millimeter-wave system. Values of β from 3 to as high as 5 are common in microcellular models [56, 282, 107], macrocellular models [141], and indoor WLAN models [96].

The linear portions of the log-distance path-loss model capture the average variation of signal strength as a function of distance and may be called the mean path loss. In measurements, though, there is substantial variation of the measured path loss around the mean path loss. For example, obstruction by buildings creates shadowing. To account for this effect, the random variable η is incorporated. It is common to select η as $\mathcal{N}(0, \sigma_{\text{shad}}^2)$. This leads to what is called log-normal shadowing, because η is added in the log domain; in the linear domain $\log_{10}(\eta)$ has a log-normal distribution. The parameter σ_{shad} would be determined from measurement data and often has a value around 6 to 8dB. In more elaborate models, it can also be a function of distance. Under the assumption that η is Gaussian, $P_r(d)$ is also Gaussian with mean $\alpha + 10\beta \log_{10}(d)$. For analytical purposes, is common to neglect shadowing and just focus on the mean path loss.

Example 5.25   Calculate the mean log-distance path loss at 100m assuming that $\beta = 4$. Compute α from the free-space equation assuming $G_t = G_r = 0\text{dB}$, $\lambda = 0.01\text{m}$, and reference distance $d_0 = 1\text{m}$.

Answer: First, compute the reference distance by finding $\alpha = P_r(1)$ from the free-space equation (5.278):

$$P_r(1) = 20\log_{10}(1) + 20\log_{10}(4\pi) - G_t - G_r - 20\log_{10}(0.01) \tag{5.282}$$

$$= 62\text{dB}. \tag{5.283}$$

Second, compute the mean path loss for the given parameters:

$$P_r(100) = 62\text{dB} + 10 \cdot 4\log_{10}(100) \tag{5.284}$$

$$= 62\text{dB} + 80\text{dB} \tag{5.285}$$

$$= 142\text{dB}. \tag{5.286}$$

Compared with Example 5.24, there is an additional 40dB of loss due to the higher path-loss exponent in this case.

Example 5.26   Consider the same setup as in Example 5.25. Suppose that the transmit power is $P_{\text{tx}} = 20\text{dBm}$ and that $\sigma_{\text{shad}} = 8\text{dB}$.

- Determine the probability that the received power $P_{\text{rx}}(100) < -110\text{dBm}$.

  Answer: The received signal power is

$$P_{\text{rx}}(100) = P_{\text{tx}} - P_r(100) \tag{5.287}$$

$$= P_{\text{tx}} - \alpha - 10\beta \log_{10}(100) - \eta \tag{5.288}$$

$$= 20\text{dBm} - 142\text{dB} - \eta \tag{5.289}$$

$$= -122\text{dBm} - \eta. \tag{5.290}$$

  Now note that

$$\mathbb{P}\left[P_{\text{rx}}(100) < -110\text{dBm}\right] = \mathbb{P}\left[-122\text{dBm} - \eta < -110\text{dBm}\right] \tag{5.291}$$

$$= \mathbb{P}\left[-12\text{dB} < \eta\right] \tag{5.292}$$

where the dBm disappears because dBm $-$ dBm $=$ dB (the 1mW cancels). Now let us rewrite this probability in terms of $x \sim \mathcal{N}(0,1)$:

$$\mathbb{P}\left[P_{\mathrm{rx}}(100) < -110\mathrm{dBm}\right] = \mathbb{P}\left[\frac{-12\mathrm{dB}}{\sigma_{\mathrm{shad}}} < x\right] \tag{5.293}$$

$$= 1 - \mathbb{P}\left[\frac{12\mathrm{dB}}{\sigma_{\mathrm{shad}}} < x\right] \tag{5.294}$$

$$= 1 - Q(1.5) \tag{5.295}$$

$$= 0.933. \tag{5.296}$$

This means that the received power is below -110dBm 93% of the time.

- Determine the value of d such that $P_{\mathrm{rx}}(d) < -110$dBm 5% of the time.

  Answer: We need to find d such that $\mathbb{P}\left[P_{\mathrm{rx}}(d) < -110\mathrm{dBm}\right] = 0.1$. Following the same steps as in the previous part of the example,

$$\mathbb{P}\left[P_{\mathrm{rx}}(d) < -110\mathrm{dBm}\right] = \mathbb{P}\left[20\mathrm{dBm} - 62\mathrm{dB} - 40\log_{10}(d) - \eta < -110\mathrm{dBm}\right] \tag{5.297}$$

$$= \mathbb{P}\left[68\mathrm{dB} - 40\log_{10}(d) < \eta\right] \tag{5.298}$$

$$= \mathbb{P}\left[\frac{68\mathrm{dB} - 40\log_{10}(d)}{8\mathrm{dB}} < x\right] \tag{5.299}$$

$$= 1 - \mathbb{P}\left[\frac{40\log_{10}(d) - 68\mathrm{dB}}{8\mathrm{dB}} < x\right] \tag{5.300}$$

$$= 1 - Q\left(\frac{40\log_{10}(d) - 68\mathrm{dB}}{8\mathrm{dB}}\right) \tag{5.301}$$

$$= 0.1 \tag{5.302}$$

where we recognize that $68\mathrm{dB} - 40\log_{10}(d)$ will be negative, so we can rearrange to get the final answer in the form of a Q-function. Using the inverse of the Q-function gives $d = 27.8$m.

There are many variations of the log-distance path-loss model, with more refinements based on measurement data and accounting for other parameters. One common model is the COST-231 extended Hata model for use between frequencies of 150MHz and 2GHz. It contains corrections for urban, suburban, and rural environments [336]. The basic equation for path loss in decibels is

$$P_{\mathrm{r}}(d) = 46.3 + 33.9\log_{10}(f_{\mathrm{c}}) - 13.82\log_{10}(h_{\mathrm{t}}) - b(h_{\mathrm{r}}) + (44.9 - 6.55\log_{10}(h_{\mathrm{t}}))\log_{10}(d) + C \quad (5.303)$$

where h_{t} is the effective transmitter (base station) antenna height ($30\mathrm{m} \leq h_t \leq 200\mathrm{m}$), h_{r} is the receiver antenna height ($1\mathrm{m} \leq h_{\mathrm{r}} \leq 10\mathrm{m}$), d is the distance between the transmitter and the receiver in kilometers ($d \geq 1\mathrm{km}$), $b(h_{\mathrm{r}})$ is the correction factor for the effective receiver antenna height, f_{c} is the carrier frequency in megahertz, the parameter $C = 0\mathrm{dB}$

for suburban or open environments, and $C = 3$dB for urban environments. The correction factor is given in decibels as

$$b(h_{\mathrm{r}}) = (1.1 \log_{10}(f_{\mathrm{c}}) - 0.7)h_{\mathrm{r}} - (1.56 \log_{10}(f_{\mathrm{c}}) - 0.8) \tag{5.304}$$

with other corrections possible for large cities. More realistic path-loss models like COST-231 are useful for simulations but are not as convenient for simple analytical calculations.

5.6.3   LOS/NLOS Path-Loss Model

The log-distance path-loss model accounts for signal blockage using the shadowing term. An alternative approach is to differentiate between blocked (NLOS) and unblocked (LOS) paths more explicitly. This allows each link to be modeled with potentially a smaller error. Such an approach is common in the analysis of millimeter-wave communication systems [20, 344, 13] and has also been used in the 3GPP standards [1, 2].

In the LOS/NLOS path-loss model, there is a distance-dependent probability that an arbitrary link of length d is LOS, which is given by $P_{\mathrm{los}}(d)$. At that same distance, $1 - P_{\mathrm{los}}(d)$ denotes the probability that the link is NLOS. There are different functions for $P_{\mathrm{los}}(d)$, which depend on the environment. For example,

$$P_{\mathrm{los}}(d) = e^{-d/C} \tag{5.305}$$

is used for suburban areas where $C = 200$m in [1]. For large distances, $P_{\mathrm{los}}(d)$ quickly converges to zero. This matches the intuition that long links are more likely to be blocked.

It is possible to find values of C for any area using concepts from random shape theory [20]. In this case, buildings are modeled as random objects that have independent shape, size, and orientation. For example, assuming the orientation of the buildings is uniformly distributed in space, it follows that

$$C = \frac{\pi}{\lambda_{\mathrm{bldg}}\mathbb{E}\left[P_{\mathrm{bldg}}\right]}, \tag{5.306}$$

where λ_{bldg} is the average number of buildings in a unit area and $\mathbb{E}\left[P_{\mathrm{bldg}}\right]$ is the average perimeter of the buildings in the investigated region. This provides a quick way to approximate the parameters of the LOS probability function without performing extensive simulations and measurements.

Let $P_{\mathrm{r}}^{\mathrm{los}}(d)$ denote the path-loss function assuming the link is LOS, and let $P_{\mathrm{r}}^{\mathrm{nlos}}(d)$ denote the path-loss function assuming the link is NLOS. While any model can be used for these functions, it is common to use the log-distance path-loss model. Shadowing is often not included in the LOS model but may be included in the NLOS model. The path-loss equation is then

$$P_{\mathrm{r}}(d) = \mathrm{I}(p_{\mathrm{los}}(d))P_{\mathrm{r}}^{\mathrm{los}}(d) + \mathrm{I}(p_{\mathrm{los}}(d))P_{\mathrm{r}}^{\mathrm{nlos}}(d) \tag{5.307}$$

where $p_{\mathrm{los}}(d)$ is a Bernoulli random variable which is 1 with probability $P_{\mathrm{los}}(d)$, and $\mathrm{I}(\cdot)$ is an indicator function that outputs 1 when the argument is true and 0 otherwise. For small values of d the LOS path-loss function dominates, whereas for large values of d the NLOS path-loss function dominates.

Example 5.27 Consider the LOS/NLOS path-loss model. Assume free-space path loss for the LOS term with $G_t = G_r = 0$dB, $\lambda = 0.01$m, and a reference distance $d_0 = 1$m. Assume a log-distance path loss for the NLOS term with $\beta = 4$ and α computed from the free-space equation with the same parameters as the LOS function. Assume LOS probability in (5.305) with $C = 200$m.

Compute the mean signal power at $d = 100$m.

Answer: First, compute the expectation of (5.307) as

$$\mathbb{E}[P_r(d)] = \mathbb{E}[I(p_{los}(d))]P_r^{los}(d) + \mathbb{E}[I(p_{los}(d))]P_r^{nlos}(d) \tag{5.308}$$

$$= P_{los}(d)P_r^{los}(d) + (1 - P_{los}(d))P_r^{nlos}(d). \tag{5.309}$$

Evaluating at $d = 100$ and using the results from Example 5.24 and Example 5.25,

$$\mathbb{E}[P_r(100)] = e^{-100/200}102\text{dB} + (1 - e^{-100/200})142\text{dB} \tag{5.310}$$

$$= 0.61 \cdot 102\text{dB} + 0.39 \cdot 142\text{dB} \tag{5.311}$$

$$= 117.6\text{dB}. \tag{5.312}$$

Example 5.28 Consider the same setup as in Example 5.27. Suppose that the transmit power is $P_{tx} = 20$dBm. Determine the probability that the received power $P_{rx}(100) < -110$dBm.

Answer: The received signal power assuming LOS is $20\text{dBm} - 102 = -82$dBm using the result of Example 5.24. The received power assuming NLOS is $20\text{dBm} - 142 = -122$dBm using the results of Example 5.25. In the LOS case, the received power is greater than -110dBm, whereas it is less than -110dBm in the NLOS case. Since there is no shadow fading, the probability that the received power $P_{rx}(100) < -110$dBm is $P_{los}(100) = 0.61$.

5.6.4 Performance Analysis Including Path Loss

Large-scale fading can be incorporated into various kinds of performance analysis. To see the impact of path loss on the algorithms proposed in this chapter, a natural approach is to include the G term in (5.275) explicitly when simulating the channel. To illustrate this concept, we consider an AWGN channel, deferring more detailed treatment until after small-scale fading has been introduced.

Consider an AWGN channel where $h[\ell] = \sqrt{G}\delta[\ell]$ and $G = E_x/P_{r,lin}(d)$. The SNR in this case is

$$\text{SNR}_{|P_{r,lin}(d)} = \frac{\text{SNR}}{P_{r,lin}(d)}. \tag{5.313}$$

Note that $\text{SNR}_{P_{\text{r,lin}}(d)}$ may be a random variable if shadowing or blockage is included in the path-loss model. Given a realization of the path loss, the probability of symbol error can then be computed by substituting $\text{SNR}_{|P_{\text{r,lin}}(d)}$ into the appropriate probability of symbol error equation $P_{\text{e}}^{\text{QAM}}\left(\frac{\text{SNR}}{P_{\text{r,lin}}(d)}\right)$, for example, for M-QAM in (4.147). The average probability of symbol error can be computed by taking the expectation with respect to the random parameters in the path loss as $\mathbb{E}\left[P_{\text{e}}\left(\frac{\text{SNR}}{P_{\text{r,lin}}(d)}\right)\right]$. In some cases, the expectation can be computed exactly. In most cases, though, an efficient way to estimate the average probability of symbol error is through Monte Carlo simulations.

Example 5.29 Compute the average probability of symbol error assuming AWGN and an LOS/NLOS path-loss model.
Answer: Let us denote the path loss in the LOS case as $P_{\text{r,lin,los}}(d)$ and in the NLOS case as $P_{\text{r,lin,nlos}}(d)$, in linear units. Based on (5.307), using similar logic to that in Example 5.27,

$$\mathbb{E}\left[P_{\text{e}}\left(\frac{\text{SNR}}{P_{\text{r,lin}}(d)}\right)\right] = P_{\text{r}}^{\text{los}}(d)P_{\text{e}}\left(\frac{\text{SNR}}{P_{\text{r,lin,los}}(d)}\right) + (1 - P_{\text{r}}^{\text{los}}(d))P_{\text{e}}\left(\frac{\text{SNR}}{P_{\text{r,lin,nlos}}(d)}\right).$$
(5.314)

Path loss is important for analyzing performance in cellular systems, or more generally any wireless system with frequency reuse. Suppose that a user is located at distance d away from the serving base station. Further suppose that there are N_{bs} interfering base stations, each a distance d_n away from the user. Assuming an AWGN channel, and treating interference as additional noise, the SINR is

$$\text{SINR}_{|P_{\text{r,lin}}(d),\{P_{\text{r,lin}}(d_n)\}_n} = \frac{\frac{E_{\text{x}}}{P_{\text{r,lin}}(d)}}{E_{\text{x}}\sum_{n=1}^{N_{\text{bs}}}\frac{1}{P_{\text{r,lin}}(d_n)} + N_{\text{o}}}$$
(5.315)

$$= \frac{\text{SNR}}{\text{SNR}\sum_{n=1}^{N_{\text{bs}}}\frac{P_{\text{r,lin}}(d)}{P_{\text{r,lin}}(d_n)} + 1}.$$
(5.316)

From this expression, it becomes clear that the ratio $P_{\text{r,lin}}(d)/P_{\text{r,lin}}(d_n)$ determines the relative significance of the interference. Ideally $P_{\text{r,lin}}(d)/P_{\text{r,lin}}(d_n) < 1$, which means that the user is associated with the strongest base station, even including shadowing or LOS/NLOS effects.

5.7 Small-Scale Fading Selectivity

In this section, we discuss small-scale fading effects. These effects, which occur on the order of a wavelength distance, are caused by the constructive and destructive interference of multipath components. These effects manifest in the time domain, leading to time selectivity or, in the frequency domain, leading to frequency selectivity. These two forms of selectivity are independent of each other, under some assumptions. Before proceeding with the mathematical details, we first develop the intuition behind time- and frequency-selective channels. Then we present the foundations for determining whether a channel is frequency selective or time selective. We conclude with showing the potential system models used in each fading regime.

5.7.1 Introduction to Selectivity

In this section, we introduce frequency selectivity and time selectivity. We pursue these in further detail in subsequent sections.

Frequency selectivity refers to the variation of the channel amplitude with respect to frequency. To understand how multipath plays an important role in frequency selectivity, consider the following two examples. Figure 5.32(a) illustrates the frequency response with a single channel tap. In the frequency domain, the Fourier transform of the single impulse function yields a flat channel; that is, the channel amplitude does not vary with frequency. We explored the flat channel in Section 5.1. Alternatively, in Figure 5.32(b), the channel impulse response has a significant amount of multipath. In the frequency

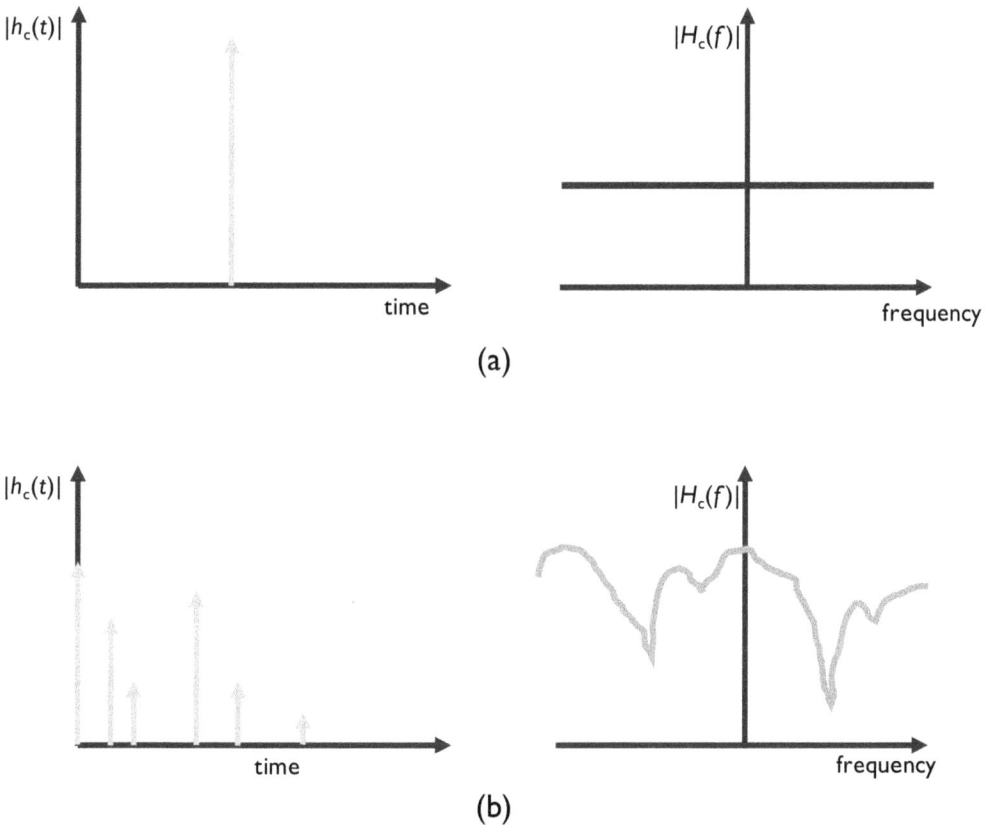

(a)

(b)

Figure 5.32 A single path channel (a) and a multipath channel (b). The single path has a frequency-flat response whereas the multipath channel has a frequency-selective response.

domain (taking the Fourier transform of the sum of shifted impulse functions), the resultant channel varies considerably with frequency.

The effect of frequency selectivity, though, depends critically on the bandwidth used to communicate in that channel. For example, illustrated in Figure 5.33 are the discrete-time complex baseband equivalent channels for two different choices of bandwidth, 1MHz and 10MHz. With the smaller bandwidth, the impulse response looks nearly like a discrete-time delta function, corresponding to a flat channel. With the larger bandwidth, there are more taps (since the symbol period is smaller) and there are many more significant taps, leading to substantial intersymbol interference. Therefore, the same channel can appear frequency flat or frequency selective depending on the signal bandwidth. The range of frequencies over which the channel amplitude remains fairly constant is the *coherence bandwidth*.

Time selectivity refers to the variation in the channel amplitude as a function of time. The degree or extent of time selectivity of a channel is measured using the *coherence time* of the channel. This simply refers to the time duration over which the channel remains fairly constant. The coherence time provides guidance on when the received signal model can be assumed to be LTI. Time selectivity is a function of the mobility present in the channel and is usually measured using the Doppler spread or the maximum Doppler shift.

For the coherence time to be useful, it must be compared with some property of the signal. For example, suppose that the receiver processes data in packets of length N_{tot}. For the LTI assumption to be valid, the channel should be roughly constant during the entire packet. Then, if TN_{tot} is less than the coherence time, we might say that the LTI assumption is good and the channel is time invariant or slow fading. This does not

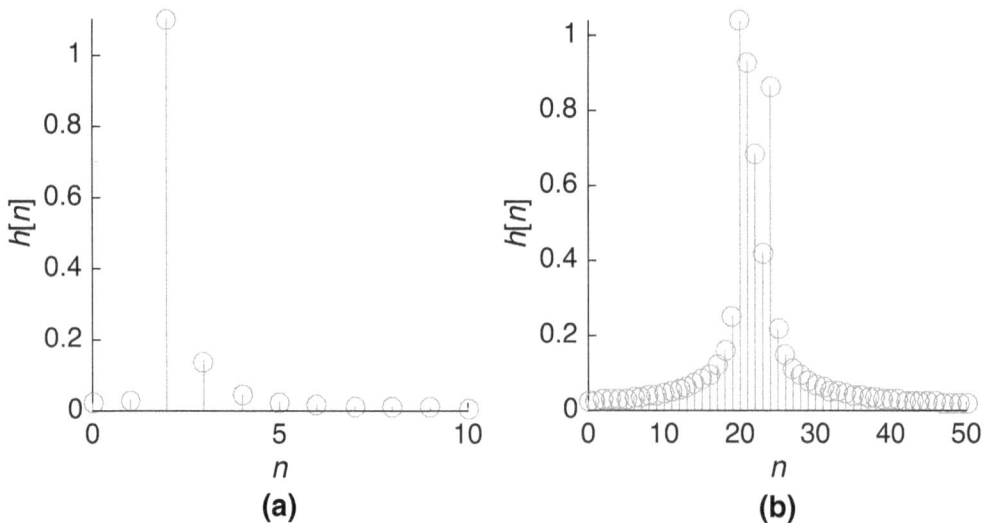

Figure 5.33 (a) Discrete-time complex baseband equivalent channel for a five-path channel with a narrow bandwidth; (b) discrete-time complex baseband equivalent channel for a five-path channel with a bandwidth ten times larger

mean that the channel does not vary at all; rather it means that during the window of interest (in this case N_{tot} symbols with period T) the channel can be considered to be time invariant. If TN_{tot} is greater than the coherence time, then we might say that the LTI assumption is not good and that the channel is time selective or fast fading.

Note that the bandwidth also plays a role in determining the coherence time of a signal. Assuming a sinc pulse-shaping filter, the symbol period $T = 1/B$. As a result, making the bandwidth larger reduces the duration of the N_{tot} symbols. Of course, increasing the bandwidth makes it more likely that the channel is frequency selective. This shows how the time and frequency variability of a channel are intertwined.

The time and frequency selectivity are coupled together. For example, if the receiver is moving and the multipaths come from different directions, each has a different Doppler shift. As a result, the time variability with several multipaths is more severe than with a single multipath. Fortunately, making some statistical assumptions about the channel allows the time selectivity to be decoupled from the frequency selectivity. Specifically, if the channel is assumed to satisfy the wide-sense stationary uncorrelated scattering assumption (WSSUS) [33], then the time-selective and frequency-selective parts of the correlation function can be decoupled, and the decision about a channel's selectivity can be made separately based on certain correlation functions, discussed in the next sections. Quantities computed from these correlation functions are compared with the bandwidth of the signal of interest, the symbol period, and the block length N_{tot} to determine the effective selectivity.

To explain this mathematically, suppose that the continuous-time complex baseband equivalent channel impulse response is $h(t, \tau)$ where t is the time and τ is the lag. This is a doubly selective channel, in that it varies both with time and with lag. The channel is measured assuming a large bandwidth, typically larger than the eventual bandwidth of the signal that will be used in this channel. For example, measurements might be made in a 500MHz channel and the results of those measurements used to decide how to partition the bandwidth among different signals. We use continuous time as this is historically how the WSSUS description was developed [33], but interpretations are also available in discrete time [152, 301].

Under the WSSUS model, a statistical description of the channel correlations is made through two functions: the power delay profile $R_{\text{delay}}(\tau)$, and the Doppler spectrum $S_{\text{Doppler}}(f)$. The power delay profile gives a measure of how the energy of the channel varies in the lag domain for two signals of very short duration (so Doppler is neglected). The Doppler spectrum measures essentially how much power is present in different Doppler shifts of two narrowband signals (so lag is neglected). Frequency selectivity is decided based on $R_{\text{delay}}(\tau)$, and time selectivity is decided based on $S_{\text{Doppler}}(f)$. In the next sections, we explore further the idea of determining the selectivity of a channel based on the power delay profile and the Doppler spectrum, or their Fourier transforms.

5.7.2 Frequency-Selective Fading

We use the power delay profile $R_{\text{delay}}(\tau)$ to determine whether a channel is frequency-selective fading. The power delay profile is typically determined from measurements; common power delay profiles can be found in different textbooks and in many standards. For example, the GSM standard specifies several different profiles, including parameters like typical urban, rural, bad urban, and others. Intuitively, in a flat channel (in the bandwidth where the power delay profile was measured) $R_{\mathrm{delay}}(\tau)$ should be close to a delta function.

The typical way to measure the severity of a power delay profile is based on the root mean square (RMS) delay spread. Define the mean excess delay as

$$\bar{\tau} = \frac{\int_0^\infty R_{\mathrm{delay}}(\tau)\tau d\tau}{\int_0^\infty R_{\mathrm{delay}}(\tau)d\tau} \qquad (5.317)$$

and the second moment as

$$\overline{\tau^2} = \frac{\int_0^\infty R_{\mathrm{delay}}(\tau)\tau^2 d\tau}{\int_0^\infty R_{\mathrm{delay}}(\tau)d\tau}. \qquad (5.318)$$

Then the RMS delay spread is the difference:

$$\sigma_{\mathrm{RMS,delay}} = \sqrt{\overline{\tau^2} - (\bar{\tau})^2}. \qquad (5.319)$$

With this definition, a channel is said to be frequency flat if the symbol period satisfies $T \gg \sigma_{\mathrm{RMS,delay}}$. This means that the effective spread is much smaller than a symbol, so there will be little ISI between adjacent symbols.

Example 5.30   Consider the exponential power delay profile $R_{\mathrm{delay}}(\tau) = e^{-\tau/\gamma}$. Determine the RMS delay spread.

Answer: The mean excess delay is

$$\bar{\tau} = \frac{\int_0^\infty \tau e^{-\tau/\gamma}d\tau}{\int_0^\infty e^{-\tau/\gamma}d\tau} \qquad (5.320)$$

$$= \frac{\gamma^2}{\gamma} \qquad (5.321)$$

$$= \gamma. \qquad (5.322)$$

The second moment is

$$\overline{\tau^2} = \frac{\int_0^\infty e^{-\tau/\gamma}\tau^2 d\tau}{\int_0^\infty e^{-\tau/\gamma}d\tau} \qquad (5.323)$$

$$= \frac{2\gamma^3}{\gamma} \qquad (5.324)$$

$$= 2\gamma^2. \qquad (5.325)$$

Therefore, the RMS delay spread is

$$\sigma_{\mathrm{RMS,delay}} = \sqrt{2\gamma^2 - \gamma^2} \qquad (5.326)$$

$$= \gamma. \qquad (5.327)$$

Therefore, the value γ is the RMS delay spread.

The Fourier transform of the power delay profile is known as the spaced-frequency correlation function:

$$S_{\text{delay}}(\Delta_{\text{lag}}) = \int_0^\infty R_{\text{delay}}(\tau)e^{-j2\pi\Delta_{\text{lag}}\tau}d\tau. \tag{5.328}$$

It measures the correlation as a function of the difference $\Delta_{\text{lag}} = f_2 - f_1$ between sinusoids sent on two different carrier frequencies. The spaced-frequency correlation function is used to define the coherence bandwidth of the channel. One definition of coherence bandwidth is the smallest value of Δ_{lag} such that $|S_{\text{delay}}(B_{\text{coh}})| = 0.5S_{\text{delay}}(0)$. Essentially this is the first point where the channel becomes decorrelated by 0.5.

It is common to define the coherence bandwidth based on the RMS delay spread. For example,

$$B_{\text{coh}} = \frac{1}{5\sigma_{\text{RMS,delay}}}. \tag{5.329}$$

The coherence bandwidth is interpreted like traditional bandwidth and is meant to give a measure over which the channel is (statistically speaking) reasonably flat. In particular, a channel is flat if the bandwidth $B \ll B_{\text{coh}}$. There are several different definitions of coherence bandwidth in the literature; all have an inverse relationship with the RMS delay spread [191, 270].

Example 5.31   Determine the spaced-frequency correlation function, the coherence bandwidth from the spaced-frequency correlation function, and the coherence bandwidth from the RMS delay spread for the exponential power delay profile in Example 5.30.

Answer: The spaced-frequency correlation function is

$$S_{\text{delay}}(\Delta_{\text{lag}}) = \int_0^\infty R_{\text{delay}}(\tau)e^{-j2\pi\Delta_{\text{lag}}\tau}d\tau \tag{5.330}$$

$$= \int_0^\infty e^{-t/\tau}e^{-j2\pi\Delta_{\text{lag}}\tau}d\tau \tag{5.331}$$

$$= \frac{1}{\gamma^{-1} + j2\pi\Delta_{\text{lag}}}. \tag{5.332}$$

To find the coherence bandwidth from the spaced-frequency correlation function:

$$|S_{\text{delay}}(\Delta_{\text{lag}})| = \sqrt{\frac{1}{\gamma^{-2} - 4\pi^2\Delta_{\text{lag}}^2}} \tag{5.333}$$

and

$$|S_{\text{delay}}(0)| = \gamma. \tag{5.334}$$

The smallest nonnegative value of Δ_{lag} that determines the coherence bandwidth is

$$B_{\text{coh}} = \frac{\sqrt{3}}{4}\frac{1}{\gamma}. \tag{5.335}$$

Based on the RMS delay spread,

$$B_{\text{coh}} = \frac{1}{5\gamma}. \tag{5.336}$$

Since $\sqrt{3}/4 \approx 0.43$ and $1/5 = 0.2$, these two expressions differ by a factor of 2. The coherence bandwidth based on the RMS delay spread is more conservative between the two measures in this case.

In practice, the power delay profile or spaced-frequency correlation function is determined from measurements. For example, suppose that a training signal is used to generate channel estimate $h[n, \ell]$ at time n. Then the discrete-time power delay profile can be estimated from M observations as $\frac{1}{N} \sum_{n=0}^{N-1} |h[n, \ell]|^2$. The spaced-frequency correlation function $\mathsf{S}_{\text{delay}}(\Delta_{\text{lag}})$ could be estimated by sending sinusoids at $\Delta f = f_2 - f_1$ and estimating the correlation between their respective channels at several different times. Or it could be computed in discrete time using an OFDM system by taking the DFT of each channel estimate $\mathsf{H}[n, k]$ for a given n, assuming K total subcarriers, then estimating the spaced-frequency correlation as a function of subcarriers as $\mathsf{S}_{\text{delay}}[k_2 - k_1] = \frac{1}{N} \sum_{n=0}^{N-1} \mathsf{H}[n, k_1]\mathsf{H}^*[n, k_2]$.

5.7.3 Time-Selective Fading

The Doppler spectrum $\mathsf{S}_{\text{Doppler}}(f)$ is used to determine if a channel is time-selective fading. The Doppler spectrum can be estimated via measurements or more commonly is based on certain analytical models. Given a general Doppler spectrum, a common approach for determining the severity of the Doppler is to define an RMS Doppler spread $\sigma_{\text{RMS,doppler}}$ in the same way the RMS delay spread is defined. Then a signal is considered to be time invariant if $B \gg \sigma_{\text{RMS,doppler}}$. In mobile channels, the maximum Doppler shift may be used instead of the RMS Doppler spread. The maximum Doppler shift is $f_{\text{m}} = f_c \nu/c$ where ν is the maximum velocity and c is the speed of light. The maximum shift occurs when the transmitter is moving either straight to or straight from the receiver with velocity ν. For many systems, the maximum Doppler shift gives a reasonable approximation of the RMS Doppler spread, typically leading to a more conservative definition of time selective.

The maximum Doppler shift varies as a function of mobility. When there is higher mobility, the velocity is higher. From a system design perspective, we often use the maximum velocity to determine the coherence time. For example, a fixed wireless system might assume only pedestrian speeds of 2mph, whereas a mobile cellular system might be designed for high-speed trains that travel at hundreds of miles per hour.

Example 5.32 Determine the maximum Doppler shift for a cellular system with $f_c = 1.9\text{GHz}$ that serves high-speed trains with a velocity of 300km/h.

Answer: The velocity of 300km/h in meters per second is 83.3m/s. The maximum Doppler shift is then

$$f_m = 1.9 \times 10^9 \frac{83.3}{2.97 \times 10^8} \tag{5.337}$$

$$= 533\text{Hz}. \tag{5.338}$$

Analytical models can also be used to determine the Doppler spectrum. Perhaps the most common is the Clarke-Jakes model. In this model, the transmitter is stationary while the receiver is moving with a velocity of ν directly toward the transmitter. There is a ring of isotropic scatterers around the receiver, meaning that multipaths arrive from all different directions and with different corresponding Doppler shifts. The Doppler spectrum under this assumption is

$$S_{\text{Doppler}}(f) = \begin{cases} \frac{1}{\pi\sqrt{f_m^2 - f^2}} & \nu \in [-f_m, f_m] \\ 0 & \nu \notin [-f_m, f_m]. \end{cases} \tag{5.339}$$

Plotting the Clarke-Jakes spectrum in Figure 5.34, we see what is known as the horned Doppler spectrum.

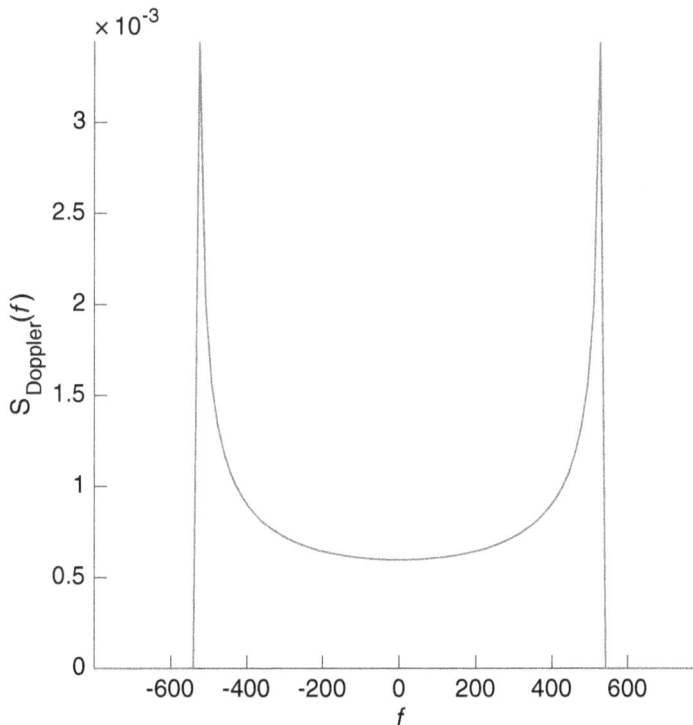

Figure 5.34 The Clarke-Jakes spectrum plotted for $f_m = 533$Hz

The time selectivity of the channel can also be determined from the spaced-time correlation function, which is the inverse Fourier transform of the Doppler spectrum:

$$R_{\text{Doppler}}(\Delta_{\text{time}}) = \int_{-\infty}^{\infty} S_{\text{Doppler}}(f)e^{j2\pi\Delta_{\text{time}}f}df. \qquad (5.340)$$

The spaced-time correlation function essentially gives the correlation between narrowband signals (so delay spread can be neglected) at two different points in time $\Delta_{\text{time}} = t_2 - t_1$. The coherence time of the channel is defined as the smallest value of Δ_{time} such that $|R_{\text{Doppler}}(\Delta_{\text{time}})| = 0.5R_{\text{Doppler}}(0)$. The resulting value is T_{coh}. It is common to define the coherence time based on either the RMS Doppler spread or the maximum Doppler shift, in the same way that the coherence bandwidth is defined. A channel is said to be LTI over block N_{tot} if $TN_{\text{tot}} \ll T_{\text{coh}}$. For the Clarke-Jakes spectrum, it is common to take $T_{\text{coh}} = 0.423/f_{\text{m}}$ [270].

Example 5.33 Determine the coherence time using the maximum Doppler shift for the same situation as in Example 5.32. Also, if a single-carrier system with 1MHz of bandwidth is used, and packets of length $N_{\text{tot}} = 100$ are employed, determine if the channel is time selective.

Answer: The coherence time is $T_{\text{coh}} = 1/(5f_{\text{m}}) = 0.375$ms. With a bandwidth of 1MHz, T is at most $1\mu s$, assuming sinc pulse shaping, less if other forms of pulse shaping are used. Comparing $N_{\text{tot}}T = 100\mu s$ with T_{coh}, we can conclude that the channel will be time invariant during the packet.

For the Clarke-Jakes spectrum, the spaced-time correlation function can be computed as

$$R_{\text{Doppler}}(\Delta_{\text{time}}) = J_0(2\pi f_{\text{m}}\Delta_{\text{time}}) \qquad (5.341)$$

where $J_0(\cdot)$ is the zeroth-order Bessel function. The spaced-time correlation function is plotted in Figure 5.35. The ripples in the temporal correlation function lead to rapid decorrelation but do show some longer-term correlations over time.

An interesting aspect of time-selective fading is that it depends on the carrier. Normally we use the baseband equivalent channel model and forget about the carrier f_{c}. Here is one place where it is important. Note that the higher the carrier, the smaller the coherence time for a fixed velocity. This means that higher-frequency signals suffer more from time variations than lower-frequency signals.

In practice, it is common to determine the spaced-time correlation function from measurements. For example, suppose that short training sequences are used to generate channel $h[n, \ell]$ at time n over N measurements. The spaced-time correlation function may then be estimated as $R_{\text{Doppler}}[n_1 - n_2] = \frac{1}{N}\sum_{n=0}^{N-1}\sum_{\ell=0}^{\infty} h[n_1, \ell]h^*[n_2, \ell]$. The Doppler spectrum could also be measured through estimating the power spectrum.

5.7.4 Signal Models for Channel Selectivity

The selectivity of a channel determines which signal processing channel model is appropriate. As a result of the decomposition of time and frequency selectivity, there are four

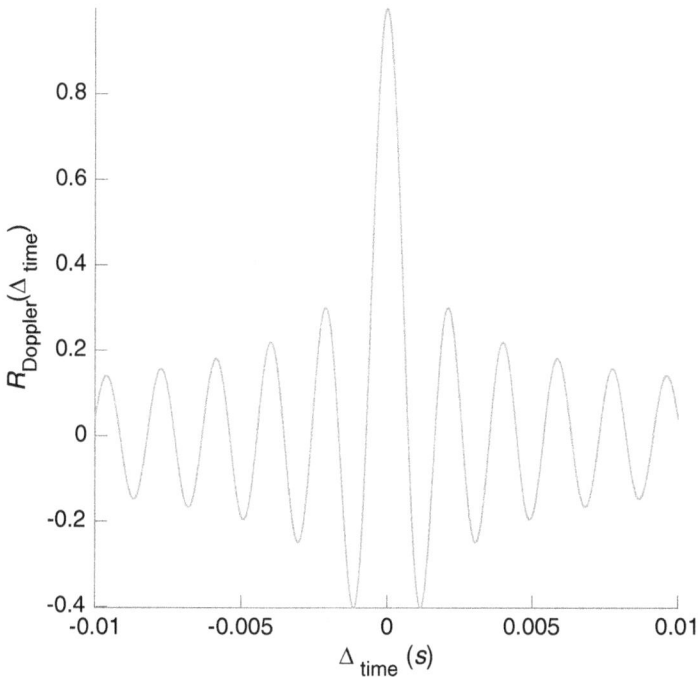

Figure 5.35 The spaced-time correlation function corresponding to the Clarke-Jakes spectrum plotted for $f_m = 533\text{Hz}$

regions of selectivity. In this section, we present typical models for each region and comment on the signal processing required at the receiver in each case.

Time Invariant/Frequency Flat In this case, the equivalent system, including the channel, carrier frequency offset, and frame delay (supposing symbol synchronization has been performed), can be written as

$$y[n] = e^{j2\pi\epsilon n} h s[n - d] + v[n] \qquad (5.342)$$

for $n = 0, 1, \ldots, N_{\text{tot}} - 1$. The signal processing steps required at the receiver were dealt with extensively in Section 5.1.

Time Invariant/Frequency Selective In this case, the equivalent system, including the channel, carrier frequency offset, and delay, is

$$y[n] = e^{j2\pi\epsilon n} \sum_{\ell=0}^{L} h[\ell] s[n - d - \ell] + v[n] \qquad (5.343)$$

for $n = 0, 1, \ldots, N_{\text{tot}} - 1$ where the impulse response $\{h[\ell]\}_{\ell=0}^{L}$ includes the effects of multipath, the transmit pulse shape, and receive matched filter, as well as any symbol synchronization errors. The signal processing steps required at the receiver were dealt with extensively in Section 5.2 through Section 5.4.

Time Variant/Frequency Flat   Assuming that the channel changes slowly with respect to the symbol period but changes faster than TN_{tot},

$$y[n] = h[n]s[n - d] + v[n] \tag{5.344}$$

for $n = 0, 1, \ldots, N_{\text{tot}} - 1$. If the channel changes too fast relative to T, then the transmit pulse shape will be significantly distorted and a more complex linear time-varying system model will be required. We have incorporated the presence of a small carrier frequency offset into the time-varying channel; the presence of larger offsets would require a different model.

One way to change the channel estimation and equalization algorithms in Section 5.1 is to include a tracking loop. The idea is to use periodically inserted pilots (at less than the coherence time) to exploit the correlation between the samples of $h[n]$. In this way, an estimate of $\widehat{h}[n]$ can be generated by using concepts of prediction and estimation, for example, a Wiener or Kalman filter [172, 143]. Another approach is to avoid channel estimation altogether and resort to differential modulation techniques like DQPSK. Generally, these methods have an SNR penalty over coherent modulations but have relaxed or no channel estimation requirements [156, 287].

Time Variant/Frequency Selective   Assuming that the channel changes slowly with respect to the symbol period but changes faster than TN_{tot},

$$y[n] = \sum_{\ell=0}^{L} h[n, \ell]s[n - d - \ell] + v[n] \tag{5.345}$$

for $n = 0, 1, \ldots, N_{\text{tot}} - 1$. The channel is described by a two-dimensional linear time-varying system with impulse response $\{h[n, \ell]\}_{\ell=0}^{L}$, which is often called a doubly selective channel. We have again incorporated the presence of carrier frequency offset into the time-varying channel. If the channel changes a lot, then some additional Nyquist assumptions may need to be made and the discrete-time model would become much more complex (typically this is required only for extremely high Dopplers).

The time- and frequency-selective channel is the most challenging from a signal processing perspective because the channel taps change over time. Operating under this kind of channel is challenging for two reasons. First, estimating the channel coefficients is difficult. Second, even with the estimated coefficients, the equalizer design is also challenging. One way to approach this problem is to use basis expansion methods [213] to represent the doubly selective channel as a function of a smaller number of more slowly varying coefficients. This parametric approach can help with channel estimation. Then modifications of OFDM modulations can be used that are more suitable for time-varying channels [320, 359]. Operating in the time- and frequency-selective region is common for underwater communication [92]. It is not yet common for terrestrial systems, though there is now growing interest by companies like Cohere Technologies [230].

5.8   Small-Scale Channel Models

Because of the challenges associated with time selectivity, wireless communication systems are normally engineered such that the channel is time invariant over a packet, frame, block, or burst (different terminology for often the same thing). This is generally called fading. Even in such systems, though, the channel may still vary from frame to frame. In this section, we describe stochastic small-scale fading models. These models are used to describe how to generate multiple realizations of a channel for analysis or simulation. First, we review some models for flat fading, in particular Rayleigh, Ricean, and Nakagami fading. Then we review some models for frequency-selective fading, including a generalization of the Rayleigh model and the Saleh-Valenzuela clustered channel model. We conclude with an explanation about how to compute a bound on the average probability of symbol error in a flat-fading channel.

5.8.1 Flat-Fading Channel Models

In this section, we present different models for flat-fading channels. We focus on the case where the channel h_s is a random variable that is drawn independently from frame to frame. We provide some explanation about how to deal with correlation over time at the end.

The most common model for flat fading is the Rayleigh channel model. In this case, h_s has distribution $\mathcal{N}_\mathbb{C}(0, 1)$. The variance is selected so that $\mathbb{E}[|h_\text{s}|^2] = 1$, so that all gain in the channel is added by the large-scale fading channel. This is called the Rayleigh model because the envelope of the channel $|h_\text{s}|$ has the Rayleigh distribution, in this case given by $f_{h_\text{s}}(x) = 2x \exp(-x^2)$. The magnitude squared $|h_\text{s}|^2$ is the sum of the squares of two $\mathcal{N}(0, 1/2)$ random variables, giving it a (scaled due to the $1/2$) chi-square distribution. The channel phase (h_s) is uniformly distributed on $[0, 2\pi]$. The Rayleigh fading channel model is said to model the case where there is a rich scattering NLOS environment. In this case, paths arrive from all different directions with slightly different phase shifts, and based on the central limit theorem, the distribution converges to a Gaussian.

Sometimes there is a dominant LOS path. In these cases, the Ricean channel model is used, parameterized by the K-factor. In the Ricean model, h_s has distribution $\mathcal{N}_\mathbb{C}(\mu, \sigma^2)$. The Rice factor is $K = |\mu|^2/\sigma^2$. Enforcing $\mathbb{E}[|h_\text{s}|^2] = 1$, then $\mu^2 + \sigma^2 = 1$. Substituting for $\mu^2 = \sigma^2 K$ and simplifying leads to $|\mu| = \sqrt{K/(1 + K)}$ and $\sigma^2 = 1/(1 + K)$. Then, in terms of K, h_s has distribution $\mathcal{N}_\mathbb{C}(e^\theta \sqrt{K/(1 + K)}, 1/(1 + K))$ where $\theta = \text{phase}(\mu)$, but in most cases $\theta = 0$ is selected. The Rice factor varies from $K = 0$ (corresponding to the Rayleigh case) to $K = \infty$ (corresponding to a non-fading channel).

Other distributions are used as well, inspired by measurement data, to give a better fit with observed data. The most common is the Nakagami-m distribution, which is a distribution for $|h_\text{s}|$ with an extra parameter m, with $m = 1$ corresponding to the Rayleigh case. The phase of h_s is taken to be uniform on $[0, 2\pi]$. The Nakagami-m distribution is

$$f(x) = \frac{2m^m x^{2m-1}}{\Gamma(m)} e^{-mx^2}. \tag{5.346}$$

When $|h_\text{s}|$ is Nakagami-m, $|h_\text{s}|^2$ has a gamma distribution $\Gamma(m, 1/m)$. The gamma distribution with two parameters is also used to give further flexibility for approximating measured data.

Time selectivity can also be incorporated into flat-fading models. This is most commonly done for the Rayleigh fading distribution. Let $h_\text{s}[n]$ denote the small-scale fading distribution as a function of n. In this case, n could index symbols or frames, depending on whether the channel is being generated in a symbol-by-symbol fashion or a frame-by-frame fashion. For purposes of illustration, we generate the channel once per N_{tot} symbols.

Suppose that we want to generate data with spatial correlation function $R_{\text{Doppler}}(\Delta_{\text{time}})$. Let $R_{\text{Doppler}}[k] = R_{\text{Doppler}}(kTN_{\text{tot}})$ be the equivalent discrete-time correlation function. We can generate a Gaussian random process with correlation function $R_{\text{Doppler}}[k]$ by generating an IID Gaussian random process with $\mathcal{N}_{\mathbb{C}}(0,1)$ and filtering that process with a filter $q[k]$ such that $R_{\text{Doppler}}[k] = q[k] * q^*[-k]$. Such a filter can be found using algorithms in statistical signal processing [143]. It is often useful to implement the convolution by representing the filters in terms of their poles and zeros. For example, the coefficients of an all-pole IIR filter can be found by solving the Yule-Walker equations using the Levinson-Durbin recursion. The resulting channel would then be considered an autoregressive random process [19]. Other filter approximations are also possible, for example, using an autoregressive moving average process [25].

There are also deterministic approaches for generating random time-selective channels. The most widely known approach is Jakes's sum-of-sinusoids approach [165], as a way to generate a channel with approximately the Clarke-Jakes Doppler spectrum. In this case, a finite number of sinusoids are summed together, with the frequency determined by the maximum Doppler shift and a certain amplitude profile. This approach was widely used with hardware simulators. A modified version with a little more randomness improves the statistical properties of the model [368].

5.8.2 Frequency-Selective Channel Models

In this section, we present different models for frequency-selective fading channels, focusing on the case where the channel coefficients are generated independently for each frame. Unlike in the previous section, we present two classes of models. Some are defined directly in discrete time. Others are physical models generated in continuous time, then converted to discrete time.

The discrete-time frequency-selective Rayleigh channel is a generalization of Rayleigh fading. In this model, the variance of the taps changes according to the specified symbol-spaced power delay profile $R_{\text{delay}}[\ell] = R_{\text{delay}}(\ell T)$. In this model $h_{\text{s}}[\ell]$ is distributed as $\mathcal{N}_{\mathbb{C}}(0, R_{\text{delay}}[\ell])$. A special case of this model is the uniform power delay profile where $h_{\text{s}}[\ell]$ is distributed as $\mathcal{N}_{\mathbb{C}}(0,1)$, which is often used for analysis because of its simplicity. Sometimes the first tap has a Ricean distribution to model an LOS component.

Note that the total power in the channel $\sum_{\ell=0}^{L} |h_{\text{s}}[\ell]|^2 = \sum_{\ell=0}^{L} R_{\text{delay}}[\ell]$, which depending on the power delay profile may be greater than 1. This is realistic because the presence of multiple paths allows the receiver to potentially capture more energy versus a flat-fading channel. Sometimes, though, the power delay profile may be normalized to have unit energy, which is useful when making comparisons with different power delay profiles.

Another class of frequency-selective channel models forms what are known as clustered channel models, the first of which is the Saleh-Valenzuela model [285]. This is a model for a continuous-time impulse response based on a set of distributions for amplitudes and delays and is a generalization of the single-path and two-path channels used as examples in Section 3.3.3. The complex baseband equivalent channel is

$$h(t) = \sum_{m=0}^{\infty} \sum_{q=0}^{\infty} \alpha_{m,q} g(t - T_m - \tau_{m,q}) \tag{5.347}$$

where T_m is the cluster delay, $\alpha_{m,q}$ is the complex path gain, and $\tau_{m,q}$ is the path delay. The discrete-time equivalent channel is then given using the calculations in Section 3.3.5 as

$$h_{\mathrm{s}}[n] = T \sum_{m=0}^{\infty} \sum_{q=0}^{\infty} \alpha_{m,q} g(nT - T_m - \tau_{m,q}). \tag{5.348}$$

The choice of $g(t)$ depends on exactly where the channel is being simulated. If the simulation only requires $\{h_{\mathrm{s}}[\ell]\}_{\ell=0}^{L}$ obtained after matched filtering and sampling, then choosing $g(t) = g_{\mathrm{tx}}(t) * g_{\mathrm{rx}}(t)$ makes sense. If the channel is being simulated with over-sampling prior to symbol synchronization and matched filtering, then alternatively it may make sense to replace $q(t)$ with a lowpass filter with bandwidth corresponding to the bandwidth of $x(t)$. In this case, the discrete-time equivalent may be obtained through oversampling, for example, by T/M_{rx}.

The Saleh-Valenzuela model is inspired by physical measurements that show that multipaths tend to arrive in clusters. The parameter T_m denotes the delay of a cluster, and the corresponding $\tau_{m,q}$ denotes the qth ray from the mth cluster.

The cluster delays are modeled as a Poisson arrival process with parameter Φ. This means that the interarrival distances $T_m - T_{m-1}$ are independent with exponential distribution $f(T_m|T_{m-1}) = \Phi \exp(-\Phi(T_m - T_{m-1}))$. Similarly, the rays are also modeled as a Poisson arrival process with parameter ϕ, giving $\tau_{m,\ell} - \tau_{m,\ell-1}$ an exponential distribution. The parameters Φ and ϕ would be determined from measurements.

The gains $\alpha_{m,q}$ are complex Gaussian with a variance that decreases exponentially as the cluster delay and the ray delay increase. Specifically, $\alpha_{m,q}$ is distributed as $\mathcal{N}_{\mathbb{C}}(0, \exp(-T_m/\bar{T}) \exp(-\tau_{m,q}/\bar{\tau})$ where $\bar{T}$ and $\bar{\tau}$ are also parameters in the model.

The Saleh-Valenzuela model is distinctive in that it facilitates simulations with two scales of randomness. The first scale is in the choice of clusters and rays. The second scale is in the variability of the amplitudes over time. A simulation may proceed as follows. First, the cluster delays $\{T_m\}$ and the associated ray delays $\tau_{m,q}$ are generated from the Poisson arrival model. For this given set of clusters and rays, several channel realizations may be generated. Conditioned on the cluster delays and ray delays, these channel coefficients are generated according to the power delay profile $\sum_{m=0}^{\infty} e^{-T_m/\Phi} e^{-\tau_{m,q}/\phi} u(t - T_m)$ where $u(t)$ is the unit step function. In practical applications of this model, the number of rays and clusters is truncated, for example, ten clusters and 50 rays per cluster.

The impulse-based channel model in (5.347) can be used in other configurations, with or without clusters. For example, given a model of a propagation environment, ray tracing could be used to determine propagation paths, gains, and delays between a transmitter and a receiver. This information could be used to generate a channel model through (5.347). The model could be updated by moving a user through an area. In another variation, the cluster locations are determined from ray tracing, but then the rays arriving from each cluster are assumed to arrive randomly with a certain distribution (usually specified by the angle spread) [37, 38]. Other variations are possible.

There are different ways to incorporate mobility into a frequency-selective channel model. In the case of the discrete-time frequency-selective Rayleigh channel, the same approaches can be used as in the flat-fading case, but applied per tap. For example, each tap could be filtered so that every tap has a specified temporal correlation function (usually assumed to be the same). In the case of the Saleh-Valenzuela model, the clusters and rays could also be parameterized with an angle in space, and a Doppler shift derived from that angle. This could be used to model a time-varying channel given a set of clusters and rays.

5.8.3   Performance Analysis with Fading Channel Models

The coherence time and the coherence bandwidth, along with the corresponding signal bandwidth and sample period, determine the type of equivalent input-output relationship used for system design. Given a fading channel operating regime, an appropriate receiver can be designed based on system considerations, including the target error rate. The impact of fading on the system performance, though, depends on the discrete-time channel and the receiver processing used to deal with fading.

In complicated system design problems, as is the case in most standards-based systems, performance of the system is typically estimated using Monte Carlo simulation techniques. Essentially, this involves generating a realization of the fading channel, generating a packet of bits to transmit, generating additive noise, and processing the corresponding received signal.

For some special cases, though, it is possible to use a stochastic description of the channel to predict the performance without requiring simulation. This is useful in making initial system design decisions, followed up by more detailed simulations. In this section, we provide an example of the analysis of the probability of symbol error.

Consider a flat-fading channel described as

$$y[n] = \sqrt{E_x} h_s s[n] + v[n] \tag{5.349}$$

where we assume the large-scale fading is $G = E_x$. When the channel is modeled as a random variable, the instantaneous probability of symbol error rate written as $P_e\left(\frac{E_x}{N_o}|h_s\right)$ is also a random variable. One measure of performance in this case is the average probability of symbol error. Other measures based on a notion of outage are also possible. The average probability of error is written as

$$P_e\left(\frac{E_x}{N_o}\right) = \mathbb{E}_{h_s}\left[P_e\left(\frac{E_x}{N_o}|h_s\right)\right] \tag{5.350}$$

where $P_e\left(\frac{E_x}{N_o}|h_s\right)$ is the probability of symbol error conditioned on a given value of h_s. The expectation is taken with respect to all channels in the distribution to give

$$P_e\left(\frac{E_x}{N_o}\right) = \int_c P_e\left(\frac{E_x}{N_o}|c\right) f_{h_s}(c) \mathrm{d}c. \tag{5.351}$$

Note that we indicate this as a single integral, but since the channel is complex, it is more general to have a double integral. Given a prescribed fading channel model and probability of symbol error expression, the expectation or a bound can sometimes be calculated in closed form.

In this section, we calculate the expectation of the union bound on the probability of symbol error in an AWGN channel. Using the union bound,

$$P_e\left(\frac{E_x}{N_o}\Big| h_s\right) \le (M-1)Q\left(\sqrt{\frac{E_x|h|^2}{N_o}\frac{d_{min}^2}{2}}\right). \tag{5.352}$$

While solutions exist to help calculate the $Q(\cdot)$ function, such as Craig's formula [80], a simpler method is to use the Chernoff bound $Q(x) \le e^{-\frac{x^2}{2}}$. This gives

$$P_e\left(\frac{E_x}{N_o}\Big| h_s\right) \le \frac{1}{2}(M-1)e^{-\frac{E_x|h_s|^2}{4N_o}d_{min}^2}. \tag{5.353}$$

To proceed, we need to specify a distribution for the channel.

Suppose that the channel has the Rayleigh distribution. A way to evaluate (5.353) is to use the fact that h_s is $\mathcal{N}_\mathbb{C}(0,1)$ and that any distribution function integrates to 1. Then, substituting in for $f_{h_s}(c) = \pi^{-1}\exp(-|c|^2)$, it follows that

$$\int_c \frac{1}{2}(M-1)e^{-\frac{E_x|c|^2}{4N_o}d_{min}^2}\frac{1}{\pi}e^{-|c|^2}dc = \frac{1}{2}(M-1)\int_c e^{-\frac{E_x|c|^2d_{min}^2}{4N_o}+|c|^2}dc \tag{5.354}$$

$$= \frac{1}{2}(M-1)\int_c e^{-|c|^2\left(\frac{E_x|c|^2d_{min}^2}{4N_o}+1\right)}dc \tag{5.355}$$

$$= \frac{1}{2}(M-1)\int_c \frac{\frac{E_x|c|^2d_{min}^2}{4N_o}+1}{\frac{E_x|c|^2d_{min}^2}{N_o}+1}dc \tag{5.356}$$

$$= \frac{1}{2}(M-1)\frac{1}{\frac{E_xd_{min}^2}{4N_o}+1}. \tag{5.357}$$

Putting it all together:

$$P_e\left(\frac{E_x}{N_o}\right) \le \frac{1}{2}(M-1)\frac{1}{\frac{E_xd_{min}^2}{4N_o}+1}. \tag{5.358}$$

Example 5.34   Compute the Chernoff upper bound on the union bound for Rayleigh fading with M-QAM.

Answer: For M-QAM, we can insert $d_{min}^2 = \frac{6}{M-1}$ into (5.358) to obtain

$$P_e\left(\frac{E_x}{N_o}\right) \le \frac{1}{2}(M-1)\frac{1}{\frac{E_x3}{2(M-1)N_o}+1}. \tag{5.359}$$

This provides somewhat more intuition than the exact solution, computed in [310] as

$$
P_e\left(\frac{E_x}{N_o}\right) = 2\left(1 - \frac{1}{\sqrt{M}}\right)\left(1 - \sqrt{\frac{\frac{E_x}{N_o}}{1 + \frac{E_x}{N_o}}}\right)
$$

$$
- \frac{1}{4}\left(1 - \frac{1}{\sqrt{M}}\right)^2\left(1 - \sqrt{\frac{\frac{E_x}{N_o}}{1 + \frac{E_x}{N_o}}}\left(\frac{4}{\pi}\tan^{-1}\sqrt{\frac{1 + \frac{E_x}{N_o}}{\frac{E_x}{N_o}}}\right)\right). \tag{5.360}
$$

A comparison of both the exact and upper bounds is provided in Figure 5.36 for 4-QAM. Here it is apparent that the upper bound is loose but has the correct slope at high SNR.

The implication of (5.358) is that the probability of error decreases as a function of the inverse of the SNR for Rayleigh fading channels. Note, though, that for the non-fading AWGN channel, the probability of error decreases exponentially (this can be visualized from the Chernoff upper bound). This means that fading channels require a much higher average SNR to achieve a given probability of error. The difference between the required SNR for an AWGN channel at a particular error rate and the target required for a fading channel is known as the small-scale fading margin. This is the extra power required to compensate for fading in the channel. An illustration is provided in Figure 5.36 for 4-QAM. For example, at a symbol error rate of 10^{-2}, 4dB of SNR are required for an AWGN channel but 14dB are required for the Rayleigh channel. This means that a 10dB small-scale fade margin would be required in Rayleigh fading compared to an AWGN channel.

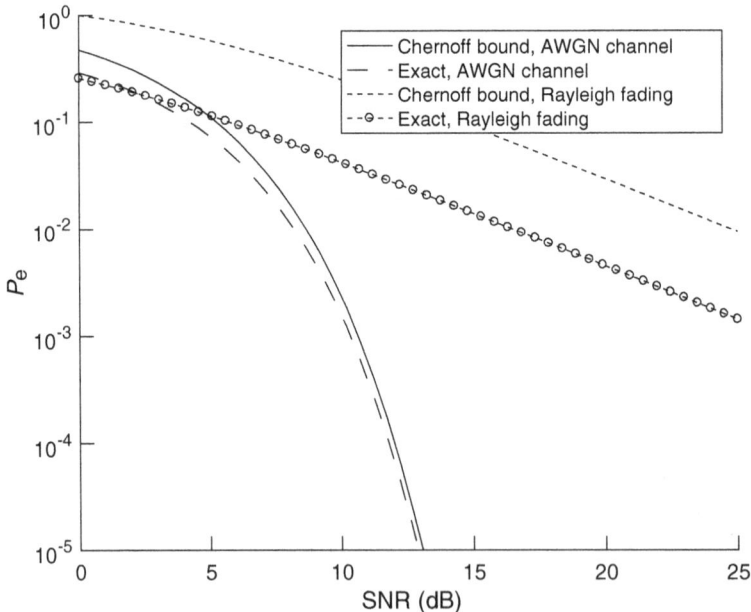

Figure 5.36 Probability of error curve comparing Gaussian and Rayleigh fading for 4-QAM

5.9   Summary

- Single-path propagation channels delay and attenuate the received signal.

- Multipath propagation channels create intersymbol interference. The discrete-time complex baseband equivalent channel includes the transmit and receive pulse shaping. It can be modeled with FIR filter coefficients $\{h[\ell]\}_{\ell=0}^{L}$.

- Equalization is a means of removing the effects of multipath propagation. Linear equalization designs a filter that is approximately the inverse of the channel filter. Equalization can also be performed in the frequency domain using OFDM and SC-FDE frameworks.

- Frame synchronization involves identifying the beginning of a transmission frame. In OFDM, frame synchronization is called symbol synchronization. Known training sequences or the periodic repetition of a training sequence can be used to estimate the beginning of the frame.

- Carrier frequency offset is created by (small) differences in the carrier frequencies used at the transmitter and the receiver. This creates a phase rotated on the received signal. Carrier frequency offset synchronization involves estimating the offset and derotating the received signal to remove it. Special signal designs are used to enable frequency offset estimation prior to channel estimation.

- Training sequences are sequences known to both the transmitter and the receiver. They are inserted to allow the receiver to estimate unknown parameters like the channel, the frame start, or the carrier frequency offset. Training sequences with good correlation properties are useful in many algorithms.

- Propagation channel models are used to evaluate the performance of signal processing algorithms. Large-scale fading captures the average characteristics of the channel over hundreds of wavelengths, whereas small-scale fading captures the channel behavior on the order of wavelengths. Channel models exist for the large-scale and the small-scale fading components.

- Path loss describes the average loss in signal power as a function of distance. It is normally measured in decibels. The log-distance path-loss model describes the loss as a function of the path-loss exponent and possibly an additional random variable to capture shadowing. The LOS/NLOS path-loss model has a distance-dependent probability function that chooses from an LOS or NLOS log-distance path-loss model.

- The small-scale fading characteristics of a channel can be described through frequency selectivity and time selectivity.

- Frequency selectivity is determined by looking at the power delay profile, computing the RMS delay spread, and seeing if it is significant relative to the symbol period. Alternatively, it can be assessed by looking at the coherence bandwidth and comparing the bandwidth of the signal.

- Time selectivity is quantified by looking at the spaced-time correlation function and comparing the frame length. Alternatively, it can be determined from the RMS Doppler spread or the maximum Doppler shift and comparing the signal bandwidth.

- There are several different flat-fading and frequency-selective channel models. Most models are stochastic, treating the channel as a random variable that changes from frame to frame. Rayleigh fading is the most common flat-fading channel model. Frequency-selective channels can be generated directly based on a description of their taps or based on a physical description of the channel.

Problems

1. Re-create the results from Figure 5.5 but with 16-QAM. Also determine the smallest value of M_{rx} such that 16-QAM has a 1dB loss at a symbol error rate of 10^{-4}.

2. Re-create the results from Example 5.3 assuming training constructed from Golay sequences as $[\mathbf{a}_8; \mathbf{b}_8]$, followed by 40 randomly chosen 4-QAM symbols. Explain how to modify the correlator to exploit properties of the Golay complementary pair. Compare your results.

3. Consider the system
$$y[n] = hs[n] + v[n] \tag{5.361}$$
where $s[n]$ is a zero-mean WSS random process with correlation $r_{ss}[n]$, $v[n]$ is a zero-mean WSS random process with correlation $r_{vv}[n]$, and $s[n]$ and $v[n]$ are uncorrelated. Find the linear MMSE equalizer g such that the mean squared error is minimized:
$$\mathbb{E}[|e[n]|^2] = \mathbb{E}[|s[n] - g^* y[n]|^2]. \tag{5.362}$$

   (a) Find an equation for the MMSE estimator g.

   (b) Find an equation for the mean squared error (substitute your estimator in and compute the expectation).

   (c) Suppose that you know $r_{ss}[n]$ and you can estimate $r_{yy}[n]$ from the received data. Show how to find $r_{vv}[n]$ from $r_{ss}[n]$ and $r_{yy}[n]$.

   (d) Suppose that you estimate $r_{yy}[n]$ through sample averaging of N samples, exploiting the ergodicity of the process. Rewrite the equation for g using this functional form.

   (e) Compare the least squares and the MMSE equalizers.

4. Consider a frequency-flat system with frequency offset. Suppose that 16-QAM modulation is used and that the received signal is
$$y[n] = e^{j2\pi \epsilon n} \sqrt{E_x} s[n] + v[n] \tag{5.363}$$

where $\epsilon = f_{\text{offset}} T_s$ and $\exp(j2\pi\epsilon n)$ is unknown to the receiver. Suppose that the SNR is 10dB and the packet size is $N = 101$ symbols. The effect of ϵ is to rotate the actual constellation.

(a) Consider $y[0]$ and $y[1]$ in the absence of noise. Illustrate the constellation plot for both cases and discuss the impact of ϵ on detection.

(b) Over the whole packet, where is the worst-case rotation?

(c) Suppose that ϵ is small enough that the worst-case rotation occurs at symbol 100. What is the value of ϵ such that the rotation is greater than $\pi/2$? For the rest of this problem assume that ϵ is less than this value.

(d) Determine the ϵ such that the symbol error rate is 10^{-3}. To proceed, first find an expression for the probability of symbol error as a function of ϵ. Make sure that ϵ is included somewhere in the expression. Set equal to 10^{-3} and solve.

5. Let $\mathbf{T}$ be a Toeplitz matrix and let $\mathbf{T}^*$ be its Hermitian conjugate. If $\mathbf{T}$ is either square or tall and full rank, prove that $\mathbf{T}^*\mathbf{T}$ is an invertible square matrix.

6. Consider the training structure in Figure 5.37. Consider the problem of estimating the channel from the first period of training data. Let $s[0], s[1], \ldots, s[N_{\text{tr}}-1]$ denote the training symbols, and let $s[N_{\text{tr}}], s[N_t + 1], \ldots, s[N - 1]$ denote the unknown QAM data symbols. Suppose that we can model the channel as a frequency-selective channel with coefficients $h[0], h[1], \ldots, h[\ell]$. The received signal (assuming synchronization has been performed) is

$$y[n] = \sum_{\ell=0}^{L} h[\ell]s[n - \ell] + v[n]. \tag{5.364}$$

(a) Write the solution for the least squares channel estimate from the training data. You can use matrices in your answer. Be sure to label the size of the matrices and their contents very carefully. Also list any critical assumptions necessary for your solution.

(b) Now suppose that the estimated channel is used to estimate an equalizer that is applied to $y[n]$, then passed to the detector to generate $\{\hat{s}[n]\}_{N_{\text{tr}}}^{N-1}$ tentative

Frame

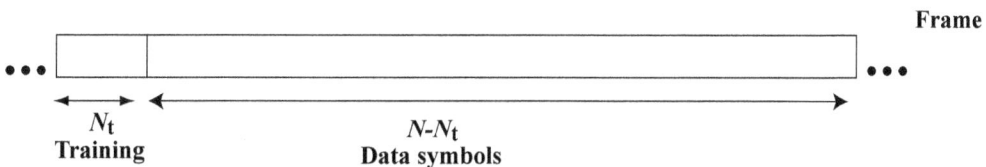

N_t
Training

$N\text{-}N_t$
Data symbols

Figure 5.37 A frame structure with a single training sequence of length N_t followed by $N - N_t$ data symbols

symbol decisions. We would like to improve the detection process by using a decision-directed channel estimate. Write the solution for the least squares channel estimate from the training data and tentative symbol decisions.

(c) Draw a block diagram for the receiver that includes synchronization, channel estimation from training, equalization, reestimation of the channel, reequalization, and the additional detection phase.

(d) Intuitively, explain how the decision-directed receiver should perform relative to only training as a function of SNR (low, high) and N_{tr} (small, large). Is there a benefit to multiple iterations? Please justify your answer.

7. Let $\widehat{\mathbf{H}}$ be the Toeplitz matrix that is used in computing the least squares equalizer $\widehat{\mathbf{f}}_{n_{\mathrm{d}}}$. Prove that if $\widehat{h}[0] \neq 0$, $\widehat{\mathbf{H}}$ is full rank.

8. Consider a digital communication system where the same transmitted symbol $s[n]$ is repeated over two different channels. This is called repetition coding. Let h_1 and h_2 be the channel coefficients, assumed to be constant during each time instance and estimated perfectly. The received signals are corrupted by $v_1[n]$ and $v_2[n]$, which are zero-mean circular symmetric complex AWGN of the same variance σ^2, that is, $v_1[n]$, $v_2[n] \sim \mathcal{N}_c(0, \sigma^2)$. In addition, $s[n]$ has zero mean and $\mathbb{E}|s[n]|^2 = 1$. We assume that $s[n]$, $v_1[n]$, and $v_2[n]$ are uncorrelated with each other. The received signals on the two channels are given by

$$y_1[n] = h_1 s[n] + v_1[n], \tag{5.365}$$
$$y_2[n] = h_2 s[n] + v_2[n]. \tag{5.366}$$

Suppose that we use the equalizers g_1 and g_2 for the two time instances, where g_1 and g_2 are complex numbers such that $|g_1|^2 + |g_2|^2 = 1$. The *combined signal*, denoted as $z[n]$, is formed by summing the equalized signals in two time instances:

$$z[n] = g_1 y_1[n] + g_2 y_2[n]. \tag{5.367}$$

If we define the following vectors:

$$\mathbf{y}[n] = \begin{bmatrix} y_1[n] \\ y_2[n] \end{bmatrix} \tag{5.368}$$

$$\mathbf{h} = \begin{bmatrix} h_1 \\ h_2 \end{bmatrix} \tag{5.369}$$

$$\mathbf{v}[n] = \begin{bmatrix} v_1[n] \\ v_2[n] \end{bmatrix} \tag{5.370}$$

$$\mathbf{g} = \begin{bmatrix} g_1 \\ g_2 \end{bmatrix}, \tag{5.371}$$

then

$$\mathbf{y}[n] = \mathbf{h}s[n] + \mathbf{v}[n] \tag{5.372}$$

and

$$z[n] = \mathbf{g}^* \, \mathbf{y}[n]. \tag{5.373}$$

(a) Write an expression for $z[n]$, first in terms of the vectors $\mathbf{h}$ and $\mathbf{g}$, then in terms of g_1, g_2, h_1, and h_2. Thus you will have two equations for $z[n]$.

(b) Compute the mean and the variance of the noise component in the combined signal $z[n]$. Remember that the noise is Gaussian and uncorrelated.

(c) Compute the SNR of the combined signal $z[n]$ as a function of σ, h_1, h_2, g_1, and g_2 (or σ, $\mathbf{h}$, and $\mathbf{g}$). In this case the SNR is the variance of the combined received signal divided by the variance of the noise term.

(d) Determine g_1 and g_2 as functions of h_1 and h_2 (or $\mathbf{g}$ as a function of $\mathbf{h}$) to maximize the SNR of the combined signal. *Hint:* Recall the Cauchy-Schwarz inequality for vectors and use the conditions when equality holds.

(e) Determine a set of equations for finding the LMMSE equalizers g_1 and g_2 to minimize the mean squared error, which is defined as follows:

$$\mathbb{E}||e[n]||^2 = \mathbb{E}|s[n] - z[n]|^2. \tag{5.374}$$

Simplify your equation by exploiting the orthogonality principle but do not solve for the unknown coefficients yet. *Hints:* Using the vector format might be useful, and you can assume you can interchange expectation and differentiation. First, expand the absolute value, then take the derivative with respect to $\mathbf{g}^*$ and set the result equal to zero. Simplify as much as possible to get an expression of $\mathbf{g}$ as a function of h_1, h_2, and σ.

(f) The autocorrelation matrix $\mathbf{R_{yy}}[0]$ of $\mathbf{y}[n]$ is defined as

$$\mathbf{R_{yy}}[0] = \mathbb{E}(\mathbf{y}[n]\mathbf{y}^*[n]) = \begin{pmatrix} r_{y_1 y_1}[0] & r_{y_1 y_2}[0] \\ r_{y_2 y_1}[0] & r_{y_2 y_2}[0] \end{pmatrix}. \tag{5.375}$$

Compute $\mathbf{R_{yy}}[0]$ as a function of σ, h_1, and h_2. *Hint:* $r_{ss}[0] = \mathbb{E}|s[n]|^2 = 1$ and $r_{ab}[0] = \mathbb{E}(a[n]b^*[n])$ for random processes $a[n]$ and $b[n]$.

(g) Now solve the set of equations you formulated in part (e).

9. Consider an order L-tap frequency-selective channel. Given a symbol period of T and after oversampling with factor M_{rx} and matched filtering, the received signal is

$$r[k] = \sum_{n=-\infty}^{\infty} s[n]h\left(\frac{kT}{M} - nT\right) + v\left(\frac{kT}{M}\right), \tag{5.376}$$

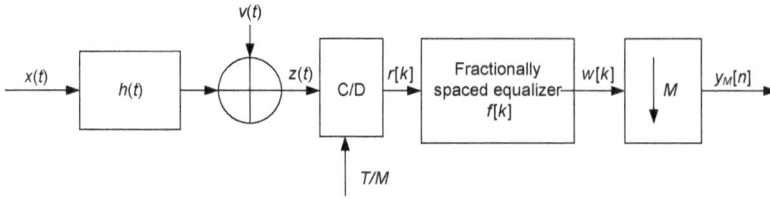

Figure 5.38 Block diagram of a receiver with FSE

where s is the transmitted symbols and v is AWGN. Suppose a *fractionally spaced equalizer* (FSE) is applied to the received signal before downsampling. The FSE $f[k]$ is an FIR filter with tap spacing of T/M_{rx} and length MN. Figure 5.38 gives the block diagram of the system.

(a) Give an expression for $w[k]$, the output of the FSE, and $y_M[n]$, the signal after downsampling.

(b) We can express the FSE filter coefficients in the following matrix form:

$$\mathbf{F} = \begin{bmatrix} f[0] & f[1] & \cdots & f[M-1] \\ f[M] & f[M+1] & \cdots & f[2M-1] \\ \vdots & \vdots & \ddots & \vdots \\ f[(N-1)M] & f[(N-1)M+1] & \cdots & f[NM-1] \end{bmatrix}. \qquad (5.377)$$

We can interpret every $(j+1)^{\text{th}}$ column of $\mathbf{F}$ as a T-spaced *subequalizer* $f_j[n]$ where $j = 0, 2, ..., M-1$. Similarly, we can express the FSE output, the channel coefficients, and the noise as T-spaced subsequences:

$$r_j[n] = r\left[nT - T\frac{M-j}{M}\right] \qquad (5.378)$$

$$h_j[n] = h\left[nT - T\frac{M-j}{M}\right] \qquad (5.379)$$

$$v_j[n] = w\left[nT - T\frac{M-j}{M}\right]. \qquad (5.380)$$

Using these expressions, show that

$$r_j[n] = s[n] \star h_j[n] + v_j[n]. \qquad (5.381)$$

(c) Now, express $y_M[n]$ as a function of $s[n]$, $f_j[n]$, $h_j[n]$, and $v_j[n]$. This expression is known as the *multichannel model* of the FSE since the output is the sum of the symbols convolved with the T-spaced subequalizers and subsequences.

(d) Draw a block diagram of the multichannel model of the FSE based on part (c).

(e) Now consider a noise-free case. Given the channel matrix $\mathbf{H}_j$ given by

$$
\begin{bmatrix}
h[0] & 0 & \cdots & \vdots \\
h[M-j] & h[j] & \cdots & 0 \\
\vdots & h[M-j] & \ddots & h[j] \\
h[(L-1)M-j] & \vdots & \cdots & h[M-j] \\
0 & h[(L-1)M-j] & \ddots & \vdots \\
\vdots & \vdots & \cdots & h[(L-1)M-j]
\end{bmatrix},
\qquad (5.382)
$$

we define

$$
\mathbf{H} = [\mathbf{H}_0\ \mathbf{H}_1\ \cdots\ \mathbf{H}_{M-1}] \qquad (5.383)
$$
$$
\mathbf{f} = \begin{bmatrix} \mathbf{f}_{M-1}^{\mathrm{T}}\ \mathbf{f}_{M-2}^{\mathrm{T}}\ \cdots\ \mathbf{f}_0^{\mathrm{T}} \end{bmatrix}, \qquad (5.384)
$$

where $\mathbf{f}_j$ is the $(j^{\mathrm{th}}+1)$ column of $\mathbf{F}$. Perfect channel equalization can occur only if the solution to $\mathbf{h} = \mathbf{Cf}$, $\mathbf{e} = \mathbf{e}_d$ lies in the column space of $\mathbf{H}$, and the channel $\mathbf{H}$ has all linearly independent rows. What are the conditions on the length of the equalizer N given M_{rx} and L to ensure that the latter condition is met? *Hint:* What are the dimensions of the matrices/vectors?

(f) Given your answer in (e), consider the case of $M = 1$ (i.e., not using an FSE). Can perfect equalization be performed in this case?

10. Consider an order L-tap frequency-selective channel. After matched filtering, the received signal is

$$
y[n] = \sum_{\ell=0}^{L} h[\ell]s[n-\ell] + v[n], \qquad (5.385)
$$

where $s[n]$ is the transmitted symbols, $h[\ell]$ is the channel coefficient, and $v[n]$ is AWGN. Suppose the received symbols corresponding to a known training sequence $\{1, -1, -1, 1\}$ are $\{0.75 + j0.75, -0.75 - j0.25, -0.25 - j0.25, 0.25 + j0.75\}$. Note that this problem requires numerical solutions; that is, when you are asked to solve the least squares problems, the results must be specific numbers. MATLAB, LabVIEW, or MathScript in LabVIEW might be useful for solving the problem.

(a) Formulate the least squares estimator of $\{h[\ell]\}_{\ell=0}^{L}$ in the matrix form. Do not solve it yet. Be explicit about your matrices and vectors, and list your assumptions (especially the assumption about the relationship between L and the length of the training sequence).

(b) Assuming $L=1$, solve for the least squares channel estimate based on part (a).

(c) Assume we use a linear equalizer to remove the effects of the channel. Let $\{f[\ell]\}_{\ell=0}^{2}$ be an FIR equalizer. Let n_{d} be the equalizer delay. Formulate the least squares estimator of $\{f[\ell]\}_{\ell=0}^{2}$ given the channel estimate in part (b). Do not solve it yet.

(d) Determine the range of values of n_{d}.

(e) Determine the n_{d} that minimizes the squared error $J_f[n_{\mathrm{d}}]$ and solve for the least squares equalizer corresponding to this n_{d}. Also provide the value of the minimum squared error.

(f) With the same assumptions as in parts (a) through (e) and with the value of n_{d} found in part (e), formulate and solve for the direct least squares estimator of the equalizer's coefficients $\{f[\ell]\}_{\ell=0}^{2}$. Also provide the value of the squared error.

11. Prove the following two properties of the DFT:

(a) Let $x[n] \leftrightarrow X[k]$ and $x_1[n] \leftrightarrow X_1[k]$. If $X_1[k] = e^{\mathrm{j}2\pi km/n} X[k]$, we have

$$x_1[n] = \begin{cases} x[((n-m))_N] & 0 \le n \le N-1 \\ 0 & \text{otherwise.} \end{cases} \tag{5.386}$$

(b) Let $y[n] \leftrightarrow Y[k]$, $h[n] \leftrightarrow H[k]$ and $s[n] \leftrightarrow S[k]$. If $Y[k] = H[k]S[k]$, we have $y[n] = \sum_{\ell=0}^{N-1} h[\ell]s[((n-\ell))_N]$.

12. Consider an OFDM system. Suppose that your buddy C.P. uses a cyclic postfix instead of a cyclic prefix. Thus

$$w[n] = \frac{1}{N} \sum_{n=0}^{N-1} s[k] e^{\mathrm{j}2\pi \frac{kn}{N}} \tag{5.387}$$

for $n = 0, 1, \ldots, N + L_{\mathrm{c}}$ where L_{c} is the length of the cyclic prefix. The values of $w[n]$ for $n < 0$ and $n > N + L_{\mathrm{c}}$ are unknown. Let the received signal be

$$y[n] = \sum_{\ell=0}^{L} h[\ell]w[n-\ell] + v[n]. \tag{5.388}$$

(a) Show that you can still recover $\{s[k]\}_{k=0}^{N-1}$ with a cyclic postfix instead of a cyclic prefix.

(b) Draw a block diagram of the system.

(c) What are the differences between using a cyclic prefix and a cyclic postfix?

13. Consider an SC-FDE system with equivalent system $y[n] = \sum_{\ell=0}^{L} h[\ell]s[n-\ell]+v[n]$. The length of the cyclic prefix is L_c. Prove that the cyclic prefix length should satisfy $L_c \geq L$.

14. Consider an OFDM system with $N = 256$ subcarriers in 5MHz of bandwidth, with a carrier of $f_c = 2$GHz and a length $L = 16$ cyclic prefix. You can assume sinc pulse shaping.

    (a) What is the subcarrier bandwidth?

    (b) What is the length of the guard interval?

    (c) Suppose you want to make the OFDM symbol periodic including the cyclic prefix. The length of the period will be 16. Which subcarriers do you need to zero in the OFDM symbol?

    (d) What is the range of frequency offsets that you can correct using this approach?

15. Consider an OFDM system with N subcarriers.

    (a) Derive a bound on the bit error probability for 4-QAM transmission. Your answer should depend on $h[k]$, N_o, and d_{min}.

    (b) Plot the error rate curves as a function of SNR for $N = 1, 2, 4$, assuming that $h[0] = E_x$, $h[1] = E_x/2$, $h[2] = -jE_x/2$, and $h[3] = E_x e^{-j2\pi/3}/3$.

16. How many pilot symbols are used in the OFDM symbols during normal data transmission (not the CEF) in IEEE 802.11a?

17. IEEE 802.11ad is a WLAN standard operating at 60GHz. It has much wider bandwidth than previous WLAN standards in lower-frequency bands. Four PHY formats are defined in IEEE 802.11ad, and one of them uses OFDM. The system uses a bandwidth of 1880MHz, with 512 subcarriers and a fixed 25% cyclic prefix. Now compute the following:

    (a) What is the sample period duration assuming sampling at the Nyquist rate?

    (b) What is the subcarrier spacing?

    (c) What is the duration of the guard interval?

    (d) What is the OFDM symbol period duration?

    (e) In the standard among the 512 subcarriers, only 336 are used as data subcarriers. Assuming we use code rate 1/2 and QPSK modulation, compute the maximum data rate of the system.

18. Consider an OFDM communication system and a discrete-time channel with taps $\{h[\ell]\}_{\ell=0}^{L}$. Show mathematically why the cyclic prefix length L_c must satisfy $L \geq L_c$.

19. In practice, we may want to use multiple antennas to improve the performance of the received signal. Suppose that the received signal for each antenna can be modeled as

$$x_1[n] = \sum_{\ell=0}^{L} h_1[\ell]s[n-\ell] + v_1[n] \qquad (5.389)$$

$$x_2[n] = \sum_{\ell=0}^{L} h_2[\ell]s[n-\ell] + v_2[n]. \qquad (5.390)$$

Essentially you have two observations of the same signal. Each is convolved by a different discrete-time channel.

In this problem we determine the coefficients of a set of equalizers $g_1^{(\Delta)}[k]$ and $g_2^{(\Delta)}[k]$ such that

$$\sum_{k=0}^{K} g_1^{(\Delta)}[k]h_1[n-k] + \sum_{k=0}^{K} g_2^{(\Delta)}[k]h_2[n-k] = \delta[n-\Delta] \qquad (5.391)$$

where Δ is a design parameter.

(a) Suppose that you send training data $\{t[n]\}_{n=0}^{N_{\text{tr}}-1}$. Formulate a least squares estimator for finding the estimated coefficients of the channel $\{\widehat{h}_1[\ell]\}_{\ell=0}^{L}$ and $\{\widehat{h}_2[\ell]\}_{\ell=0}^{L}$.

(b) Formulate the least squares equalizer design problem given your channel estimate. *Hint:* You need a squared error. Do not solve it yet. Be explicit about your matrices and vectors.

(c) Solve for the least squares equalizer estimate using the formulation in part (b). You can use matrices in your answer. List your assumptions about dimensions and so forth.

(d) Draw a block diagram of a QAM receiver that includes this channel estimator and equalizer.

(e) Now formulate and solve the direct equalizer estimation problem. List your assumptions about dimensions and so forth.

(f) Draw a block diagram of a QAM receiver that includes this equalizer.

20. Consider a wireless communication system with a frame structure as illustrated in Figure 5.39. Training is interleaved around (potentially different) bursts of

| ... | training | data 1 | training | data 2 | ... |

Figure 5.39 Frames with training data

data. The same training sequence is repeated. This structure was proposed relatively recently and is used in the single-carrier mode in several 60GHz wireless communication standards.

Let the training sequence be $\{t[n]\}_{n=0}^{N_{tr}-1}$ and the data symbols be $\{s[n]\}$. Just to make the problem concrete, suppose that

$$w[n] = \begin{cases} t[n] & n \in [0, N_{tr} - 1] \\ s[n - N_{tr}] & n \in [N_t, N_t + N - 1] \\ t[n - (N_t + N)] & n \in [N_t + N, 2N_{tr} + N - 1] \\ s[n - (2N_{tr} + N)] & n \in [2N_{tr} + N, N_t + N - 1] \\ \text{etc.} \end{cases} \tag{5.392}$$

Suppose that the channel is linear and time invariant. After matched filtering, synchronization, and sampling, the received signal is given by the usual relationship

$$y[n] = \sum_{\ell=0}^{L} h[\ell]w[n - \ell] + v[n]. \tag{5.393}$$

Assume that N is much greater than N_{tr}, and that $N_{tr} \geq L$.

(a) Consider the sequence $\{w[n]\}_{n=0}^{2N_{tr}+N-1}$. Show that there exists a cyclic prefix of length N_{tr}.

(b) Derive a single-carrier frequency-domain equalization structure that exploits the cyclic prefix we have created. Essentially, show how we can recover $\{s[n]\}_{n=0}^{N-1}$ from $\{y[n]\}$.

(c) Draw a block diagram for the transmitter. You need to be as explicit as possible in indicating how the training sequence gets incorporated.

(d) Draw a block diagram for the receiver. Be careful.

(e) Suppose that we want to estimate the channel from the training sequence using the training on either side of the data. Derive the least squares channel estimator and determine the minimum value of N_{tr} required for this estimator to satisfy the conditions required by least squares.

(f) Can you use this same trick with OFDM, using training for the cyclic prefix? Explain why or why not.

21. Consider an OFDM communication system. Suppose that the system uses a bandwidth of 40MHz, 128 subcarriers, and a length 32 cyclic prefix. Suppose the carrier frequency is 5.785GHz.

(a) What is the sample period duration assuming sampling at the Nyquist rate?

(b) What is the OFDM symbol period duration?

(c) What is the subcarrier spacing?

(d) What is the duration of the guard interval?

(e) How much frequency offset can be corrected, in hertz, using the Schmidl-Cox method with all odd subcarriers zeroed, assuming only fine offset correction? Ignore the integer offset.

(f) Oscillators are specified in terms of their offset in parts per million. Determine how many parts per million of variation are tolerable given the frequency offset in part (e).

22. Consider an OFDM communication system with cyclic prefix of length L_c. The samples conveyed to the pulse-shaping filter are

$$w[n] = \frac{1}{N} \sum_{n=0}^{N-1} s[m] e^{j2\pi \frac{m(n-L_c)}{N}}, \qquad n = 0, \ldots, N + L_c - 1. \tag{5.394}$$

Recall that

$$w[n] = w[n+N], \text{ for } n = 0, \ldots, L_c - 1 \tag{5.395}$$

is the cyclic prefix. Suppose that the cyclic prefix is designed such that the channel order L satisfies $L_c = 2L$. In this problem, we use the cyclic prefix to perform carrier frequency offset estimation. Let the receive signal after match filtering, frame offset correction, and downsampling be written as

$$y[n] = e^{j2\pi \epsilon n} \sum_{\ell=0}^{L} h[\ell] w[n - \ell] + v[n], \text{ for } n = 0, \ldots, N + L_c - 1. \tag{5.396}$$

(a) Consider first a single OFDM symbol. Use the redundancy in the cyclic prefix to derive a carrier frequency offset estimator. *Hint:* Exploit the redundancy in the cyclic prefix but remember that $L_c = 2L$. You need to use the fact that the cyclic prefix is longer than the channel.

(b) What is the correction range of your estimator?

(c) Incorporate multiple OFDM symbols into your estimator. Does it work if the channel changes for different OFDM symbols?

23. **Synchronization Using Repeated Training Sequences and Sign Flipping**
Consider the framing structure illustrated in Figure 5.40. This system uses a repetition of four training sequences. The goal of this problem is to explore the impact of multiple repeated training sequences on frame synchronization, frequency offset synchronization, and channel estimation.

(a) Suppose that you apply the frame synchronization, frequency offset estimation, and channel estimation algorithms using only two length N_{tr} training sequences. In other words, ignore the two additional repetitions of the

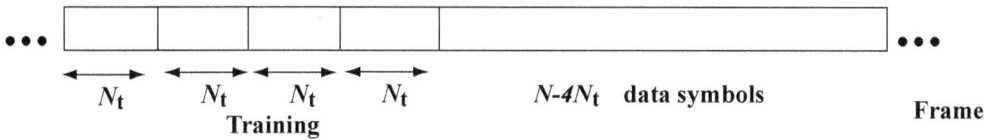

Figure 5.40 A communication frame with four repeated training sequences followed by data symbols

training signal. Please comment on how frame synchronization, frequency off-set estimation, and channel estimation work on the aforementioned packet structure.

(b) Now focus on frequency offset estimation. Treat the framing structure as a repetition of two length $2N_{\mathrm{tr}}$ training signals. Present a correlation-based frequency offset estimator that uses two length $2N_{\mathrm{tr}}$ training signals.

(c) Treat the framing structure as a repetition of four length N_{tr} training signals. Propose a frequency offset estimator that uses correlations of length N_{tr} and exploits all four training signals.

(d) What range of frequency offsets can be corrected in part (b) versus part (c)? Which is better in terms of accuracy versus range? Overall, which approach is better in terms of frame synchronization, frequency offset synchronization, and channel estimation? Please justify your answer.

(e) Suppose that we flip the sign of the third training sequence. Thus the training pattern becomes $T, T, -T, T$ instead of T, T, T, T. Propose a frequency offset estimator that uses correlations of length N_{tr} and exploits all four training signals.

(f) What range of frequency offsets can be corrected in this case? From a frame synchronization perspective, what is the advantage of this algorithm versus the previoius algorithm you derived?

24. **Synchronization in OFDM Systems**   Suppose that we would like to implement the frequency offset estimation algorithm. Suppose that our OFDM systems operate with $N = 128$ subcarriers in 2MHz of bandwidth, with a carrier of $f_{\mathrm{c}} = 1\mathrm{GHz}$ and a length $L = 16$ cyclic prefix.

(a) What is the subcarrier bandwidth?

(b) We would like to design a training symbol that has the desirable periodic correlation properties that are useful in the discussed algorithm. What period should you choose and why? Be sure to consider the effect of the cyclic prefix.

(c) For the period you suggest in part (b), which subcarriers do you need to zero in the OFDM symbol?

(d) What is the range of frequency offsets that you can correct using this approach without requiring the modulo correction?

25. Consider the IEEE 802.11a standard. By thinking about the system design, provide plausible explanations for the following:

    (a) Determine the amount of carrier frequency offset correction that can be obtained from the short training sequence.

    (b) Determine the amount of carrier frequency offset correction that can be obtained from the long training sequence.

    (c) Why do you suppose the long training sequence has a double-length guard interval followed by two repetitions?

    (d) Why do you suppose that the training comes at the beginning instead of at the middle of the transmission as in the mobile cellular system GSM?

    (e) Suppose that you use 10MHz instead of 20MHz in IEEE 802.11a. Would you rather change the DFT size or change the number of zero subcarriers?

26. Answer the following questions for problem 17:

    (a) How much frequency offset (in Hertz) can be corrected using the Schmidl-Cox method with all odd subcarriers zeroed, assuming only fine offset correction? Ignore the integer offset.

    (b) Oscillator frequency offsets are specified in parts per million (ppm). Determine how many parts per million of variation are tolerable given the result in part (a).

27. GMSK is an example of continuous-phase modulation (CPM). Its continuous-time baseband transmitted signal can be written as

$$s(t) = e^{j\frac{\pi}{2} \sum_{n=-\infty}^{\infty} a[n]\phi(t-nT)} \tag{5.397}$$

where $a[n]$ is a sequence of BPSK symbols and $\phi(t)$ is the CPM pulse. For GSM, $T = 6/1.625e6 \approx 3.69\mu s$.

The BPSK symbol sequence $a[n]$ is generated from a differentially binary (0 or 1) encoded data sequence $d[n]$ where

$$a[n] = 1 - 2(d[n] \oplus d[n-1]) \tag{5.398}$$

where $\oplus$ denotes modulo-2 addition. Let us denote the BPSK encoded data sequence as

$$b[n] = 1 - 2(d[n]). \tag{5.399}$$

Then (5.398) can be rewritten in a simpler form as

$$a[n] = b[n]b[n-1]. \tag{5.400}$$

Consider the Gaussian pulse response

$$g(t) = B\sqrt{\frac{2\pi}{\ln 2}} e^{-\frac{2\pi^2 B^2 t^2}{\ln 2}}. \tag{5.401}$$

Denote the rectangle function of duration T in the usual way as

$$\operatorname{rect}\left(\frac{t}{T}\right) = \begin{cases} 1 & |t| \leqslant \frac{T}{2} \\ 0 & \text{elsewhere.} \end{cases} \tag{5.402}$$

For GSM, $BT = 0.3$. This means that $B = 81.25\text{kHz}$. The B here is not the bandwidth of the signal; rather it is the 3dB bandwidth of the pulse $g(t)$.

The combined filter response is

$$h(t) = g(t) * \operatorname{rect}(t/T) = \frac{1}{T} \int_{-T/2}^{T/2} g(t - \tau) \mathrm{d}\tau. \tag{5.403}$$

Then the CPM pulse is given by

$$\phi(t) = \int_{-\infty}^{t} h(\tau) \mathrm{d}\tau. \tag{5.404}$$

Because of the choice of BT product, it is recognized that

$$\phi(t) \approx \begin{cases} 0, & t \leq 0h \\ 1, & t \geq 4T. \end{cases} \tag{5.405}$$

This means that the current GMSK symbol $a[n]$ depends most strongly on the three previous symbols $a[n-1], a[n-2]$, and $a[n-3]$.

Note that the phase output $\phi(t)$ depends on all previous bits $d[n]$ because of the infinite integral. Consequently, the GSM modulator is initialized by the state give in Figure 5.41. The state is reinitialized for every transmitted burst.

The classic references for linearization are [171, 173], based on the pioneering work on linearizing CPM signals in [187]. Here we summarize the linearization approach of [171, 173], which was used in [87] for the purpose of blind channel equalization of GMSK signals. We sketch the idea of the derivation from [87] here.

(a) Exploiting the observation in (5.405), argue that

$$s(t) = e^{j\frac{\pi}{2} \sum_{n=-\infty}^{\infty} a[n]\phi(t-nT)} \tag{5.406}$$

$$\approx e^{j\frac{\pi}{2} \sum_{k=-\infty}^{n-4} a[k]} \prod_{k=n-3}^{n} e^{j\frac{\pi}{2} a[k]\phi(t-kT)} \tag{5.407}$$

for $t \in [nT, (n+1)T)$. This makes the dependence of $s(t)$ on the current symbol and past three symbols clearer.

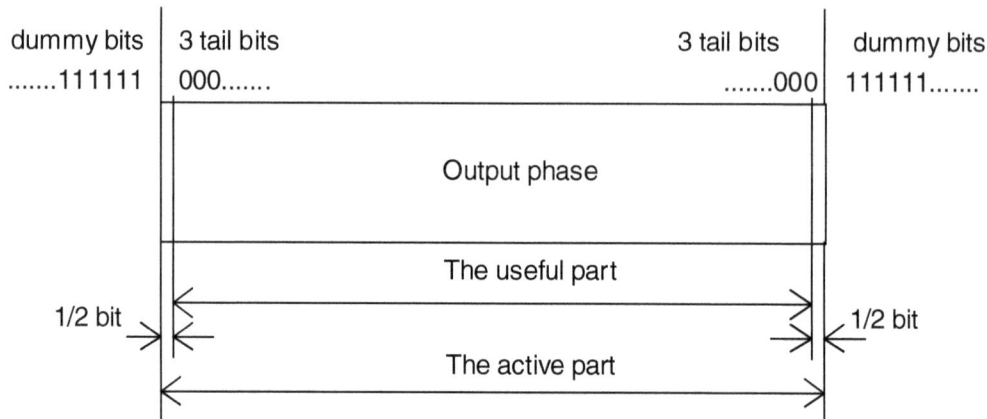

Figure 5.41 GMSK is a modulation with memory. This figure based on [99] shows how each burst is assumed to be initialized.

(b) Explain how to obtain

$$e^{j\frac{\pi}{2}a[k]\phi(t-kT)} = \cos\left(\frac{\pi}{2}a[k]\phi(t-kT)\right) + j\sin\left(\frac{\pi}{2}a[k]\phi(t-kT)\right). \quad (5.408)$$

(c) Using the fact that $a[n]$ are BPSK modulated with values $+1$ or -1, and the even property of cosine and the odd property of sine, show that

$$e^{j\frac{\pi}{2}a[k]\phi(t-kT)} = \cos\left(\frac{\pi}{2}\phi(t-kT)\right) + ja[k]\sin\left(\frac{\pi}{2}\phi(t-kT)\right). \quad (5.409)$$

(d) Now let

$$\text{sgn}(t) = \begin{cases} -1 & t < 0 \\ 0 & t = 0 \\ 1 & t > 0 \end{cases}. \quad (5.410)$$

Note that for any real number $t = \text{sgn}(t)\,|t|$. Now define

$$\beta(t) = \cos\left(\frac{\pi}{2}\text{sgn}(t)\phi(t)\right). \quad (5.411)$$

It can be shown exploiting the symmetry of $\phi(t)$ that

$$\cos\left(\frac{\pi}{2}\phi(t)\right) = \beta(t) \quad (5.412)$$

$$\sin\left(\frac{\pi}{2}\phi(t)\right) = \beta(t-4T). \quad (5.413)$$

Substitute in for β to obtain

$$s(t) \approx e^{j\frac{\pi}{2}\sum_{k=-\infty}^{n-4} a[k]} \prod_{k=n-3}^{n} (\beta(t-kT) + ja[n]\beta(t-kT-4T)). \quad (5.414)$$

(e) Show that

$$s(t) \approx \sum_{n=-\infty}^{\infty} a_0[n]c_0(t - nT) \qquad (5.415)$$

where

$$a_0[n] := e^{j\frac{\pi}{2}\sum_{k=-\infty}^{n} a[k]} = ja[n]a_0[n-1] = -a[n]a[n-1]a_0[n-2] \qquad (5.416)$$

and

$$c_0(t) = \beta(t - T)\beta(t - 2T)\beta(t - 3T)\beta(t - 4T) \qquad (5.417)$$

for $t \in [0, 5T]$. This is called the one-term approximation.

(f) Modify the typical transmit block diagram to implement the transmit waveform in (5.415).

(g) Modify the typical receiver block diagram to implement the receiver processing in a frequency-selective channel.

28. Plot the path loss for a distance from 1m to 200m for the following channel models and given parameters. The path loss should be plotted in decibels.

(a) Free space assuming $G_t = G_r = 3$dB, and $\lambda = 0.1$m

(b) Mean log distance using a 1m reference with free-space parameters as in part (a) with path-loss exponent $\beta = 2$

(c) Mean log distance using a 1m reference with free-space parameters as in part (a) with path-loss exponent $\beta = 3$

(d) Mean log distance using a 1m reference with free-space parameters as in part (a) with path-loss exponent $\beta = 4$

29. Plot the path loss at a distance of 1km for wavelengths from 10m to 0.01m for the following channel models and given parameters. The path loss should be plotted in decibels and the distance plotted on a log scale.

(a) Free space assuming $G_t = G_r = 3$dB

(b) Mean log distance using a 1m reference with free-space parameters as in part (a) with path-loss exponent $\beta = 2$

(c) Mean log distance using a 1m reference with free-space parameters as in part (a) with path-loss exponent $\beta = 3$

(d) Mean log distance using a 1m reference with free-space parameters as in part (a) with path-loss exponent $\beta = 4$

30. Consider the path loss for the log-distance model with shadowing for $\sigma_{\text{shad}} = 8\text{dB}$, $\eta = 4$, a reference distance of 1m with reference loss computed from free space, assuming $G_{\text{r}} = G_{\text{r}} = 3\text{dB}$, for a distance from 1m to 200m. Plot the mean path loss. For every value of d you plot, also plot the path loss for ten realizations assuming shadow fading. Explain what you observe. To resolve this issue, some path-loss models have a distance-dependent σ_{shad}.

31. Consider the LOS/NLOS path-loss model with $P_{\text{los}}(d) = e^{-d/200}$, free space for the LOS path loss, log distance without shadowing for the NLOS with $\beta = 4$, reference distance of 1m, $G_{\text{t}} = G_{\text{r}} = 0\text{dB}$, and $\lambda = 0.1\text{m}$. Plot the path loss in decibels for distances from 1 to 400m.

    (a) Plot the LOS path loss.

    (b) Plot the NLOS path loss.

    (c) Plot the mean path loss.

    (d) For each value of distance in your plot, generate ten realizations of the path loss and overlay them on your plot.

    (e) Explain what happens as distances become larger.

32. Consider the path loss for the log-distance model assuming that $\beta = 3$, with reference distance 1m, $G_{\text{t}} = G_{\text{r}} = 0\text{dB}$ and $\lambda = 0.1\text{m}$, and $\sigma_{\text{shad}} = 6\text{dB}$. The transmit power is $P_{\text{tx}} = 10\text{dBm}$. Generate your plots from 1 to 500m.

    (a) Plot the received power based on the mean path-loss equation.

    (b) Plot the probability that $P_{\text{rx}}(d) < -110\text{dBm}$.

    (c) Determine the maximum value of d such that $P_{\text{rx}}(d) > -110\text{dBm}$ 90% of the time.

33. Consider the LOS/NLOS path-loss model with $P_{\text{los}}(d) = e^{-d/200}$, free space for the LOS path loss, log distance without shadowing for the NLOS with $\beta = 4$, reference distance of 1m, $G_{\text{t}} = G_{\text{r}} = 0\text{dB}$, and $\lambda = 0.1\text{m}$. The transmit power is $P_{\text{tx}} = 10\text{dBm}$. Generate your plots from 1 to 500m.

    (a) Plot the received power based on the mean path-loss equation.

    (b) Plot the probability that $P_{\text{rx}}(d) < -110\text{dBm}$.

    (c) Determine the maximum value of d such that $P_{\text{rx}}(d) > -110\text{dBm}$ 90% of the time.

34. Consider the same setup as in the previous problem but now with shadowing on the NLOS component with $\sigma_{\text{shad}} = 6\text{dB}$. This problem requires some extra work to include the shadowing in the NLOS part.

(a) Plot a realization of the received path loss here. Do not forget about the shadowing.

(b) Plot the probability that $P_{\text{rx}}(d) < -110\text{dBm}$.

(c) Determine the maximum value of d such that $P_{\text{rx}}(d) > -110\text{dBm}$ 90% of the time.

35. Consider a simple cellular system. Seven base stations are located in a hexagonal cluster, with one in the center and six surrounding. Specifically, there is a base station at the origin and six base stations located 400m away at $0°$, $60°$, $120°$, and so on. Consider the path loss for the log-distance model assuming that $\beta = 4$, with reference distance 1m, $G_t = G_r = 0\text{dB}$, and $\lambda = 0.1\text{m}$. You can use 290K and a bandwidth of $B = 10\text{MHz}$ to compute the thermal noise power as kTB. The transmit power is $P_{\text{tx}} = 40\text{dBm}$.

(a) Plot the SINR for a user moving along the $0°$ line from a distance of 1m from the base station to 300m. Explain what happens in this curve.

(b) Plot the mean SINR for a user from a distance of 1m from the base station to 300m. Use Monte Carlo simulations. For every distance, generate 100 user locations randomly on a circle of radius d and average the results. How does this compare with the previous curve?

36. **Link Budget Calculations** In this problem you will compute the acceptable transmission range by going through a link budget calculation. You will need the fact that noise variance is kTB where k is Boltzmann's constant (i.e., $k = 1.381 \times 10^{-23}\text{W/Hz/K}$), T is the effective temperature in kelvins (you can use $T = 300\text{K}$), and B is the signal bandwidth. Suppose that field measurements were made inside a building and that subsequent processing revealed that the data fit the log-normal model. The path-loss exponent was found to be $n = 4$. Suppose that the transmit power is 50mW, 10mW is measured at a reference distance $d_0 = 1\text{m}$ from the transmitter, and $\sigma = 8\text{dB}$ for the log-normal path-loss model. The desired signal bandwidth is 1MHz, which is found to be sufficiently less than the coherence bandwidth; thus the channel is well modeled by the Rayleigh fading channels. Use 4-QAM modulation.

(a) Calculate the noise power in decibels referenced to 1 milliwatt (dBm).

(b) Determine the SNR required for the Gaussian channel to have a symbol error rate of 10^{-2}. Using this value and the noise power, determine the minimum received signal power required for at least 10^{-2} operation. Let this be P_{min} (in decibels as usual).

(c) Determine the largest distance such that the received power is above P_{min}.

(d) Determine the small-scale link margin L_{Rayleigh} for the flat Rayleigh fading channel. The fade margin is the difference between the SNR required for the

Rayleigh case and the SNR required for the Gaussian case. Let $P_{\text{Rayleigh}} = L_{\text{Rayleigh}} + P_{\text{min}}$.

(e) Determine the large-scale link margin for the log-normal channel with a 90% outage. In other words, find the received power $P_{\text{large-scale}}$ required such that $P_{\text{large-scale}}$ is greater than P_{Rayleigh} 90% of the time.

(f) Using $P_{\text{large-scale}}$, determine the largest acceptable distance in meters that the system can support.

(g) Ignoring the effect of small-scale fading, what is the largest acceptable distance in meters that the system can support?

(h) Describe the potential impact of diversity on the range of the system.

37. Suppose that field measurements were made inside a building and that subsequent processing revealed that the data fit the path-loss model with shadowing. The path-loss exponent was found to be $n = 3.5$. If 1mW was measured at $d_0 = 1$m from the transmitter and at a distance of $d_0 = 10$m, 10% of the measurements were stronger than -25dBm, find the standard deviation σ_{shad} for the log-normal model.

38. Through real measurement, the following three received power measurements were obtained at distances of 100m, 400m, and 1.6km from a transmitter:

Distance from Transmitter	Received Power
100m	0dBm
400m	-20dBm
1.6km	-50dBm

(a) Suppose that the path loss follows the log-distance model, without shadowing, with reference distance $d_0 = 100$m. Use least squares to estimate the path-loss exponent.

(b) Suppose that the noise power is -90dBm and the target SNR is 10dB. What is the minimum received signal power required, also known as the receiver sensitivity?

(c) Using the value calculated in part (b), what is the approximate acceptable largest distance in kilometers that the system can support?

39. Classify the following as slow or fast and frequency selective or frequency flat. Justify your answers. In some cases you will have to argue why your answer is correct by making reasonable assumptions based on the application. Assume that the system occupies the full bandwidth listed.

(a) A cellular system with carrier frequency of 2GHz, bandwidth of 1.25MHz, that provides service to high-speed trains. The RMS delay spread is $2\mu s$.

(b) A vehicle-to-vehicle communication system with carrier frequency of 800MHz and bandwidth of 100kHz. The RMS delay spread is 20ns.

(c) A 5G communication system with carrier frequency of 3.7GHz and bandwidth of 200MHz.

(d) A 60GHz wireless personal area network with a bandwidth of 2GHz and an RMS delay spread of 40ns. The main application is high-speed multimedia delivery.

(e) Police-band radio. Vehicles move at upwards of 100mph and communicate with a base station. The bandwidth is 50kHz at 900MHz carrier.

40. Consider a wireless communication system with a frequency carrier of $f_c = 1.9$GHz and a bandwidth of $B = 5$MHz. The power delay profile shows the presence of three strong paths and is plotted in Figure 5.42.

(a) Compute the RMS delay spread.

(b) Consider a single-carrier wireless system. Classify the system as *frequency flat* or *frequency selective* and *slow* or *fast*. Assume the system provides service to vehicle-to-vehicle communication at the speed of $v = 120$km/h and not packet transmission but symbol transmission. You should use the stricter relationship between coherence time T_{coh} and maximum Doppler frequency f_{m} by $T_{\text{coh}} \approx \frac{0.5}{f_{\text{m}}}$.

(c) Consider an OFDM system with IFFT/FFT size of $N = 1024$. What is the maximum-length cyclic prefix needed to allow for efficient equalization in the frequency domain with OFDM?

(d) For the specified OFDM system, which also provides services to vehicle-to-vehicle communication at urban speeds of 30mph, is the channel slow or fast fading? Justify your answer.

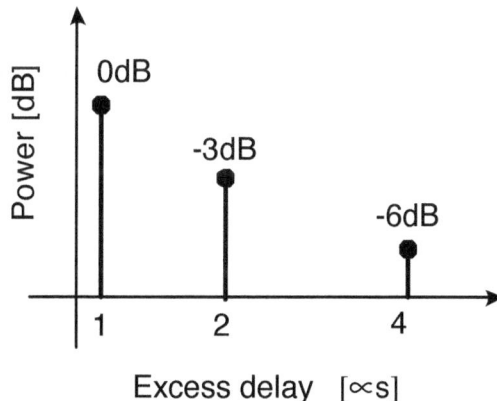

Figure 5.42 The power delay profile of a three-path channel

41. Suppose that $R_{\text{delay}}(\tau) = 1$ for $\tau \in [0, 20\mu s]$ and is zero otherwise.

    (a) Compute the RMS delay spread.

    (b) Compute the spaced-frequency correlation function $S_{\text{delay}}(\Delta_{\text{lag}})$.

    (c) Determine if the channel is frequency selective or frequency flat for a signal with bandwidth $B = 1\text{MHz}$.

    (d) Find the largest value of bandwidth such that the channel can be assumed to be frequency flat.

42. Suppose that $R_{\text{delay}}(\tau) = e^{-\tau/10\mu s}u(\tau) + \frac{1}{4}e^{-(\tau-8\mu s)/5\mu s}u(\tau - 8\mu s)$ for $\tau \in [0, 20\mu s]$.

    (a) Compute the RMS delay spread.

    (b) Compute the spaced-frequency correlation function $S_{\text{delay}}(\Delta_{\text{lag}})$.

    (c) Determine if the channel is frequency selective or frequency flat for a signal with bandwidth $B = 1\text{MHz}$.

    (d) Find the largest value of bandwidth such that the channel can be assumed to be frequency flat.

43. Sketch the design of an OFDM system with 80MHz channels, supporting pedestrian speeds of 3km/h, an RMS delay spread of $8\mu s$, operating at $f_c = 800\text{MHz}$. Explain how you would design the preamble to permit synchronization and channel estimation and how you would pick the parameters such as the number of subcarriers and the cyclic prefix. Justify your choice of parameters and explain how you deal with mobility, delay spread, and how much frequency offset you can tolerate. You do not need to do any simulation for this problem; rather make calculations as required and then justify your answers.

44. Sketch the design of an SC-FDE system with 2GHz channels, supporting pedestrian speeds of 3km/h, and an RMS delay spread of 50ns, operating at $f_c = 64\text{GHz}$. Explain how you would design the preamble to permit synchronization and channel estimation and how you would pick the parameters. Justify your choice of parameters and explain how you deal with mobility, delay spread, and how much frequency offset you can tolerate. You do not need to do any simulation for this problem; rather make calculations as required and then justify your answers.

45. Consider an OFDM system with a bandwidth of $W = 10\text{MHz}$. The power delay profile estimates the RMS delay spread to be $\sigma_{\text{RMS,delay}} = 5\mu s$. Possible IFFT/FFT sizes for the system are $N = \{128, 256, 512, 1024, 2048\}$. Possible cyclic prefix sizes (in terms of fractions of N) are $\{1/4, 1/8, 1/16, 1/32\}$.

    (a) What combination of FFT and cyclic prefix sizes provides the minimum amount of overhead?

(b) Calculate the fraction of loss due to this overhead.

(c) What might be some issues associated with minimizing the amount of overhead in this system?

46. Compute the maximum Doppler shift for the following sets of parameters:

    (a) 40MHz of bandwidth, carrier of 2.4GHz, and supporting 3km/h speeds

    (b) 2GHz of bandwidth, carrier of 64GHz, and supporting 3km/h speeds

    (c) 20MHz of bandwidth, carrier of 2.1GHz, and supporting 300km/h speeds

47. In fixed wireless systems, the Clarke-Jakes spectrum may not be appropriate. A better model is the Bell Doppler spectrum where $S_{\text{Doppler}}(f) = \frac{1}{1+K_{\text{B}}\left(\frac{f}{f_{\text{m}}}\right)^2}$ has been found to fit to data [363] where K_{B} is a constant so that $\int_f S_{\text{Doppler}}(f)df = 1$. In this case, the maximum Doppler shift is determined from the Doppler due to moving objects in the environment.

    (a) Determine K_{B} for a maximum velocity of 160km/h and a carrier frequency of 1.9GHz.

    (b) Determine the RMS Doppler spread.

    (c) Is a signal with bandwidth of $B = 20$MHz and blocks of length 1000 symbols well modeled as time invariant?

48. Look up the basic parameters of the GSM cellular system. Consider a dual-band phone that supports $f_{\text{c}} = 900$MHz and $f_{\text{c}} = 1.8$GHz. What is the maximum velocity that can be supported on each carrier frequency so that the channel is sufficiently constant during a burst period? Explain your work and list your assumptions.

49. Look up the basic parameters of an IEEE 802.11a system. What is the maximum amount of delay spread that can be tolerated by an IEEE 802.11a system? Suppose that you are in a square room and that the access point is located in the middle of the room. Consider single reflections. How big does the room need to be to give this maximum amount of channel delay?

50. Consider 4-QAM transmission. Plot the following for error rates of 1 to 10^{-5} and SNRs of 0dB to 30dB. This means that for higher SNRs do not plot values of the error rate below 10^{-5}.

    • Gaussian channel, exact expression

    • Gaussian channel, exact expression, neglecting the quadratic term

    • Gaussian channel, exact expression, neglecting the quadratic term and using a Chernoff upper bound on the Q-function

    • Rayleigh fading channel, Chernoff upper bound on the Q-function

51. **Computer** In this problem, you will create a flat-fading channel simulator function. The input to your function is $x(nT/M_{tx})$, the sampled pulse-amplitude modulated sequence sampled at T/M_{tx} where M_{tx} is the oversampling factor. The output of your function is $r(nT/M_{rx})$, which is the sampled receive pulse-amplitude modulated signal (5.7). The parameters of your function should be the channel h (which does not include $\sqrt{E_x}$), the delay τ_d, M_{tx}, M_{rx}, and the frequency offset ϵ. Essentially, you need to determine the discrete-time equivalent channel, convolve it with $x(nT/M_{tx})$, resample, incorporate frequency offset, add AWGN, and resample to produce $r(nT/M_{rx})$. Be sure that the noise is added correctly so it is correlated. Design some tests and demonstrate that your code works correctly.

52. **Computer** In this problem, you will create a symbol synchronization function. The input is $y(nT/M_{rx})$, which is the matched filtered receive pulse-amplitude modulated signal, M_{rx} the receiver oversampling factor, and the block length for averaging. You should implement both versions of the MOE algorithm. Demonstrate that your code works. Simulate $x(t)$ obtained from 4-QAM constellations with 100 symbols and raised cosine pulse shaping with $\alpha = 0.25$. Pass your sampled signal with $M_{tx} \geq 2$ through your channel simulator to produce the sampled output with M_{rx} samples per symbol. Using a Monte Carlo simulation, with an offset of $\tau_d = T/3$ and $h = 0.3e^{j\pi/3}$, perform 100 trials of estimating the symbol timing with each symbol synchronization algorithm for SNRs from 0dB to 10dB. Which one gives the best performance? Plot the received signal after correction; that is, plot $\tilde{y}[nM_{rx}+k^\star]$ for the entire frame with real on the x-axis and imaginary on the y-axis for 0dB and 10dB SNR. This should give a rotated set of 4-QAM constellation symbols. Explain your results.

53. **Computer** In this problem, you will implement the transmitter and receiver for a flat-fading channel. The communication system has a training sequence of length 32 that is composed of length 8 Golay sequences and its complementary pairs as $[a_8, b_8, a_8, b_8]$. The training sequence is followed by 100 M-QAM symbols. No symbols are sent before or after this packet. Suppose that $M_{rx} = 8$, $\tau_d = 4T+T/3$, and $h = e^{j\pi/3}$. There is no frequency offset. Generate the matched filtered receive signal. Pass it through your symbol timing synchronizer and correct for the timing error. Then perform the following:

    (a) Develop a frame synchronization algorithm using the given training sequence. Demonstrate the performance of your algorithm by performing 100 Monte Carlo simulations for a range of SNR values from 0dB to 10dB. Count the number of frame synchronization errors as a function of SNR.

    (b) After correction for frame synchronization error, estimate the channel. Plot the estimation error averaged over all the Monte Carlo trials as a function of SNR.

    (c) Compute the probability of symbol error. Perform enough Monte Carlo trials to estimate the symbol error reliably to 10^{-4}.

54. **Computer** Consider the same setup as in the previous problem but now with a frequency offset of $\epsilon = 0.03$. Devise and implement a frequency offset correction algorithm.

(a) Assuming no delay in the channel, demonstrate the performance of your algorithm by performing 100 Monte Carlo simulations for a range of SNR values from 0dB to 10dB. Plot the average frequency offset estimation error.

(b) After correction for frequency synchronization error, estimate the channel. Plot the estimation error averaged over all the Monte Carlo trials as a function of SNR.

(c) Perform enough Monte Carlo trials to estimate the symbol error reliably to 10^{-4}.

(d) Repeat each of the preceding tasks by now including frame synchronization and symbol timing with τ_d as developed in the previous problem. Explain your results.

55. **Computer** In this problem, you will create a multipath channel simulator function where the complex baseband equivalent channel has the form

$$h(t) = T\,\mathrm{sinc}(t/T) * \sum_{k=1}^{K} \alpha_k \delta(t - \tau_k) \qquad (5.418)$$

and the channel has multiple taps, each consisting of complex amplitude α_k and delay τ_k. The input to your function is $x(nT/M_\mathrm{tx})$, the sampled pulse-amplitude modulated sequence sampled at T/M_tx where M_tx is the oversampling factor. The output of your function is $r(nT/M_\mathrm{rx})$, which is sampled from

$$r(t) = e^{\mathrm{j}2\pi f_c t} \sum_{k=1}^{K} \alpha_k x(t - \tau_k) + v(t). \qquad (5.419)$$

Your simulator should include the channel parameters M_tx, M_rx, and the frequency offset ϵ. Check that your code works correctly by comparing it with the flat-fading channel simulator. Design some other test cases and demonstrate that your code works.

Chapter 6

MIMO Communication

Thus far the book has focused on wireless communication systems that use a single antenna for transmission or reception. Many wireless systems, however, make use of multiple antennas. These antennas may be used to obtain diversity against fading in the channel, to support higher data rates, or to provide resilience to interference. This chapter considers communication in systems with multiple transmit and/or receive antennas. Broadly this is known as multiple-input multiple-output (MIMO) communication. The chapter provides background on different techniques for using multiple transmit and receive antennas, including beamforming, spatial multiplexing, and space-time coding. It also explains how algorithms developed in prior chapters including channel estimation and equalization extend when multiple antennas are present.

6.1   Introduction to Multi-antenna Communication

Multiple antennas have been used in wireless communication systems since the very beginning, even in Marconi's early transatlantic experiments at the beginning of the 1900s. Wireless communication systems have been making use of simple applications of multiple antennas for several decades. In cellular systems they have been used for sectoring (to focus the energy of the base station on a particular geographic sector), downtilt (to focus the energy away from the horizon), and receive diversity (to improve signal reception from low-power mobile stations). In wireless local area networks or Wi-Fi, two antennas have often been used for diversity in both the client (e.g., laptop) and the access point. These antennas would often be configured to be spatially far apart, for example, on either side of the laptop screen. These early applications did not require redesigning the communication signal to support multiple antennas.

Several commercial wireless systems have integrated multiple antennas into their design and deployment. Techniques like transmit beamforming and precoding are supported by cellular systems (3GPP since release 7) and wireless local area networks (IEEE 802.11n, IEEE 802.11ac, and IEEE 802.11ad). Systems that exploit multiple antennas for both transmit and receive are said to employ MIMO communication [256].

The use of MIMO comes from signal processing and control theory. It is used to refer to a system that has multiple inputs and multiple outputs. In the case of a wireless system, a MIMO system is created from the propagation channel that results from having multiple inputs (from different transmit antennas) and multiple outputs (from different receive antennas). There are several other related terms, illustrated in Figure 6.1, that are used to refer to different configurations of antennas. Single-input single-output (SISO) refers to a communication link with one transmit and one receive antenna. This is the usual communication link considered thus far in this book. The other configurations are summarized in this section using examples from flat fading and are generalized in subsequent sections.

The emphasis in this chapter is explicitly on the use of multiple antennas to create a MIMO system. In every direction of propagation using a transverse wave, though, there can be two orthogonal polarizations. Usually these are known as horizontal and vertical, circular left and circular right, or ±45°. Every antenna has two patterns, one for its primary polarization and one for its orthogonal polarization. Some physical antenna structures, like patch antennas, can be used to send two different polarized signals by exciting different modes. In other cases, different antennas are used to send different polarizations. Channel models are more complicated with polarization. For example, the path loss when using vertical polarization is more adversely affected by rain than with horizontal polarizations. Also, the small-scale fading model becomes more complicated because of cross-pole decoupling (the leakage of one polarization into another) [39]. Polarization is important in cellular systems where the random orientation of the cellular phone creates

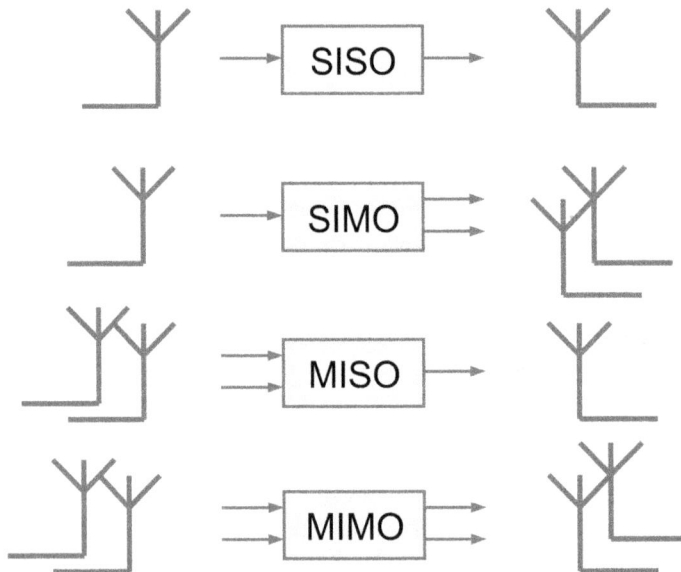

Figure 6.1 The terminology resulting from different antenna configurations: single-input single-output (SISO), single-input multiple-output (SIMO), multiple-input single-output (MISO), and multiple-input multiple-output (MIMO)

different polarized signals. We defer explicit treatment of polarization in this chapter, but note that polarization is an effective complement to MIMO communication.

6.1.1 Single-Input Multiple-Output (SIMO)

SIMO refers to a communication link with multiple receive antennas and one transmit antenna. This is often called receive diversity and is perhaps the oldest use of multiple antennas [53, 261]. The basic idea is to add extra receive antennas and somehow combine the different outputs. The processing can be performed in analog or in digital. For example, an analog switch may be used to switch between two receive antennas. Alternatively, multiple RF chains and multiple continuous-to-discrete converters can be used so that all the antenna outputs can be processed together in digital. The potential benefits of SIMO operation include array gain (resistance to noise) and diversity gain (resistance to fading).

Diversity is an important concept in wireless communications. In essence, diversity means having multiple avenues to get symbols from the transmitter to the receiver. If the system is well designed, each avenue sees a different propagation channel, which increases the likelihood that at least one of those channels is good. Diversity can be obtained in many ways in wireless systems. For example, with the use of coding and interleaving, coded data can experience channels separated by the coherence time, in what is called time diversity. Alternatively, a signal with a large bandwidth can be used to take advantage of the multiple paths between the transmitter and the receiver, in what is called frequency-selective diversity. Diversity can also be achieved using multiple antennas, in what is called spatial diversity.

As an example, consider a flat-fading SIMO system with digital receiver processing as illustrated in Figure 6.2. After matched filtering, symbol timing, frequency offset correction, and frame synchronization, the output corresponding to each receive antenna can be written as

$$y_1[n] = h_1 s[n] + v_1[n] \tag{6.1}$$
$$y_2[n] = h_2 s[n] + v_2[n] \tag{6.2}$$

or in matrix form as

$$\mathbf{y}[n] = \mathbf{h}s[n] + \mathbf{v}[n]. \tag{6.3}$$

The scalars h_1 and h_2 are the channels from the transmit antenna to receive antennas 1 and 2, respectively. If the antennas are spaced far enough apart—greater than the coherence distance—then the channels are said to be uncorrelated. This provides the receiver with diversity since it has multiple opportunities to extract the same information sent through different fading channels and augmented by a different additive noise term. Good receiver signal processing algorithms take advantage of both observations in the equalization and detection process.

Selection diversity is one example of a way to process signals from multiple receive antennas [261]. The channel with the largest magnitude is selected (given by $\bar{k}$) and then $y_{\bar{k}}$ is processed. This system obtains diversity from fading because if h_1 and h_2 are independent (which follows if the channels are jointly Gaussian and uncorrelated), then

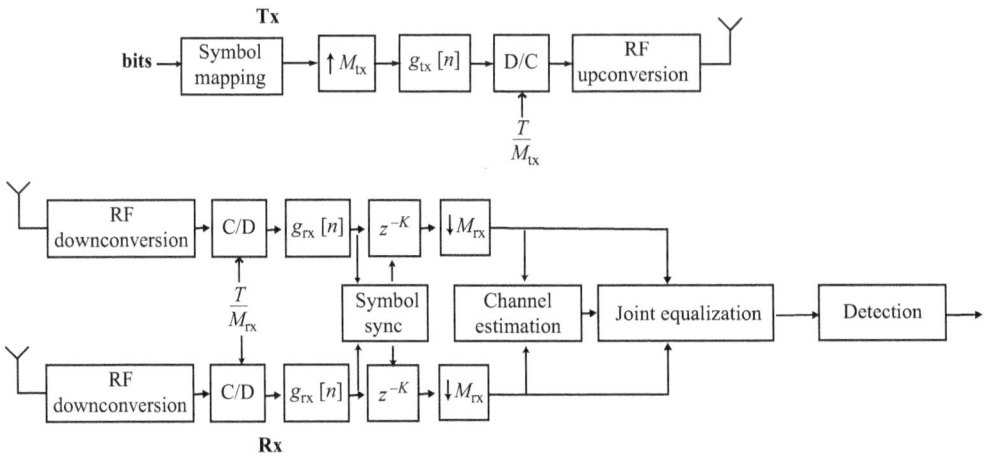

Figure 6.2 Block diagram for a QAM system exploiting SIMO with receive processing in digital and multi-antenna linear equalization. The same transmitter structure is employed as in a SISO scenario. The receiver, though, includes two RF chains. Some processing steps may be performed jointly to leverage common information received on both receive antennas. This example shows symbol synchronization being done jointly. The channel estimation may be done independently for each receiver, but the equalization operation should combine both received signals prior to detection. Frame synchronization and frequency offset synchronization are not shown but can also leverage multiple receive antennas.

the probability that $|h_1|$ and $|h_2|$ are both small is much smaller than the probability that either one is small.

There are many other kinds of receiver processing techniques, including maximum ratio combining (MRC) [52], equal gain combining (which used to be known as the linear adder [10]), and joint space-time equalization [90, 351, 241]. An illustration of the impact of receive diversity is provided in Figure 6.3.

Virtually every technique studied thus far in this book can be extended to the case of multiple receive antennas without dramatic changes to the resulting signal processing algorithms. Consider how functions like symbol synchronization, frame synchronization, and frequency offset correction might incorporate observations from multiple antennas. SIMO is discussed in more detail in Section 6.2.

6.1.2 Multiple-Input Single-Output (MISO)

MISO refers to a communication link with multiple transmit antennas and one receive antenna. This is often called transmit diversity or transmit beamforming. Multiple transmit antennas have been in wide use only recently in commercial wireless systems as they often require additional signal processing at the transmitter to extract the most benefit from the antennas.

A MISO system uses multiple transmit antennas to convey a signal to one receive antenna. Generally, communication in the MISO case is much more challenging than in the SIMO case because the transmit signals are combined by the channel. Obtaining

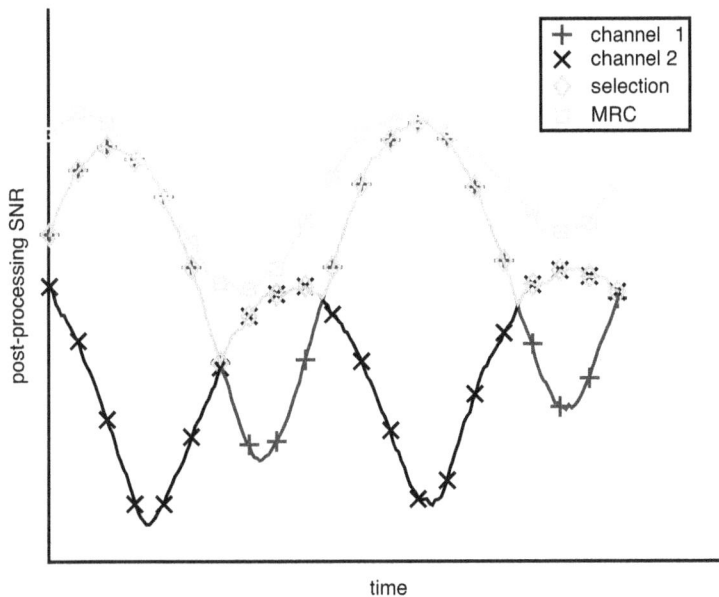

Figure 6.3 The impact of diversity in a two-antenna SIMO system. One realization for each channel is shown. The effective channels created using antenna selection diversity and maximum ratio combining are also shown. Receive antenna diversity reduces the occurrences of deep fades.

some advantage from spatial diversity is much more difficult. From a signal processing perspective, either some knowledge about the channel is required at the transmitter or the information needs to be spread in a special way across the transmit antennas to obtain a diversity advantage.

One approach for using multiple transmit antennas that does not require channel state information at the transmitter is known as transmit delay diversity [357]. A variation of delay diversity (called cyclic delay diversity [83]) is used in IEEE 802.11n. The basic idea, as illustrated in Figure 6.4, is to send a successively delayed version of the transmitted signal on each transmit antenna. The effect is to create a frequency-selective channel from a flat-fading channel. For example, if the delay is exactly one symbol period and there are two transmit antennas (after matched filtering, symbol timing, frequency offset correction, and frame synchronization), then

$$y[n] = h_1 s_1[n] + h_2 s_2[n] + v[n] \tag{6.4}$$

$$= h_1 s[n] + h_2 s[n-1] + v[n] \tag{6.5}$$

where $y[n]$ experiences intersymbol interference! At first it seems like it is bad that now an equalizer is required. But note that $s[n]$ contributes to both $y[n]$ and $y[n+1]$; thus there are two places to extract meaningful information about $s[n]$. This is known as frequency-selective diversity. Unfortunately, linear time-domain equalizers do not reap much of an advantage from this sort of diversity. To obtain the most benefits, SC-FDE is required, or

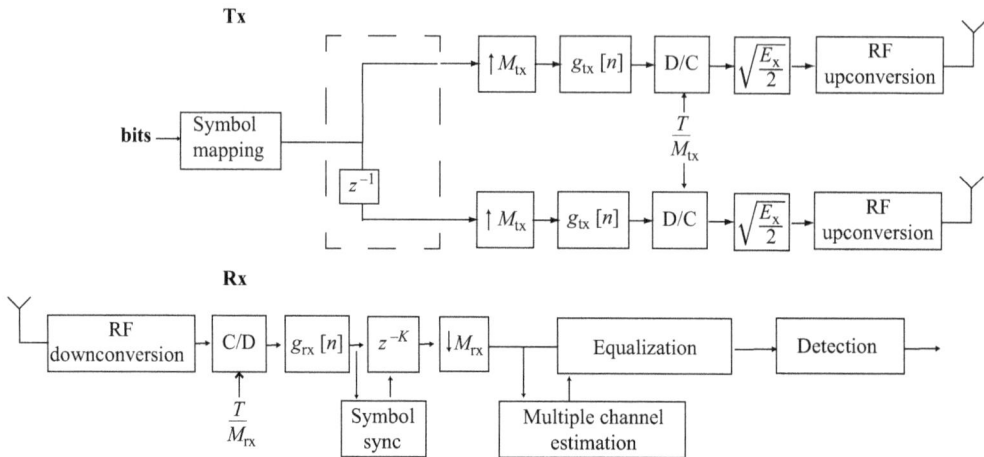

Figure 6.4 Block diagram for a QAM system exploiting MISO with delay diversity at the transmitter. The transmitter structure is modified over the SISO case by the inclusion of a space-time coding block (the dashed box), which in this case implements the delay operation. The receiver structure is similar to the SISO case, except that estimation of the channels from each antenna is required in general and the equalization may be more complex. With delay diversity, the receiver can use processing similar to that for frequency-selective channels. Frame synchronization and frequency offset synchronization are not shown but can also leverage multiple receive antennas.

coding and interleaving with OFDM, or a nonlinear receiver like the maximum likelihood sequence detector.

There are a number of different approaches for using multiple transmit antennas. Approaches that do not use channel state information at the transmitter are known as space-time codes [326, 8, 327, 125]. Delay diversity is an example of a space-time code. Essentially, with space-time coding, information is spread across all of the transmit antennas. More sophisticated receiver algorithms are then used to detect the transmitted signal. The main advantage of space-time coding is that it can obtain a diversity advantage without requiring information about channel. The main disadvantage is that (except in special cases) achieving the best performance requires higher-complexity algorithms at the receiver.

There are several ways to use channel state information at the transmitter, the most common being for transmit beamforming. The idea is to change the amplitude and phase of the signal sent from each antenna so that the signals combine favorably at the receiver. This results in better performance than is achieved when the channel is unknown but requires additional system design considerations to obtain an estimate of the channel at the transmitter [207].

An illustration of the impact of different diversity combining techniques is shown in Figure 6.5. Approaches that use channel state information at the transmitter outperform approaches that do not exploit knowledge of the channel. Learning the channel at the transmitter requires additional support in the communication system, for example, the use of feedback. MIS, including space-time coding and beamforming, is discussed in more detail in Section 6.3.

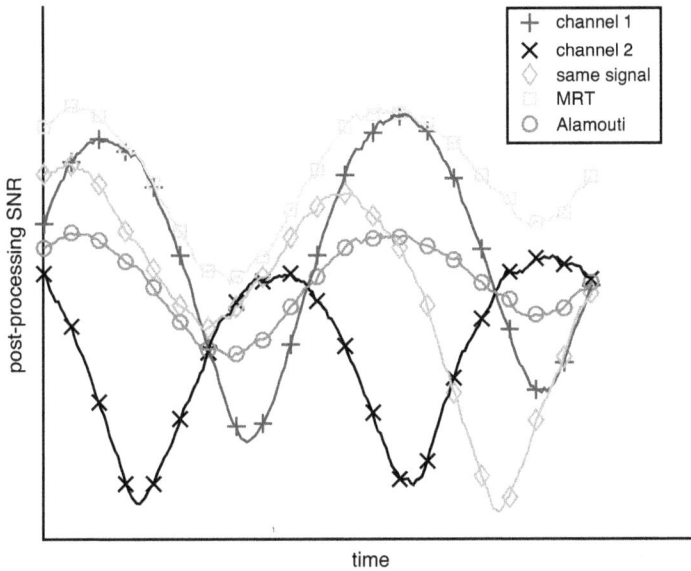

Figure 6.5 The impact of diversity in a two-transmit-antenna MISO system. One realization for each channel is shown. The effective channels created by sending the same signal across both antennas (spatial repetition) is also shown, along with maximum ratio transmission (which uses channel state information at the transmitter) and the Alamouti space-time block code. Sending the same information across both transmit antennas results in a channel that is no better than either of the original channels. Maximum ratio transmission obtains better performance compared with the Alamouti code but requires knowledge of the channel at the transmitter.

6.1.3 Multiple-Input Multiple-Output (MIMO)

MIMO wireless communication refers to the general case of a communication link that has multiple transmit and multiple receive antennas. SISO, SIMO, and MISO are also considered special cases of MIMO communication. The principle of MIMO communication is the joint design of transceiver techniques that make simultaneous use of multiple communication channels.

MIMO communication is an established technology in wireless communication systems. It has been widely deployed in wireless local area networks through IEEE 802.11n and is a key feature of various third- and fourth-generation cellular standards and next-generation wireless local area network protocols. At this point, MIMO is incorporated into millions of laptops, phones, and tablets.

Spatial multiplexing, also called V-BLAST [115], is one of the most widely used transmission techniques for MIMO systems [255]. The idea of spatial multiplexing is to send different data from each transmit antenna, using the same spectrum and with the same total power as in a SISO system. The transmitted data streams are mixed together by the MIMO channel, creating cross-talk or self-interference. Because each receiver sees a different combination of signals from all the transmitters, joint processing can be used to separate out the signals. Essentially, the cross-talk between different transmitted signals can be canceled. This allows multiple interference-free channels to coexist, using the same spectral resources, thus increasing the channel capacity [130, 328].

The idea of spatial multiplexing is illustrated in Figure 6.6 for the case of two transmit and two receive antennas. In this example, two symbols are sent at the same time and the transmit power is split between them. This allows the symbol rate to be doubled without increasing the bandwidth. Assuming a flat-fading channel (after matched filtering, symbol timing, frequency offset correction, and frame synchronization), the received signals can be written as

$$y_1[n] = h_{1,1}s_1[n] + h_{1,2}s_2[n] + v_1[n] \tag{6.6}$$
$$y_2[n] = h_{2,1}s_1[n] + h_{2,2}s_2[n] + v_2[n] \tag{6.7}$$

or in matrix form

$$\mathbf{y}[n] = \mathbf{H}\mathbf{s}[n] + \mathbf{v}[n]. \tag{6.8}$$

The coefficients $h_{k,\ell}$ denote the channel between the ℓth transmit antenna and the kth receive antenna. Assuming that the channel coefficients are different, each receiver observes a different linear combination of the symbols $s_1[n]$ and $s_2[n]$ and the channel

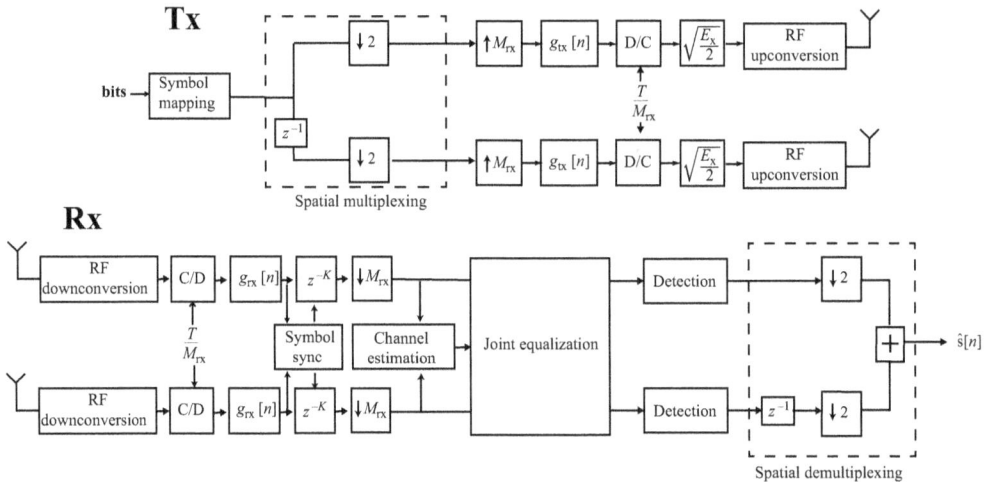

Figure 6.6 Block diagram for a QAM system exploiting MIMO spatial multiplexing with two transmitter and two receiver antennas. The spatial multiplexing operation is illustrated in the dashed box on the transmitter. This is essentially a 1:2 serial-to-parallel converter (also correctly called a demultiplexer). The demultiplexing operation takes one symbol stream $s[n]$ and produces two symbol streams $s_1[n] = s[2n]$ and $s_2[n] = s[2n+1]$. The power splitting is shown explicitly. The receiver involves the usual synchronization operations (frame and frequency offset synchronization not shown) followed by joint channel estimation, joint equalization, and independent symbol detection. Sometimes the detection and joint equalization are combined. After the detection is the spatial demultiplexing operation, which is essentially a 2:1 parallel-to-serial converter (also correctly called a multiplexer).

matrix will be invertible. A simple way to remove the effects of the channel is to invert the matrix in (6.8) and apply this inverse to the received signal to obtain

$$\widehat{\mathbf{s}} = \mathbf{H}^{-1}\mathbf{y}[n] \tag{6.9}$$

followed by independent detection.

MIMO systems with spatial multiplexing are much more complex to implement than SISO systems because of the self-interference created by the transmit antennas. This complicates channel estimation, synchronization, and equalization among other system functions. MIMO systems are presented in more detail in Section 6.4, followed by treatment of MIMO-OFDM because of its wide use in Section 6.5.

6.2 Receiver Diversity for Flat-Fading SIMO Systems

Multiple receive antennas provide a means of receiving multiple copies of the same signal. Each copy sees a different noise realization and a different channel realization. In this section, we focus on two algorithms for combining the outputs of all the antennas in a SIMO system and provide some analysis of those algorithms. We begin with a brief review of SIMO flat-fading channel models, explaining how Rayleigh fading generalizes to the SIMO case. Then we describe two algorithms for processing the antenna outputs in a SIMO system: antenna selection and maximum ratio combining. We provide some analysis of these algorithms for Rayleigh fading. We defer treatment of frequency-selective channel estimation and synchronization to the MIMO case (which specializes to SIMO).

6.2.1 SIMO Flat-Fading Channel Models

The received signal in a SIMO flat-fading channel model with block fading is given in (6.3), but with all the vector dimensions extended to $N_{\mathrm{r}} \times 1$. The channel for the SIMO flat-fading case is represented by the $N_{\mathrm{r}} \times 1$ vector $\mathbf{h}$. As in the SISO case, it is common to decompose multi-antenna channels into their large-scale and small-scale components. Assuming that the antennas are colocated (in an area separated by a few wavelengths), then each antenna sees the same large-scale component and only the small-scale coefficient differs. Then the channel may be decomposed as

$$\mathbf{h} = \sqrt{G}\mathbf{h}_{\mathrm{s}} \tag{6.10}$$

where the large-scale gain is $G = E_{\mathrm{x}}/P_{\mathrm{r,lin}}(d)$, $P_{\mathrm{r,lin}}(d)$ is the distance-dependent pathloss term in linear (to distinguish it from the more common decibel measure), and the small-scale fading coefficients are collected in the vector $\mathbf{h}_{\mathrm{s}}$.

There are many multi-antenna stochastic channel models for $\mathbf{h}_{\mathrm{s}}$. Two conceptual models are illustrated in Figure 6.7. The most common model used in MIMO systems is the Rayleigh model with spatial correlation where $\mathbf{h}_{\mathrm{s}}$ has vector complex Gaussian distribution $\mathcal{N}_{\mathbb{C}}(0, \mathbf{R}_{\mathrm{rx}})$ where the main parameter is the receive spatial correlation matrix given by $\mathbf{R}_{\mathrm{rx}} = \mathbb{E}\left[\mathbf{h}_{\mathrm{s}}\mathbf{h}_{\mathrm{s}}^{*}\right]$. In most small-scale models, it is common for the average energy of the channel to be normalized to 1; in the vector case this means that $[\mathbf{R}_{\mathrm{rx}}]_{k,k} = 1$. The off-diagonal terms in the spatial correlation matrix $[\mathbf{R}_{\mathrm{rx}}]_{k,\ell}$ give the spatial correlation

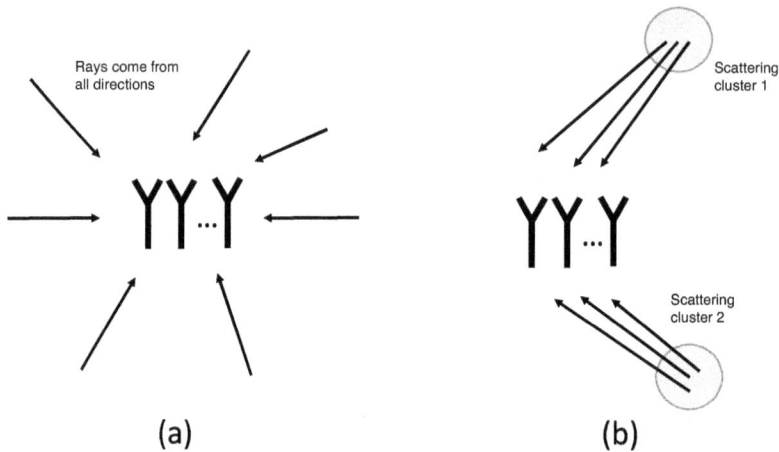

Figure 6.7 Two different channel-modeling scenarios. In (a), there is rich scattering, which means that multipaths arrive from many different directions. Alternatively, in (b), there is clustered scattering. Multipaths are clustered together in potentially different-size clusters.

between the channel seen by antenna k and the channel seen by antenna ℓ. This function is defined similarly to the spaced-time correlation function or the spaced-frequency correlation function. The spatial case of $\mathbf{R}_{\mathrm{rx}} = \mathbf{I}$ is known as IID Rayleigh fading or sometimes just Rayleigh fading. In this case, the channels experienced by each receive antenna are completely uncorrelated. In cases where $\mathbf{R}_{\mathrm{rx}}$ is non-identity, the channel coefficients may be correlated. The exponentially correlated model is often used for analysis where $[\mathbf{R}_{\mathrm{rx}}]_{k,\ell} = \rho^{k-\ell}$; ρ is the complex correlation coefficient between two adjacent antennas.

Most of the analysis in this chapter is based on the IID Rayleigh fading channel. This is normally justified by the rich scattering assumption where there are many multipaths in the environment and the antennas are spaced far enough apart. To be uncorrelated, the antennas should be spaced by the coherence distance, which is a selectivity quantity that can be related to the coherence time. For WLAN and cellular applications, the coherence distance is anywhere from $\lambda/2$ on devices surrounded by clutter to 10λ for antennas that are high up on a tower, where λ is the wavelength. Further decorrelation can be achieved by making the patterns of the antennas different [111, 112, 263].

The clustered channel model [313], a spatial generalization of the Saleh-Valenzuela channel model [285], is also used in multi-antenna systems for channel modeling. Suppose that there are C clusters in the propagation environment, each with R rays. Let the vector $\mathbf{a}(\theta)$ denote the array response vector, which is

$$\mathbf{a}(\theta) = [1, e^{-\mathrm{j}\frac{2\pi\Delta}{\lambda}\sin(\theta)}, \ldots, e^{-\mathrm{j}(N_{\mathrm{r}}-1)\frac{2\pi\Delta}{\lambda}\sin(\theta)}]^T \tag{6.11}$$

for the case of a uniform linear array with isotropic elements. The term Δ/λ is the spacing between two adjacent antenna elements in terms of wavelength. The array is oriented on the z-axis, and the angle θ is called the azimuth direction. Let $\theta_{c,r}$ denote the angle of arrival of the rth ray from the cth cluster. Let $\alpha_{c,r}$ denote the complex gain of that ray, often modeled as Rayleigh. The clustered channel is given as the sum of the rays

$$\mathbf{h}_{\text{s}} = \frac{1}{\sqrt{RC}} \sum_{c=0}^{C-1} \sum_{r=0}^{R-1} \alpha_{c,r} \mathbf{a}(\theta_{c,r}). \tag{6.12}$$

The factor of $\sqrt{RC}$ is used to satisfy the power constraints. This clustered model formulation assumes that all the clusters are equidistant. Further modifications account for differences in path loss for each cluster.

The clustered channel model is is often generated stochastically by making assumptions about the distribution of clusters and angles. Suppose that each cluster is associated with a mean angle of arrival θ_c. It is common for the mean angle of arrival to be uniformly distributed. Suppose that the rays are generated around each cluster with a certain power distribution, called the power azimuth spectrum $P_{\text{az}}(\theta)$. Let σ_{as} be the angle spread associated with the distribution. The most common choice for MIMO channel modeling is the Laplacian distribution, which is

$$P_{\text{az}}(\theta) = \frac{C_{\text{lap}}}{\sqrt{2}\sigma_{\text{as}}} e^{-\left|\frac{\sqrt{2}\theta}{\sigma_{\text{as}}}\right|}, \quad \theta \in [-\pi, \pi) \tag{6.13}$$

where $C_{\text{lap}} = \frac{1}{1-\exp(-\sqrt{2}\pi/\sigma_{\text{as}})}$ is a normalization factor needed to make the function integrate to 1. The power azimuth spectrum for each cluster is then shifted by the cluster's mean angle of arrival as $P_{\text{az}}((\theta - \theta_c)_{2\pi})$.

Using these distributions, it is possible to generate (6.12) via simulation. It is also possible to derive an equivalent $\mathbf{R}_{\text{rx}}$ for the clustered channel model [298, 349, 178], though the expressions have high computational complexity. Lower-complexity solutions follow using a small angle approximation [113]. In this sense, the clustered channel model provides a way to generate the spatial correlation matrix for a correlated Rayleigh channel, based on parameters of the propagation environment. In general, having fewer clusters and/or a small angle spread leads to significant correlation ($\mathbf{R}_{\text{rx}}$ is far away from an identity matrix). Alternatively, having many clusters and/or a large angle spread tends to reduce correlation and converges eventually to the IID Rayleigh channel.

A special case of the clustered model is the single-path channel where

$$\mathbf{h}_{\text{s}} = \alpha \mathbf{a}(\theta). \tag{6.14}$$

In this case, the elements of $\mathbf{h}_{\text{s}}$ are fully correlated since α is common for all antennas. This kind of channel model has been widely used in statistical array processing and in smart antenna communication systems. It is most practical in cases where there is only one significant propagation path, making it more suitable for LOS links for higher-frequency communication systems like millimeter wave [145, 268].

6.2.2 Antenna Selection

One simple technique for receive diversity in a SIMO system is called *selection combining*, wherein the receiver measures the SNR on each antenna and chooses to decode the signal

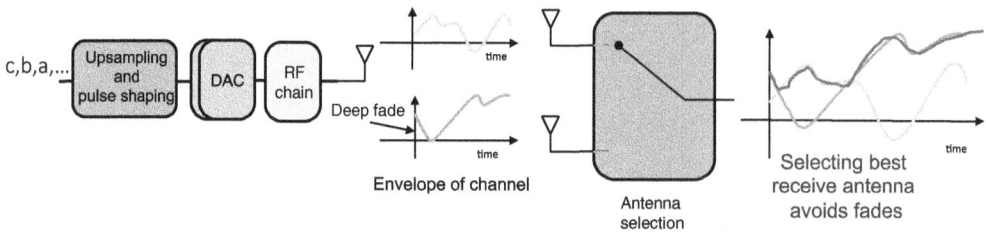

Figure 6.8 Block diagram of a SIMO system implementing antenna selection

from the antenna with the highest SNR [261]. An illustration of a receiver implementing selection combining is shown in Figure 6.8. Consider the received signal on the ith antenna

$$y_i[n] = h_i s[n] + v_i[n] \tag{6.15}$$

for $i = 1, \ldots, N_{\text{r}}$. Let $i^\star$ be the solution to $\arg\max_{i=1,2,\ldots,N_{\text{r}}} |h_i|^2$. Since the SNR of the ith branch given h_i is $|h_i|^2/N_{\text{o}}$, this selection results in choosing the m that gives the highest SNR.

The signal corresponding to $i^\star$,

$$y_{i^\star}[n] = h_{i^\star} s[n] + v_{i^\star}[n], \tag{6.16}$$

is selected for symbol detection. The post-selection SNR is the effective SNR after the selection, given knowledge of the channel:

$$\text{SNR}_{\mathbf{h}}^{\text{sel}} = \frac{|h_{i^\star}|^2}{N_{\text{o}}}. \tag{6.17}$$

A good post-selection SNR is achieved if the channel from at least one antenna is good.

The performance achieved by selection diversity depends on the assumptions about the fading distribution. In Example 6.1, Example 6.2, and Example 6.3, the Chernoff bound on the probability of symbol error is calculated for uncorrelated Rayleigh fading.

Example 6.1   Consider a SIMO system with N_{r} receive antennas, and suppose that the channel coefficients are IID $\mathcal{N}_{\mathbb{C}}(0, G)$. Compute the CDF and PDF of $|h_{\bar{k}}|^2$.

Answer: The CDF of $|h_{i^\star}|^2$ is the CDF of $\max_i |h_i|^2$ given by $\mathbb{P}\left(\max_i |h_{i^\star}|^2 \leq x\right)$

$$\mathbb{P}\left(\max_i |h_i|^2 \leq x\right) = \mathbb{P}\left(|h_1|^2 \leq x\right) \mathbb{P}\left(|h_2|^2 \leq x\right) \cdots \mathbb{P}\left(|h_{N_{\text{r}}}|^2 \leq x\right) \tag{6.18}$$

since the h_i are independent.

In general, the sum of K $\mathcal{N}(0, 1)$ squared random variables has a chi-square distribution with K degrees of freedom. Because each h_i is $\mathcal{N}_{\mathbb{C}}(0, G)$, $|h_i|^2$ is a scaled chi-square distribution with 2 degrees of freedom (because h_i is complex). In this case, the distribution is especially simple and is

$$\mathbb{P}\left(|h_i|^2 \leq x\right) = 1 - e^{-x/G}. \tag{6.19}$$

Therefore,

$$\mathbb{P}\left(\max_i |h_i|^2 \leq x\right) = \left(1 - e^{-x/G}\right)^{N_{\mathrm{r}}}. \tag{6.20}$$

The PDF is computed from

$$\frac{\mathrm{d}}{\mathrm{d}x}\mathbb{P}\left(\max_i |h_i|^2 \leq x\right) = \frac{1}{G}e^{-x/G}(1 - e^{-x/G})^{N_{\mathrm{r}}-1}N_{\mathrm{r}}. \tag{6.21}$$

Example 6.2   For the same setting as in Example 6.1, assuming $G = E_{\mathrm{x}}/P_{\mathrm{r,lin}}(d)$, determine the average post-selection SNR given by $\mathbb{E}[\max_i |h_i|^2]N_{\mathrm{o}}$.

  Answer:

$$\frac{1}{N_{\mathrm{o}}}\mathbb{E}\left[\max_i |h_i|^2\right] = \frac{1}{N_{\mathrm{o}}}\int_0^\infty \frac{1}{G}xe^{-x/G}(1 - e^{-x/G})^{N_{\mathrm{r}}-1}N_{\mathrm{r}}\mathrm{d}x \tag{6.22}$$

$$= \frac{G}{N_{\mathrm{o}}}\sum_{i=1}^{N_{\mathrm{r}}}\frac{1}{i} \tag{6.23}$$

$$= \frac{E_{\mathrm{x}}}{N_{\mathrm{o}}P_{\mathrm{r,lin}}(d)}\sum_{i=1}^{N_{\mathrm{r}}}\frac{1}{i}. \tag{6.24}$$

Note that increasing N_{r} provides a diminishing gain to the average SNR, a result known as *SNR hardening*.

Example 6.3   For the same setting as in Example 6.1, compute the Chernoff bound on the union bound on the average probability of symbol error, neglecting path loss, that is, $G = E_{\mathrm{x}}$.

  Answer:

$$P_{\mathrm{e}} \leq (M-1)\mathbb{E}\left[Q\left(\sqrt{\frac{1}{2N_{\mathrm{o}}}d_{\min}^2 \max_i |h_i|^2}\right)\right] \tag{6.25}$$

$$\leq \frac{M-1}{2}\mathbb{E}\left[e^{-\frac{1}{4N_{\mathrm{o}}}d_{\min}^2 \max_i |h_i|^2}\right] \tag{6.26}$$

$$= \frac{M-1}{2}\int_0^\infty e^{-\frac{1}{4N_{\mathrm{o}}}d_{\min}^2 x}e^{-x/G}(1 - e^{-x/G})^{N_{\mathrm{r}}-1}N_{\mathrm{r}}\mathrm{d}x \tag{6.27}$$

$$= \frac{M-1}{2}\frac{N_{\mathrm{r}}\Gamma(\frac{G}{4N_{\mathrm{o}}}d_{\min}^2 + 1)\Gamma(N_{\mathrm{r}})}{\Gamma(\frac{G}{4N_{\mathrm{o}}}d_{\min}^2 + N_{\mathrm{r}} + 1)} \tag{6.28}$$

where $\Gamma(x) = \int_0^\infty t^{x-1} e^{-t} dt$ is the gamma function. Performing a series expansion around G/N_o and substituting for $G = E_x/P_{r,\text{lin}}(d)$, neglecting higher-order terms, it can be concluded that

$$P_e \lesssim \frac{(M-1)N_r \Gamma(N_r)}{2} \left(\frac{1}{\frac{E_x}{4N_o} d_{\text{min}}^2} \right)^{N_r}. \tag{6.29}$$

The factor of $\left(\frac{1}{\frac{E_x}{N_o}} \right)^{N_r}$ is common in many calculations for Rayleigh fading channels involving the probability of symbol error. Recall that for the SISO case from Section 5.8.3, the probability of symbol error was proportional to $\left(\frac{1}{\frac{E_x}{N_o}} \right)$ for high SNR. The additional exponent of N_r in the case of selection diversity is called the *diversity gain*. A comparison of the exact probability of symbol error is provided in Figure 6.9 for different values of N_r for QPSK modulations. Exact expressions for the probability of error are available in the literature; see, for example, [310] and the references therein.

There are other variations and extensions of antenna selection. One variation is known as scanning diversity [142, 53]. The idea is to scan the received signal successively on different antennas, stopping on the first antenna that is above a predetermined threshold. That antenna is used until the signal drops below a threshold, when the scanning process begins again. In this way the receiver does not have to measure the signals from all the receive antennas, allowing the use of an analog antenna switch in hardware, at the expense of worse performance.

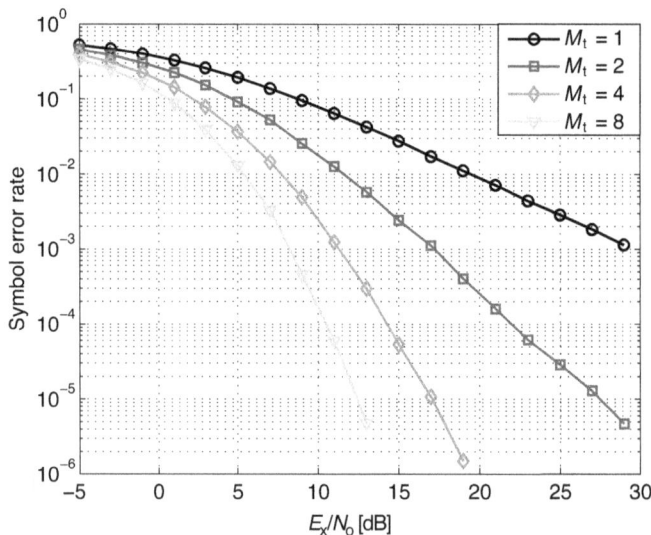

Figure 6.9 Performance of selection combining in Rayleigh fading channels for QPSK modulation and different values of N_r

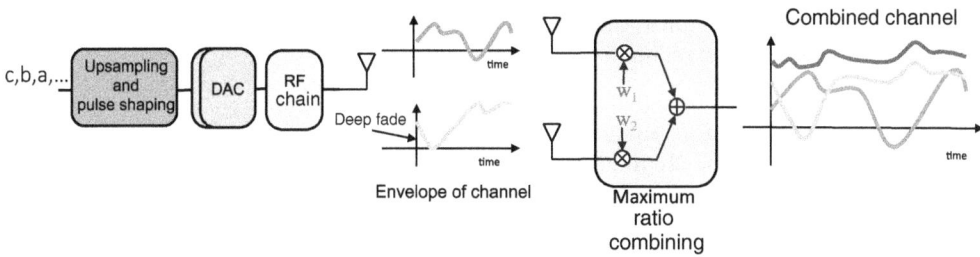

Figure 6.10 Block diagram of a SIMO system implementing maximum ratio combining

Antenna selection can be extended to frequency-selective channels [21, 22, 23, 242]. In a single-carrier system, the receiver would need a better metric for selecting the best antenna, for example, based on the mean squared error of the equalizer. In an OFDM system, antenna selection could be performed independently as described on each separate subcarrier. It might make more sense, however, to perform the switching in analog prior to the OFDM demodulator to reduce the hardware requirements. This would require a different metric to choose the antenna that is good for many subcarriers [189, 289].

A main disadvantage of antenna selection is that it throws away the signals on the receive antennas that were not selected for processing. This prompts more sophisticated algorithms that combine the received signals from several antennas.

6.2.3 Maximum Ratio Combining

Maximum ratio combining (MRC) is a receiver signal processing technique for SIMO systems that combines the signals from the receiver antenna to maximize the post-combining SNR [52]. The objective of MRC is to determine the coefficients of a vector $\mathbf{w}$, called a beamforming vector, such that the combined signal

$$r[n] = \mathbf{w}^*\mathbf{y}[n] \tag{6.30}$$
$$= \mathbf{w}^*\mathbf{h}s[n] + \mathbf{w}^*\mathbf{v}[n] \tag{6.31}$$

has the largest possible post-processing SNR.

The post-processing SNR is the effective SNR after beamforming, conditioned on knowledge of $\mathbf{h}$:

$$\text{SNR}_{|\mathbf{h}} = \frac{|\mathbf{w}^*\mathbf{h}|^2}{N_o\|\mathbf{w}\|^2}. \tag{6.32}$$

Expanding the norm and absolute values,

$$\text{SNR}_{|\mathbf{h}} = \frac{(\mathbf{w}^*\mathbf{h})(\mathbf{w}^*\mathbf{h})^*}{N_o\mathbf{w}^*\mathbf{w}}. \tag{6.33}$$

The objective is to find $\mathbf{w}$ to maximize the post-processing SNR:

$$\mathbf{w} = \arg\max_{\mathbf{w}\in\mathbb{C}^{N_r}} \frac{(\mathbf{w}^*\mathbf{h})(\mathbf{w}^*\mathbf{h})^*}{N_o\mathbf{w}^*\mathbf{w}}. \tag{6.34}$$

To solve this problem, recognize that $\mathbf{w}^*\mathbf{h}$ is just the inner product between two vectors. From the Cauchy-Schwarz inequality, it follows that

$$|\mathbf{w}^*\mathbf{h}|^2 \le \|\mathbf{w}\|^2 \|\mathbf{h}\|^2 \tag{6.35}$$

where equality holds if $\mathbf{w} = c\mathbf{h}$ and c is an arbitrary nonzero complex value. Consequently,

$$\frac{(\mathbf{w}^*\mathbf{h})(\mathbf{w}^*\mathbf{h})^*}{N_\mathrm{o}\mathbf{w}^*\mathbf{w}} \le \frac{(c^*\mathbf{h}^*\mathbf{h})(c\mathbf{h}^*\mathbf{h})^*}{N_\mathrm{o}|c|^2\mathbf{h}^*\mathbf{h}} \tag{6.36}$$

$$= \frac{1}{N_\mathrm{o}}\mathbf{h}^*\mathbf{h} \tag{6.37}$$

$$= \frac{1}{N_\mathrm{o}}\|\mathbf{h}\|^2. \tag{6.38}$$

This reveals that the choice of c is immaterial, so it can be set to $c = 1$ without loss of generality. This also provides a convenient expression for the post-combining SNR that can be used for subsequent performance analysis. The intuitive interpretation is that $\mathbf{w}$ is a spatial matched filter, matched to the vector $\mathbf{h}$.

The resulting post-combining SNR with the matched filter is then

$$\mathrm{SNR}_{|\mathbf{h}}^{\mathrm{MRC}} = \frac{\|\mathbf{h}\|^2}{N_\mathrm{o}} \tag{6.39}$$

$$= \frac{G\|\mathbf{h}_\mathrm{s}\|^2}{N_\mathrm{o}}. \tag{6.40}$$

Since $\|\mathbf{h}\|^2 \ge \max_i |h_i|^2$, MRC outperforms antenna selection.

Example 6.4, Example 6.5, and Example 6.6 consider the average SNR and the probability of symbol error after optimum combining, assuming an IID Rayleigh fading channel.

Example 6.4   Consider a SIMO system with N_r receive antennas, and suppose that the channel coefficients h_i are IID $\mathcal{N}_\mathbb{C}(0, G)$. Compute the CDF and PDF of $\mathrm{SNR}_{|\mathbf{h}}^{\mathrm{MRC}}$.

Answer: Since h_i are $\mathcal{N}_\mathbb{C}(0, G)$, it follows that $\sum_{i=1}^{N_\mathrm{r}} |h_i|^2$ is the sum of $2N_\mathrm{r}$ IID $\mathcal{N}(0, G/2)$ random variables. This is a scaled (because of the $1/2$) chi-square distribution denoted $\chi_{2N_\mathrm{r}}^2$ with $2N_\mathrm{r}$ degrees of freedom. As a result, $\|\mathbf{h}\|^2/N_\mathrm{o}$ is also a scaled chi-square distribution, or equivalently a gamma distribution with $\Gamma(N_\mathrm{r}, G/N_\mathrm{o})$. One form of the CDF is given by

$$\mathbb{P}\left(\frac{\|\mathbf{h}\|^2}{N_\mathrm{o}} \le x\right) = 1 - \sum_{m=0}^{N_\mathrm{r}-1} \frac{1}{m!}\left(x\frac{N_\mathrm{o}}{G}\right)^m e^{-x\frac{N_\mathrm{o}}{G}} \tag{6.41}$$

whereas the PDF is

$$\frac{\mathrm{d}}{\mathrm{d}x}\mathbb{P}\left(\frac{\|\mathbf{h}\|^2}{N_\mathrm{o}} \le x\right) = \frac{1}{N_\mathrm{r}!}\left(\frac{N_\mathrm{o}}{G}\right)^{N_\mathrm{r}} e^{-x\frac{N_\mathrm{o}}{G}} x^{N_\mathrm{r}-1}. \tag{6.42}$$

Example 6.5   For the same setting as in Example 6.4, determine the average post-combining SNR.

Answer:

$$\mathbb{E}\left[\frac{\|\mathbf{h}\|^2}{N_{\mathrm{o}}}\right] = \int_{x=0}^{\infty} \frac{1}{N_{\mathrm{r}}!}\left(\frac{N_{\mathrm{o}}}{G}\right)^{N_{\mathrm{r}}} e^{-x\frac{N_{\mathrm{o}}}{G}} x^{N_{\mathrm{r}}} \mathrm{d}x \tag{6.43}$$

$$= \frac{G}{N_{\mathrm{o}}} N_{\mathrm{r}}. \tag{6.44}$$

In this case the SNR increases linearly with N_{r}, unlike the case with antenna selection. The increase in effective SNR here is also known as the maximum array gain.

Example 6.6   For the same setting as in Example 6.4, compute the Chernoff bound on the probability of symbol error assuming $G = E_{\mathrm{x}}$.

Answer:

$$P_{\mathrm{e}} \leq (M-1)\mathbb{E}\left[Q\left(\sqrt{\frac{E_{\mathrm{x}}}{2N_{\mathrm{o}}} d_{\mathrm{min}}^2 \|\mathbf{h}\|^2}\right)\right] \tag{6.45}$$

$$\leq \frac{(M-1)}{2}\mathbb{E}\left[e^{-\frac{E_{\mathrm{x}}}{4N_{\mathrm{o}}}\|\mathbf{h}\|^2}\right] \tag{6.46}$$

$$= \frac{(M-1)}{2}\int_0^{\infty} e^{-\frac{E_{\mathrm{x}}}{4N_{\mathrm{o}}} d_{\mathrm{min}}^2 x} \frac{1}{N_{\mathrm{r}}!} e^{-x} x^{N_{\mathrm{r}}-1} \mathrm{d}x \tag{6.47}$$

$$= \frac{(M-1)}{2}\frac{1}{\left(\frac{E_{\mathrm{x}}}{4N_{\mathrm{o}}} d_{\mathrm{min}}^2 + 1\right)^{N_{\mathrm{r}}}} \tag{6.48}$$

$$\approx \frac{(M-1)}{2}\frac{1}{\left(\frac{E_{\mathrm{x}}}{4N_{\mathrm{o}}} d_{\mathrm{min}}^2\right)^{N_{\mathrm{r}}}}. \tag{6.49}$$

The performance analysis in Example 6.6 shows that MRC also achieves a diversity gain of N_{r}. A comparison of the exact probability of symbol error is provided in Figure 6.11 for different values of N_{r} for QPSK modulations. Exact expressions for the probability of error are available in the literature (see, for example, [310] and the references therein). Comparing Figure 6.9 and Figure 6.11, it can be seen that at high SNR, MRC provides about a 2dB gain over antenna selection. This is the extra benefit achieved from optimum combining.

It is possible to view antenna selection through the lens of MRC. Essentially, antenna selection chooses combining weights from the set

$$\mathcal{S}_{\mathrm{sel}} = \left\{\begin{bmatrix}1\\0\\\vdots\\0\end{bmatrix}, \begin{bmatrix}0\\1\\\vdots\\0\end{bmatrix}, \cdots \begin{bmatrix}0\\0\\\vdots\\1\end{bmatrix}\right\}, \tag{6.50}$$

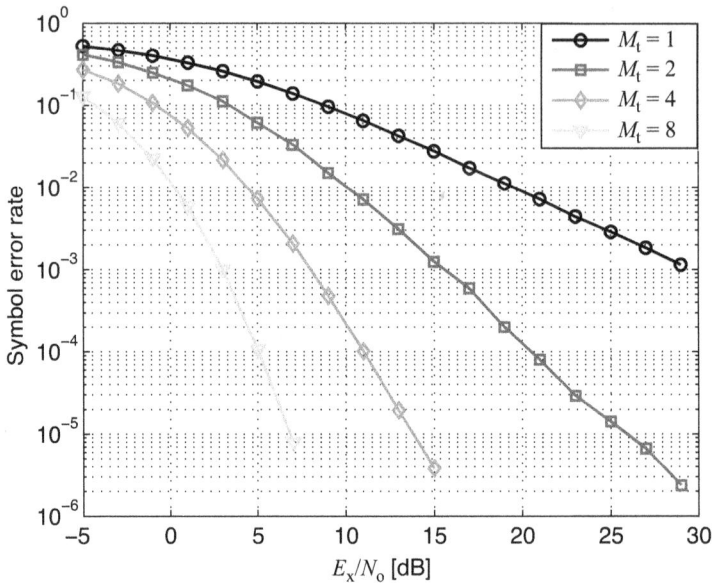

Figure 6.11 Performance of MRC in Rayleigh fading channels for QPSK modulation and different values of N_{r}

which is smaller than $\mathbb{C}^{N_{\mathrm{r}}}$. Thus it is expected that MRC should perform better than antenna selection because the antenna selection is included in the possible set of beamforming vectors used by MRC.

MRC can be extended to frequency-selective channels. In a single-carrier system, more complex beamforming would be employed. For example, the vector beamformer might be replaced with a combined equalizer and combiner [90, 351, 241]. Essentially each receive antenna would be filtered and the results combined. The weights for such an equalizing solution can be derived using the least squares approach in Chapter 5; variations are described for MIMO systems in Section 6.4.7. MRC can be applied in a straightforward fashion in an OFDM system by performing the combining individually on each subcarrier.

There are other kinds of beamforming besides MRC. For example, the beamforming weights can be chosen to minimize the mean squared error at the receiver. This would allow for the beamforming to deal with spatially correlated noise, created, for example, if there is also interference. Such beamforming techniques provide better performance in the presence of correlated noise or interference.

A beamforming technique related to MRC is equal gain combining (EGC) [10, 53, 165]. With EGC, the beamforming vector is constrained to have the form $\mathbf{w}^{T} = [e^{\mathrm{j}\theta_1}, e^{\mathrm{j}\theta_2}, \ldots, e^{\mathrm{j}\theta_{N_{\mathrm{r}}}}]^{T}$. Unlike MRC, with EGC only phase shifts of the received signal are allowed. Let $\phi_i = \mathrm{phase}(h_i)$. Solving a similar optimization to the one in (6.34) over the space of constrained beamforming vectors gives the solution $\mathbf{w}^{T} = e^{\mathrm{j}\theta}[e^{\mathrm{j}\phi_1}, e^{\mathrm{j}\phi_2}, \ldots, e^{\mathrm{j}\phi_{N_{\mathrm{r}}}}]^{T}$. The arbitrary phase θ is usually selected to be either 0 or $-\theta_1$ (phase synchronizing to the first antenna branch). EGC matches the phases of the channels so that they add coherently at the receiver. The resulting SNR is $\frac{E_x}{N_o} \left(\sum_i |h_i| \right)^2$. It can be shown (though the calculation is not trivial [14]) that EGC achieves a diversity gain of N_r. At high SNR, in Rayleigh fading channels, EGC typically performs about 1dB worse than MRC and 1dB better than antenna selection.

6.3 Transmit Diversity for MISO Systems

Multiple transmit antennas can be used to send potentially different signals into the propagation environment. Unlike multiple receive antennas, with multiple transmit antennas, the signals are combined in the air by the channel. This subtle difference leads to substantial algorithmic changes in the system design. This section reviews approaches for exploiting transmit antennas both with and without channel state information at the transmitter in MISO systems. When properly designed, transmit diversity techniques can achieve diversity gains in Rayleigh fading channels similar to those achieved with receive diversity in SIMO systems.

6.3.1 MISO Flat-Fading Channel Models

The received signal model in a MISO system depends on the transmission strategy, for example, space-time coding or transmit beamforming. The channel for the MISO flat-fading case, though, can be represented similarly to the SIMO case, except as a $1 \times N_t$ vector given by $\mathbf{h}^*$. The conjugate transpose is used because all vectors in this book are column vectors; the conjugate transpose is used instead of the transpose just for convenience. With a transpose, all the channel models for SIMO flat fading in (6.2.1) also apply to the MISO case.

   Assuming that the antennas are colocated (in an area separated by a few wavelengths), each antenna sees the same large-scale component and only the small-scale coefficient differs. Then the channel may be decomposed as in the SIMO case as $\mathbf{h} = \sqrt{G_{\mathrm{MIMO}}} \mathbf{h}_s$ where in this case $G_{\mathrm{MIMO}} = \frac{E_x}{P_{r,\mathrm{lin}}(d) N_t} = \frac{G}{N_t}$. The extra factor of N_t comes from the scaling of the transmit signal by $\sqrt{E_x / N_t}$ and will be clear later. All the large-scale models for $P_{r,\mathrm{lin}}(d)$ can be used in the MISO case without change. The multi-antenna stochastic channel models for $\mathbf{h}_s$ are similar to the SIMO case. There is a difference in terminology, though. The Rayleigh model with spatial correlation where $\mathbf{h}_s$ has vector complex Gaussian distribution $\mathcal{N}_{\mathbb{C}}(0, \mathbf{R}_{tx})$ where $\mathbf{R}_{tx} = \mathbb{E}\left[\mathbf{h}_s \mathbf{h}_s^* \right]$ is the transmit spatial correlation matrix. In the clustered channel model, $\theta_{c,r}$ becomes the angle of arrival (not the angle of departure) of the rth ray to the cth cluster. Now we proceed to explain the performance of different communication strategies in the MISO channel.

6.3.2 Why Spatial Repetition Does Not Work

The naive approach to using multiple transmit antennas is to send the same signal from all the antennas. This section explains the fallacy of this "intuitive" approach, as it breaks down in fading channels.

Suppose that the same symbol $s[n]$ is sent from each transmit signal. The pulse shaping is the same for all the transmitted signals. Assuming perfect synchronization and after matched filtering, the received signal is

$$y[n] = h_1 s[n] + h_2 s[n] + \cdots + h_{N_t} s[n] + v[n]. \tag{6.51}$$

Collecting terms together,

$$y[n] = s[n] \left(h_1 + h_2 + \cdots + h_{N_t} \right) + v[n]. \tag{6.52}$$

At this point, it is tempting to be optimistic, because the effective channel here is the sum $\sum_{i=1}^{N_r} h_i$. Unfortunately, the channel coefficients are complex numbers and the addition is not coherent.

To expand further, we consider the IID Rayleigh fading channel where h_i is IID $\mathcal{N}_\mathbb{C}(0, G_{\text{MIMO}})$. Because sums of Gaussian random variables are Gaussians, $(h_1 + h_2 + \cdots + h_{N_t})$ is Gaussian with mean

$$\mathbb{E}\left[(h_1 + h_2 + \cdots + h_{N_t})\right] = 0 \tag{6.53}$$

and variance

$$\mathbb{E}\left[(h_1 + h_2 + \cdots + h_{N_t})^2\right] = \mathbb{E}[|h_1|^2] + \mathbb{E}[|h_2|^2] + \cdots \mathbb{E}[|h_{N_t}|^2] \tag{6.54}$$

$$= N_t G_{\text{MIMO}} \tag{6.55}$$

$$= \frac{1}{N_t} G N_t \tag{6.56}$$

$$= G. \tag{6.57}$$

Therefore, this sum channel $(h_1 + h_2 + \cdots + h_{N_t})$ is $\mathcal{N}_\mathbb{C}(0, G)$. This means that (6.51) is equivalent (in the sense that the distribution of the effective channel is the same) to

$$y[n] = hs[n] + v[n], \tag{6.58}$$

which is just the usual SISO fading channel. Consequently, there is no diversity if the symbol is repeated from multiple transmit antennas. This was illustrated in Figure 6.5, where it was shown that the effective post-processing SNR resulting from spatial repetition is no better than that from individual channels.

The failure of spatial repetition to coherently combine in the channel encourages the development of other transmission techniques. In the next subsections, we introduce two such approaches. One main approach is transmit beamforming, where the objective is to adjust the phase of the signal sent on each antenna so that the resulting signals combine more coherently. The other approach is space-time coding, where the transmitted signals are encoded in such a way that the coded symbols experience many different combinations of the channel coefficients. Both approaches, when properly designed, lead to a diversity of N_t in Rayleigh fading channels.

6.3.3   Transmit Beamforming

Exploiting channel state information is one means to avoid the pitfalls in Section 6.3.2. One way to exploit channel state information at the transmitter is through transmit beamforming as illustrated in Figure 6.12. In this case, the same symbol is sent from each transmit antenna but is scaled by a complex weight f_j that is allowed to depend on the channel state. The role of f_j is to change the phase and amplitude of the transmitted signal to better match the channel and lead to a good coherently combined over-the-air signal.

Define the beamforming weight vector as $\mathbf{f}^T = [f_1, f_2, \ldots, f_{N_t}]^T$. The received symbol, after matched filtering and synchronization, is

$$y[n] = \mathbf{h}^* \mathbf{f} s[n] + v[n]. \tag{6.59}$$

The effective received channel $\mathbf{h}^* \mathbf{f}$ is simply the inner product between $\mathbf{h}$ and $\mathbf{f}$. To maintain the transmit power constraint, $\mathbf{f}$ is scaled such that $\sum_{m=1}^{N_t} |f_j|^2 \mathbb{E}[|s[n]|^2] = N_t$, which is achieved by beamforming vectors that satisfy $\|\mathbf{f}\|^2 = N_t$. The choice of N_t is the result of our assumption that each antenna branch is scaled by E_x/N_t. Unlike the case of receive beamforming, the choice of the transmit beamforming vector does not impact the noise variance.

As in the case of MRC, it makes sense to choose the beamformer $\mathbf{f}$ to maximize the received SNR. The objective is to maximize $|\mathbf{h}^* \mathbf{f}|^2 / N_o$ subject to $\|\mathbf{f}\|^2 = N_t$. The resulting beamformer is known as the maximum ratio transmission (MRT) solution. Using the same arguments as in Section 6.2.3, the optimizer has the form $\mathbf{f} = \alpha \mathbf{h}$. To satisfy the power constraint with maximum power, observe that

$$\mathbf{f}^* \mathbf{f} = |\alpha|^2 \mathbf{h}^* \mathbf{h} \tag{6.60}$$
$$= N_t. \tag{6.61}$$

Therefore, the scaling factor α must satisfy $|\alpha|^2 = N_t/\|\mathbf{h}\|^2$. The simple solution is to take $\alpha = \sqrt{N_t}/\|\mathbf{h}\|$. In this way, the transmit beamformer acts as a spatial matched filter to the channel and is given by $\mathbf{f} = \sqrt{N_t}\mathbf{h}/\|\mathbf{h}\|$.

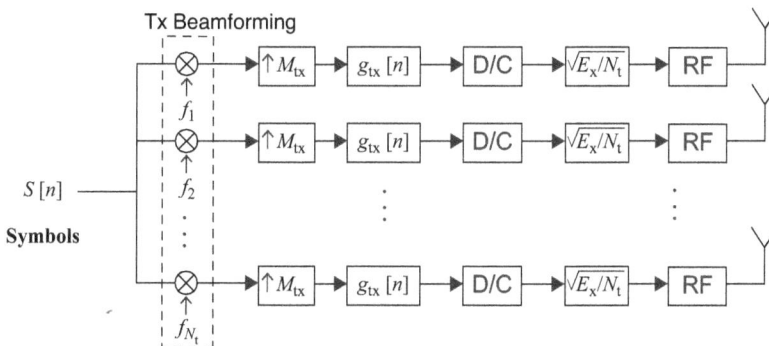

Figure 6.12 The operation of transmit beamforming. The weights f_j may be determined based on the channel state. After multiplication by the weight, the resulting signal $f_j s[n]$ is passed through the usual pulse-shaping, discrete-to-continuous conversion, and RF operations.

The received SNR given MRT is

$$\text{SNR}_{\mathbf{h}}^{\text{MRT}} = \frac{|\mathbf{h}^*\mathbf{f}|^2}{N_{\text{o}}} \tag{6.62}$$

$$= \frac{N_{\text{t}}|\mathbf{h}^*\mathbf{h}|^2}{\|\mathbf{h}\|^2 N_{\text{o}}} \tag{6.63}$$

$$= \|\mathbf{h}\|^2 \frac{N_{\text{t}}}{N_{\text{o}}}. \tag{6.64}$$

$$= \|\mathbf{h}_{\text{s}}\|^2 \frac{G_{\text{MIMO}} N_{\text{t}}}{N_{\text{o}}} \tag{6.65}$$

$$= \|\mathbf{h}_{\text{s}}\|^2 \frac{G}{N_{\text{o}}}. \tag{6.66}$$

Comparing (6.40) and (6.64) reveals that the SNR is the same for both MRT and MRC! The performance analysis in Example 6.4 through Example 6.6 therefore also applies to the case of MRT. In particular, we can conclude that MRT offers a diversity order of N_{t} in IID Rayleigh fading channels.

Other constraints on the beamforming weights are also possible. For example, forcing each weight to have the form $\exp(j\theta_k)/\sqrt{N_{\text{t}}}$ would yield something equivalent to EGC, resulting in the equal gain transmission solution [205, 237]. This approach is attractive because it allows each transmit antenna to send a fixed fraction of the power $E_{\text{x}}/N_{\text{t}}$ instead of being designed for full power E_{x}. It is also possible to implement phase shifting in analog using digitally controlled phase shifters. Selecting the weight vector from the set $\mathcal{S}_{\text{sel}}$ (scaled by $\sqrt{E_{\text{x}}}$) leads to transmit antenna selection [283, 146, 228, 288]. Transmit antenna selection seems to be most interesting when the switching operation can be performed in analog, sharing one RF chain among many possible transmit antennas [167].

The main disadvantage of transmit beamforming based on the channel state is the requirement that $\mathbf{h}$ be known (perfectly) at the transmitter. This leads to developing strategies that make use of only information about $\mathbf{h}$ that is acquired from quantized channel state information sent from the receiver back to the transmitter.

6.3.4 Limited Feedback Beamforming

Limited feedback describes a methodology in wireless communication systems where the transmitter is informed about the channel state through a low-rate (or limited) feedback channel from the receiver back to the transmitter [211, 207].

Feedback channels are widely used in wireless communication systems to support many other system functions, including power control and packet acknowledgments. Because feedback channels consume system resources, their capacity is limited by design. This means that they must convey as little information as possible to accomplish the assigned task.

The idea of limited feedback beamforming [208, 235, 207] is for the receiver to select the best possible transmit beamforming vector from a codebook of possible beamforming vectors. It turns out that this approach is more efficient than quantizing the channel directly because it leads to further feedback reductions.

Let $\mathcal{F} = \{\mathbf{f}_1, \mathbf{f}_2, \ldots, \mathbf{f}_{N_{\mathrm{LF}}}\}$ denote a codebook of beamforming vectors. Each entry $\mathbf{f}_j$ is an $N_{\mathrm{t}} \times 1$ vector. All the vectors satisfy $\|\mathbf{f}_j\| = N_{\mathrm{t}}$ to maintain the power constraint. There is a total of N_{LF} vectors in the codebook; typically N_{LF} is chosen to be a power of 2. In the case of MISO transmit beamforming, a reasonable approach to selecting a good transmit beamforming vector is to find the index of the vector that maximizes the received SNR:

$$j^{\star} = \underset{j \in \{1,2,\ldots,N_{\mathrm{LF}}\}}{\arg\max} \frac{1}{N_{\mathrm{o}}} \|\mathbf{h}^* \mathbf{f}_j\|^2 \tag{6.67}$$

$$= \underset{j \in \{1,2,\ldots,N_{\mathrm{LF}}\}}{\arg\max} \|\mathbf{h}^* \mathbf{f}_j\|^2. \tag{6.68}$$

Essentially the best beamforming vector is $\mathbf{f}_{j^\star}$, which has the strongest inner product with the channel $\mathbf{h}$. The receiver sends the index $j^\star$ back to the transmitter through the limited feedback channel.

The performance of limited feedback beamforming depends on the specific choice of the codebook $\mathcal{F}$. Typically the codebook is designed ahead of time and specified in a standard. The most common design criterion is to find a codebook such that the minimum inner product is maximized, making the beamforming vectors far apart in some sense. This leads to a Grassmannian codebook (exploiting the connection between this beamforming problem and packings of lines on the Grassmann manifold [208, 235]). Other codebook designs are possible to satisfy other objectives, for example, making the codebook easier to search or reducing the precision required for its storage [358, 284, 150, 195]. To provide a concrete example, note that 3GPP LTE release 8 uses the following beamforming codebook for two antennas with $N_{\mathrm{LF}} = 4$:

$$\mathcal{F} = \left\{ \begin{bmatrix} 1 \\ 1 \end{bmatrix}, \begin{bmatrix} 1 \\ -1 \end{bmatrix}, \begin{bmatrix} 1 \\ j \end{bmatrix}, \begin{bmatrix} 1 \\ -j \end{bmatrix} \right\}. \tag{6.69}$$

Analyzing the performance of limited feedback beamforming in fading channels is difficult, since it requires the distribution of $\|\mathbf{h}^* \mathbf{f}_{j^\star}\|$, which depends on the codebook. It has been shown that in Rayleigh fading channels, a diversity order of N_{t} can be achieved for Grassmannian codebooks provided that $N_{\mathrm{LF}} \geqslant N_{\mathrm{t}}$ [208, Theorem 4]. More detailed analysis of limited feedback beamforming in other channel settings is also available [199, 209, 278, 203, 155, 150].

6.3.5 Reciprocity-Based Beamforming

An alternative to limited feedback beamforming is to use reciprocity in the propagation environment. In the case of a MISO system, this involves deriving the transmit beamforming vector from the channel observed in a SIMO system. To explain, suppose that user A with N antennas used for both transmit and receive communicates with a user with one antenna for transmit and receive. Denote the SIMO channel as $\mathbf{h}_{B \to A}$ and the MISO channel as $\mathbf{h}_{A \to B}^*$. The principle of reciprocity is that the propagation channel from A to B is the same as the propagation channel from B to A. This means the path gains, phases, and delays are the same. Using our notation, this means that $\mathbf{h}_{A \to B} = \mathbf{h}_{B \to A}^{\mathrm{c}}$. Using reciprocity, user A can measure $\mathbf{h}_{B \to A}$ based on training data sent from user B

and can use this information to design the transmit beamformer $\mathbf{f} = \mathbf{h}^c_{B \to A}/\|\mathbf{h}_{B \to A}\|$. Of course, this requires the channel to be constant during the total time for transmission from B to A to B, called the ping-pong time. With reciprocity, channel state information from A to B is obtained for free from the channel measurement from B to A.

Wireless systems that exploit reciprocity typically use time-division duplex systems where the same f_c is used for transmission from A to B and from B to A. It may be possible to use different carriers if the frequencies are close together [199], but this is still an ongoing topic of research.

The main advantages of reciprocity-based beamforming are (a) that no quantization of the channel or beamforming vector is required and (b) the training overhead is reduced since estimating the channel from N_t transmit antennas takes approximately N_t times more training than estimating the channel from one antenna. The main disadvantage is that a calibration phase is required because of differences in the analog front ends between the transmit and receive paths, especially differences between the power amplifier output impedance and the input impedance of the local noise amplifier [57].

There are two approaches for calibration: self-calibration [201, 202] and over-the-air calibration; see, for example, [305]. With self-calibration, an extra transceiver is used to perform the calibration [201, 202]. The extra transceiver is used to determine differences between the transmit and receive paths successively for each antenna during a special calibration transmission phase. In the over-the-air approach, the received channel is fed back to the transmitter, so that the transmitter can determine the required calibration parameters in the baseband [305]. Note that the required channel accuracy is quite high when performing calibration, versus what is required for limited feedback beamforming, but the calibration occurs less frequently (on the order of seconds, minutes, or longer).

6.3.6 The Alamouti Code

It is possible to obtain diversity from multiple transmit antennas without channel state information when the transmitted signal is specially designed. Such a construction is generally known as space-time coding [325, 326]. The Alamouti code [8] is perhaps the most elegant and commercially successful space-time code, and it appears in the WCDMA standard [153].

The Alamouti code is a space-time code designed for two transmit antennas. To simplify the exposition, we focus on the time instants $n = 0$ and $n = 1$. At time instant $n = 0$, transmit antenna 1 sends $s[0]$ and transmit antenna 2 sends $s[1]$. The corresponding received signal is

$$y[0] = (h_1 s[0] + h_2 s[1]) + v[0]. \tag{6.70}$$

At time instant $n = 1$, transmit antenna 1 sends $(-s^*[1])$ and transmit antenna 2 sends $s^*[0]$. The corresponding received signal is

$$y[1] = (-h_1 s^*[1] + h_2 s^*[0]) + v[1]. \tag{6.71}$$

Taking the conjugate of (6.71) leads to an equation in the non-conjugated symbols:

$$y^*[1] = (-h_1^* s[1] + h_2^* s[0]) + v^*[1]. \tag{6.72}$$

Combining (6.70) and (6.72) into a vector input-output relationship gives

$$\underbrace{\begin{bmatrix} y[0] \\ y^*[1] \end{bmatrix}}_{\mathbf{y}} = \underbrace{\begin{bmatrix} h_1 & h_2 \\ h_2^* & -h_1^* \end{bmatrix}}_{\mathbf{H}} \underbrace{\begin{bmatrix} s[0] \\ s[1] \end{bmatrix}}_{\mathbf{s}} + \underbrace{\begin{bmatrix} v[0] \\ v^*[1] \end{bmatrix}}_{\mathbf{v}} \tag{6.73}$$

or in matrix form

$$\mathbf{y} = \mathbf{Hs} + \mathbf{v}. \tag{6.74}$$

Multiplying both sides on the left by $\mathbf{H}^*$ gives

$$\mathbf{H}^*\mathbf{y} = \mathbf{H}^*\mathbf{Hs} + \mathbf{H}^*\mathbf{v}. \tag{6.75}$$

The resulting $\mathbf{H}$ has a special structure. Observe that

$$\mathbf{H}^*\mathbf{H} = \begin{bmatrix} h_1^* & h_2 \\ h_2^* & -h_1 \end{bmatrix} \begin{bmatrix} h_1 & h_2 \\ h_2^* & -h_1^* \end{bmatrix} \tag{6.76}$$

$$= \begin{bmatrix} |h_1|^2 + |h_2|^2 & h_1^*h_2 - h_1^*h_2 \\ h_2^*h_1 - h_1h_2^* & |h_1|^2 + |h_2|^2 \end{bmatrix} \tag{6.77}$$

$$= \begin{bmatrix} |h_1|^2 + |h_2|^2 & 0 \\ 0 & |h_1|^2 + |h_2|^2 \end{bmatrix}. \tag{6.78}$$

The columns of $\mathbf{H}$ are orthogonal!

Because of the structure of $\mathbf{H}$, the filtered noise term $\mathbf{H}^*\mathbf{v}$ also has a special structure. The entries are complex Gaussian (because linear combinations of Gaussians are Gaussian). The mean is $\mathbb{E}[\mathbf{H}^*\mathbf{v}] = \mathbf{0}$ and the covariance is

$$\mathbb{E}\left[\mathbf{H}^*\mathbf{v}\mathbf{v}^*\mathbf{H}\right] = \mathbf{H}^*\mathbb{E}\left[\mathbf{v}\mathbf{v}^*\right]\mathbf{H} \tag{6.79}$$

$$= N_o\mathbf{H}^*\mathbf{H} \tag{6.80}$$

$$= N_o(|h_1|^2 + |h_2|^2)\mathbf{I}. \tag{6.81}$$

Therefore, the entries of $\mathbf{H}^*\mathbf{v}$ are IID with $\mathcal{N}_\mathbb{C}(0, N_o|h_1|^2 + |h_2|^2)$. Given the structure of $\mathbf{H}$, (6.75) simplifies as

$$\mathbf{H}^*\mathbf{y} = (|h_1|^2 + |h_2|^2)\mathbf{s} + \mathbf{H}^*\mathbf{v}. \tag{6.82}$$

Because the noise terms are independent, the decoding of (6.82) can proceed independently on each entry (this property is called single-symbol decoding) without sacrificing performance. The symbols can then be found by solving the ML detection problems:

$$\widehat{s_1} = \underset{s\in\mathcal{C}}{\arg\min} \left\| \begin{bmatrix} h_1^* & h_2 \end{bmatrix} \mathbf{y} - (|h_1|^2 + |h_2|^2)s \right\|^2 \tag{6.83}$$

$$\widehat{s_2} = \underset{s\in\mathcal{C}}{\arg\min} \left\| \begin{bmatrix} h_2^* & -h_1 \end{bmatrix} \mathbf{y} - (|h_1|^2 + |h_2|^2)s \right\|^2. \tag{6.84}$$

To analyze the performance of the Alamouti transmit diversity technique, consider the post-processing SNR derived from (6.82) and (6.81):

$$\text{SNR}_{\mathbf{h}}^{\text{Ala}} = \frac{(|h_1|^2 + |h_2|^2)^2}{N_o|h_1|^2 + |h_2|^2} \tag{6.85}$$

$$= \frac{1}{N_o}(|h_1|^2 + |h_2|^2) \tag{6.86}$$

$$= \frac{G_{\text{MIMO}}}{N_o}\|\mathbf{h}_s\|^2 \tag{6.87}$$

$$= \frac{G}{N_o N_t}\|\mathbf{h}_s\|^2. \tag{6.88}$$

This SNR is the same as the post-processing SNR achieved with MRT in (6.64) with a factor of $N_t = 2$ in the denominator. Because $10\log_{10} 2 \approx 3\text{dB}$, the performance achieved by the Alamouti code is 3dB worse than that achieved by MRT. The penalty is a result of not using channel state information at the transmitter and may be considered a fair trade-off. An illustration of the performance differences between the Alamouti code and MRT is provided in Figure 6.5.

The average SNR and bounds on the probability of symbol error for the Alamouti code can be characterized using the derivations in Section 6.2.3. It achieves a diversity order of 2 in Rayleigh fading channels. Characterizations of performance in other fading channels are widely available [346, 59, 245].

The Alamouti code achieves full diversity (2 when $N_t = 2$ and $N_r = 1$), sends two symbols in two time periods (called the full-rate property in space-time parlance), and has single-symbol decodability (the fact that the detection of each symbol can be performed separately without any loss of optimality as in (6.83) and (6.84)). Generalizations of the Alamouti code to larger numbers of antennas, for example, orthogonal designs [327] or quasi-orthogonal designs [163], sacrifice one or more of these properties.

6.3.7 Space-Time Coding

The Alamouti code is a special case of what is known as a space-time code [326]. A space-time code is a type of code that is usually designed to obtain diversity advantage from multiple transmit antennas. This section explains some general concepts of space-time codes and their performance analysis in IID Rayleigh fading channels using the pairwise error probability.

A codebook is a generalization of a constellation. With a constellation, a sequence of B bits is mapped to one of the $M = 2^B$ complex symbols in the constellation $\mathcal{C}$. In a general space-time code, the codebook takes the place of the constellation and codewords take the place of symbols. In a space-time code, each codeword can be visualized as an $N_t \times N_{\text{code}}$ matrix where N_{code} is the number of temporal symbol periods used by the code. Let us denote this codebook as $\mathcal{S}$ and the kth entry of the codebook as $\mathbf{S}_k$. We assume a normalization of the codewords such that $\mathbb{E}[\text{tr}(\mathbf{S}_k^*\mathbf{S}_k)] = N_t N_{\text{code}}$.

Example 6.7   Determine the codebook for the Alamouti code when BPSK symbols are used.

Answer: The general form of the Alamouti codeword is

$$\mathbf{S} = \begin{bmatrix} s_1 & -s_2^* \\ s_2 & s_1^* \end{bmatrix} \tag{6.89}$$

where s_1 ands s_2 are constellation points. If s_1 and s_2 come from a BPSK constellation, then there is a total of four possible codewords. Enumerating each possible value of s_1 and s_2 leads to

$$\mathcal{S} = \left\{ \begin{bmatrix} 1 & -1 \\ 1 & 1 \end{bmatrix}, \begin{bmatrix} 1 & 1 \\ -1 & 1 \end{bmatrix}, \begin{bmatrix} -1 & -1 \\ 1 & -1 \end{bmatrix}, \begin{bmatrix} -1 & 1 \\ -1 & -1 \end{bmatrix} \right\}. \tag{6.90}$$

Now consider the received signal assuming that codeword $\mathbf{S} = [\mathbf{s}[0], \mathbf{s}[1], \ldots, \mathbf{s}[N_{\text{code}} - 1]]$ is transmitted over a MISO flat-fading channel:

$$y[n] = \mathbf{h}^* \mathbf{s}[n] + v[n] \quad n = 0, 1, \ldots, N_{\text{code}} - 1. \tag{6.91}$$

This can be written compactly in matrix form by collecting adjacent observations as columns

$$\mathbf{Y} = \mathbf{h}^* \mathbf{S} + \mathbf{V} \tag{6.92}$$

where $\mathbf{Y}$ and $\mathbf{V}$ are $1 \times N_{\text{code}}$ matrices. We use this approach because the derivations below extend to MIMO channels directly using $\mathbf{H}$ in place of $\mathbf{h}^*$ and making $\mathbf{Y}$ and $\mathbf{V}$ $N_{\text{r}} \times N_{\text{code}}$ matrices.

An alternative vector form is achieved by stacking all the columns using the $\text{vec}(\mathbf{Y})$ operator. The vec operator generates a vector by stacking the columns of a matrix on top of each other. The vec often shows up with the Kronecker product $\otimes$. The Kronecker product of an $N \times M$ matrix $\mathbf{A}$ and a $P \times Q$ matrix $\mathbf{B}$ is

$$\mathbf{A} \otimes \mathbf{B} = \begin{bmatrix} a_{11}\mathbf{B} & \cdots & a_{1M}\mathbf{B} \\ \vdots & \ddots & \vdots \\ a_{N1}\mathbf{B} & \cdots & a_{NM}\mathbf{B} \end{bmatrix}. \tag{6.93}$$

A useful identity is the fact that

$$\text{vec}(\mathbf{ABC}) = (\mathbf{C}^T \otimes \mathbf{A})\text{vec}(\mathbf{B}). \tag{6.94}$$

Using these definitions to rewrite (6.92) gives

$$\mathbf{y} = (\mathbf{I}_{N_t} \otimes \mathbf{h}^*)\mathbf{s} + \mathbf{v} \tag{6.95}$$

where $\mathbf{y} = \text{vec}(\mathbf{Y})$ and $\mathbf{s} = \text{vec}(\mathbf{S})$.

The maximum likelihood detector for a space-time code is derived in a similar way to the scalar decoder from Chapter 4, except using the multivariate Gaussian distribution from Chapter 3 and the vector equation in (6.95). A related derivation for spatial multiplexing is provided in Section 6.4.3, which can be applied directly to (6.95). The detailed derivation for the space-time-coded case is presented as a problem. The resulting detector is

$$\widehat{\mathbf{S}} = \underset{\mathbf{Q} \in \mathcal{S}}{\arg \min} \, \|\mathbf{y} - (\mathbf{I} \otimes \mathbf{h}^*) \mathrm{vec}(\mathbf{Q})\|^2 \,. \tag{6.96}$$

It can be rewritten more compactly using (6.92) as

$$\widehat{\mathbf{S}} = \underset{\mathbf{Q} \in \mathcal{S}}{\arg \min} \, \|\mathbf{Y} - \mathbf{h}^* \mathbf{Q}\|_F^2 \,. \tag{6.97}$$

The optimum detector searches over all possible transmitted codewords to find the one that was most likely transmitted. The complexity of the brute-force search over N_{code} entries can be reduced when there is structure in the code, such as orthogonal structure as found in the Alamouti code or trellis structure [326].

The performance of a space-time code does not just depend on the channel through the post-processing SNR as in the case of beamforming; it also depends on the spatial structure of the codebook. Given a realization of $\mathbf{h}$, the conditional probability of codeword error $\mathbb{P}(E_{\mathrm{x}}/N_{\mathrm{o}}|\mathbf{h})$ can be upper bounded using similar arguments to those in Chapter 4 using the pairwise error probability. The detailed derivation is provided in the literature [326, 134]. The pairwise error probability, in this case taking $G = E_{\mathrm{x}}$, is

$$\mathbb{P}(\mathbf{Q}_k \rightarrow \mathbf{Q}_\ell | \mathbf{h}) = Q\left(\sqrt{\frac{1}{2N_{\mathrm{o}}} \|\mathbf{h}^*(\mathbf{Q}_k - \mathbf{Q}_\ell)\|_F^2}\right) \,. \tag{6.98}$$

Looking for the worst error event and inserting into the union bound gives

$$\mathbb{P}(E_{\mathrm{x}}/N_{\mathrm{o}}|\mathbf{h}) \leq (N_{\mathrm{code}} - 1)Q\left(\sqrt{\frac{1}{2N_{\mathrm{o}}} \underset{\mathbf{Q}_k \neq \mathbf{Q}_\ell \in \mathcal{S}}{\min} \|\mathbf{h}^*(\mathbf{Q}_k - \mathbf{Q}_\ell)\|_F^2}\right) \,. \tag{6.99}$$

In the SISO case studied in Chapter 4, the upper bound depended only on the minimum distance of the constellation. The same is true for beamforming as seen in Example 6.6. In (6.99), however, the channel and the codewords are coupled together. This is a main difference between space-time coding and other diversity techniques. As a result, performance depends in particular on the structure and design of the codebook.

The average probability of error can be used to devise criteria for good codebooks in fading channels. We consider the specific case of IID Rayleigh fading channels with $G_{\mathrm{MIMO}} = E_{\mathrm{x}}/N_{\mathrm{t}}$ to neglect the large-scale fading contribution. Taking the expectation of (6.98) and writing the Chernoff upper bound:

$$\mathbb{E}_{\mathbf{h}}\left[\mathbb{P}(\mathbf{Q}_k \rightarrow \mathbf{Q}_\ell | \mathbf{h})\right] \leq \frac{(N_{\mathrm{code}} - 1)}{2} \mathbb{E}_{\mathbf{h}}\left[e^{-\frac{1}{4N_{\mathrm{o}}} \|\mathbf{h}^*(\mathbf{Q}_k - \mathbf{Q}_\ell)\|_F^2}\right] \tag{6.100}$$

$$= \frac{(N_{\mathrm{code}} - 1)}{2} \mathbb{E}_{\mathbf{h}}\left[e^{-\frac{1}{4N_{\mathrm{o}}} \mathbf{h}^*(\mathbf{Q}_k - \mathbf{Q}_\ell)(\mathbf{Q}_k - \mathbf{Q}_\ell)^* \mathbf{h}}\right] \,. \tag{6.101}$$

Let $\mathbf{R}_{k,\ell} = (\mathbf{Q}_k - \mathbf{Q}_\ell)(\mathbf{Q}_k - \mathbf{Q}_\ell)^*$, which is called the error covariance matrix. Using the definition of the moment-generating function of a multivariate Gaussian distribution (related to the characteristic function; see [332] for more details), and substituting in for G_{MIMO}, it can be shown that

$$\mathbb{E}_{\mathbf{h}}\left[\mathbb{P}(\mathbf{Q}_k \to \mathbf{Q}_\ell|\mathbf{h})\right] \leq \frac{(N_{\text{code}} - 1)}{2} \frac{1}{\left|\mathbf{I} + \frac{E_{\mathrm{x}}}{4N_{\mathrm{t}}N_{\mathrm{o}}}\mathbf{R}_{k,\ell}\right|} \tag{6.102}$$

$$= \frac{(N_{\text{code}} - 1)}{2} \frac{1}{\prod_{m=1}^{\text{rank}(\mathbf{R}_{k,\ell})} \left(1 + \frac{E_{\mathrm{x}}}{4N_{\mathrm{t}}N_{\mathrm{o}}}\lambda_m(\mathbf{R}_{k,\ell})\right)} \tag{6.103}$$

where the second step uses the fact that the determinant of a positive definite matrix is the product of its eigenvalues and that the eigenvalues of a positive definite matrix of the form $\mathbf{I} + \mathbf{R}$ are $1 + \lambda_m(\mathbf{R})$. The worst-case error is used to upper bound the average probability of error as

$$\mathbb{E}_{\mathbf{h}}\left[\mathbb{P}(E_{\mathrm{x}}/N_{\mathrm{o}}|\mathbf{h})\right] \leq \frac{(N_{\text{code}} - 1)}{2} \max_{k,\ell,k\neq\ell} \frac{N_{\text{code}} - 1}{\prod_{m=1}^{\text{rank}(\mathbf{R}_{k,\ell})} \left(1 + \frac{E_{\mathrm{x}}}{4N_{\mathrm{t}}N_{\mathrm{o}}}\lambda_m(\mathbf{R}_{k,\ell})\right)} \tag{6.104}$$

$$\approx \frac{(N_{\text{code}} - 1)}{2} \max_{k,\ell,k\neq\ell} \frac{N_{\text{code}} - 1}{\left(\frac{E_{\mathrm{x}}}{4N_{\mathrm{t}}N_{\mathrm{o}}}\right)^{\text{rank}(\mathbf{R}_{k,\ell})} \prod_{m=1}^{\text{rank}(\mathbf{R}_{k,\ell})} \lambda_m(\mathbf{R}_{k,\ell})}. \tag{6.105}$$

Essentially, what can be gleaned from (6.105) is that the diversity performance of the space-time code depends on the worst-case error covariance rank($\mathbf{R}_{k,\ell}$). A space-time code with a full-rank error covariance for all possible error pairs is known as a full-rank space-time code. The coding gain of the code is related to the term $\prod_{m=1}^{\text{rank}(\mathbf{R}_{k,\ell})} \lambda_m(\mathbf{R}_{k,\ell})$. For two codes of similar diversity performance, the one with the larger coding gain for that minimum-rank codeword will generally have better performance.

Example 6.8   Determine the diversity performance of the Alamouti space-time code using the pairwise error probability (PEP) approach.

   Answer: Evaluating the diversity performance using PEP requires computing the error covariance matrix. In general this is challenging, but for special codes like the Alamouti code the task can be accomplished. Because of the structure of the Alamouti code, using (6.89), the codeword difference has the form

$$\mathbf{Q}_k - \mathbf{Q}_\ell = \frac{1}{\sqrt{2}} \begin{bmatrix} s_1^{(k)} - s_1^{(\ell)} & -(s_2^{(k)} - s_2^{(\ell)})^* \\ s_2^{(k)} - s_2^{(\ell)} & (s_1^{(k)} - s_1^{(\ell)})^* \end{bmatrix}. \tag{6.106}$$

Following a similar derivation to that in (6.78),

$$\mathbf{R}_{k,\ell} = \frac{1}{2}(\mathbf{Q}_k - \mathbf{Q}_\ell)(\mathbf{Q}_k - \mathbf{Q}_\ell)^* \tag{6.107}$$

$$= \frac{1}{2}\left(|s_1^{(k)} - s_1^{(\ell)}|^2 + |s_2^{(k)} - s_2^{(\ell)}|^2\right)\mathbf{I}. \tag{6.108}$$

Because rank($\mathbf{I}$) = 2, as long there is at least one error difference, the error covariance matrix is always full rank and the Alamouti code achieves a diversity of 2.

The design of space-time codes has been an active area of interest. The problem is nontrivial since the properties of the code depend on different properties of the error covariance matrices. The rank and determinant criteria for code design were proposed in [326] where space-time trellis codes were derived. Delay diversity [357] (see Figure 6.4) turns out to be a special case in [326]. A variation of delay diversity called cyclic delay diversity [83, 133] is used in IEEE 802.11n.

6.4   MIMO Transceiver Techniques

MIMO communication makes full use of multiple transmit antennas and receive antennas. MIMO provides a generalization of both SIMO and MISO systems. It also introduces the spatial multiplexing mode of communication, where the objective is to send many symbols simultaneously, taking advantage of the higher capacity that is generally present in MIMO channels. In this section, we introduce the spatial multiplexing concept. We explain how to generalize maximum likelihood symbol detection and equalization to the MIMO flat-fading channel and provide some performance analysis for IID Rayleigh fading channels. Using knowledge of the channel at the transmitter, we show how to generalize transmit beamforming to the MIMO case, in what is called transmit precoding, and its extension to limited feedback. We develop channel estimators for MIMO flat fading, as detection, equalization, and precoding all require an estimate of the channel. Finally, we provide some extensions of time-domain equalization and channel estimation to MIMO frequency-selective channels.

6.4.1   Spatial Multiplexing

Spatial multiplexing is a communication technique where independent symbols are sent from each transmit antenna. A description of spatial multiplexing was provided in Section 6.1.3, emphasizing the application to two transmit antennas. Now we explain the extension to N_t transmit antennas and describe how it performs with different receiver algorithms.

An illustration of spatial multiplexing is provided in Figure 6.13. Consider a symbol stream $s[n]$ as in earlier chapters. The spatial multiplexer operates as a $1 : N_t$ serial-to-parallel converter. The outputs of the spatial multiplexer are the symbol streams denoted as $s_j[n] = s[N_t n + (j-1)]$ for $j = 1, \ldots, N_t$. These symbols are pulse shaped and scaled for transmission to generate a signal

$$x_j(t) = \sqrt{\frac{E_x}{N_t}} \sum_{n=-\infty}^{\infty} s_j[n] g_{tx}(t - nT), \tag{6.109}$$

which is analogous to $x(t)$ in (4.3) in Chapter 4. The only difference is the amplitude scaling. To keep the total transmitted power constant, the symbols must be scaled by $\sqrt{E_x/N_t}$ instead of by $\sqrt{E_x}$. In this way, $\sum_{j=1}^{N_t} \mathbb{E}[|x_j(t)|^2] = E_x$.

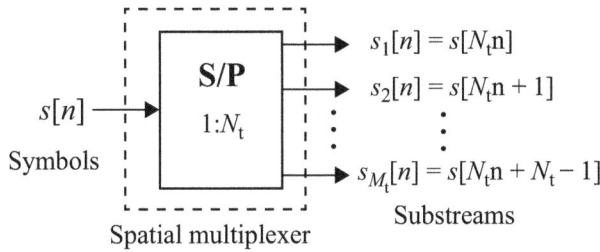

Figure 6.13 The operation of spatial multiplexing at the transmitter. Spatial multiplexing is essentially a serial-to-parallel conversion that takes symbols and produces subsymbols or streams. The output symbols are then pulse shaped, converted to continuous time, and upconverted.

Let $h_{i,j}$ denote the complex baseband equivalent flat-fading channel between the jth transmit antenna and the ith receive antenna. Suppose that this channel includes large-scale fading, small-scale fading, and thus the scaling by $E_{\text{x}}/N_{\text{t}}$. The baseband received signal, assuming perfect synchronization and sampling, is

$$y_i[n] = (h_{i,1}s_1[n] + h_{i,2}s_2[n] + \cdots + h_{i,N_{\text{t}}}s_{N_{\text{t}}}[n]) + v_i[n] \qquad (6.110)$$

where $v_i[n]$ is the usual AWGN with $\mathcal{N}_{\mathbb{C}}(0, N_{\text{o}})$. Stacking all the received observations into a vector gives

$$\underbrace{\begin{bmatrix} y_1[n] \\ y_2[n] \\ \vdots \\ y_{N_{\text{r}}}[n] \end{bmatrix}}_{\mathbf{y}[n]} = \underbrace{\begin{bmatrix} h_{1,1} & h_{1,2} & \cdots & h_{1,N_{\text{t}}} \\ h_{2,1} & h_{2,2} & \cdots & h_{2,N_{\text{t}}} \\ \vdots & \ddots & \ddots & \vdots \\ h_{N_{\text{r}},1} & h_{N_{\text{r}},2} & \cdots & h_{N_{\text{r}},N_{\text{t}}} \end{bmatrix}}_{\mathbf{H}} \underbrace{\begin{bmatrix} s_1[n] \\ s_2[n] \\ \vdots \\ s_{N_{\text{t}}}[n] \end{bmatrix}}_{\mathbf{s}[n]} + \underbrace{\begin{bmatrix} v_1[n] \\ v_2[n] \\ \vdots \\ v_{N_{\text{r}}}[n] \end{bmatrix}}_{\mathbf{v}[n]} \qquad (6.111)$$

leading to the classic MIMO equation

$$\mathbf{y}[n] = \mathbf{H}\mathbf{s}[n] + \mathbf{v}[n]. \qquad (6.112)$$

A simplified system block diagram of a MIMO communication system is provided in Figure 6.14.

The receiver is tasked with detecting the symbol vector $\mathbf{s}[n]$ from the observation $\mathbf{y}[n]$. This can be performed based on different criteria as described in the next section.

6.4.2   MIMO Flat-Fading Channel Models

A MIMO flat-fading channel is represented by the $N_{\text{r}} \times N_{\text{t}}$ matrix $\mathbf{H}$. Each column of $[\mathbf{H}]_{:,j}$ corresponds to the SIMO channel experienced by the transmission from the jth transmit antenna. Each row of $[\mathbf{H}]_{i,:}$ is likewise the MISO channel from the transmitter to the ith receive antenna. MIMO channel models therefore share similarities with SIMO and MISO models.

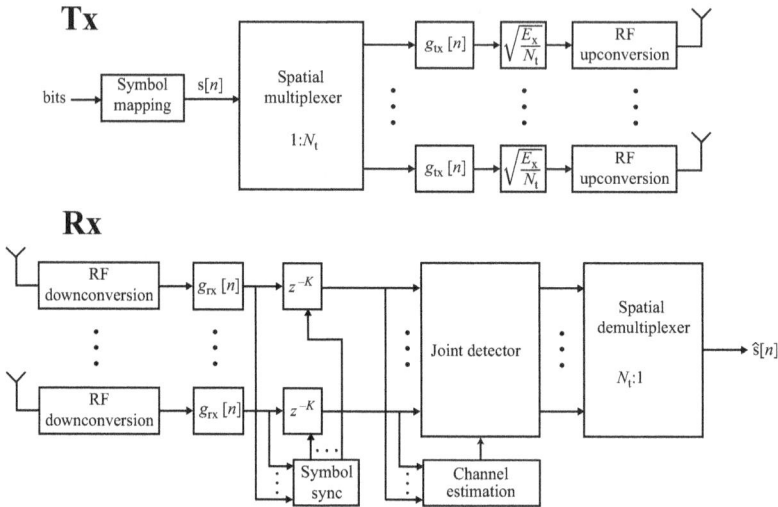

Figure 6.14 A block diagram for a complete spatial multiplexing system including the transmitter and the receiver. The bits are mapped to symbols and then stacked into vectors at the transmitter. The entries of each vector are then transmitted separately on each transmit antenna. At the receiver, the outputs from each antenna may be synchronized separately, and then joint equalization and detection are used to generate an estimate of the transmitted symbol vector. The upsampling and downsampling are not shown explicitly in the figure.

Note that the dimensions of $\mathbf{H}$ are $N_r \times N_t$. It is common, though, to refer to a MIMO system with N_t transmit antennas and N_r receive antennas as an $N_t \times N_r$ system. For example, a 2×3 system has two transmit antennas and three receive antennas.

Assuming that the antennas are colocated (in an area separated by a few wavelengths), the path loss between each transmit and receive antenna can be assumed to be the same factor $P_{r,lin}(d)$. As a result, all the large-scale path-loss models apply to the MIMO case as well without any changes. There are some further generalizations, for example, to LOS MIMO, distributed antennas, or cloud radio access networks, where antennas may have different geographic locations and a more general model might be required.

Then the channel may be decomposed as in the SIMO and MISO cases as a product of large-scale and small-scale components as $\mathbf{H} = \sqrt{G_{\mathrm{MIMO}}}\mathbf{H}_s$. As in the MISO case, $G_{\mathrm{MIMO}} = \frac{E_x}{P_{r,lin}(d)N_t}$ with $P_{r,lin}(d)$ capturing the large-scale path-loss effects. The small-scale fading matrix $\mathbf{H}_s$ captures all the effects of spatial selectivity in the channel.

There are many stochastic models for MIMO communication systems, which treat $\mathbf{H}_s$ as a random variable or construct $\mathbf{H}_s$ as a function of random variables. Often we describe the distribution of $\mathbf{H}_s$ in terms of $\mathbf{h}_s = \mathrm{vec}(\mathbf{H}_s)$, to leverage existing multivariate distributions (which apply to vectors and not matrices).

The most widely used model for analysis of MIMO communication systems is the IID Rayleigh fading channel. In this case, $[\mathbf{H}_s]_{i,j}$ is $\mathcal{N}_{\mathbb{C}}(0,1)$ and is IID. In other words, $\mathbf{h}_s$ has a multivariate Gaussian distribution $\mathcal{N}_{\mathbb{C}}(\mathbf{0}, \mathbf{I}_{N_r N_t})$. The IID Rayleigh model is usually justified by the "rich scattering assumption" [115]. The main argument is that there are many scatters in the environment, which leads to Gaussianity by the central limit theorem, and the antennas are spaced far enough apart, which justifies the uncorrelated (and thus independent in a Gaussian channel) assumption. We use the IID Rayleigh fading channel for the analysis in this chapter.

The most common generalization of the IID Rayleigh fading channel is to incorporate spatial correlation, in what is called the Kronecker channel model or separable correlation model. Let $\mathbf{R}_{\text{tx}}$ denote the transmit spatial correlation matrix, let $\mathbf{R}_{\text{rx}}$ be the receive spatial correlation matrix, and let $\mathbf{H}_{\text{w}}$ be an IID Rayleigh fading matrix. With the Kronecker model,

$$\mathbf{H}_{\text{s}} = \mathbf{R}_{\text{rx}}^{1/2} \mathbf{H}_{\text{w}} \mathbf{R}_{\text{tx}}^{1/2}. \tag{6.113}$$

The transmit correlation matrix captures the effects of scattering around the transmit antenna array, and the receive spatial correlation matrix captures scattering around the receive antenna array. This is called the Kronecker channel model because the resulting $\mathbf{h}_{\text{s}}$ is distributed as $\mathcal{N}_{\mathbb{C}}(\mathbf{0}, \mathbf{R}_{\text{tx}}^{\text{T}} \otimes \mathbf{R}_{\text{rx}})$, which gives a Kronecker structure to the spatial correlation matrix of the vectorized channel $\mathbf{E}[\mathbf{h}_{\text{s}}\mathbf{h}_{\text{s}}^*]$. Further generalizations are possible where the correlation is not restricted to having Kronecker structure [295, 350].

The clustered channel model can be used to separately generate the transmit and receive covariance matrices using, for example, closed form expressions like those found in [298] or [113]. Alternatively, the cluster parameters can be generated randomly, and $\mathbf{H}_{\text{s}}$ can be simulated as a generalization of (6.12) as

$$\mathbf{H}_{\text{s}} = \frac{1}{\sqrt{RC}} \sum_{c=0}^{C-1} \sum_{r=0}^{R-1} \alpha_{c,r} \mathbf{a}_{\text{rx}}(\theta_{c,r}) \mathbf{a}_{\text{tx}}^*(\phi_{c,r}). \tag{6.114}$$

There are other generalizations that involve asymmetric numbers of clusters on the transmit and receive sides. A variation of this sum-of-rays-type channel model is used, for example, in the 3GPP spatial channel model. The approach based on the exact computation of the spatial correlation matrix is used in the IEEE 802.11n standard [96].

6.4.3 Detection and Equalization for Spatial Multiplexing

In this section we review the maximum likelihood and zero-forcing (ZF) detectors and analyze their performance in IID Rayleigh fading channels.

Maximum Likelihood Detector for Spatial Multiplexing Suppose that all the symbols $s_j[n]$ come from the same constellation $\mathcal{C}$. It is possible to generalize to multiple constellations (though uncommon in practice) as explored in the problems at the end of this chapter. The constellation formed by the resulting $\mathbf{s}[n]$ is called the vector constellation. It consists of all the vectors with every possible symbol and is denoted as $\mathcal{S}$. Essentially, $\mathcal{S} = \mathcal{C} \times \mathcal{C} \cdots \times \mathcal{C}$. The cardinality of $\mathcal{S}$ is

$$|\mathcal{S}| = |\mathcal{C}|^{N_{\text{t}}}. \tag{6.115}$$

The size of the vector constellation thus grows exponentially with N_{t}, leading to very large constellations!

Example 6.9   Illustrate the vector constellation for $N_t = 2$ and BPSK modulation.
   Answer: With BPSK, $\mathcal{C} = \{-1, 1\}$. Forming all combinations of symbols,

$$\mathcal{S} = \left\{ \begin{bmatrix} 1 \\ 1 \end{bmatrix}, \begin{bmatrix} 1 \\ -1 \end{bmatrix}, \begin{bmatrix} -1 \\ 1 \end{bmatrix}, \begin{bmatrix} -1 \\ -1 \end{bmatrix} \right\}. \tag{6.116}$$

The maximum likelihood detection problem is to determine the best $\widehat{\mathbf{s}}[n] \in \mathcal{S}$ that maximizes the likelihood of $\mathbf{y}[n]$ given a candidate vector symbol $\bar{\mathbf{s}}[n]$ and $\mathbf{H}$. The derivation is similar to that in Chapter 4. First note that $\mathbf{v}[n]$ is multivariate Gaussian with distribution $\mathcal{N}_{\mathbb{C}}(0, N_o\mathbf{I})$. Therefore, the likelihood function is given by (using (3.300))

$$f_{\mathbf{y}|\mathbf{H},\mathbf{s}}(\mathbf{y}[n]|s[n] = \bar{\mathbf{s}}, \mathbf{H}) = \frac{1}{\pi^{N_r} N_o^{N_r}} e^{-\frac{1}{N_o}(\mathbf{y}[n]-\mathbf{H}\bar{\mathbf{s}})^*(\mathbf{y}[n]-\mathbf{H}\bar{\mathbf{s}})}. \tag{6.117}$$

The maximum likelihood detector solves

$$\widehat{s}[n] = \arg\max_{\bar{\mathbf{s}}\in\mathcal{S}} f_{\mathbf{y}|\mathbf{H},\mathbf{s}}(\mathbf{y}[n]|s[n] = \bar{\mathbf{s}}, \mathbf{H}) \tag{6.118}$$

$$= \arg\max_{\bar{\mathbf{s}}\in\mathcal{S}} \frac{1}{\pi^{N_r} N_o^{N_r}} e^{-\frac{1}{N_o}(\mathbf{y}[n]-\mathbf{H}\bar{\mathbf{s}})^*(\mathbf{y}[n]-\mathbf{H}\bar{\mathbf{s}})}. \tag{6.119}$$

Since the $\frac{1}{\pi^{N_r} N_o^{N_r}}$ does not impact the maximizer, and recalling that the exponential function is monotonically increasing,

$$\widehat{s}[n] = \arg\max_{\bar{\mathbf{s}}\in\mathcal{S}} -\frac{1}{N_o}(\mathbf{y}[n] - \mathbf{H}\bar{\mathbf{s}})^*(\mathbf{y}[n] - \mathbf{H}\bar{\mathbf{s}}) \tag{6.120}$$

$$= \arg\min_{\bar{\mathbf{s}}\in\mathcal{S}} (\mathbf{y}[n] - \mathbf{H}\bar{\mathbf{s}})^*(\mathbf{y}[n] - \mathbf{H}\bar{\mathbf{s}}). \tag{6.121}$$

Rewriting the inner product in (6.121) as a norm gives the main result:

$$\widehat{\mathbf{s}}[n] = \arg\min_{\bar{\mathbf{s}}\in\mathcal{S}} \|\mathbf{y}[n] - \mathbf{H}\bar{\mathbf{s}}\|^2. \tag{6.122}$$

Based on (6.122), the ML decoder for spatial multiplexing performs a brute-force search over all possible vector symbols $|\mathcal{C}|^{N_t}$. Each step involves a matrix multiplication, vector difference, and vector norm operation. The complexity can be reduced somewhat if the channel remains constant over many symbol periods. Then a distorted vector constellation can be computed ahead of time for the entire block by precomputing $\mathbf{Hs}$ for all $\mathbf{s} \in \mathcal{S}$ and using this constellation to avoid the product in (6.122). The disadvantage is the increase in storage required. There are several low-complexity algorithms that approximate the solution to (6.122), including the sphere decoder [140, 347, 24].

The performance of spatial multiplexing can be evaluated using pairwise error probability in a similar way as is performed for space-time codes in Section 6.3.7. First we consider the case of AWGN. For this performance analysis we consider just small-scale fading and let $G = E_x/N_t$. Let $\mathbf{s}^{(k)} \in \mathcal{S}$ denote the codeword transmitted, and let $\mathbf{s}^{(\ell)}$

denote the codeword that is decoded. Assuming the ML decoder is used, the pairwise error probability in the AWGN channel is

$$P(\mathbf{s}^{(k)} \rightarrow \mathbf{s}^{(\ell)}|\mathbf{H}) = Q\left(\sqrt{\frac{E_\mathrm{x}}{2N_\mathrm{o}N_\mathrm{t}}\frac{\|\mathbf{Hs}^{(k)} - \mathbf{Hs}^{(\ell)}\|^2}{2}}\right). \tag{6.123}$$

The probability of vector symbol error is then bounded as

$$P(E_\mathrm{x}/N_\mathrm{o}|\mathbf{H}) \leq (|\mathcal{S}| - 1)\max_{k,\ell,k\neq\ell} Q\left(\sqrt{\frac{E_\mathrm{x}}{2N_\mathrm{o}N_\mathrm{t}}\|\mathbf{Hs}^{(k)} - \mathbf{Hs}^{(\ell)}\|^2}\right) \tag{6.124}$$

$$\leq (|\mathcal{S}| - 1)Q\left(\sqrt{\frac{E_\mathrm{x}}{2N_\mathrm{o}N_\mathrm{t}}\min_{k,\ell,k\neq\ell}\|\mathbf{Hs}^{(k)} - \mathbf{Hs}^{(\ell)}\|^2}\right). \tag{6.125}$$

The performance of spatial multiplexing depends on the minimum distance of the distorted vector constellation given by $\min_{k,\ell,k\neq\ell}\|\mathbf{Hs}^{(k)} - \mathbf{Hs}^{(\ell)}\|^2$. In a SISO system, the minimum distance is a function only of the constellation; in the MIMO case, it is also a function of the channel. The channel distorts the vector constellation and changes the distance properties. Having performance that depends on the channel realization makes performance analysis more challenging since it is not simply a function of a post-processing SNR.

To see how the channel impacts performance, let $\mathbf{e}^{(k,\ell)} = \mathbf{s}^{(k)} - \mathbf{s}^{(\ell)}$. Then the error term becomes

$$\|\mathbf{Hs}^{(k)} - \mathbf{Hs}^{(\ell)}\|^2 = \|\mathbf{He}^{(k,\ell)}\|^2. \tag{6.126}$$

If the channel $\mathbf{H}$ is low rank, then it is possible that one of the error vectors lies in the null space of $\mathbf{H}$, which would make $\|\mathbf{He}^{(k,\ell)}\|^2 = 0$. This makes the upper bound equal to $(|\mathcal{S}| - 1)0.5$. Alternatively, suppose that the channel is a scaled unitary matrix that satisfies $\mathbf{H}^*\mathbf{H} = c\mathbf{I}$. Then, because of unitary invariance, it follows that $\|\mathbf{He}^{(k,\ell)}\|^2 = |c|^2\|\mathbf{e}^{(k,\ell)}\|^2$ and good performance is guaranteed, as long as the error vector is nonzero. In this case, the channel rotates all the error vectors by the same amount, preserving their distance properties.

Now we compute an upper bound on the probability of error for spatial multiplexing with a maximum likelihood receiver in an IID Rayleigh fading channel. To solve this using a similar calculation to that in (6.100) it is useful to rewrite the error term as

$$\|\mathbf{He}^{(k,\ell)}\|^2 = \|\mathrm{vec}(\mathbf{He}^{(k,\ell)})\|^2 \tag{6.127}$$

where the last step follows from the fact that for a vector $\mathbf{x}$, $\mathrm{vec}(\mathbf{x}) = \mathbf{x}$. Using the identity in (6.94),

$$\|\mathbf{He}^{(k,\ell)}\|^2 = \|(\mathbf{e}^{(k,\ell)T} \otimes \mathbf{I}_{N_\mathrm{r}})\mathrm{vec}(\mathbf{H})\|^2. \tag{6.128}$$

Defining $\mathbf{h} = \mathrm{vec}(\mathbf{H})$, and expanding the inner product and using the distributive property of the Kronecker product,

$$\|\mathbf{He}^{(k,\ell)}\|^2 = \mathbf{h}^*(\mathbf{e}^{(k,\ell)c}\mathbf{e}^{(k,\ell)\ T} \otimes \mathbf{I}_{N_\mathrm{r}})\mathbf{h}. \tag{6.129}$$

Let the error covariance matrix for spatial multiplexing be defined as $\mathbf{R}_{k,\ell} = \mathbf{e}^{(k,\ell)c}\mathbf{e}^{(k,\ell)\ T}$. Following the same steps as in (6.100)–(6.102):

$$\mathbb{E}_{\mathbf{H}}\left[P(\mathbf{s}^{(k)} \to \mathbf{s}^{(\ell)}|\mathbf{H})\right] \le \frac{1}{2} \frac{1}{\left|\mathbf{I} + \frac{E_{\mathrm{x}}}{4N_{\mathrm{o}}N_{\mathrm{t}}}\mathbf{R}_{k,\ell} \otimes \mathbf{I}_{N_{\mathrm{r}}}\right|} \tag{6.130}$$

$$= \frac{1}{2} \frac{1}{\left(1 + \frac{E_{\mathrm{x}}}{4N_{\mathrm{o}}N_{\mathrm{t}}}\|\mathbf{e}^{(k,\ell)}\|^2\right)^{N_{\mathrm{r}}}}. \tag{6.131}$$

The last step follows from two facts. First, $\mathrm{rank}(\mathbf{R}_{k,\ell}) = 1$ and its only nonzero eigenvalue is $\|\mathbf{e}^{(k,\ell)}\|^2$. Second, the eigenvalues of $\mathbf{A} \otimes \mathbf{B}$ are $\lambda_k(\mathbf{A})\lambda_m(\mathbf{B})$. Finally, the smallest value of $\|\mathbf{e}^{(k,\ell)}\|^2$ is the $d_{\min}$ of the component constellation $\mathcal{C}$.

The main intuition derived from (6.131) is that the spatial multiplexing with a maximum likelihood receiver obtains at most a diversity gain of N_{r}. There is no diversity obtained from having multiple transmit antennas. The reason is that the information sent across each transmit antenna is effectively independent since the constituent symbols $s[n]$ are all IID. Since there is no redundancy in space, the rank of the error covariance matrix $\mathbf{R}_{k,\ell}$ is 1. It is possible to derive codes that combine space-time coding and spatial multiplexing to achieve a diversity as high as $N_{\mathrm{t}}N_{\mathrm{r}}$ [149, 214, 31].

Spatial multiplexing with a maximum likelihood receiver is usually the benchmark receiver for performance comparisons. In Rayleigh fading channels it achieves a diversity order of N_{r}, which is the maximum that can be achieved by lower-complexity receivers (without channel state information from the transmitter or more sophisticated kinds of space-time coding). The main drawback of the maximum likelihood receiver is the requirement for a brute-force search over all the transmitted symbol vectors. This leads to the study of other low-complexity receiver techniques like the zero-forcing receiver.

Zero-Forcing Detector for Spatial Multiplexing The zero-forcing detector is one of the simplest yet most effective receivers for a spatial multiplexing system. Consider the system in (6.112). Now suppose that $\mathbf{H}$ is full rank and that $N_{\mathrm{r}} \ge N_{\mathrm{t}}$. From Chapter 3, the least squares estimate is

$$\widehat{\mathbf{s}}[n] = (\mathbf{H}^*\mathbf{H})^{-1}\mathbf{H}^*\mathbf{y}[n] \tag{6.132}$$

where $(\mathbf{H}^*\mathbf{H})^{-1}\mathbf{H}^*$ is simply the pseudo-inverse of $\mathbf{H}$ given by $\mathbf{H}^\dagger$.

In the presence of noise, the equalized received signal is

$$\mathbf{H}^\dagger\mathbf{y}[n] = \mathbf{s}[n] + \mathbf{H}^\dagger\mathbf{v}[n]. \tag{6.133}$$

The zero-forcing receiver detects each symbol stream separately, neglecting the fact that the noise is now spatially correlated. Let $\mathbf{z} = \mathbf{H}^\dagger\mathbf{y}[n]$. The zero-forcing detector then computes for $j = 1, 2, \ldots, N_{\mathrm{t}}$

$$\widehat{s}_j[n] = \arg\min_{c \in \mathcal{C}} |\mathbf{z}_j[n] - c|^2 \tag{6.134}$$

separately for each entry. In terms of search complexity, the zero-forcing detector computes N_{t} searches over the $|\mathcal{C}|$ constellation symbols, computing a scalar magnitude operation for each entry. In contrast, the maximum likelihood detector computes a search over $|\mathcal{C}|^{N_t}$ vector symbols and has to compute a difference and norm of length N_r for each entry. The complexity for zero forcing is much reduced compared with that of the maximum likelihood solution.

Performance of the zero-forcing detector can be analyzed through the probability of stream symbol error. A vector symbol error occurs when there is one or more stream symbol errors. The post-processing SNR for the jth symbol stream can be computed from

$$\mathbf{z}_j[n] = \mathbf{s}_j[n] + [\mathbf{H}^\dagger \mathbf{v}[n]]_j. \tag{6.135}$$

The noise is zero-mean complex Gaussian with covariance matrix $N_o \mathbf{H}^\dagger \mathbf{H}^{\dagger\,*}$. Using the pseudo-inverse definition,

$$\mathbf{H}^\dagger \mathbf{H}^{\dagger\,*} = (\mathbf{H}^* \mathbf{H})^{-1} \mathbf{H}^* \mathbf{H} (\mathbf{H}^* \mathbf{H})^{-1} \tag{6.136}$$

$$= (\mathbf{H}^* \mathbf{H})^{-1}. \tag{6.137}$$

Therefore, the variance of $[\mathbf{H}^\dagger \mathbf{v}[n]]_j$ is $[(\mathbf{H}^* \mathbf{H})^{-1}]_{j,j}$ and the post-processing SNR is given by

$$\text{SNR}_j^{\text{ZF}} = \frac{1}{N_o [(\mathbf{H}^* \mathbf{H})^{-1}]_{j,j}}. \tag{6.138}$$

This shows that the performance of the ZF receiver depends on the "invertibility" of the channel. If the channel was ill conditioned, then the inverse would "blow up," creating a very small post-processing SNR.

Example 6.10 Consider a 2×2 MIMO system with channel matrices given as follows:

$$\mathbf{H}_1 = \begin{bmatrix} e^{j\pi/3} & e^{j\pi/2} \\ e^{j\pi/7} & e^{-j\pi/4} \end{bmatrix} \tag{6.139}$$

$$\mathbf{H}_2 = \begin{bmatrix} e^{j\pi/6} & 0.5e^{j\pi} \\ e^{j\pi/5} & 0.1e^{j\pi/8} \end{bmatrix}. \tag{6.140}$$

First, compute the zero-forcing equalizers for the two channels, and after that calculate the post-processing SNR for each channel, assuming $1/N_o = 10\text{dB}$.

Answer: The ZF equalizer is just the pseudo-inverse of the channel matrix, which is the same as the normal inverse for the given 2×2 matrices. For the first channel matrix,

$$\mathbf{H}_1^{-1} = \frac{1}{e^{j\pi/3}e^{-j\pi/4} - e^{j\pi/7}e^{j\pi/2}} \begin{bmatrix} e^{-j\pi/4} & -e^{j\pi/2} \\ -e^{j\pi/7} & e^{j\pi/3} \end{bmatrix} \tag{6.141}$$

$$= \frac{1}{e^{j\pi/12} - e^{j9\pi/14}} \begin{bmatrix} e^{-j\pi/4} & -e^{j\pi/2} \\ -e^{j\pi/7} & e^{j\pi/3} \end{bmatrix}. \tag{6.142}$$

Similarly,

$$\mathbf{H}_2^{-1} = \frac{1}{0.1e^{\mathrm{j}7\pi/24} + 0.5e^{\mathrm{j}\pi/5}} \begin{bmatrix} 0.1e^{\mathrm{j}\pi/8} & 0.5 \\ -e^{\mathrm{j}\pi/5} & e^{\mathrm{j}\pi/6} \end{bmatrix}. \tag{6.143}$$

The post-processing SNR for the jth stream is given by (6.138). Plugging in $N_t = 2$ and $1/N_o = 10\text{dB}$, for $\mathbf{H}_1$,

$$\text{SNR}^{\text{ZF}}(\mathbf{H}_1) = \begin{bmatrix} 11.8591 \\ 11.8591 \end{bmatrix} = \begin{bmatrix} 10.74\text{dB} \\ 10.74\text{dB} \end{bmatrix} \tag{6.144}$$

$$\text{SNR}^{\text{ZF}}(\mathbf{H}_2) = \begin{bmatrix} 13.6878 \\ 1.7794 \end{bmatrix} = \begin{bmatrix} 11.36\text{dB} \\ 2.50\text{dB} \end{bmatrix}. \tag{6.145}$$

The probability of error in an AWGN channel is obtained by inserting the post-processing SNR in (6.138) into the expressions in Section 4.4.5. Analyzing the average probability of error is more complicated. It can be shown for Rayleigh fading channels that (6.138) is a (scaled) chi-squared random variable with $2(N_r - N_t + 1)$ degrees of freedom [132]. This result can be used to show that SNR_j^{ZF} has an equivalent distribution as a SIMO system with $N_r - N_t + 1$ antennas but with G replaced by G_{MIMO}. As a result, the diversity advantage is $N_r - N_t + 1$. If $N_r = N_t$, then all the dimensions of the channel are used in the inverse and there is no "excess" for diversity advantage. Interestingly, if there is even just one excess antenna $N_r = N_t + 1$, then zero forcing achieves second-order diversity. The diversity from just a single extra antenna is shared by all the symbol streams to improve performance. Because diversity effects offer dimensioning returns, a ZF receiver with a few excess antennas can perform close to the maximum likelihood receiver despite its much lower complexity.

The performance analysis results for spatial multiplexing show that neither maximum likelihood nor zero forcing benefits from having an excess number of transmit antennas. With ML detection, the diversity advantage depends only on N_r. With the ZF receiver, the diversity gain is reduced if large N_t is chosen and $N_t > N_r$ is not feasible. Additional gains from multiple transmit antennas can be obtained by using, for example, transmit precoding.

6.4.4 Linear Precoding

It is possible to exploit channel state information at the transmitter in MIMO spatial multiplexing systems using linear transmit precoding. A linear precoder is a matrix that maps N_s streams onto N_t transmit antennas. Most applications of linear precoding focus on the case where $N_s < N_t$ and use channel state information to design the precoder. The special case of $N_s = 1$ corresponds to sending a single stream using beamforming.

A spatial multiplexing system with linear precoding is illustrated in Figure 6.15. Let $\mathbf{F}$ denote the $N_t \times N_s$ precoding matrix. The precoding matrix is applied to the $N_s \times 1$ symbol vector $\mathbf{s}[n]$. There are different ways to normalize the precoder $\mathbf{F}$. In this section, we consider the case where $\text{tr}(\mathbf{FF}^*) = N_t$ to match our assumptions earlier where the

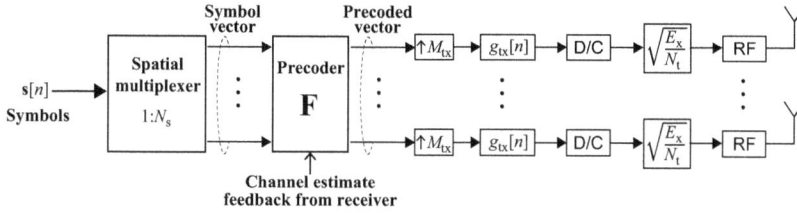

Figure 6.15 Spatial multiplexing at the transmitter with linear precoding. The spatial multiplexer produces a symbol vector of dimension $N_s \times 1$. The precoding operation multiplies the symbol vector by a precoding matrix $\mathbf{F}$, usually determined by the channel state, to produce the precoded output. The outputs are subsequently pulse shaped, converted to continuous time, and upconverted.

scaling per antenna branch is E_x/N_t. Under this assumption, the classed MIMO received signal equation is

$$\mathbf{y}[n] = \mathbf{HFs}[n] + \mathbf{v}[n]. \tag{6.146}$$

Other normalizations of the precoder are also possible if, for example, different powers are allocated to the different symbols [71, 370]. Additional constraints might also be imposed, for example, a per-antenna power constraint [71].

A special case of linear precoding is known generally as eigenbeamforming. To inform the construction, it is useful to have some background on the singular value decomposition (SVD) of a matrix. Any nonsquare $N \times M$ matrix $\mathbf{A}$ can be factored as

$$\mathbf{A} = \mathbf{U}_{\text{full}} \mathbf{\Sigma} \mathbf{V}_{\text{full}}^*. \tag{6.147}$$

The matrix $\mathbf{U}_{\text{full}}$ is unitary $N \times N$, $\mathbf{V}_{\text{full}}$ is unitary $M \times M$, and $\mathbf{\Sigma}$ is an $N \times M$ matrix with *ordered* (from largest to smallest) nonnegative entries $\{\sigma_k\}$ (called singular values) on the main diagonal and zero elsewhere. The columns of $\mathbf{U}_{\text{full}}$ are called the left singular vectors of $\mathbf{A}$ and the columns of $\mathbf{V}_{\text{full}}$ are called the right singular vectors of $\mathbf{A}$. The squared singular values of $\mathbf{A}$ are the eigenvalues of $\mathbf{AA}^*$ and $\mathbf{A}^*\mathbf{A}$.

Suppose that the SVD of $\mathbf{H} = \mathbf{U\Sigma V}^*$. Let $\mathbf{F} = [\mathbf{V}]_{:,1:N_s}$. Substituting into (6.146),

$$\mathbf{y}[n] = \mathbf{U\Sigma V}^*[\mathbf{V}]_{:,1:N_s}\mathbf{s}[n] + \mathbf{v}[n]. \tag{6.148}$$

Multiplying the received signal vector by $\mathbf{U}^*$ gives

$$\mathbf{U}^*\mathbf{y}[n] = \mathbf{U}^*\mathbf{U\Sigma V}^*[\mathbf{V}]_{:,1:N_s}\mathbf{s}[n] + \mathbf{U}^*\mathbf{v}[n] \tag{6.149}$$

$$= \mathbf{\Sigma V}^*[\mathbf{V}]_{:,1:N_s}\mathbf{s}[n] + \mathbf{U}^*\mathbf{v}[n] \tag{6.150}$$

$$= [\mathbf{\Sigma}]_{:,1:N_s}\mathbf{s}[n] + \mathbf{U}^*\mathbf{v}[n] \tag{6.151}$$

where (6.150) follows because $\mathbf{U}$ is unitary; and (6.151) follows because $\mathbf{V}$ is unitary; thus $\mathbf{V}^*[\mathbf{V}]_{:,1:N_s} = [\mathbf{I}]_{:,1:N_s}$. Also because $\mathbf{U}$ is unitary, $\mathbf{U}^*\mathbf{v}[n]$ remains $\mathcal{N}_\mathbb{C}(0, N_o\mathbf{I})$ since $\mathbb{E}[\mathbf{U}^*\mathbf{v}[n]\mathbf{v}^*[n]\mathbf{U}] = N_o\mathbf{I}$. Unlike the zero-forcing case, the noise remains IID. Defining $\mathbf{z}[n] = \mathbf{U}^*\mathbf{y}[n]$ and using $\mathbf{v}[n]$ in place of $\mathbf{U}^*\mathbf{v}[n]$ since they have the same distribution gives the equivalent received signal

$$\mathbf{z}[n] = [\mathbf{\Sigma}]_{:,1:N_s}\mathbf{s}[n] + \mathbf{v}[n]. \tag{6.152}$$

It is easy to decode the ith stream because all the symbols are decoupled:

$$\mathbf{z}_j[n] = \sigma_j \mathbf{s}_j[n] + \mathbf{v}_j[n]. \tag{6.153}$$

Unlike the case of zero forcing, there is no loss in performing the stream detection independently since the noise remains IID in this case. Consequently, the performance of the optimum detector is dramatically simplified.

Linear precoding as described in this section has a nice interpretation. Effectively with the chosen precoder each symbol rides along its own eigenmode (singular vector) of the channel using a beamforming vector $\mathbf{f}_j = [\mathbf{F}]_{:,j}$ chosen to excite that mode. Linear precoding effectively couples symbols into the MIMO channel along its preferred "directions."

The performance of the jth stream depends on the post-processing SNR

$$\text{SNR} = \frac{1}{N_\text{o}} \sigma_j^2, \tag{6.154}$$

which depends only on the singular values of the channel. If $N_\text{t} = N_\text{s}$, linear precoding provides performance that is similar to that of the zero-forcing decoder. Precoding, though, can be tuned to give better performance by adapting the rate and power per stream. The benefits of precoding appear when used with $N_\text{s} < N_\text{t}$. For example, consider a 4×4 channel $\mathbf{H}$ that has only two significant singular values. By choosing $N_\text{s} = 2$, the transmitter can send information on the two best eigenmodes and can avoid wasting effort on the two bad ones. When N_s is chosen dynamically as the channel changes, this is known as rank adaptation or multimode precoding [206, 148, 179].

The advantages of precoding are also realized in terms of its diversity performance. If $N_\text{s} \le N_\text{t}$ and $N_\text{s} \le N_\text{r}$, then in Rayleigh fading channels precoding achieves a diversity order of $(N_\text{r} - N_\text{s} + 1)(N_\text{t} - N_\text{s} + 1)$ [210].

In this case diversity is obtained both from excess transmit and excess receive antennas. If multimode precoding is employed, then a diversity of $N_\text{t} N_\text{r}$ can be achieved, which is what is achieved using single-stream beamforming. Overall, precoding makes use of the channel to provide better performance, make use of extra transmit and receive antennas, and allow optimal low-complexity decoding at the receiver.

Requiring channel state information at the transmitter is a main limitation of linear precoding as described in this section. A solution to this problem is to use limited feedback precoding, where the transmit precoder is selected from a codebook of possible precoders known to the receiver and the transmitter.

6.4.5   Extensions to Limited Feedback

Limited feedback precoding is a generalization of limited feedback beamforming. It provides a framework for providing channel state information (in the form of a codeword recommendation) from the receiver to the transmitter. Limited feedback precoding is found in IEEE 802.11n and 3GPP LTE, among other wireless systems.

A spatial multiplexing system with limited feedback is illustrated in Figure 6.16. Let $\mathcal{F} = \{\mathbf{F}_1, \mathbf{F}_2, \dots, \mathbf{F}_{N_\text{LF}}\}$ denote a codebook of precoding matrices. Each entry $\mathbf{F}_k$ is an $N_\text{t} \times N_\text{s}$ precoding matrix that satisfies the appropriate precoding design constraints (in our case it satisfies the constraint that the columns be unit norm and orthogonal).

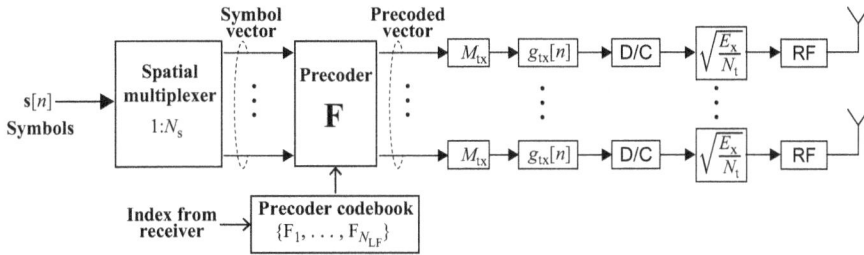

Figure 6.16 Spatial multiplexing at the transmitter with limited feedback linear precoding. The spatial multiplexer produces a symbol vector of dimension $N_s \times 1$. The precoding operation multiplies the symbol vector by a precoding matrix $\mathbf{F}$, which comes from a codebook of precoding vectors. An index sent from the receiver determines which precoder in the codebook is selected.

Selecting the optimum codeword from the codebook depends on the chosen receiver design. There are different criteria, and each is reasonable under different assumptions [210]. For example, if a zero-forcing receiver is employed, one approach would be to maximize the minimum post-processing SNR given the composite channel $\mathbf{HF}_k$ as

$$k^\star = \underset{k=1,2,\ldots,N_{\mathrm{LF}}}{\arg\max} \ \underset{m=1,2,\ldots,N_s}{\min} \frac{E_{\mathrm{x}}}{N_t N_0 [(\mathbf{F}_k^* \mathbf{H}^* \mathbf{HF}_k)^{-1}]_{m,m}} \tag{6.155}$$

$$= \underset{k=1,2,\ldots,N_{\mathrm{LF}}}{\arg\min} \ \underset{m=1,2,\ldots,N_s}{\max} [(\mathbf{F}_k^* \mathbf{H}^* \mathbf{HF}_k)^{-1}]_{m,m}. \tag{6.156}$$

Another approach, inspired by (6.154), is to maximize the minimum singular value of $\mathbf{HF}_k$.

The diversity obtained using limited feedback precoding is the same as that achieved with perfect channel state information if the codebook is large enough. With a good design, choosing $N_{\mathrm{LF}} \geq N_t$ is all that is required [210].

Antenna subset selection is a special case of limited feedback precoding [146, 228, 207]. With antenna subset selection, the precoding matrices are chosen from subsets of columns of the identity matrix. The maximum codebook size is $\binom{N_t}{N_s}$. Antenna subset selection can also be viewed as a generalization of MISO antenna selection.

Example 6.11   Illustrate the limited feedback codebook derived from antenna subset selection, assuming $N_s = 2$ and $N_t = 3$.

Answer: There is a total of $\binom{3}{2} = 3$ entries in this codebook. The resulting codebook is

$$\mathcal{F} = \left\{ \sqrt{\frac{3}{2}} \begin{bmatrix} 1 & 0 \\ 0 & 1 \\ 0 & 0 \end{bmatrix}, \sqrt{\frac{3}{2}} \begin{bmatrix} 1 & 0 \\ 0 & 0 \\ 0 & 1 \end{bmatrix}, \sqrt{\frac{3}{2}} \begin{bmatrix} 0 & 0 \\ 1 & 0 \\ 0 & 1 \end{bmatrix} \right\}. \tag{6.157}$$

The scaling factor is chosen to ensure that $\mathrm{tr}(\mathbf{F}_k^* \mathbf{F}_k) = N_t$.

There are a number of different possible codebook designs for $\mathcal{F}$. Perhaps the classic design is Grassmannian precoding [210] where the codewords correspond to points on the Grassmann manifold. The 3GPP LTE standard uses a nested codebook design that allows the transmitter to select a smaller N_s than indicated by the receiver [127, Section 7.2.2]. IEEE 802.11n has three feedback options: one where the receiver feeds back the quantized channel coefficients directly (full channel state information feedback) and two where the receiver first computes beamforming feedback matrices and sends the non-compressed version (noncompressed beamforming matrix feedback) or the compressed version using Givens rotation (compressed beamforming matrix feedback) [158, Section 20.3.12.2].

Example 6.12 In this example, we compare the performance of different MIMO configurations and performance with two transmit antennas and two receive antennas. To make comparisons fair, we keep the rate (size of the vector constellation) fixed within each strategy. We compute the average probability of symbol error rate assuming an IID Rayleigh channel. We consider four possible cases:

- SIMO with maximum ratio combining (only one transmit antenna is used) with a 16-QAM constellation.

- MIMO with transmit and receive antenna selection (select the transmit and receive antennas with the best channel) with a 16-QAM constellation. Antenna selection is a special case of limited feedback precoding.

- Spatial multiplexing with a zero-forcing receiver and 4-QAM constellations.

- Spatial multiplexing with a maximum likelihood receiver and 4-QAM constellations.

Answer the following questions based on the performance plot in Figure 6.17.

- Why does antenna selection in the MIMO channel outperform optimum combining in the SIMO channel?

  Answer: In the 1×2 SIMO channel, there is only second-order diversity. In contrast, since there are two transmit antennas and two receive antennas in a 2×2 MIMO channel, antenna selection provides a diversity advantage of four, which can be seen through similar arguments to those for antenna selection in Section 6.2.2. Therefore, at high SNR, the performance curves cross and the system with higher diversity has better performance.

- For what SNRs is the antenna selection technique preferred to spatial multiplexing with a zero-forcing receiver?

  Answer: Antenna selection has both transmit and receive diversity with a diversity advantage of four, whereas spatial multiplexing with a zero-forcing receiver has a diversity advantage of only one. From Figure 6.17, antenna selection outperforms zero forcing for all the SNR values considered.

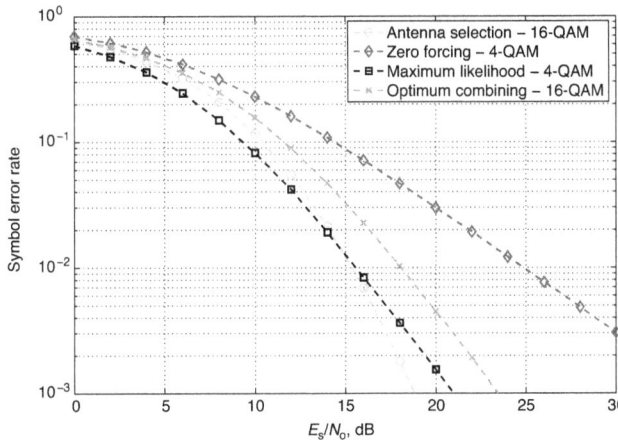

Figure 6.17 Symbol error probability comparison for different 2×2 MIMO techniques

- For what SNRs is the antenna selection technique preferred to spatial multiplexing with an ML receiver?

  Answer: Figure 6.17 shows that the antenna selection technique is preferred to spatial multiplexing with an ML receiver when SNR is (approximately) greater than 14.5dB. When the SNR is smaller than 14.5dB, spatial multiplexing with an ML receiver is preferred over antenna selection.

- What is the performance difference between zero forcing and the maximum likelihood receiver?

  Answer: The maximum likelihood receiver is better than the zero-forcing receiver in terms of symbol error rate for all SNR values because of its higher diversity advantage. The difference between ML and ZF increases as SNR increases.

- What is the complexity difference between zero forcing and the maximum likelihood receiver?

  Answer: First, there are some calculations that are performed on the block of symbols. Zero forcing requires the calculation of a matrix inverse. Maximum likelihood does not require initial calculations, but more efficient implementation uses a modified constellation $\{\mathbf{H}s | s \in \mathcal{S}\}$. The complexity is essentially then one matrix inverse versus 16 matrix multiplies.

  Second, with zero forcing, two independent detections of 4-QAM require one matrix multiply (requiring four multiplies and two adds) and eight scalar minimum distance calculations. With maximum likelihood detection, the joint symbol detection requires 16 vector minimum distance calculations, which is equivalent to 32 scalar minimum distance calculations. The complexity with maximum likelihood detection is at least twice as high as for zero forcing, not including the first set of calculations.

6.4.6 Channel Estimation in MIMO Systems

All of the SIMO, MISO, and MIMO approaches described thus far require knowledge of the channel at the receiver. This section provides some generalizations of the channel estimators described in Section 5.1.4 to estimate the MIMO channel. Through suitable choice of N_t and N_r, these approaches can be applied to estimate SIMO and MISO channels.

Let $\{\mathbf{t}[n]\}_{n=0}^{N_{tr}-1}$ denote a vector of training symbols. We take the vector of training symbols and create a training matrix to simplify the exposition:

$$\mathbf{T} = \begin{bmatrix} \mathbf{t}[0] & \mathbf{t}[1] & \cdots & \mathbf{t}[N_{tr}-1] \end{bmatrix}. \tag{6.158}$$

The simplest approach for channel estimation is to take turns training each antenna. Suppose that N_{tr} is divisible by N_t, and let $\{t[n]\}_{n=0}^{N_{tr}/N_t-1}$ denote a scalar training sequence. Sending this sequence successively on each antenna gives

$$\mathbf{T} = \begin{bmatrix} t[0] & \cdots & t[N_{tr}-1] & 0 & \cdots & 0 & \cdots \\ 0 & \cdots & 0 & t[0] & \cdots & t[N_{tr}-1] & \cdots \\ \vdots & \cdots & \vdots & 0 & \cdots & 0 & \cdots \end{bmatrix}. \tag{6.159}$$

Consider the received signal during the transmission of the training sequence on transmit antenna j at receive antenna i:

$$\mathbf{y}_i[n+(i-1)N_{tr}] = h_{i,j}t[n] + \mathbf{v}_i[n+(i-1)N_{tr}] \tag{6.160}$$

for $n = 0, 1, \ldots, N_{tr}/N_t - 1$. Stacking into a vector gives

$$\mathbf{y}_{i,j} = h_{i,j}\mathbf{t} + \mathbf{v}_{i,j}. \tag{6.161}$$

The least squares solution is

$$\widehat{h}_{i,j} = (\mathbf{t}^*\mathbf{t})^{-1}\mathbf{t}^*\mathbf{y}_{m,k}. \tag{6.162}$$

When the training signal is sent from transmit antenna j, the signal received on antenna i is used to estimate $h_{i,j}$. After cycling through all the transmit antennas, the receiver can put together an estimate of the entire channel matrix $\mathbf{H}$.

The problem with this simple approach is that it requires transmit antennas to power off and on. This may create challenges for the analog front end. It also results in reduced performance when there are per-antenna power constraints (when all the transmit antennas are not used, not all the power is delivered). A solution is to use a more general training structure. Suppose that a general $\mathbf{T}$ is used with columns $\mathbf{t}[n]$. Then

$$\mathbf{y}[n] = \mathbf{H}\mathbf{t}[n] + \mathbf{v}[n]. \tag{6.163}$$

Now, collecting successive vectors together for $n = 0, 1, \ldots, N_{tr} - 1$,

$$\mathbf{Y} = \mathbf{H}\mathbf{T} + \mathbf{V}. \tag{6.164}$$

The least squares solution for $\mathbf{H}$ is

$$\widehat{\mathbf{H}} = \mathbf{YT}^*(\mathbf{TT}^*)^{-1}. \tag{6.165}$$

Further details about this derivation are provided in Example 6.13 and Example 6.14.

A good design for the training sequence is to have orthogonal sequences sent on each antenna. If this is the case, then $\mathbf{TT}^* = N_{\mathrm{tr}}\mathbf{I}$ and the least squares channel estimate simplifies as

$$\widehat{\mathbf{H}} = \mathbf{YT}^*, \tag{6.166}$$

which is essentially a matrix correlation operation.

Example 6.13 Consider a MISO frequency-flat slow-fading channel with two transmit antennas and one receive antenna. Let $h_{1,1}$ and $h_{1,2}$ be the channels from the transmit antennas to the receive antenna. Assuming that the independent symbols $s_1[n]$ and $s_2[n]$ are transmitted, the received signal at this antenna is given by

$$y_1[n] = h_{1,1}s_1[n] + h_{1,2}s_2[n] + v_1[n], \tag{6.167}$$

where $v_1[n]$ is the noise at the receive antenna.

Formulate and solve a least squares channel estimation problem to estimate the channel coefficients of the MISO channel, assuming that we send two training sequences $\{t_1[n]\}_{n=0}^{N_{\mathrm{tr}}-1}$ and $\{t_2[n]\}_{n=0}^{N_{\mathrm{tr}}-1}$ through the MISO channel at the same time. Be specific about the expression of the squared error and the resulting matrix and vector definitions, to provide a point of comparison for the MIMO case in the next example.

Answer: The input-output relationship in the matrix form in the absence of noise is given by

$$\underbrace{\begin{bmatrix} y_1[0] & \cdots & y_1[N_{\mathrm{tr}} - 1] \end{bmatrix}}_{\mathbf{y}_1^T} = \underbrace{\begin{bmatrix} h_{1,1} & h_{1,2} \end{bmatrix}}_{\mathbf{h}_1^T} \underbrace{\begin{bmatrix} t_1[0] & t_1[1] & \cdots & t_1[N_{\mathrm{tr}} - 1] \\ t_2[0] & t_2[1] & \cdots & t_2[N_{\mathrm{tr}} - 1] \end{bmatrix}}_{\mathbf{T}}. \tag{6.168}$$

We write using row vectors to be consistent with the derivation in (6.164). The squared error between the observations and the known data for dummy vector $\mathbf{a}^T = [a_1, a_2]$ is

$$J(\mathbf{a}) = \sum_{n=0}^{N_{\mathrm{tr}}-1} |y_1[n] - a_1 t_1[n] - a_2 t_2[n]|^2 \tag{6.169}$$

$$= ||\mathbf{y}_1 - \mathbf{T}^T\mathbf{a}||^2. \tag{6.170}$$

The least squares channel estimate is the solution to

$$\widehat{\mathbf{h}}_1 = \arg\min_{\mathbf{a} \in \mathbb{C}^{2 \times 1}} ||\mathbf{y}_1 - \mathbf{T}^T\mathbf{a}||^2. \tag{6.171}$$

Assuming that $\mathbf{T}$ is full rank, the least squares MISO channel estimate is

$$\widehat{\mathbf{h}}_1^T = \mathbf{y}_1^T\mathbf{T}^*(\mathbf{TT}^*)^{-1} \tag{6.172}$$

where we have written the final solution in the form as in (6.165).

Example 6.14   Consider the system in Example 6.13 where the receiver is equipped with one additional antenna to form a MIMO 2×2 system. The channels from the transmit antennas to the second receive antenna are denoted as h_{21} and h_{22}. The received signal at the second receive antenna is given by

$$y_2[n] = h_{2,1}s_1[n] + h_{2,2}s_2[n] + v_2[n], \tag{6.173}$$

where $v_2[n]$ is the noise at the second receive antenna. Formulate and solve for the least squares channel estimator using the same two training sequences $\{t_1[n]\}_{n=0}^{N_{\text{tr}}-1}$ and $\{t_2[n]\}_{n=0}^{N_{\text{tr}}-1}$, chosen such that $\mathbf{T}$ is full rank.

Answer: We define $\mathbf{y}_2$ in a similar fashion as $\mathbf{y}_1$, and $\mathbf{h}_2$ in a similar way as $\mathbf{h}_1$. The input-output relationship in the matrix form in the absence of noise is given by

$$\underbrace{\begin{bmatrix} y_1[0] & \cdots & y_1[N_{\text{tr}}-1] \\ y_2[0] & \cdots & y_2[N_{\text{tr}}-1] \end{bmatrix}}_{\mathbf{Y}} = \underbrace{\begin{bmatrix} h_{1,1} & h_{2,1} \\ h_{1,2} & h_{2,2} \end{bmatrix}}_{\mathbf{H}} \underbrace{\begin{bmatrix} t_1[0] & t_1[1] & \cdots & t_1[N_{\text{tr}}-1] \\ t_2[0] & t_2[1] & \cdots & t_2[N_{\text{tr}}-1] \end{bmatrix}}_{\mathbf{T}} \tag{6.174}$$

or equivalently

$$\begin{bmatrix} \mathbf{y}_1^{\mathsf{T}} \\ \mathbf{y}_2^{\mathsf{T}} \end{bmatrix} = \begin{bmatrix} \mathbf{h}_1^{\mathsf{T}} \\ \mathbf{h}_2^{\mathsf{T}} \end{bmatrix} \mathbf{T}. \tag{6.175}$$

Now consider the squared error term, which in the matrix case is written using the Frobenius norm $\|\mathbf{A}\|_F^2 = \sum_i \sum_j [\mathbf{A}]_{i,j}$ as

$$\left\| \begin{bmatrix} \mathbf{y}_1^{\mathsf{T}} \\ \mathbf{y}_2^{\mathsf{T}} \end{bmatrix} - \begin{bmatrix} \mathbf{h}_1^{\mathsf{T}} \\ \mathbf{h}_2^{\mathsf{T}} \end{bmatrix} \mathbf{T} \right\|_F^2 = \text{tr} \left(\left(\begin{bmatrix} \mathbf{y}_1^{\mathsf{T}} \\ \mathbf{y}_2^{\mathsf{T}} \end{bmatrix} - \begin{bmatrix} \mathbf{h}_1^{\mathsf{T}} \\ \mathbf{h}_2^{\mathsf{T}} \end{bmatrix} \mathbf{T} \right)^* \left(\begin{bmatrix} \mathbf{y}_1^{\mathsf{T}} \\ \mathbf{y}_2^{\mathsf{T}} \end{bmatrix} - \begin{bmatrix} \mathbf{h}_1^{\mathsf{T}} \\ \mathbf{h}_2^{\mathsf{T}} \end{bmatrix} \mathbf{T} \right) \right) \tag{6.176}$$

$$= \|\mathbf{y}_1 - \mathbf{T}^{\mathsf{T}}\mathbf{h}_1\|^2 + \|\mathbf{y}_2 - \mathbf{T}^{\mathsf{T}}\mathbf{h}_2\|^2. \tag{6.177}$$

Because the cost function is separable (the minimizer of the first term is independent of the minimizer of the second term), it is possible to solve for each channel separately as

$$\widehat{\mathbf{h}}_i = (\mathbf{T}^c\mathbf{T}^T)^{-1}\mathbf{T}^c\mathbf{y}_i, \text{ for } i = 1, 2. \tag{6.178}$$

Equivalently,

$$\mathbf{h}_i = \underset{\mathbf{a}_i \in \mathbb{C}^{2 \times 1}}{\arg \min} \|\mathbf{y}_i - \mathbf{T}\mathbf{a}_i\|^2, \text{ for } i = 1, 2. \tag{6.179}$$

Assume that $\mathbf{T}$ is full rank. The MIMO least squares channel estimate is given by

$$\widehat{\mathbf{h}}_i = (\mathbf{T}^c\mathbf{T}^T)^{-1}\mathbf{T}^c\mathbf{y}_i, \text{ for } i = 1, 2. \tag{6.180}$$

Writing in matrix form using (6.165), then

$$\widehat{\mathbf{H}} = \mathbf{Y}\mathbf{T}^*(\mathbf{T}\mathbf{T}^*)^{-1}. \tag{6.181}$$

In general, for a given SISO channel estimation error achieved with a length N_{SISO} training sequence, a MIMO system requires a length $N_{\mathrm{tr}}N_{\mathrm{SISO}}$ matrix sequence to achieve the same estimation error performance. The overhead is only a function of the number of transmit antennas. For a given coherence time, the number of symbols during which the channel is constant is fixed. As the number of transmit antennas increases, the required training grows accordingly. For a large number of antennas, little time may be left to actually send data. This creates a trade-off between different system parameters where there is tension between the number of transmit antennas used, the time devoted to training, and the number of receive antennas.

6.4.7 Going Beyond the Flat-Fading Channel to Frequency-Selective Channels

The exposition in this section has focused so far on MIMO communication in flat-fading channels. MIMO communication is also possible in frequency-selective channels. This section reviews the MIMO frequency-selective channel model and describes how concepts like equalization are generalized in the MIMO setting. Emphasis is placed on spatial multiplexing. At present, commercial systems employing MIMO communication in frequency-selective channels use MIMO-OFDM, which is explored in more detail in Section 6.5.

Frequency-Selective Channel Equalization Let $\{h_{i,j}[\ell]\}_{\ell=0}^{L}$ denote the complex baseband discrete-time equivalent channel between the jth transmit antenna and the ith receive antenna. The channel includes all scaling factors. Spatial multiplexing is used at the transmitter where the symbol stream $\{s_j[n]\}$ is sent on antenna j. The discrete-time baseband received signal (assuming synchronization has been performed) is

$$y_i[n] = \sum_{\ell=0}^{L} h_{i,1}[\ell]s_1[n-\ell] + h_{i,2}[\ell]s_2[n-\ell] + \cdots + h_{i,N_t}[\ell]s_{N_t}[n-\ell]. \qquad (6.182)$$

Compared with the SISO channel model in (5.87), in the MIMO case, each receive antenna observes a linear combination of all the signals from the transmit antennas, filtered by their respective antennas. This self-interference among the transmitted signals makes it hard to apply directly all the algorithms from Chapter 5 without substantial modifications.

Define the multivariate impulse response $\{\mathbf{H}[\ell]\}_{\ell=0}^{L}$ where $[\mathbf{H}[\ell]]_{i,j} = h_{i,j}[\ell]$. Then the received signal can be written in matrix form as

$$\mathbf{y}[n] = \sum_{\ell=0}^{L} \mathbf{H}[\ell]\mathbf{s}[n-\ell] + \mathbf{v}[n]. \qquad (6.183)$$

This is the standard MIMO input-output equation for a frequency-selective fading channel.

Because the channel is a multivariate filter, a linear equalization should also be a multivariate filter. Let $\{\mathbf{G}[k]\}_{k=0}^{K}$ denote an order K multivariate impulse response that corresponds to the equalizer. A least squares equalizer would find a $\{\mathbf{G}[k]\}_{k=0}^{K}$ such that

$$\sum_{k=0}^{K} \mathbf{G}[k]\mathbf{H}[n-k] \approx \delta[n-n_{\mathrm{d}}]\mathbf{I}. \tag{6.184}$$

The least squares equalizer can be found using a similar approach to that in Section 5.2.2 using block Toeplitz matrices. Let $\bar{\mathbf{H}}_{k,m}$ denote the $(L+K+1) \times (K+1)$ Toeplitz matrix constructed from $h_{k,m}[n]$ as in (5.91). Furthermore, let $\mathbf{g}_{k,m} = [g_{k,m}[0], g_{k,m}[1], \ldots, g_{k,m}[K]]^{T}$. Then the equalizer for recovering the mth stream can be found by solving

$$\begin{bmatrix} \mathbf{H}_{1,m} & \mathbf{H}_{2,m} & \cdots & \mathbf{H}_{N_{\mathrm{r}},m} \end{bmatrix} \begin{bmatrix} \mathbf{g}_{1,m} \\ \mathbf{g}_{2,m} \\ \vdots \\ \mathbf{g}_{N_{\mathrm{r}},m} \end{bmatrix} = \mathbf{e}_{n_{\mathrm{d}}} \tag{6.185}$$

with $\mathbf{e}_{n_{\mathrm{d}}}$ as defined in (5.91).

An interesting fact about MIMO equalizers is that, under certain circumstances, it is possible to find a perfect FIR matrix inverse. In other words, the least squares approach can have zero error. The exact conditions for finding a perfect inverse are described in [184, 367] and involve conditions on K, L, N_{r}, and the coprimeness of certain polynomials. The intuition is that if N_{r} is large enough, then the block Toeplitz matrix in (6.189) will become square or even fat. In this case, assuming the matrix is full rank, either one (if square) or an infinite (if fat) number of solutions can exist. When more than one solution is available, common convention is to take the minimum norm solution, which gives the equalizer coefficients with the smallest norm and thus the least noise enhancement. These two solutions are explored in Example 6.15.

Example 6.15   Consider a frequency-selective slow-fading SIMO system where the transmitter has a single antenna and the receiver has two antennas. Assume that the channels from the transmit antenna to each receive antenna are frequency selective with $L+1$ taps. The received signals at the receive antennas are given by

$$y_1[n] = \sum_{\ell=0}^{L} h_1[\ell]s[n-\ell] + v_1[n] \tag{6.186}$$

$$y_2[n] = \sum_{\ell=0}^{L} h_2[\ell]s[n-\ell] + v_2[n]. \tag{6.187}$$

Note that the receiver has two observations of the same signal. Each is convolved by a different frequency-selective channel. The receiver then applies a set of equalizers

$\{g_1[k]\}_{k=0}^K$ and $\{g_2[k]\}_{k=0}^K$ such that

$$\sum_{k=0}^K g_1[k]h_1[n-k] + \sum_{k=0}^K g_2[k]h_2[n-k] = \delta[n-n_d]. \tag{6.188}$$

Formulate and solve the least squares equalizer problem given knowledge of $\{\widehat{h}_1[\ell]\}_{\ell=0}^L$ and $\{\widehat{h}_2[\ell]\}_{\ell=0}^L$.

Answer: Based on (6.189), the coefficients of equalizer $\mathbf{g}_{1,1}$ and $\mathbf{g}_{2,1}$ are obtained by solving

$$\underbrace{\begin{bmatrix} \mathbf{H}_{1,1} & \mathbf{H}_{2,1} \end{bmatrix}}_{\bar{\mathbf{H}}} \underbrace{\begin{bmatrix} \mathbf{g}_{1,1} \\ \mathbf{g}_{2,1} \end{bmatrix}}_{\mathbf{g}} = \mathbf{e}_{n_d}. \tag{6.189}$$

The solution depends on the rank and dimension of $\bar{\mathbf{H}}$. The dimensions are $(L+K+1) \times (2(K+1))$. The matrix is square or fat if $K \geq L-1$. The matrix is full rank in this case if the two channels are coprime, which means that they do not share any zeros in their Z transforms [184]. If $K < L-1$, then the solution is the usual least squares equalizer estimate given by

$$\mathbf{g} = (\bar{\mathbf{H}}^*\bar{\mathbf{H}})^{-1}\bar{\mathbf{H}}^*\mathbf{e}_{n_d} \tag{6.190}$$

where $\mathbf{g}_{1,1} = [\mathbf{g}]_{:,1:K+1}$ and $\mathbf{g}_{2,1} = [\mathbf{g}]_{:,K+2:2K+2}$. The optimum delay can be selected by finding the delay that minimizes the residual error $\mathbf{e}_{n_d}^*(\mathbf{I} - \bar{\mathbf{H}}(\bar{\mathbf{H}}^*\bar{\mathbf{H}})^{-1}\bar{\mathbf{H}}^*)\mathbf{e}_{n_d}$.

If K is chosen large enough such that $K \geq L$, then $\bar{\mathbf{H}}$ is a fat matrix and there are an infinite number of possible solutions. It is common to choose from among those solutions the one that has the minimum norm $\|\mathbf{g}\|$, which is

$$\mathbf{g} = \bar{\mathbf{H}}^* \left(\bar{\mathbf{H}}\bar{\mathbf{H}}^*\right)^{-1} \mathbf{e}_{n_d}. \tag{6.191}$$

The delay can be further optimized by finding the value of n_d such that $\mathbf{e}_{n_d}^* \left(\bar{\mathbf{H}}\bar{\mathbf{H}}^*\right)^{-1} \mathbf{e}_{n_d}$ is minimized.

Frequency-Selective Channel Estimation   Channel estimation also generalizes to the MIMO case by forming a suitable least squares problem. Suppose that training sequence $\{t_j[n]\}_{n=0}^{N_{tr}-1}$ is sent from antenna j. We focus on estimating the set of channels $\{h_{i,j}[\ell]\}_{j=1}^{N_t}$ from the observations at antenna i. Writing the observed data as a function of the unknowns in matrix form following the approach in (5.173), let

$$\mathbf{T}_j = \begin{bmatrix} t_j[L] & \cdots & t_j[0] \\ t_j[L+1] & \ddots & t_j[1] \\ \vdots & & \vdots \\ t_j[N_{tr}-1] & \cdots & t_j[N_{tr}-1-L] \end{bmatrix} \tag{6.192}$$

$$\mathbf{y}_i = \begin{bmatrix} y_i[L] \\ y_i[L+1] \\ \vdots \\ y_i[N_{\text{tr}} - 1] \end{bmatrix} \tag{6.193}$$

$$\mathbf{h}_{i,j} = \begin{bmatrix} h_{i,j}[0] \\ h_{i,j}[1] \\ \vdots \\ h_{i,j}[L] \end{bmatrix} \tag{6.194}$$

$$\mathbf{v}_i = \begin{bmatrix} v_i[L] \\ v_i[L+1] \\ \vdots \\ v_i[N_{\text{tr}} - 1] \end{bmatrix}. \tag{6.195}$$

Then the observation can be written as

$$\mathbf{y}_i = \underbrace{\begin{bmatrix} \mathbf{T}_1 & \mathbf{T}_2 & \cdots & \mathbf{T}_{N_t} \end{bmatrix}}_{\bar{\mathbf{T}}} \underbrace{\begin{bmatrix} \mathbf{h}_{i,1} \\ \mathbf{h}_{i,2} \\ \vdots \\ \mathbf{h}_{i,N_t} \end{bmatrix}}_{\mathbf{h}_i} + \mathbf{v}_i. \tag{6.196}$$

The least squares channel estimate follows as

$$\widehat{\mathbf{h}}_i = (\bar{\mathbf{T}}^*\bar{\mathbf{T}})^{-1}\bar{\mathbf{T}}^*\mathbf{y}_i. \tag{6.197}$$

Compared with the SISO case, the MIMO case estimates all the channels seen by receive antenna k at once using the fact that training sequences are sent simultaneously from each transmit antenna.

Example 6.16   Consider a frequency-selective slow-fading MISO system where the transmitter has two antennas and the receiver has a single antenna. Assume that the channels from each transmit antenna to the receive antenna are frequency selective with $L+1$ taps. The received signal is given by

$$y_1[n] = \sum_{\ell=0}^{L} h_{1,1}[\ell]s_1[n-\ell] + h_{1,2}[\ell]s_2[n-\ell] + v_1[n]. \tag{6.198}$$

The transmitter uses training sequences $\{t_1[n]\}_{n=0}^{N_{\text{tr}}-1}$ and $\{t_2[n]\}_{n=0}^{N_{\text{tr}}-1}$. Solve for the least squares channel estimate.

Answer: The squared error can be written as follows:

$$J(\mathbf{h}_{1,1}, \mathbf{h}_{1,2}) = \sum_{n=L}^{N_{\text{tr}}-1} \left| y_1[n] - \sum_{\ell=0}^{L} h_{1,1}[\ell] t_1[n-\ell] + h_{1,2}[\ell] t_2[n-\ell] \right|^2 \quad (6.199)$$

$$= \| \mathbf{y}_1 - (\mathbf{T}_1 \mathbf{h}_{1,1} + \mathbf{T}_2 \mathbf{h}_{1,2}) \|^2. \quad (6.200)$$

Defining $\mathbf{h}_1 = [\mathbf{h}_{1,1}^T, \mathbf{h}_{1,2}^T]^T$ and $\bar{\mathbf{T}} = [\mathbf{T}_1, \mathbf{T}_2]$, then the solution is

$$\widehat{\mathbf{h}}_1 = (\bar{\mathbf{T}}^* \bar{\mathbf{T}})^{-1} \bar{\mathbf{T}}^* \mathbf{y}_1. \quad (6.201)$$

Generalizations are possible for each of the impairments and the algorithms for dealing with them from Chapter 5 to MIMO frequency-selective channels. Direct equalizers can be formulated and solved using appropriately defined block matrices. Carrier frequency offset and frame synchronization are also important in MIMO systems. We defer discussion of MIMO-specific synchronization algorithms to the specific case of MIMO dealt with in Section 6.5.

A block diagram of a MIMO communication system is provided in Figure 6.18. Many of the functional blocks serve the same purpose as in the SISO communication system but with added complexity because there are multiple inputs and multiple outputs. One of the functions with the highest complexity is the linear equalizer. Both estimating and applying the equalizer require a significant amount of computation. An alternative that has been widely successful in commercial systems is to use OFDM for its easy equalization properties. This is discussed in more detail in the next section.

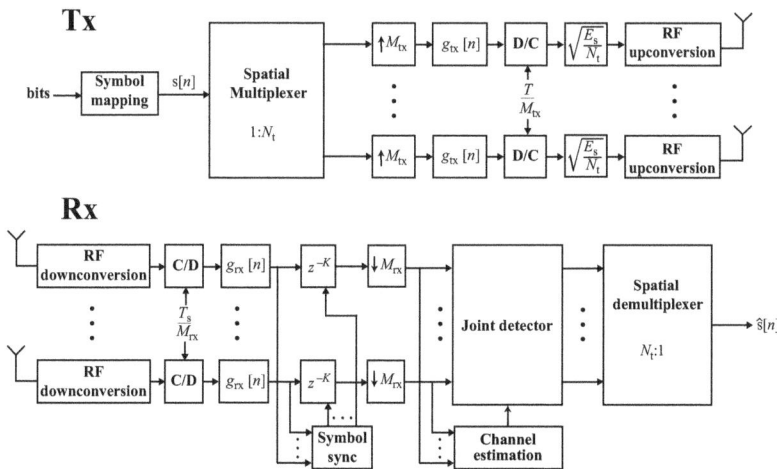

Figure 6.18 A possible transmit block diagram for a MIMO system with spatial multiplexing

6.5 MIMO-OFDM Transceiver Techniques

MIMO-OFDM combines the spatial multiplexing and diversity features of MIMO communication with the ease of equalization when using OFDM modulation. Indeed, MIMO-OFDM is currently the de facto approach for MIMO communication. It is used in IEEE 802.11n [158] and IEEE 802.11ac [161]. A variation known as MIMO-OFDMA (orthogonal frequency-division multiple access) is used in WiMAX [157], 3GPP LTE [127], and 3GPP LTE Advanced. This section reviews MIMO-OFDM with an emphasis on its application with spatial multiplexing. After a review of the system model, the operation of different receiver functions is explained in more detail. Equalization and precoding are explained in MIMO-OFDM, where it is shown that they follow from the narrowband MIMO exposition. Different approaches for channel estimation are reviewed based on training data sent in the frequency domain. Least squares estimators are derived for the channel estimates and are found to have additional structure. Finally, a generalization of the Moose algorithm for carrier frequency offset synchronization and frame synchronization to MIMO-OFDM is described, assuming a single offset among all the transmit antennas.

6.5.1 System Model

Consider a MIMO-OFDM system with spatial multiplexing as illustrated in Figure 6.19 and Figure 6.20. Denote the symbol stream (assumed to originate in the frequency domain) as $\mathsf{s}[n]$, the $N_t \times 1$ vector symbol as $\mathbf{s}[n]$, and the corresponding subsymbols on antenna j as $\mathsf{s}_j[n]$. After the spatial multiplexing operation, each subsymbol stream is passed into a SISO-OFDM transmitter operation that consists of a $1:N$ serial-to-parallel operation followed by an N-IDFT and the addition of a cyclic prefix of length L_c. Let $\mathbf{w}[n]$ denote the $N_t \times 1$ time-domain vector output of the cyclic prefix addition block and $w_j[n]$ the samples to be sent on the jth transmit antenna.

Consider the signal at the rth receive antenna assuming perfect synchronization:

$$y_i[n] = \sum_{j=1}^{N_t} \sum_{\ell=0}^{L} h_{i,j}[\ell] w_j[n-\ell] + \mathsf{v}_i[n]. \tag{6.202}$$

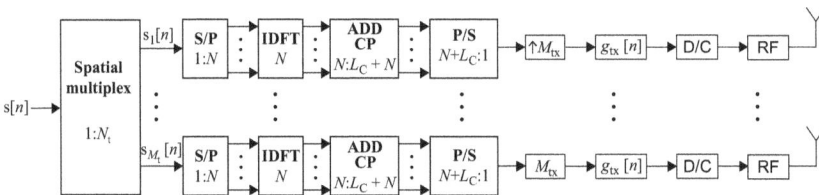

Figure 6.19 Block diagram of a MIMO-OFDM system transmitter. The spatial multiplexing operation happens prior to the usual OFDM transmitter operations. The output of each OFDM modulator is possibly pulse shaped, converted to continuous time, then upconverted.

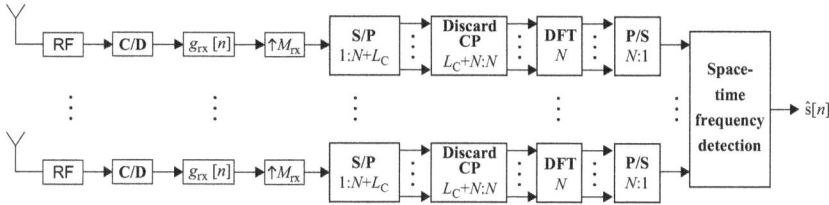

Figure 6.20 Block diagram of a MIMO-OFDM system receiver

We proceed by following the steps in (5.161)–(5.167), exploiting the linearity of the DFT operation. Let

$$h_{i,j}[k] = \sum_{\ell=0}^{L} h_{i,j}[\ell] e^{-j\frac{2\pi k \ell}{N}} \tag{6.203}$$

denote the N-DFT of the (zero-padded) channel between the rth receive antenna and the mth transmit antenna. Discarding the first L_c samples and taking the N-DFT of the result gives

$$y_i[n] = \sum_{j=1}^{N_t} h_{i,j}[k] s_m[k] + v_i[n]. \tag{6.204}$$

Now define the matrix response $[\mathbf{H}[k]]_{i,j} = h_{i,j}[k]$. Equivalently, $\mathbf{H}[k]$ is related to $\mathbf{H}[n]$ through the N-DFT of the matrix channel:

$$\mathbf{H}[k] = \sum_{\ell=0}^{L} \mathbf{H}[\ell] e^{-j\frac{2\pi k \ell}{N}}. \tag{6.205}$$

Stacking the observations in (6.204) for $i = 1, 2, \ldots, N_r$ gives the canonical MIMO-OFDM system equation

$$\mathbf{y}[k] = \mathbf{H}[k]\mathbf{s}[k] + \mathbf{v}[k]. \tag{6.206}$$

Compared to the case where OFDM is not used in Section 6.4.7, (6.206) trades the multivariate convolution in (6.183) for a per-subcarrier matrix multiply. This leads to reductions in equalizer complexity and better performance in most cases. Compared with the special case of SISO-OFDM in (5.168), per-subcarrier equalization has higher complexity in the MIMO case. Furthermore, there are additional trade-offs to be made in terms of the choice of equalizer and detector and their resulting performance.

The similarity of the MIMO-OFDM signal model in (6.206) and the flat-fading MIMO signal model in (6.112) is one of the primary reasons that so much emphasis in research is placed on flat-fading MIMO channel models. The main observation is that many flat-fading results (equalization, detection, and precoding) can be readily extended to MIMO-OFDM with a suitable change in notation. This underlies the relevance of the emphasis on flat-fading signal models in this chapter.

6.5.2   Equalization and Detection

The equalization options for MIMO-OFDM are similar to those of their flat-fading counterparts. A maximum likelihood detector would solve

$$\widehat{\mathbf{s}}[k] = \arg\min_{\bar{\mathbf{s}} \in \mathcal{S}} \|\mathbf{y}[k] - \mathbf{H}[k]\bar{\mathbf{s}}\|^2 \tag{6.207}$$

for $k = 0, 1, \ldots, N - 1$. The brute-force search complexity is the same as in the flat-fading case. The main difference is that in the case of MIMO-OFDM, the candidate symbol vectors are multiplied by $\mathbf{H}[k]$, which is different for every subcarrier. This may make it harder to precompute all the possible distorted symbol vectors because of higher storage requirements.

The zero-forcing detector operates in an analogous fashion as well. With $\mathbf{G}[k] = \mathbf{H}[k]^\dagger$, the receiver first computes

$$\mathbf{z}[k] = \mathbf{G}[k]\mathbf{y}[k], \tag{6.208}$$

then applies a separate detector for each stream to compute

$$\widehat{\mathbf{s}}_k[n] = \arg\min_{c \in \mathcal{C}} |\mathbf{z}_k[n] - c|^2. \tag{6.209}$$

Complexity can be reduced in a couple of different ways. First, instead of computing an inverse and then performing the product in (6.208), the solution can be found through a QR factorization [131]. Second, the relationship between adjacent inverses can be exploited to reduce the computational complexity need to find an inverse for every subcarrier [49].

Performance analysis of these detectors is more complicated in a MIMO-OFDM system as it depends on the channel model. Both delay spread and more generally the power delay profile impact the performance [46]. If each $\mathbf{H}[n]$ is IID with IID zero-mean Gaussian entries with the same variance, then $\mathbf{H}[k]$ are also IID with the same distribution. As a result, the maximum likelihood and zero-forcing detectors for spatial multiplexing achieve the diversity performance as predicted in Section 6.4.1. In theory, though, the total amount of diversity available is $N_t N_r(L + 1)$; achieving this much diversity requires either coding and interleaving [6] or space-frequency codes [319].

6.5.3   Precoding

Precoding generalizes in a natural way to MIMO-OFDM based on the relationship in (6.206). Let $\mathbf{F}[k]$ denote a precoder derived for channel $\mathbf{H}[k]$, and suppose now that $\mathbf{s}[k]$ is $N_s \times 1$. Then the received signal with precoding is

$$\mathbf{x}[k] = \mathbf{H}[k]\mathbf{F}[k]\mathbf{s}[k] + \mathbf{v}[k] \tag{6.210}$$

for $k = 0, 1, \ldots, N - 1$. The receiver performs equalization and detection based on the combined channel $\mathbf{H}[k]\mathbf{F}[k]$.

The concept of limited feedback can be used to convey quantized precoders from the receiver back to the transmitter using a shared codebook and applying quantization to each $\mathbf{H}[k]$. The feedback overhead, however, grows with N since each subcarrier needs its own precoder. There are different approaches for avoiding this requirement. For example, a single precoder can be used for several adjacent subcarriers, exploiting the coherence bandwidth of the channel [229]. Alternatively, every K^{th} precoder can be fed back and then interpolation used at the transmitter to recover the missing ones [68, 67, 251]. Another approach is to quantize the lower-dimensional temporal channel matrix [307], but this requires different quantization strategies. Despite the potential overheads associated with precoding, it is widely deployed with MIMO-OFDM in IEEE 802.11n, IEEE 802.11ac, WiMAX, 3GPP LTE, and 3GPP LTE Advanced.

6.5.4 Channel Estimation

One of the critical receiver operations in a MIMO-OFDM system is channel estimation. This section reviews different approaches for estimating the channel based on known symbols inserted into the transmitted sequence. These strategies use either the time-domain received waveform $\mathbf{y}[n]$ or the frequency-domain received waveform $\mathbf{y}[k]$ to estimate $\{\mathbf{H}[k]\}_{k=0}^{N-1}$ or $\{\mathbf{H}[\ell]\}_{\ell=0}^{L}$ and then determine $\{\mathbf{H}[k]\}_{k=0}^{N-1}$ through application of the N-DFT to each $\{h_{i,j}[\ell]\}_{\ell=0}^{L}$. More information on channel estimation for MIMO-OFDM is found in the literature [198].

For simplicity of exposition, we focus on training data inserted into a single OFDM symbol. Multiple symbols may be exploited by augmenting the appropriate matrices. Suppose that $\{\mathbf{t}[n]\}_{n=0}^{N-1}$ is a known training sequence and that $\mathbf{s}[n] = \mathbf{t}[n]$ for $n = 0, 1, \ldots, N-1$. In this case $N_{\text{tr}} = N$. The time-domain signal after the cyclic prefix addition is $\{\mathbf{w}[n]\}_{n=0}^{N+L-1}$. Treating $\{\mathbf{w}[n]\}_{n=0}^{N+L-1}$ as known information, the approaches in Section 6.4.7 can be exploited at the receiver to estimate the channel based on $\{\mathbf{y}[n]\}_{n=0}^{N+L-1}$. The block size of $N+L$ in the MIMO-OFDM case is the only significant difference. The resulting $\mathbf{T}_i$ matrix (constructed in this case from $\{\mathbf{w}[n]\}_{n=0}^{N+L-1}$) is $N \times (L+1)$ and $\mathbf{y}_i$ is $N \times 1$. The time-domain least squares channel estimate follows from (6.197). Perhaps the only other notable difference in the least squares formulation in this case is that the performance of the estimator depends on the matrix constructed from the DFT of the training sequences. Having good correlation properties in $\{\mathbf{t}[n]\}_{n=0}^{N-1}$ does not necessarily translate into good correlation properties in the resulting $\{\mathbf{w}[n]\}_{n=0}^{N+L-1}$. Consequently, other specially designed sequences may be of interest [26, 126].

Now we consider algorithms that directly exploit the training sequence in the frequency domain. The received signal is

$$\mathbf{y}[k] = \mathbf{H}[k]\mathbf{t}[k] + \mathbf{v}[k] \quad k = 0, 1, \ldots, N-1. \tag{6.211}$$

It is not possible to estimate $\mathbf{H}[k]$ from only $\mathbf{t}[k]$; estimating $\mathbf{H}[k]$ directly requires training in multiple OFDM symbol periods. Here we focus on only one OFDM symbol; therefore, we consider methods that estimate $\{\mathbf{H}[\ell]\}_{\ell=0}^{L}$ instead. This is motivated by the usual choice of $N \gg L$, which means there are N frequency-domain channel coefficients but only $L+1$ time-domain coefficients.

Writing (6.211) as a function of the time-domain channel requires a bit of care. First, we rewrite $\mathbf{H}[k]$ as a function of $\mathbf{H}[\ell]$ in matrix form:

$$\mathbf{H}[k] = \sum_{\ell=0}^{L} \mathbf{H}[\ell] e^{-j\frac{2\pi k\ell}{N}} \tag{6.212}$$

$$= \begin{bmatrix} \mathbf{H}[0] & \mathbf{H}[1] & \cdots & \mathbf{H}[L] \end{bmatrix} \begin{bmatrix} \mathbf{I}_{N_t} \\ e^{-j\frac{2\pi k}{N}}\mathbf{I}_{N_t} \\ \vdots \\ e^{-j\frac{2\pi kL}{N}}\mathbf{I}_{N_t} \end{bmatrix}. \tag{6.213}$$

Now define the vector

$$\mathbf{e}[k]^T = \begin{bmatrix} 1 & e^{-j\frac{2\pi k}{N}} & \cdots & e^{-j\frac{2\pi kL}{N}} \end{bmatrix}. \tag{6.214}$$

Rewrite the channel using the Kronecker product

$$\mathbf{H}[k] = \begin{bmatrix} \mathbf{H}[0] & \mathbf{H}[1] & \cdots & \mathbf{H}[L] \end{bmatrix} (\mathbf{e}[k] \otimes \mathbf{I}_{N_t}) \tag{6.215}$$

and compute

$$\text{vec}(\mathbf{H}[k]) = \left((\mathbf{e}[k]^T \otimes \mathbf{I}_{N_t}) \otimes \mathbf{I}_{N_r}\right) \underbrace{\begin{bmatrix} \text{vec}(\mathbf{H}[0]) \\ \text{vec}(\mathbf{H}[1]) \\ \vdots \\ \text{vec}(\mathbf{H}[L]) \end{bmatrix}}_{\mathbf{h}}. \tag{6.216}$$

Now we use Kronecker product identities to rewrite (6.211) as

$$\mathbf{y}[k] = \text{vec}(\mathbf{y}[k]) \tag{6.217}$$

$$= \text{vec}\left(\mathbf{H}[k]\mathbf{t}[k]\right) + \mathbf{v}[k] \tag{6.218}$$

$$= \left(\mathbf{t}[k]^T \otimes \mathbf{I}_{N_r}\right) \text{vec}(\mathbf{H}[k]) + \mathbf{v}[k] \tag{6.219}$$

$$= \left(\mathbf{t}[k]^T \otimes \mathbf{I}_{N_r}\right) \left((\mathbf{e}[k]^T \otimes \mathbf{I}_{N_t}) \otimes \mathbf{I}_{N_r}\right)\mathbf{h} + \mathbf{v}[k] \tag{6.220}$$

$$= \underbrace{\left(\mathbf{e}[k]^T \otimes \mathbf{t}[k]^T \otimes \mathbf{I}_{N_r}\right)}_{\bar{\mathbf{T}}[k]}\mathbf{h} + \mathbf{v}[k]. \tag{6.221}$$

The known values of the training can be used to build a least squares estimator. Suppose that pilot subcarriers are used. This means that training is known only at subcarriers $\mathcal{K} = \{k_1, k_2, \ldots, k_t\}$. Training over the entire OFDM symbol $\mathcal{K} = \{0, 1, \ldots, N-1\}$ is a special case. Stacking the observations in (6.221) gives

$$\underbrace{\begin{bmatrix} \mathbf{y}[k_1] \\ \mathbf{y}[k_2] \\ \vdots \\ \mathbf{y}[k_t] \end{bmatrix}}_{\bar{\mathbf{y}}} = \underbrace{\begin{bmatrix} \bar{\mathbf{T}}[k_1] \\ \bar{\mathbf{T}}[k_2] \\ \vdots \\ \bar{\mathbf{T}}[k_t] \end{bmatrix}}_{\bar{\mathbf{T}}} \mathbf{h} + \begin{bmatrix} \mathbf{v}[k_1] \\ \mathbf{v}[k_2] \\ \vdots \\ \mathbf{v}[k_t] \end{bmatrix}. \tag{6.222}$$

The dimensions of $\bar{\mathbf{T}}$ are $|\mathcal{K}|N_{\mathrm{r}} \times N_{\mathrm{t}}N_{\mathrm{r}}(L+1)$. With enough pilot subcarriers, $|\mathcal{K}|$ can be made large enough to ensure that $\bar{\mathbf{T}}$ is square or tall. Assuming a good training sequence design so that $\bar{\mathbf{T}}$ is full rank, the least squares estimate is then computed in the usual way as $\hat{\mathbf{h}} = (\bar{\mathbf{T}}^*\bar{\mathbf{T}})^{-1}\bar{\mathbf{T}}^*\bar{\mathbf{y}}$. The final step would be to reform $\{\mathbf{H}[\ell]\}_{\ell=0}^{L}$ from $\mathbf{h}$ and then to take the N-DFT to find $\{\mathbf{H}[k]\}_{k=0}^{N-1}$. While the complexity of the least squares solution seems high, keep in mind that $(\bar{\mathbf{T}}^*\bar{\mathbf{T}})^{-1}\bar{\mathbf{T}}^*$ can be precomputed so only a matrix multiplication is required to generate the estimate. A final takeaway from this derivation is that the least squares estimator in Chapter 5 is powerful and can be applied to a variety of sophisticated settings in the context of MIMO-OFDM.

There are many generalizations and extensions of least squares channel estimators for MIMO-OFDM that result in better performance and/or lower complexity. A general approach for channel estimates is described in [198], including using multiple OFDM symbols and simplified estimators. Achieving good performance requires choosing good training sequences on each antenna. Training sequence optimality and designs are proposed [198, 26, 126]. It has been found, for example, that rotated versions of Frank-Zadoff-Chu sequences [119, 70] have good performance because they satisfy certain optimality criteria and are constant modulus in both time and frequency domains.

6.5.5 Carrier Frequency Synchronization

Synchronization is an important operation for MIMO-OFDM communication systems. This section reviews MIMO-OFDM carrier frequency offset estimation, also providing some guidance on frame synchronization. The main approach is to generalize the Moose and Schmidl-Cox approaches from Chapter 5 that leverage periodicity in the transmitted signal.

The extent of synchronization required in a MIMO-OFDM system depends on whether the transmit antennas and receive antennas are each (respectively) locally synchronized. Consider a system where each RF chain independently generates an imperfect version of f_{c}. The transmit signal generated by the jth transmit antenna is created by upconversion using a carrier frequency $f_{\mathrm{c},j}^{(\mathrm{tx})}$, whereas the downconverted signal at the ith receive antenna is demodulated using carrier frequency $f_{\mathrm{c},i}^{(\mathrm{rx})}$. Let the normalized frequency offset be $\epsilon_{i,j} = (f_{\mathrm{c},i}^{(\mathrm{rx})} - f_{\mathrm{c},j}^{(\mathrm{tx})})T$. Using the same logic as in Section 5.4.1, (6.202) with carrier frequency offset becomes

$$y_i[n] = \sum_{j=1}^{N_{\mathrm{t}}} e^{\mathrm{j}2\pi\epsilon_{i,j}n} \sum_{\ell=0}^{L} h_{i,j}[\ell]w_j[n-\ell] + v_i[n]. \tag{6.223}$$

Example 6.17   Simplify the received signal equation in (6.223) for the case of flat fading with single-carrier modulation, and write it in matrix form. Explain how to correct for carrier frequency offset.

Answer: Simplifying the case where $L = 0$ and assuming single-carrier modulation,

$$y_i[n] = \sum_{j=1}^{N_{\mathrm{t}}} e^{\mathrm{j}2\pi\epsilon_{i,j}n} h_{i,j}s_j + v_i[n]. \tag{6.224}$$

Stacking this results in

$$\mathbf{y}[n] = \underbrace{\begin{bmatrix} h_{11}e^{j2\pi\epsilon_{11}n} & h_{12}e^{j2\pi\epsilon_{12}n} & \dots & h_{1N_t}e^{j2\pi\epsilon_{1N_t}n} \\ h_{21}e^{j2\pi\epsilon_{21}n} & h_{22}e^{j2\pi\epsilon_{22}n} & \dots & h_{2N_t}e^{j2\pi\epsilon_{2N_t}n} \\ \dots & \dots & \dots & \dots \\ h_{N_r1}e^{j2\pi\epsilon_{N_r1}n} & h_{N_r2}e^{j2\pi\epsilon_{N_r2}n} & \dots & h_{N_rN_t}e^{j2\pi\epsilon_{N_rN_t}n} \end{bmatrix}}_{\mathbf{H}[n]} \mathbf{s}[n] + \mathbf{v}[n]. \quad (6.225)$$

If estimates of the channel $\mathbf{H}$ and all the frequency offsets $\{\epsilon_{i,j}\}_{i=1,j=1}^{N_r,N_t}$ are available, then an effective time-varying matrix $\mathbf{H}[n]$ can be constructed and used for joint carrier frequency offset correction and equalization. It should be clear that this approach is more sensitive to offset error, since even small errors can create big matrix inverse fluctuations.

Even if the offsets $\{\epsilon_{i,j}\}$ can be estimated, they cannot be easily removed with a multiplication by $e^{-j2\pi\epsilon_{i,j}n}$ because the offsets show up inside the summation in (6.223). The problem is further enhanced by the observation that the offsets are different for each $i = 1, 2, \dots, N_r$, making it challenging to apply any joint correction to $\mathbf{y}[n]$. The problem is simplified somewhat if a common reference is used at the receiver, using joint equalization and carrier frequency offset synchronization [324] or more sophisticated estimation methods [181], but the resulting receiver complexity is still high. For this reason, it is important that a common reference be used to generate the carriers at both the transmitter and the receiver.

If a common reference is used at each transmitter and receiver, then $\epsilon_{i,j} = \epsilon$ for all i and j, and there is only a single offset. Then the offset can be pulled out of the sum in (6.223) since it no longer depends on j. Since the offset does not depend on i, a matrix equation can be written as

$$\mathbf{y}[n] = e^{j2\pi\epsilon n} \sum_{\ell=0}^{L} \mathbf{H}[\ell]\mathbf{w}[n - \ell] + \mathbf{v}[n]. \quad (6.226)$$

Given an estimate of the offset $\hat{\epsilon}$, correction of the offset in (6.226) amounts to generating the signal $\exp(-j2\pi\hat{\epsilon}n)\mathbf{y}[n]$. The remaining challenge is then to develop good algorithms for estimating the offset that work with multiple transmit and multiple receive antennas.

The simplest generalization of the Schmidl-Cox approach is to send the same training sequence simultaneously from every antenna. If this is the case, then $\mathbf{w}[n] = \mathbf{1}w[n]$ where $\mathbf{1}$ is an $N_t \times 1$ vector of all 1s and the receive signal in (6.226) simplifies as

$$\mathbf{y}[n] = e^{j2\pi\epsilon n} \sum_{\ell=0}^{L} \underbrace{\mathbf{H}[\ell]\mathbf{1}}_{\widetilde{\mathbf{h}}[\ell]} w[n - \ell] + \mathbf{v}[n] \quad (6.227)$$

so that each received signal has the form

$$y_i[n] = e^{j2\pi\epsilon n} \sum_{\ell=0}^{L} \widetilde{h}_i[\ell]w[n - \ell] + v_i[n]. \quad (6.228)$$

Assuming for simplicity of exposition that the odd subcarriers are zeroed such that

$$w[n + L_c] = w[n + L_c + N/2],\tag{6.229}$$

then with $\bar{y}_i[n] = y_i[n + L_c]$ (in the absence of noise)

$$\bar{y}_i[n + N/2] = e^{j2\pi\epsilon N/2}\bar{y}_i[n]\tag{6.230}$$

for $n = 0, 1, \ldots, N/2 - 1$ and $i = 1, 2, \ldots, N_r$. A simple estimator using least squares follows as

$$\hat{\epsilon} = \frac{\sum_{n=0}^{N/2-1}\sum_{i=1}^{N_r}\bar{y}_i^*[n + N/2]\bar{y}_i[n]}{\pi N}.\tag{6.231}$$

The integer offset can be estimated in a similar manner with a second training sequence. While this approach is simple, sending the same training sequence on all the antennas means that the approaches employed in Section 6.5.4 cannot be applied to estimate the channel.

An easy generalization of the periodic training structure is to send training data separately on the transmit antennas orthogonally in time. For example, one periodic OFDM symbol could be sent successively from each transmit antenna. Assuming the training is sent successively from each transmit antenna j, then

$$y_i[n + (j-1)(N + L_c)]$$
$$= e^{j2\pi\epsilon[n+(j-1)(N+L_c)]}\sum_{\ell=0}^{L}h_{i,j}[\ell]w[n - \ell] + v_i[n + (j-1)(N + L_c)].\tag{6.232}$$

Using the $\bar{y}_i[n]$ notation as before (in the absence of noise),

$$\bar{y}_i[n + N/2 + (j-1)(N + L_c)] = e^{j2\pi\epsilon N/2}\bar{y}_r[n + (j-1)(N + L_c)]\tag{6.233}$$

for $n = 0, 1, \ldots, N/2 - 1$, $i = 1, 2, \ldots, N_r$, and $j = 1, 2, \ldots, N_t$. Again, a least squares estimator can be formulated as

$$\hat{\epsilon} = \frac{\sum_{n=0}^{N/2-1}\sum_{i=1}^{N_r}\sum_{j=1}^{N_t}\bar{y}_i^*[n + N/2 + (j-1)(N + L_c)]\bar{y}_i[n + (j-1)(N + L_c)]}{\pi N}.$$
$$\tag{6.234}$$

Compared to (6.231), the approach in (6.234) offers improved performance through additional averaging over time. Note that the indexing would change slightly if a second OFDM symbol were sent from each antenna for the purpose of performing the integer offset, but the idea remains the same.

The carrier frequency offset estimator can be further generalized by allowing simultaneous transmission of training sequences from each transmit antenna. This ensures that each antenna is operating at its peak power (there are per-antenna power constraints in practical MIMO systems) but adds further complexity in the resulting estimators. Joint design of training sequences for channel estimation and carrier frequency offset estimation is provided in [215, 365, 323, 65, 225, 126]. Discussions of system implementations with carrier frequency offset estimation are found in [318, 342, 47].

While we did not consider frame synchronization explicitly, the approach for carrier frequency offset estimation can be leveraged to perform frame synchronization as well. Performance can improve over the SISO case because noise is averaged across different receive antennas, leading to a better estimate.

6.6 Summary

- SIMO communication systems use a single transmit antenna and multiple receive antennas. There are different ways to combine the outputs of each antenna, including antenna selection and maximum ratio combining. These approaches achieve second-order diversity in IID Rayleigh channels. Most receiver processing algorithms for SISO systems extend without major changes to SIMO.

- MISO communication systems use multiple transmit antennas and a single receive antenna. Taking advantage of the transmit antennas is more difficult than in the SIMO case. To achieve the most benefits, either space-time coding or transmit beamforming is required. Space-time coding spreads signals across the antennas in a special way but usually requires higher-complexity decoding. Transmit beamforming can be implemented efficiently using limited feedback, where the best beamforming vector is selected from a codebook known to both the transmitter and the receiver. Transmit beamforming can also be implemented using channel reciprocity, where the receive and transmit channels are reciprocal because of the use of frequency-division duplexing.

- MIMO communication systems use multiple transmit and receive antennas. Spatial multiplexing is the most common transmission technique in MIMO systems.

- Precoding is the generalization of transmit beamforming to the MIMO setting. The optimum precoder comes from the dominant right singular vectors of the channel. The precoder can also be designed using limited feedback from a precoder codebook known to both the transmitter and the receiver. Antenna subset selection is a special case of limited feedback where the codebook consists of different subsets of transmit antennas.

- Channel estimation in MIMO systems requires sending different training sequences on each transmit antenna. When least squares is performed, the channels to each receive antenna can be estimated separately.

- Equalization is more challenging in MIMO systems. Linear equalization in the time domain would require implementing a multivariate equalizer, which usually has high complexity. As a result, MIMO-OFDM is common. With MIMO-OFDM, equalization and detection strategies for flat-fading MIMO systems can be used per subcarrier.

- Carrier frequency offset is a more significant problem in MIMO systems compared with SISO systems. Locking the transmit carriers and receive carriers simplifies the problem. Synchronization techniques like the Schmidl-Cox approach and the Moose algorithm extend to the MIMO setting with some modifications.

Problems

1. Look up cyclic delay diversity as used in IEEE 802.11n and explain how it works.

2. Consider a flat Rayleigh fading MISO system that employs delay diversity with $N_t = 2$ $N_r = 1$ to send data $\{s[n]\}_{n=0}^{N-1}$.

   (a) Supposing that $s[-1] = 0$ and $s[N] = 0$, form the 2×101 space-time codeword matrix $\mathbf{S}$. The length is 101 so that symbol $s[99]$ sees both transmit antennas.

   (b) Show that this space-time code has full diversity.

3. Consider a SIMO system with flat fading described by

   $$\mathbf{y}[n] = \mathbf{h}s[n] + \mathbf{v}[n]. \tag{6.235}$$

   Suppose that the $\mathbf{v}[n]$ is zero mean with covariance $\mathbf{R}_v$. Determine the receive beamforming vector $\mathbf{w}$ that maximizes

   $$\text{SNR}_{|\mathbf{h}} = \frac{|\mathbf{w}^*\mathbf{h}|^2}{\mathbb{E}[|\mathbf{w}^*\mathbf{v}[n]|^2]}. \tag{6.236}$$

   You may assume that $\mathbf{R}_v$ is invertible.

4. Consider MMSE receiver beamforming in a SIMO system with flat fading described by

   $$\mathbf{y}[n] = \mathbf{h}s[n] + \mathbf{v}[n]. \tag{6.237}$$

   Suppose that $\mathbf{v}[n]$ is zero mean with covariance $\mathbf{R}_v$, $s[n]$ is a zero-mean unit variance IID sequence, and $s[n]$ and $\mathbf{v}[n]$ are independent. Determine the receive beamforming vector $\mathbf{w}$ that minimizes the MMSE defined as

   $$\mathbb{E}\left[|\mathbf{w}^*\mathbf{y}[n] - s[n]|^2\right]. \tag{6.238}$$

   Show how the result is simplified in the case where $\mathbf{R}_v = \mathbf{I}$, and explain the differences with MRC.

5. Consider a SIMO system with a single-path channel with $\theta = \pi/4$ and $\alpha = 0.25e^{j\pi/3}$.

   (a) Determine the MRC beamforming solution.

   (b) Determine an expression for $|\mathbf{w}^*\mathbf{h}_s|^2$.

   (c) Generate a polar plot of $|\mathbf{a}(\phi)^*\mathbf{h}_s|^2$. Interpret your results.

6. Three well-known diversity combining methods are maximum ratio combining (MRC), selection combining (SC), and switch-and-stay combining (SSC). Compare and contrast these techniques. In a flat-fading channel, what is the impact of each method on the symbol error rate; that is, how is the probability of symbol error different from the symbol error rate for a SISO channel? Note: You may need to consult additional references to solve this problem.

7. Consider the groundbreaking space-time code devised by Alamouti, used in a MIMO system with $N_t = 2$ and N_r receive antennas. Derive the pairwise error probability for the Alamouti code and show that the diversity order is $2N_r$.

8. Derive the expression for the pairwise error probability of space-time coding with N_r receive antennas. *Hint:* Write (6.92) for an $N_r \times N_t$ matrix $\mathbf{H}$ instead of a $1 \times N_t$ matrix $\mathbf{h}^*$, then use properties of the Kronecker product.

9. Consider 4-QAM transmission. Using Monte Carlo simulations, plot the probability of symbol error for the following transmission strategies up to error rates of 10^{-3}:

   • Gaussian channel

   • Rayleigh fading channel

   • SIMO Rayleigh fading channel with antenna selection (select the antenna with the best channel) with two antennas

   • SIMO Rayleigh fading channel with optimum combining with two, three, and four antennas

   • MISO Rayleigh fading channel with the Alamouti code

   Hint: To perform this simulation, you only need to simulate the discrete-time system. For example, for the Gaussian channel you could simulate

   $$y[n] = \sqrt{E_x} s[n] + v[n]. \tag{6.239}$$

   You should generate approximately $N = 100 \times 1/P_e$ symbols where P_e is the minimum target error probability. Then scale the symbols and add to each symbol an independently generated noise relation. Perform detection and count the errors. The average number of errors divided by N is your estimate of the probability of symbol error.

   Based on your results, answer the following questions:

   (a) Determine the small-scale fade margin for each technique at a symbol error rate of 10^{-2}.

   (b) What is the benefit of optimum combining versus antenna selection?

   (c) What is the benefit of antenna selection versus the Alamouti code?

   (d) What is the impact of increasing the number of receive antennas? How does the increased diversity benefit the system?

10. Consider the Alamouti code with two transmit and two receive antennas. Derive the optimum combiner for the case where two receive antennas are used. Essentially, write the equations for the received signal on each antenna. Stack them together and perform the spatial matched filter. Then show how the observations on each antenna can be combined to get better performance. Explain the diversity order.

11. Derive the maximum likelihood decoder for a spatial multiplexing system with $N_t = 2$ and where different constellations $\mathcal{C}_1$ and $\mathcal{C}_2$ are used.

12. **Computer** Develop a flat-fading MIMO channel simulator. Your simulator should generate a flat-fading matrix channel of dimensions $N_r \times N_t$ for the following channel models:

    (a) IID Rayleigh fading

    (b) Spatially correlated Rayleigh fading with input covariances $\mathbf{R}_{tx}$ and $\mathbf{R}_{rx}$

    (c) Rank-one LOS channel where $\mathbf{H} = \alpha \mathbf{a}(\theta)\mathbf{a}^*(\phi)$ where α is $\mathcal{N}_{\mathbb{C}}(0, 1)$ and $\mathbf{a}(\cdot)$ is the uniform linear array response vector

    (d) Clustered channel model in (6.114) with inputs C, R, and Laplacian power azimuth spectrum in (6.13)

13. **Computer** Using Monte Carlo simulations, plot the probability of symbol error for a 4-QAM SIMO flat-fading channel with given distribution down to a probability of symbol error of 10^{-3} for $N_t = 1, 2, 4, 8$. Compute the following and interpret your results. What is the effect of correlation? How does it behave for larger numbers of antennas?

    (a) IID Rayleigh fading

    (b) Spatially correlated Rayleigh fading using the exponential correlation model with $\rho = 0.5e^{j\pi/3}$

14. **Computer** Using Monte Carlo simulations, plot the probability of symbol error for a 4-QAM MISO flat-fading channel down to a probability of symbol error of 10^{-3} for $N_t = 2$ for the following signaling strategies. Assume IID Rayleigh fading. Interpret your results.

    (a) Maximum ratio transmission

    (b) Equal gain transmission (take the phase of the MRT solution)

    (c) Transmit antenna selection (choose the best transmit antenna and put all your power there)

    (d) Limited feedback transmit beamforming with the LTE codebook in (6.69)

    (e) Alamouti code

15. **Computer** In this problem, you compare the performance of different MIMO configurations and performance with *two* transmit antennas and *two* receive antennas. To make comparisons fair, we keep the rate (size of the vector constellation) fixed within each strategy. Using Monte Carlo simulations, plot the probability of symbol error for the following transmission strategies up to error rates of 10^{-3} in an IID Rayleigh fading channel:

    - Alamouti code with 16-QAM

- Spatial multiplexing with 4-QAM for each data stream with a zero-forcing receiver

- Spatial multiplexing with 4-QAM for each data stream with a maximum likelihood receiver

Hint: To perform this simulation, you only need to simulate the discrete-time system. For example, for spatial multiplexing, you need to simulate

$$\mathbf{y}[n] = \mathbf{H}\mathbf{s}[n] + \mathbf{v}[n] \qquad (6.240)$$

where $\mathbf{H}$ has IID Rayleigh fading and path loss is neglected. You should generate about 100 noise realizations and for each noise realization 100 channel realizations for each simulation point. This means 100×100 simulations per SNR point. You might get by with less to get a smooth curve. Also, just as a reminder:

- Spatial multiplexing with a zero-forcing receiver involves computing $(\mathbf{H})^{-1}\mathbf{y}[n]$ and then detecting each substream separately. Note that you need to compute the vector error; that is, an error is counted if either or both of the symbols sent at the same time differ from what was transmitted. In other words, there is an error if $\widehat{\mathbf{s}}[n] \neq \mathbf{s}[n]$.

- Spatial multiplexing with a maximum likelihood receiver involves computing $\arg\min_{\mathbf{s}\in\mathcal{S}} \|\mathbf{y}[n] - \sqrt{\frac{E_x}{N_t}}\mathbf{H}\mathbf{s}\|^2$ where $\mathcal{S}$ is the set of all possible vector symbols, for example, all possible symbols sent from antenna 1 and all possible symbols sent from antenna 2. In this case it has 16 entries. Again, compute the vector error rate.

Because we compare 16-QAM transmission and spatial multiplexing with 4-QAM transmission, the rate comparisons are fair.

(a) For what SNRs is the Alamouti technique preferred to spatial multiplexing?

(b) What is the performance difference between zero forcing and the maximum likelihood receiver?

(c) What is the complexity difference between zero forcing and the maximum likelihood receiver?

References

[1] 3GPP TR 36.814, "Further advancements for E-UTRA physical layer aspects (Release 9)," March 2010. Available at https://mentor.ieee.org/802.11/dcn/03/11-03-0940-04-000n-tgn-channel-models.doc.

[2] 3GPP TR 36.873, "Technical Specification Group radio access network; Study on 3D channel model for LTE (release 12)," September 2014.

[3] T. J. Abatzoglou, J. M. Mendel, and G. A. Harada, "The constrained total least squares technique and its applications to harmonic superresolution," *IEEE Transactions on Signal Processing*, vol. 39, no. 5, pp. 1070–1087, May 1991.

[4] N. Abramson, "Internet access using VSATs," *IEEE Communications Magazine*, vol. 38, no. 7, pp. 60–68, July 2000.

[5] O. Akan, M. Isik, and B. Baykal, "Wireless passive sensor networks," *IEEE Communications Magazine*, vol. 47, no. 8, pp. 92–99, August 2009.

[6] E. Akay, E. Sengul, and E. Ayanoglu, "Bit interleaved coded multiple beamforming," *IEEE Transactions on Communications*, vol. 55, no. 9, pp. 1802–1811, 2007.

[7] I. Akyildiz, W. Su, Y. Sankarasubramaniam, and E. Cayirci, "A survey on sensor networks," *IEEE Communications Magazine*, vol. 40, no. 8, pp. 102–114, August 2002.

[8] S. M. Alamouti, "A simple transmit diversity technique for wireless communications," *IEEE Journal on Selected Areas in Communications*, pp. 1451–1458, October 1998.

[9] K. Ali and H. Hassanein, "Underwater wireless hybrid sensor networks," in *IEEE Symposium on Computers and Communications Proceedings*, July 2008, pp. 1166–1171.

[10] F. Altman and W. Sichak, "A simplified diversity communication system for beyond-the-horizon links," *IRE Transactions on Communications Systems*, vol. 4, no. 1, pp. 50–55, March 1956.

[11] J. G. Andrews, S. Buzzi, W. Choi, S. V. Hanly, A. Lozano, A. C. K. Soong, and J. C. Zhang, "What will 5G be?" *IEEE Journal on Selected Areas in Communications*, vol. 32, no. 6, pp. 1065–1082, June 2014.

[12] J. G. Andrews, A. Ghosh, and R. Muhamed, *Fundamentals of WiMAX: Understanding Broadband Wireless Networking*. Prentice Hall, 2007.

[13] J. G. Andrews, T. Bai, M. Kulkarni, A. Alkhateeb, A. Gupta, and R. W. Heath, Jr., "Modeling and analyzing millimeter wave cellular systems," Submitted to *IEEE Transactions on Communications*, 2016. Available at http://arxiv.org/abs/1605.04283.

[14] A. Annamalai, C. Tellambura, and V. Bhargava, "Equal-gain diversity receiver performance in wireless channels," *IEEE Transactions on Communications*, vol. 48, no. 10, pp. 1732–1745, October 2000.

[15] E. Arikan, "Channel polarization: A method for constructing capacity-achieving codes for symmetric binary-input memoryless channels," *IEEE Transactions on Information Theory*, vol. 55, no. 7, pp. 3051–3073, July 2009.

[16] D. Astély, E. Dahlman, A. Furuskär, Y. Jading, M. Lindström, and S. Parkvall, "LTE: The evolution of mobile broadband—[LTE Part II: 3GPP release 8]," *IEEE Communications Magazine*, vol. 47, no. 4, pp. 44–51, April 2009.

[17] M. E. Austin, "Decision-feedback equalization for digital communication over dispersive channels," MIT Lincoln Labs, Lexington, MA, Technical Report 437, August 1967.

[18] D. Avagnina, F. Dovis, A. Ghiglione, and P. Mulassano, "Wireless networks based on high-altitude platforms for the provision of integrated navigation/communication services," *IEEE Communications Magazine*, vol. 40, no. 2, pp. 119–125, February 2002.

[19] K. E. Baddour and N. C. Beaulieu, "Autoregressive modeling for fading channel simulation," *IEEE Transactions on Wireless Communications*, vol. 4, no. 4, pp. 1650–1662, July 2005.

[20] T. Bai, R. Vaze, and R. Heath, "Analysis of blockage effects on urban cellular networks," *IEEE Transactions on Wireless Communications*, vol. 13, no. 9, pp. 5070–5083, September 2014.

[21] P. Balaban and J. Salz, "Dual diversity combining and equalization in digital cellular mobile radio," *IEEE Transactions on Vehicular Technology*, vol. 40, no. 2, pp. 342–354, May 1991.

[22] ——, "Optimum diversity combining and equalization in digital data transmission with applications to cellular mobile radio. I. Theoretical considerations," *IEEE Transactions on Communications*, vol. 40, no. 5, pp. 885–894, May 1992.

[23] ——, "Optimum diversity combining and equalization in digital data transmission with applications to cellular mobile radio. II. Numerical results," *IEEE Transactions on Communications*, vol. 40, no. 5, pp. 895–907, May 1992.

[24] L. Barbero and J. Thompson, "Fixing the complexity of the sphere decoder for MIMO detection," *IEEE Transactions on Wireless Communications*, vol. 7, no. 6, pp. 2131–2142, June 2008.

[25] A. Barbieri, A. Piemontese, and G. Colavolpe, "On the ARMA approximation for fading channels described by the Clarke model with applications to Kalman-based receivers," *IEEE Transactions on Wireless Communications*, vol. 8, no. 2, pp. 535–540, February 2009.

[26] I. Barhumi, G. Leus, and M. Moonen, "Optimal training design for MIMO OFDM systems in mobile wireless channels," *IEEE Transactions on Signal Processing*, vol. 51, no. 6, pp. 1615–1624, 2003.

[27] R. Barker, "Group synchronization of binary digital systems," in *Communication Theory*, W. Jackson, ed. Butterworth, 1953, pp. 273–287.

[28] J. Barry, E. Lee, and D. Messerschmitt, *Digital Communication*. Springer, 2004.

[29] O. Bejarano, E. W. Knightly, and M. Park, "IEEE 802.11ac: From channelization to multi-user MIMO," *IEEE Communications Magazine*, vol. 51, no. 10, pp. 84–90, October 2013.

[30] C. A. Belfiore and J. H. Park, "Decision feedback equalization," *Proceedings of the IEEE*, vol. 67, no. 8, pp. 1143–1156, August 1979.

[31] J. C. Belfiore, G. Rekaya, and E. Viterbo, "The Golden code: A 2×2 full-rate space-time code with nonvanishing determinants," *IEEE Transactions on Information Theory*, vol. 51, no. 4, pp. 1432–1436, April 2005.

[32] R. Bellman, *Introduction to Matrix Analysis: Second Edition*, Classics in Applied Mathematics series. Society for Industrial and Applied Mathematics, 1997.

[33] P. Bello, "Characterization of randomly time-variant linear channels," *IEEE Transactions on Communications*, vol. 11, no. 4, pp. 360–393, December 1963.

[34] A. Berg and W. Mikhael, "A survey of techniques for lossless compression of signals," in *Proceedings of the Midwest Symposium on Circuits and Systems*, vol. 2, August 1994, pp. 943–946.

[35] T. Berger and J. Gibson, "Lossy source coding," *IEEE Transactions on Information Theory*, vol. 44, no. 6, pp. 2693–2723, October 1998.

[36] C. Berrou, A. Glavieux, and P. Thitimajshima, "Near Shannon limit error-correcting coding and decoding: Turbo codes," in *Proceedings of the IEEE International Conference on Communications*, Geneva, Switzerland, May 1993, pp. 1064–1070.

[37] R. Bhagavatula, R. W. Heath, Jr., and S. Vishwanath, "Optimizing MIMO antenna placement and array configurations for multimedia delivery in aircraft," in *65th IEEE Vehicular Technology Conference Proceedings*, April 2007, pp. 425–429.

[38] R. Bhagavatula, R. W. Heath, Jr., S. Vishwanath, and A. Forenza, "Sizing up MIMO arrays," *IEEE Vehicular Technology Magazine*, vol. 3, no. 4, pp. 31–38, December 2008.

[39] R. Bhagavatula, C. Oestges, and R. W. Heath, "A new double-directional channel model including antenna patterns, array orientation, and depolarization," *IEEE Transactions on Vehicular Technology*, vol. 59, no. 5, pp. 2219–2231, June 2010.

[40] S. Bhashyam and B. Aazhang, "Multiuser channel estimation and tracking for long-code CDMA systems," *IEEE Transactions on Communications*, vol. 50, no. 7, pp. 1081–1090, July 2002.

[41] S. Biswas, R. Tatchikou, and F. Dion, "Vehicle-to-vehicle wireless communication protocols for enhancing highway traffic safety," *IEEE Communications Magazine*, vol. 44, no. 1, pp. 74–82, January 2006.

[42] E. Björnson, E. G. Larsson, and T. L. Marzetta, "Massive MIMO: Ten myths and one critical question," *IEEE Communications Magazine*, vol. 54, no. 2, pp. 114–123, February 2016.

[43] R. Blahut, *Algebraic Codes for Data Transmission*. Cambridge University Press, 2003.

[44] Bluetooth Special Interest Group, "Specification of the Bluetooth System." Available at http://grouper.ieee.org/groups/802/15/Bluetooth/.

[45] F. Boccardi, R. W. Heath, A. Lozano, T. L. Marzetta, and P. Popovski, "Five disruptive technology directions for 5G," *IEEE Communications Magazine*, vol. 52, no. 2, pp. 74–80, February 2014.

[46] H. Bolcskei, D. Gesbert, and A. Paulraj, "On the capacity of OFDM-based spatial multiplexing systems," *IEEE Transactions on Communications*, vol. 50, no. 2, pp. 225–234, 2002.

[47] H. Bolcskei, "MIMO-OFDM wireless systems: Basics, perspectives, and challenges," *IEEE Wireless Communications*, vol. 13, no. 4, pp. 31–37, August 2006.

[48] P. K. Bondyopadhyay, "The first application of array antenna," in *Proceedings of the IEEE International Conference on Phased Array Systems and Technology*, Dana Point, CA, May 21–25, 2000, pp. 29–32.

[49] M. Borgmann and H. Bolcskei, "Interpolation-based efficient matrix inversion for MIMO-OFDM receivers," in *Conference Record of the Thirty-eighth Asilomar Conference on Signals, Systems and Computers*, vol. 2, 2004, pp. 1941–1947.

[50] R. N. Bracewell, *The Fourier Transform and Its Applications*. McGraw-Hill, 1986.

[51] D. Brandwood, "A complex gradient operator and its application in adaptive array theory," *IEEE Proceedings F Communications, Radar and Signal Processing*, vol. 130, no. 1, pp. 11–16, February 1983.

[52] D. G. Brennan, "On the maximal signal-to-noise ratio realizable from several noisy signals," *Proceedings of the IRE*, vol. 43, 1955.

[53] ——, "Linear diversity combining techniques," *Proceedings of the IRE*, vol. 47, no. 6, pp. 1075–1102, June 1959.

[54] M. Briceno, I. Goldberg, and D. Wagner, "A pedagogical implementation of A5/1." Available at www.scard.org/gsm/a51.html.

[55] J. Brooks, *Telephone: The First Hundred Years*. London: HarperCollins, 1976.

[56] R. J. C. Bultitude and G. K. Bedal, "Propagation characteristics on microcellular urban mobile radio channels at 910 MHz," *IEEE Journal on Selected Areas in Communications*, vol. 7, no. 1, pp. 31–39, January 1989.

[57] N. E. Buris, "Reciprocity calibration of TDD smart antenna systems," in *2010 IEEE Antennas and Propagation Society International Symposium (APSURSI) Proceedings*, July 2010, pp. 1–4.

[58] S. F. Bush, *Smart Grid: Communication-Enabled Intelligence for the Electric Power Grid*. Wiley/IEEE, 2014.

[59] S. Caban and M. Rupp, "Impact of transmit antenna spacing on 2x1 Alamouti radio transmission," *Electronics Letters*, vol. 43, no. 4, pp. 198–199, February 2007.

[60] G. Caire, G. Taricco, and E. Biglieri, "Bit-interleaved coded modulation," *IEEE Transactions on Information Theory*, vol. 44, no. 3, pp. 927–946, May 1998.

[61] J. Camp and E. Knightly, "The IEEE 802.11s extended service set mesh networking standard," *IEEE Communications Magazine*, vol. 46, no. 8, pp. 120–126, August 2008.

[62] A. Cangialosi, J. Monaly, and S. Yang, "Leveraging RFID in hospitals: Patient life cycle and mobility perspectives," *IEEE Communications Magazine*, vol. 45, no. 9, pp. 18–23, September 2007.

[63] R. Chang and R. Gibby, "A theoretical study of performance of an orthogonal multiplexing data transmission scheme," *IEEE Transactions on Communications*, vol. 16, no. 4, pp. 529–540, August 1968.

[64] V. Chawla and D. S. Ha, "An overview of passive RFID," *IEEE Communications Magazine*, vol. 45, no. 9, pp. 11–17, September 2007.

[65] J. Chen, Y.-C. Wu, S. Ma, and T.-S. Ng, "Joint CFO and channel estimation for multiuser MIMO-OFDM systems with optimal training sequences," *IEEE Transactions on Signal Processing*, vol. 56, no. 8, pp. 4008–4019, August 2008.

[66] H.-K. Choi, O. Qadan, D. Sala, J. Limb, and J. Meyers, "Interactive web service via satellite to the home," *IEEE Communications Magazine*, vol. 39, no. 3, pp. 182–190, March 2001.

[67] J. Choi, B. Mondal, and R. W. Heath, Jr., "Interpolation based unitary precoding for spatial multiplexing MIMO-OFDM with limited feedback," *IEEE Transactions on Signal Processing*, vol. 54, no. 12, pp. 4730–4740, December 2006.

[68] J. Choi and R. W. Heath, Jr., "Interpolation based transmit beamforming for MIMO-OFDM with limited feedback," *IEEE Transactions on Signal Processing*, vol. 53, no. 11, pp. 4125–4135, 2005.

[69] J. Choi, N. Gonzalez-Prelcic, R. Daniels, C. R. Bhat, and R. W. Heath, Jr., "Millimeter-wave vehicular communication to support massive automotive sensing," *IEEE Communications Magazine*, vol. 54, no. 12, pp. 160–167, December 2016.

[70] D. Chu, "Polyphase codes with good periodic correlation properties (corresp.)," *IEEE Transactions on Information Theory*, vol. 18, no. 4, pp. 531–532, July 1972.

[71] S. T. Chung, A. Lozano, and H. Huang, "Approaching eigenmode BLAST channel capacity using V-BLAST with rate and power feedback," in *Vehicular Technology Conference Proceedings*, vol. 2, 2001, pp. 915–919.

[72] L. Cimini, "Analysis and simulation of a digital mobile channel using orthogonal frequency division multiplexing," *IEEE Transactions on Communications*, vol. 33, no. 7, pp. 665–675, July 1985.

[73] J. M. Cioffi, "EE 379A—Digital Communication: Signal Processing," http://web.stanford.edu/group/cioffi/ee379a/contents.html.

[74] A. C. Clarke, "Extra-terrestrial relays: Can rocket stations give world-wide radio coverage?" *Wireless World*, pp. 305–308, October 1945.

[75] F. Classen and H. Meyr, "Frequency synchronization algorithms for OFDM systems suitable for communication over frequency selective fading channels," in *IEEE 44th Vehicular Technology Conference Proceedings*, June 1994, pp. 1655–1659.

[76] M. Colella, J. Martin, and F. Akyildiz, "The HALO network," *IEEE Communications Magazine*, vol. 38, no. 6, pp. 142–148, June 2000.

[77] S. Coleri, M. Ergen, A. Puri, and A. Bahai, "Channel estimation techniques based on pilot arrangement in OFDM systems," *IEEE Transactions on Broadcasting*, vol. 48, no. 3, pp. 223–229, September 2002.

[78] P. Cosman, A. Kwasinski, and V. Chande, *Joint Source-Channel Coding*. Wiley/IEEE, 2016.

[79] C. Cox, *Essentials of UMTS*. Cambridge University Press, 2008.

[80] J. W. Craig, "A new, simple and exact result for calculating the probability of error for two-dimensional signal constellations," in *Conference Record of IEEE Military Communications Conference "Military Communications in a Changing World,"* McLean, VA, November 4–7, 1991, pp. 571–575.

[81] B. Crow, I. Widjaja, L. Kim, and P. Sakai, "IEEE 802.11 wireless local area networks," *IEEE Communications Magazine*, vol. 35, no. 9, pp. 116–126, September 1997.

[82] M. Cudak, A. Ghosh, T. Kovarik, R. Ratasuk, T. A. Thomas, F. W. Vook, and P. Moorut, "Moving towards mmwave-based beyond-4G (B-4G) technology," in *IEEE 77th Vehicular Technology Conference Proceedings*, June 2013, pp. 1–5.

[83] A. Dammann and S. Kaiser, "Standard conformable antenna diversity techniques for OFDM and its application to the DVB-T system," in *IEEE Global Telecommunications Conference Proceedings*, vol. 5, 2001, pp. 3100–3105.

[84] F. Davarian, "Sirius satellite radio: Radio entertainment in the sky," in *IEEE Aerospace Conference Proceedings*, vol. 3, 2002, pp. 3-1031–3-1035.

[85] G. Davidson, M. Isnardi, L. Fielder, M. Goldman, and C. Todd, "ATSC video and audio coding," *Proceedings of the IEEE*, vol. 94, no. 1, pp. 60–76, January 2006.

[86] F. De Castro, M. De Castro, M. Fernandes, and D. Arantes, "8-VSB channel coding analysis for DTV broadcast," *IEEE Transactions on Consumer Electronics*, vol. 46, no. 3, pp. 539–547, August 2000.

[87] Z. Ding and G. Li, "Single-channel blind equalization for GSM cellular systems," *IEEE Journal on Selected Areas in Communications*, vol. 16, no. 8, pp. 1493–1505, October 1998.

[88] S. DiPierro, R. Akturan, and R. Michalski, "Sirius XM satellite radio system overview and services," in *5th Advanced Satellite Multimedia Systems Conference and the 11th Signal Processing for Space Communications Workshop Proceedings*, September 2010, pp. 506–511.

[89] K. Doppler, M. Rinne, C. Wijting, C. B. Ribeiro, and K. Hugl, "Device-to-device communication as an underlay to LTE-advanced networks," *IEEE Communications Magazine*, vol. 47, no. 12, pp. 42–49, December 2009.

[90] A. Duel-Hallen, "Equalizers for multiple input/multiple output channels and PAM systems with cyclostationary input sequences," *IEEE Journal on Selected Areas in Communications*, vol. 10, no. 3, pp. 630–639, April 1992.

[91] J. Eberspächer, H. Vögel, C. Bettstetter, and C. Hartmann, *GSM—Architecture, Protocols and Services*, Wiley InterScience online books series. Wiley, 2008.

[92] T. Ebihara and K. Mizutani, "Underwater acoustic communication with an orthogonal signal division multiplexing scheme in doubly spread channels," *IEEE Journal of Oceanic Engineering*, vol. 39, no. 1, pp. 47–58, January 2014.

[93] H. Ekstrom, A. Furuskar, J. Karlsson, M. Meyer, S. Parkvall, J. Torsner, and M. Wahlqvist, "Technical solutions for the 3G long-term evolution," *IEEE Communications Magazine*, vol. 44, no. 3, pp. 38–45, March 2006.

[94] R. Emrick, P. Cruz, N. B. Carvalho, S. Gao, R. Quay, and P. Waltereit, "The sky's the limit: Key technology and market trends in satellite communications," *IEEE Microwave Magazine*, vol. 15, no. 2, pp. 65–78, March 2014.

[95] V. Erceg, L. J. Greenstein, S. Y. Tjandra, S. R. Parkoff, A. Gupta, B. Kulic, A. A. Julius, and R. Bianchi, "An empirically based path loss model for wireless channels in suburban environments," *IEEE Journal on Selected Areas in Communications*, vol. 17, no. 7, pp. 1205–1211, July 1999.

[96] V. Erceg, L. Schumacher, P. Kyritsi, A. Molisch, D. S. Baum, A. Y. Gorokhov, C. Oestges, Q. Li, K. Yu, N. Tal, B. Dijkstra, A. Jagannatham, C. Lanzl, V. J. Rhodes, J. Medbo, D. Michelson, M. Webster, E. Jacobsen, D. Cheung, C. Prettie, M. Ho, S. Howard, B. Bjerke, L. Jengx, H. Sampath, S. Catreux, S. Valle, A. Poloni, A. Forenza, and R. W. Heath, "TGn channel models," *IEEE 802.11-03/940r4*, May 2004. Available at https://mentor.ieee.org/802.11/dcn/03/11-03-0940-04-000n-tgn-channel-models.doc.

[97] K. Etemad, *CDMA2000 Evolution: System Concepts and Design Principles*. Wiley, 2004.

[98] ——, "Overview of mobile WiMAX technology and evolution," *IEEE Communications Magazine*, vol. 46, no. 10, pp. 31–40, October 2008.

[99] ETSI, "Modulation," ETSI TS 100 959 V8.4.0; also 5.04.

[100] ——, "GSM Technical Specification. Digital cellular telecommunications system (Phase 2+); Physical layer on the radio path; General description," ETSI GSM 05.01, 1996.

[101] ETSI Security Algorithms Group of Experts (SAGE) Task Force, "General report on the design, specification and evaluation of 3GPP standard confidentiality and integrity algorithms (3G TR 33.908 version 3.0.0 release 1999)," 3GPP, Technical Report, 1999.

[102] D. Falconer, S. L. Ariyavisitakul, A. Benyamin-Seeyar, and B. Eidson, "Frequency domain equalization for single-carrier broadband wireless systems," *IEEE Communications Magazine*, vol. 40, no. 4, pp. 58–66, April 2002.

[103] E. Falletti, M. Laddomada, M. Mondin, and F. Sellone, "Integrated services from high-altitude platforms: A flexible communication system," *IEEE Communications Magazine*, vol. 44, no. 2, pp. 85–94, February 2006.

[104] G. Faria, J. Henriksson, E. Stare, and P. Talmola, "DVB-H: Digital broadcast services to handheld devices," *Proceedings of the IEEE*, vol. 94, no. 1, pp. 194–209, January 2006.

[105] J. Farserotu and R. Prasad, "A survey of future broadband multimedia satellite systems, issues and trends," *IEEE Communications Magazine*, vol. 38, no. 6, pp. 128–133, June 2000.

[106] K. Feher, "1024-QAM and 256-QAM coded modems for microwave and cable system applications," *IEEE Journal on Selected Areas in Communications*, vol. 5, no. 3, pp. 357–368, April 1987.

[107] M. J. Feuerstein, K. L. Blackard, T. S. Rappaport, S. Y. Seidel, and H. H. Xia, "Path loss, delay spread, and outage models as functions of antenna height for microcellular system design," *IEEE Transactions on Vehicular Technology*, vol. 43, no. 3, pp. 487–498, August 1994.

[108] P. Fire, *A Class of Multiple-Error-Correcting Binary Codes for Non-independent Errors*, SEL: Stanford Electronics Laboratories series. Department of Electrical Engineering, Stanford University, 1959.

[109] M. P. Fitz, "Further results in the fast estimation of a single frequency," *IEEE Transactions on Communications*, vol. 42, no. 234, pp. 862–864, February 1994.

[110] G. Fodor, E. Dahlman, G. Mildh, S. Parkvall, N. Reider, G. Miklós, and Z. Turányi, "Design aspects of network assisted device-to-device communications," *IEEE Communications Magazine*, vol. 50, no. 3, pp. 170–177, March 2012.

[111] A. Forenza and R. W. Heath, Jr., "Benefit of pattern diversity via two-element array of circular patch antennas in indoor clustered MIMO channels," *IEEE Transactions on Communications*, vol. 54, no. 5, pp. 943–954, 2006.

[112] A. Forenza and R. W. Heath, "Optimization methodology for designing 2-CPAs exploiting pattern diversity in clustered MIMO channels," *IEEE Transactions on Communications*, vol. 56, no. 10, pp. 1748–1759, October 2008.

[113] A. Forenza, D. J. Love, and R. W. Heath, "Simplified spatial correlation models for clustered MIMO channels with different array configurations," *IEEE Transactions on Vehicular Technology*, vol. 56, pp. 1924–1934, July 2007.

[114] G. Forney and D. Costello, "Channel coding: The road to channel capacity," *Proceedings of the IEEE*, vol. 95, no. 6, pp. 1150–1177, June 2007.

[115] G. J. Foschini, "Layered space-time architecture for wireless communication in a fading environment when using multiple antennas," *Bell Lab Technical Journal*, vol. 1, no. 2, pp. 41–59, 1996.

[116] R. Frank, "Polyphase codes with good nonperiodic correlation properties," *IEEE Transactions on Information Theory*, vol. 9, no. 1, pp. 43–45, January 1963.

[117] ——, "Comments on 'Polyphase codes with good periodic correlation properties' by Chu, David C.," *IEEE Transactions on Information Theory*, vol. 19, no. 2, pp. 244–244, March 1973.

[118] R. Frank, S. Zadoff, and R. Heimiller, "Phase shift pulse codes with good periodic correlation properties (corresp.)," *IRE Transactions on Information Theory*, vol. 8, no. 6, pp. 381–382, October 1962.

[119] R. L. Frank and S. A. Zadoff, "Phase shift pulse codes with good periodic correlation properties," *IRE Transactions on Information Theory*, vol. IT-8, pp. 381–382, 1962.

[120] A. Furuskar, S. Mazur, F. Muller, and H. Olofsson, "EDGE: Enhanced data rates for GSM and TDMA/136 evolution," *IEEE Personal Communications*, vol. 6, no. 3, pp. 56–66, June 1999.

[121] W. A. Gardner, A. Napolitano, and L. Paura, "Cyclostationarity: Half a century of research," *Signal Processing*, vol. 86, no. 4, pp. 639–697, 2006.

[122] V. K. Garg, *IS-95 CDMA and cdma2000: Cellular/PCS Systems Implementation.* Prentice Hall, 2000.

[123] V. K. Garg and J. E. Wilkes, *Principles and Applications of GSM*, Prentice Hall, 1999.

[124] C. Georghiades, "Maximum likelihood symbol synchronization for the direct-detection optical on-off-keying channel," *IEEE Transactions on Communications*, vol. 35, no. 6, pp. 626–631, June 1987.

[125] D. Gesbert, M. Shafi, D.-shan Shiu, P. Smith, and A. Naguib, "From theory to practice: An overview of MIMO space-time coded wireless systems," *IEEE Journal on Selected Areas in Communications*, vol. 21, no. 3, pp. 281–302, 2003.

[126] M. Ghogho and A. Swami, "Training design for multipath channel and frequency-offset estimation in MIMO systems," *IEEE Transactions on Signal Processing*, vol. 54, no. 10, pp. 3957–3965, October 2006.

[127] A. Ghosh, J. Zhang, J. G. Andrews, and R. Muhamed, *Fundamentals of LTE.* Prentice Hall, 2010.

[128] R. D. Gitlin and S. B. Weinstein, "Fractionally-spaced equalization: An improved digital transversal equalizer," *Bell System Technical Journal*, vol. 60, no. 2, pp. 275–296, February 1981.

[129] M. Golay, "Complementary series," *IRE Transactions on Information Theory*, vol. 7, no. 2, pp. 82–87, April 1961.

[130] A. Goldsmith, S. Jafar, N. Jindal, and S. Vishwanath, "Capacity limits of MIMO channels," *IEEE Journal on Selected Areas in Communications*, vol. 21, no. 5, pp. 684–702, 2003.

[131] G. H. Golub and C. F. V. Loan, *Matrix Computations, Third Edition*. The Johns Hopkins University Press, 1996.

[132] D. Gore, R. Heath, and A. Paulraj, "Statistical antenna selection for spatial multiplexing systems," *IEEE International Conference on Communications Proceedings*, vol. 1, 2002, pp. 450–454.

[133] D. Gore, S. Sandhu, and A. Paulraj, "Delay diversity codes for frequency selective channels," in *IEEE International Conference on Communications Proceedings*, vol. 3, 2002, pp. 1949–1953.

[134] J.-C. Guey, M. P. Fitz, M. R. Bell, and W.-Y. Kuo, "Signal design for transmitter diversity wireless communication systems over Rayleigh fading channels," *IEEE Transactions on Communications*, vol. 47, no. 4, pp. 527–537, April 1999.

[135] T. T. Ha, *Theory and Design of Digital Communication Systems*. Cambridge University Press, 2010.

[136] R. W. Hamming, "Error detecting and error correcting codes," *Bell System Technical Journal*, vol. 29, pp. 147–160, 1950.

[137] S. Hara and R. Prasad, "Overview of multicarrier CDMA," *IEEE Communications Magazine*, vol. 35, no. 12, pp. 126–133, December 1997.

[138] F. J. Harris and M. Rice, "Multirate digital filters for symbol timing synchronization in software defined radios," *IEEE Journal on Selected Areas in Communications*, vol. 19, no. 12, pp. 2346–2357, December 2001.

[139] L. Harte, *CDMA IS-95 for Cellular and PCS*. McGraw-Hill, 1999.

[140] B. Hassibi and H. Vikalo, "On the sphere-decoding algorithm. I. Expected complexity," *IEEE Transactions on Signal Processing*, vol. 53, no. 8, pp. 2806–2818, August 2005.

[141] M. Hata, "Empirical formula for propagation loss in land mobile radio services," *IEEE Transactions on Vehicular Technology*, vol. 29, no. 3, pp. 317–325, August 1980.

[142] A. Hausman, "An analysis of dual diversity receiving systems," *Proceedings of the IRE*, vol. 42, no. 6, pp. 944–947, June 1954.

[143] M. Hayes, *Statistical Digital Signal Processing and Modeling*. Wiley, 1996.

[144] R. Headrick and L. Freitag, "Growth of underwater communication technology in the U.S. Navy," *IEEE Communications Magazine*, vol. 47, no. 1, pp. 80–82, January 2009.

[145] R. W. Heath, N. González-Prelcic, S. Rangan, W. Roh, and A. M. Sayeed, "An overview of signal processing techniques for millimeter wave MIMO systems," *IEEE Journal of Selected Topics in Signal Processing*, vol. 10, no. 3, pp. 436–453, April 2016.

[146] R. Heath, Jr., S. Sandhu, and A. Paulraj, "Antenna selection for spatial multiplexing systems with linear receivers," *IEEE Communications Letters*, vol. 5, no. 4, pp. 142–144, April 2001.

[147] R. W. Heath, Jr., *Digital Wireless Communication: Physical Layer Exploration Lab Using the NI USRP*. National Technology and Science Press, 2012.

[148] R. W. Heath, Jr. and D. Love, "Multimode antenna selection for spatial multiplexing systems with linear receivers," *IEEE Transactions on Signal Processing*, vol. 53, pp. 3042–3056, 2005.

[149] R. W. Heath, Jr. and A. Paulraj, "Linear dispersion codes for MIMO systems based on frame theory," *IEEE Transactions on Signal Processing*, vol. 50, no. 10, pp. 2429–2441, 2002.

[150] R. W. Heath, Jr., T. Wu, and A. C. K. Soong, "Progressive refinement of beamforming vectors for high-resolution limited feedback," *EURASIP Journal on Advances in Signal Processing*, 2009.

[151] H. Hertz, *Electric Waves: Being Researches on the Propagation of Electric Action with Finite Velocity through Space*. Macmillan and Company, 1893.

[152] P. Hoeher, "A statistical discrete-time model for the WSSUS multipath channel," *IEEE Transactions on Vehicular Technology*, vol. 41, no. 4, pp. 461–468, November 1992.

[153] H. Holma and A. Toskala, *WCDMA for UMTS: Radio Access for Third Generation Mobile Communications*. Wiley, 2000.

[154] M.-H. Hsieh and C.-H. Wei, "Channel estimation for OFDM systems based on comb-type pilot arrangement in frequency selective fading channels," *IEEE Transactions on Consumer Electronics*, vol. 44, no. 1, pp. 217–225, February 1998.

[155] K. Huang, R. W. Heath, Jr., and J. Andrews, "Limited feedback beamforming over temporally-correlated channels," *IEEE Transactions on Signal Processing*, vol. 57, no. 5, pp. 1959–1975, May 2009.

[156] B. L. Hughes, "Differential space-time modulation," *IEEE Transactions on Information Theory*, vol. 46, no. 7, pp. 2567–2578, November 2000.

[157] IEEE, "IEEE Standard for Local and Metropolitan Area Networks. Part 16: Air Interface for Fixed and Mobile Broadband Wireless Access Systems—Amendment 2: Physical and Medium Access Control Layers for Combined Fixed and Mobile Operation in Licensed Bands and Corrigendum 1," pp. 1–822, 2006.

[158] IEEE, "IEEE Standard for Information Technology—Local and Metropolitan Area Networks—Specific Requirements—Part 11: Wireless LAN Medium Access Control (MAC) and Physical Layer (PHY) Specifications Amendment 5: Enhancements for Higher Throughput," *IEEE Std 802.11n-2009*, pp. 1–565, October 2009.

[159] IEEE, "IEEE Standard for Information Technology—Telecommunications and Information Exchange between Systems—Local and Metropolitan Area Networks—Specific Requirements—Part 15.3: Wireless Medium Access Control (MAC) and Physical Layer (PHY) Specifications for High Rate Wireless Personal Area Networks (WPANs). Amendment 2: Millimeter-Wave-Based Alternative Physical Layer Extension," December 2009.

[160] IEEE, "IEEE Standard for Information Technology—Telecommunications and Information Exchange between Systems—Local and Metropolitan Area Networks, Wireless LAN Medium Access Control (MAC) and Physical Layer (PHY) Specifications," March 2012.

[161] IEEE, "IEEE Standard for Information Technology—Telecommunications and Information Exchange between Systems—Local and Metropolitan Area Networks—Specific Requirements—Part 11: Wireless LAN Medium Access Control (MAC) and Physical Layer (PHY) Specifications—Amendment 4: Enhancements for Very High Throughput for Operation in Bands below 6 GHz." *IEEE Std 802.11ac-2013*, pp. 1–425, December 2013.

[162] ISO/IEC/IEEE, "ISO/IEC/IEEE International Standard for Information Technology—Telecommunications and Information Exchange between Systems—Local and Metropolitan Area Networks—Specific Requirements—Part 11: Wireless LAN Medium Access Control (MAC) and Physical Layer (PHY) Specifications, Amendment 3: Enhancements for Very High Throughput in the 60 GHz Band (adoption of IEEE Std 802.11ad-2012)," *ISO/IEC/IEEE 8802-11:2012/Amd.3:2014(E)*, pp. 1–634, March 2014.

[163] H. Jafarkhani, "A quasi-orthogonal space-time block code," *IEEE Transactions on Communications*, vol. 49, no. 1, pp. 1–4, 2001.

[164] A. Jahn, M. Holzbock, J. Muller, R. Kebel, M. de Sanctis, A. Rogoyski, E. Trachtman, O. Franzrahe, M. Werner, and F. Hu, "Evolution of aeronautical communications for personal and multimedia services," *IEEE Communications Magazine*, vol. 41, no. 7, pp. 36–43, July 2003.

[165] W. C. Jakes, ed., *Microwave Mobile Communications, Second Edition.* Wiley/IEEE, 1994.

[166] T. Jiang and Y. Wu, "An overview: Peak-to-average power ratio reduction techniques for OFDM signals," *IEEE Transactions on Broadcasting*, vol. 54, no. 2, pp. 257–268, June 2008.

[167] Y. Jiang and M. K. Varanasi, "The RF-chain limited MIMO system: Part I: Optimum diversity-multiplexing tradeoff," *IEEE Transactions on Wireless Communications*, vol. 8, no. 10, pp. 5238–5247, October 2009. Available at http://dx.doi.org/10.1109/TWC.2009.081385.

[168] J. Joe and S. Toh, "Digital underwater communication using electric current method," in *IEEE OCEANS 2007*, June 2007, pp. 1–4.

[169] C. R. Johnson and W. A. Sethares, *Telecommunications Breakdown: Concepts of Communication Transmitted via Software-Defined Radio.* Pearson, 2003.

[170] E. Jorg, *GSM: Architecture, Protocols and Services, Third Edition.* John Wiley and Sons, 2009.

[171] P. Jung, "Laurent's representation of binary digital continuous phase modulated signals with modulation index 1/2 revisited," *IEEE Transactions on Communications*, vol. 42, no. 234, pp. 221–224, 1994.

[172] T. Kailath, A. H. Sayed, and B. Hassibi, *Linear Estimation.* Pearson, 2000.

[173] G. Kaleh, "Simple coherent receivers for partial response continuous phase modulation," *IEEE Journal on Selected Areas in Communications*, vol. 7, no. 9, pp. 1427–1436, December 1989.

[174] S. Kay, "A fast and accurate single frequency estimator," *IEEE Transactions on Acoustics, Speech, and Signal Processing*, vol. 37, no. 12, pp. 1987–1990, December 1989.

[175] S. M. Kay, *Fundamentals of Statistical Signal Processing, Volume I: Estimation Theory.* Prentice Hall, 1993.

[176] S. Kay, *Fundamentals of Statistical Signal Processing, Volume II: Detection Theory.* Prentice Hall, 1998.

[177] J. B. Kenney, "Dedicated short-range communications (DSRC) standards in the United States," *Proceedings of the IEEE*, vol. 99, no. 7, pp. 1162–1182, July 2011.

[178] J. Kermoal, L. Schumacher, K. Pedersen, P. Mogensen, and F. Frederiksen, "A stochastic MIMO radio channel model with experimental validation," *IEEE Journal on Selected Areas in Communications*, vol. 20, no. 6, pp. 1211–1226, 2002.

[179] N. Khaled, B. Mondal, G. Leus, R. W. Heath, and F. Petre, "Interpolation-based multi-mode precoding for MIMO-OFDM systems with limited feedback," *IEEE Transactions on Wireless Communication*, vol. 6, no. 3, pp. 1003–1013, March 2007.

[180] D. Kidston and T. Kunz, "Challenges and opportunities in managing maritime networks," *IEEE Communications Magazine*, vol. 46, no. 10, pp. 162–168, October 2008.

[181] K. J. Kim, M. O. Pun, and R. A. Iltis, "Joint carrier frequency offset and channel estimation for uplink MIMO-OFDMA systems using parallel Schmidt Rao-Blackwellized particle filters," *IEEE Transactions on Communications*, vol. 58, no. 9, pp. 2697–2708, September 2010.

[182] S. Kotz, N. Balakrishnan, and N. Johnson, *Continuous Multivariate Distributions, Models and Applications*, Continuous Multivariate Distributions series. Wiley, 2004.

[183] P. Krishna and D. Husalc, "RFID infrastructure," *IEEE Communications Magazine*, vol. 45, no. 9, pp. 4–10, September 2007.

[184] S.-Y. Kung, Y. Wu, and X. Zhang, "Bezout space-time precoders and equalizers for MIMO channels," *IEEE Transactions on Signal Processing*, vol. 50, no. 10, pp. 2499–2514, October 2002.

[185] E. G. Larsson, O. Edfors, F. Tufvesson, and T. L. Marzetta, "Massive MIMO for next generation wireless systems," *IEEE Communications Magazine*, vol. 52, no. 2, pp. 186–195, February 2014.

[186] B. P. Lathi, *Linear Signals and Systems, Second Edition*. Oxford University Press, 2004.

[187] P. Laurent, "Exact and approximate construction of digital phase modulations by superposition of amplitude modulated pulses (AMP)," *IEEE Transactions on Communications*, vol. 34, no. 2, pp. 150–160, February 1986.

[188] T. Le-Ngoc, V. Leung, P. Takats, and P. Garland, "Interactive multimedia satellite access communications," *IEEE Communications Magazine*, vol. 41, no. 7, pp. 78–85, July 2003.

[189] D. Lee, G. Saulnier, Z. Ye, and M. Medley, "Antenna diversity for an OFDM system in a fading channel," in *IEEE Military Communications Conference Proceedings*, vol. 2, 1999, pp. 1104–1109.

[190] D. Lee, "JPEG 2000: Retrospective and new developments," *Proceedings of the IEEE*, vol. 93, no. 1, pp. 32–41, January 2005.

[191] W. Lee, *Mobile Cellular Telecommunications Systems*. McGraw-Hill, 1989.

[192] M. Lei, C.-S. Choi, R. Funada, H. Harada, and S. Kato, "Throughput comparison of multi-Gbps WPAN (IEEE 802.15.3c) PHY layer designs under nonlinear 60-GHz power amplifier," in *IEEE 18th International Symposium on Personal, Indoor and Mobile Radio Communications Proceedings*, September 2007, pp. 1–5.

[193] P. Lescuyer, *UMTS: Origins, Architecture and the Standard*. Springer, 2004.

[194] B. Li, Y. Qin, C. P. Low, and C. L. Gwee, "A survey on mobile WiMAX," *IEEE Communications Magazine*, vol. 45, no. 12, pp. 70–75, December 2007.

[195] Q. Li, X. Lin, and J. Zhang, "MIMO precoding in 802.16e WiMAX," *Journal of Communications and Networks*, vol. 9, no. 2, pp. 141–149, June 2007.

[196] Y. Li, L. J. J. Cimini, and N. R. Sollenberger, "Robust channel estimation for OFDM systems with rapid dispersive fading channels," *IEEE Transactions on Communications*, vol. 46, no. 7, pp. 902–915, July 1998.

[197] Y. Li, H. Minn, and R. Rajatheva, "Synchronization, channel estimation, and equalization in MB-OFDM systems," *IEEE Transactions on Wireless Communications*, vol. 7, no. 11, pp. 4341–4352, November 2008.

[198] Y. Li, "Simplified channel estimation for OFDM systems with multiple transmit antennas," *IEEE Transactions on Wireless Communications*, vol. 1, no. 1, pp. 67–75, January 2002.

[199] Y.-C. Liang and F. P. S. Chin, "Downlink channel covariance matrix (DCCM) estimation and its applications in wireless DS-CDMA systems," *IEEE Journal on Selected Areas in Communications*, vol. 19, no. 2, pp. 222–232, February 2001.

[200] S. Lin and D. J. Costello, *Error Control Coding*. Pearson, 2004.

[201] J. Liu, A. Bourdoux, J. Craninckx, P. Wambacq, B. Come, S. Donnay, and A. Barel, "OFDM-MIMO WLAN AP front-end gain and phase mismatch calibration," in *2004 IEEE Radio and Wireless Conference Proceedings*, September 2004, pp. 151–154.

[202] J. Liu, G. Vandersteen, J. Craninckx, M. Libois, M. Wouters, F. Petre, and A. Barel, "A novel and low-cost analog front-end mismatch calibration scheme for MIMO-OFDM WLANs," in *2006 IEEE Radio and Wireless Symposium Proceedings*, January 2006, pp. 219–222.

[203] L. Liu and H. Jafarkhani, "Novel transmit beamforming schemes for time-selective fading multiantenna systems," *IEEE Transactions on Signal Processing*, vol. 54, no. 12, pp. 4767–4781, December 2006.

[204] W. C. Liu, T. C. Wei, Y. S. Huang, C. D. Chan, and S. J. Jou, "All-digital synchronization for SC/OFDM mode of IEEE 802.15.3c and IEEE 802.11ad," *IEEE Transactions on Circuits and Systems I: Regular Papers*, vol. 62, no. 2, pp. 545–553, February 2015.

[205] D. J. Love and R. W. Heath, Jr., "Equal gain transmission in multiple-input multiple-output wireless systems," *IEEE Transactions on Communications*, vol. 51, no. 7, pp. 1102–1110, July 2003.

[206] D. J. Love and R. W. Heath, Jr., "Multimode precoding for MIMO wireless systems," *IEEE Transactions on Signal Processing*, vol. 53, pp. 3674–3687, 2005.

[207] D. J. Love, R. W. Heath, Jr., V. K. N. Lau, D. Gesbert, B. Rao, and M. Andrews, "An overview of limited feedback in wireless communication systems," *IEEE Journal on Selected Areas in Communications*, vol. 26, no. 8, pp. 1341–1365, October 2008.

[208] D. J. Love, R. W. Heath, Jr., and T. Strohmer, "Grassmannian beamforming for multiple-input multiple-output wireless systems," *IEEE Transactions on Information Theory*, vol. 49, no. 10, pp. 2735–2747, 2003.

[209] D. Love and R. W. Heath, Jr., "Limited feedback diversity techniques for correlated channels," *IEEE Transactions on Vehicular Technology*, vol. 55, no. 2, pp. 718–722, 2006.

[210] D. Love and R. W. Heath, Jr., "Limited feedback unitary precoding for spatial multiplexing systems," *IEEE Transactions on Information Theory*, vol. 51, no. 8, pp. 2967–2976, 2005.

[211] D. Love, R. W. Heath, Jr., W. Santipach, and M. Honig, "What is the value of limited feedback for MIMO channels?" *IEEE Communications Magazine*, vol. 42, no. 10, pp. 54–59, 2004.

[212] M. Luise and R. Reggiannini, "Carrier frequency recovery in all-digital modems for burst-mode transmissions," *IEEE Transactions on Communications*, vol. 43, no. 2/3/4, pp. 1169–1178, February 1995.

[213] X. Ma, G. B. Giannakis, and S. Ohno, "Optimal training for block transmissions over doubly selective wireless fading channels," *IEEE Transactions on Signal Processing*, vol. 51, no. 5, pp. 1351–1366, May 2003.

[214] X. Ma and G. Giannakis, "Full-diversity full-rate complex-field space-time coding," *IEEE Transactions on Signal Processing*, vol. 51, no. 11, pp. 2917–2930, November 2003.

[215] X. Ma, M.-K. Oh, G. Giannakis, and D.-J. Park, "Hopping pilots for estimation of frequency-offset and multiantenna channels in MIMO-OFDM," *IEEE Transactions on Communications*, vol. 53, no. 1, pp. 162–172, January 2005.

[216] V. H. MacDonald, "The cellular concept," *The Bell System Technical Journal*, vol. 58, no. 2, pp. 15–43, January 1979.

[217] D. J. C. MacKay and R. M. Neal, "Near Shannon limit performance of low-density parity-check codes," *Electronics Letters*, vol. 32, pp. 1645–1646, August 1996.

[218] M. D. Macleod, "Fast nearly ML estimation of the parameters of real or complex single tones or resolved multiple tones," *IEEE Transactions on Signal Processing*, vol. 46, no. 1, pp. 141–148, January 1998.

[219] U. Madhow, *Fundamentals of Digital Communication*. Cambridge University Press, 2008.

[220] J. Magnus and H. Neudecker, *Matrix Differential Calculus with Applications in Statistics and Econometrics*, Wiley Series in Probability and Statistics: Texts and References Section. Wiley, 1999.

[221] A. K. Maini and V. Agrawal, *Satellite Technology: Principles and Applications*. John Wiley and Sons, 2007.

[222] G. Maral and M. Bousquet, *Satellite Communication Systems: Systems, Techniques, and Technology, Fourth Edition*. John Wiley and Sons, 2002.

[223] T. L. Marzetta, "Noncooperative cellular wireless with unlimited numbers of base station antennas," *IEEE Transactions on Wireless Communications*, vol. 9, no. 11, pp. 3590–3600, November 2010.

[224] J. Massey, "Optimum frame synchronization," *IEEE Transactions on Communications*, vol. 20, no. 2, pp. 115–119, April 1972.

[225] H. Minn, N. Al-Dhahir, and Y. Li, "Optimal training signals for MIMO OFDM channel estimation in the presence of frequency offset and phase noise," *IEEE Transactions on Communications*, vol. 54, no. 10, pp. 1754–1759, October 2006.

[226] H. Minn, V. K. Bhargava, and K. B. Letaief, "A robust timing and frequency synchronization for OFDM systems," *IEEE Transactions on Wireless Communications*, vol. 2, no. 4, pp. 822–839, July 2003.

[227] J. Misic, *Wireless Personal Area Networks: Performance, Interconnection, and Security with IEEE 802.15.4*. John Wiley and Sons, 2008.

[228] A. F. Molisch and M. Z. Win, "MIMO systems with antenna selection," *IEEE Microwave Magazine*, vol. 5, no. 1, pp. 46–56, March 2004.

[229] B. Mondal and R. W. Heath, Jr., "Algorithms for quantized precoded MIMO-OFDM systems," in *Conference Record of the Thirty-ninth Asilomar Conference on Signals, Systems and Computers*, 2005, pp. 381–385.

[230] A. Monk, R. Hadani, M. Tsatsanis, and S. Rakib, "OTFS—Orthogonal time frequency space: A novel modulation technique meeting 5G high mobility and massive MIMO challenges." Available at http://arxiv.org/pdf/1608.02993v1.pdf.

[231] P. Moose, "A technique for orthogonal frequency division multiplexing frequency offset correction," *IEEE Transactions on Communications*, vol. 42, no. 10, pp. 2908–2914, 1994.

[232] Y. L. Morgan, "Notes on DSRC & WAVE standards suite: Its architecture, design, and characteristics," *IEEE Communications Surveys Tutorials*, vol. 12, no. 4, pp. 504–518, December 2010.

[233] D. Morton, "Viewing television's history," *Proceedings of the IEEE*, vol. 87, no. 7, pp. 1301–1304, July 1999.

[234] M. Motro, A. Chu, J. Choi, A. Pinjari, C. Bhat, J. Ghosh, and R. Heath, Jr., "Vehicular ad-hoc network (VANET) simulations of overtaking maneuvers on two-lane rural highways," 2016. Submitted to *Transportation Research Part C*.

[235] K. Mukkavilli, A. Sabharwal, E. Erkip, and B. Aazhang, "On beamforming with finite rate feedback in multiple-antenna systems," *IEEE Transactions on Information Theory*, vol. 49, no. 10, pp. 2562–2579, 2003.

[236] B. Muquet, Z. Wang, G. Giannakis, M. de Courville, and P. Duhamel, "Cyclic prefixing or zero padding for wireless multicarrier transmissions?" *IEEE Transactions on Communications*, vol. 50, no. 12, pp. 2136–2148, December 2002.

[237] C. Murthy and B. Rao, "Quantization methods for equal gain transmission with finite rate feedback," *IEEE Transactions on Signal Processing*, vol. 55, no. 1, pp. 233–245, January 2007.

[238] Y. H. Nam, B. L. Ng, K. Sayana, Y. Li, J. Zhang, Y. Kim, and J. Lee, "Full-dimension MIMO (FD-MIMO) for next generation cellular technology," *IEEE Communications Magazine*, vol. 51, no. 6, pp. 172–179, June 2013.

[239] F. Nebeker, *Signal Processing: The Emergence of a Discipline, 1948–1998*. IEEE History Center, 1998.

[240] R. Negi and J. Cioffi, "Pilot tone selection for channel estimation in a mobile OFDM system," in *IEEE Transactions on Consumer Electronics*, vol. 44, no. 3, pp. 1122–1128, 1998.

[241] B. Ng, J.-T. Chen, and A. Paulraj, "Space-time processing for fast fading channels with co-channel interferences," in *IEEE 46th Vehicular Technology Conference Proceedings, 1996. "Mobile Technology for the Human Race,"* vol. 3, April 1996, pp. 1491–1495.

[242] J. C. L. Ng, K. Letaief, and R. Murch, "Antenna diversity combining and finite-tap decision feedback equalization for high-speed data transmission," *IEEE Journal on Selected Areas in Communications*, vol. 16, no. 8, pp. 1367–1375, October 1998.

[243] D. Noble, "The history of land-mobile radio communications," *IEEE Transactions on Vehicular Technology*, pp. 1406–1416, 1962.

[244] L. Nuaymi, *WiMAX: Technology for Broadband Wireless Access*. Wiley, 2007.

[245] C. Oestges, B. Clerckx, M. Guillaud, and M. Debbah, "Dual-polarized wireless communications: From propagation models to system performance evaluation," *IEEE Transactions on Wireless Communications*, vol. 7, no. 10, pp. 4019–4031, October 2008.

[246] T. Okumura, E. Ohmori, and K. Fukuda, "Field strength and its variability in VHF and UHF land mobile service," *Review of the Electrical Communication Laboratory*, vol. 16, no. 9–10, pp. 825–873, September–October 1968.

[247] D. O'Neil, "The rapid deployment digital satellite network," *IEEE Communications Magazine*, vol. 30, no. 1, pp. 30–35, January 1992.

[248] A. V. Oppenheim and R. W. Schafer, *Discrete-Time Signal Processing, Third Edition*. Pearson, 2009.

[249] A. V. Oppenheim, A. S. Willsky, and S. Hamid, *Signals and Systems, Second Edition*. Pearson, 1996.

[250] J. Oppermann and B. S. Vucetic, "Complex spreading sequences with a wide range of correlation properties," *IEEE Transactions on Communications*, vol. 45, no. 3, pp. 365–375, March 1997.

[251] T. Pande, D. J. Love, and J. Krogmeier, "Reduced feedback MIMO-OFDM precoding and antenna selection," *IEEE Transactions on Signal Processing*, vol. 55, no. 5, pp. 2284–2293, May 2007.

[252] A. Papoulis, *Probability, Random Variables, and Stochastic Processes*. McGraw-Hill, 1991.

[253] S. Parkvall, E. Dahlman, A. Furuskar, Y. Jading, M. Olsson, S. Wanstedt, and K. Zangi, "LTE-Advanced—Evolving LTE towards IMT-Advanced," in *IEEE Vehicular Technology Conference Proceedings*, September 2008, pp. 1–5.

[254] R. Parot and F. Harris, "Resolving and correcting gain and phase mismatch in transmitters and receivers for wideband OFDM systems," in *Conference Record of the Thirty-sixth Asilomar Conference on Signals, Systems and Computers*, vol. 2, November 2002, pp. 1005–1009.

[255] A. Paulraj and T. Kailath, "U.S. #5345599: Increasing capacity in wireless broadcast systems using distributed transmission/directional reception (DTDR)," September 1994.

[256] A. Paulraj, D. Gore, R. Nabar, and H. Bolcskei, "An overview of MIMO communications—A key to gigabit wireless," *Proceedings of the IEEE*, vol. 92, no. 2, pp. 198–218, 2004.

[257] E. Perahia, "IEEE 802.11n development: History, process, and technology," *IEEE Communications Magazine*, vol. 46, no. 7, pp. 48–55, July 2008.

[258] E. Perahia, C. Cordeiro, M. Park, and L. L. Yang, "IEEE 802.11ad: Defining the next generation multi-Gbps Wi-Fi," in *2010 7th IEEE Consumer Communications and Networking Conference Proceedings*, January 2010, pp. 1–5.

[259] E. Perahia and M. X. Gong, "Gigabit wireless LANs: An overview of IEEE 802.11ac and 802.11ad," *SIGMOBILE Mobile Computing and Communications Review*, vol. 15, no. 3, pp. 23–33, November 2011.

[260] S. W. Peters and R. W. Heath, Jr., "The future of WiMAX: Multihop relaying with IEEE 802.16j," *IEEE Communications Magazine*, vol. 47, no. 1, pp. 104–111, January 2009.

[261] H. O. Peterson, H. Beverage, and J. Moore, "Diversity telephone receiving system of R.C.A. Communications, Inc." *Proceedings of the Institute of Radio Engineers*, vol. 19, no. 4, pp. 562–584, April 1931.

[262] Z. Pi and F. Khan, "An introduction to millimeter-wave mobile broadband systems," *IEEE Communications Magazine*, vol. 49, no. 6, pp. 101–107, June 2011.

[263] D. Piazza, N. J. Kirsch, A. Forenza, R. W. Heath, and K. R. Dandekar, "Design and evaluation of a reconfigurable antenna array for MIMO systems," *IEEE Transactions on Antennas and Propagation*, vol. 56, no. 3, pp. 869–881, March 2008.

[264] T. Pollet, M. Van Bladel, and M. Moeneclaey, "BER sensitivity of OFDM systems to carrier frequency offset and Wiener phase noise," *IEEE Transactions on Communications*, vol. 43, no. 234, pp. 191–193, February/March/April 1995.

[265] D. Pompili and I. Akyildiz, "Overview of networking protocols for underwater wireless communications," *IEEE Communications Magazine*, vol. 47, no. 1, pp. 97–102, January 2009.

[266] I. Poole, "What exactly is ... HD Radio?" *IEEE Communications Engineering*, vol. 4, no. 5, pp. 46–47, October–November 2006.

[267] B. M. Popovic, "Generalized chirp-like polyphase sequences with optimum correlation properties," *IEEE Transactions on Information Theory*, vol. 38, no. 4, pp. 1406–1409, July 1992.

[268] T. Rappaport, R. W. Heath, Jr., R. C. Daniels, and J. Murdock, *Millimeter Wave Wireless Communications*. Prentice Hall, 2015.

[269] T. S. Rappaport, S. Sun, R. Mayzus, H. Zhao, Y. Azar, K. Wang, G. N. Wong, J. K. Schulz, M. Samimi, and F. Gutierrez, "Millimeter wave mobile communications for 5G cellular: It will work!" *IEEE Access*, vol. 1, pp. 335–349, 2013.

[270] T. S. Rappaport, *Wireless Communications: Principles and Practice, Second Edition*. Prentice Hall, 2002.

[271] B. Razavi, *RF Microelectronics*. Prentice Hall, 1997.

[272] I. S. Reed and G. Solomon, "Polynomial codes over certain finite fields," *Journal of the Society for Industrial and Applied Mathematics*, vol. 8, pp. 300–304, June 1960.

[273] J. H. Reed, *Software Radio: A Modern Approach to Radio Engineering*. Prentice Hall, 2002.

[274] U. Reimers, "Digital video broadcasting," *IEEE Communications Magazine*, vol. 36, no. 6, pp. 104–110, June 1998.

[275] ——, "DVB—the family of international standards for digital video broadcasting," *Proceedings of the IEEE*, vol. 94, no. 1, pp. 173–182, January 2006.

[276] M. Richer, G. Reitmeier, T. Gurley, G. Jones, J. Whitaker, and R. Rast, "The ATSC digital television system," *Proceedings of the IEEE*, vol. 94, no. 1, pp. 37–43, January 2006.

[277] U. Rizvi, G. Janssen, and J. Weber, "Impact of RF circuit imperfections on multi-carrier and single-carrier based transmissions at 60 GHz," in *2008 IEEE Radio and Wireless Symposium Proceedings*, January 2008, pp. 691–694.

[278] J. C. Roh and B. D. Rao, "Transmit beamforming in multiple-antenna systems with finite rate feedback: A VQ-based approach," *IEEE Transactions on Information Theory*, vol. 52, no. 3, pp. 1101–1112, March 2006.

[279] P. Roshan, *802.11 Wireless LAN Fundamentals*. Cisco Press, 2004.

[280] S. Ross, *A First Course in Probability, Ninth Edition*. Pearson, 2012.

[281] S. Roy, J. R. Foerster, V. S. Somayazulu, and D. G. Leeper, "Ultrawideband radio design: The promise of high-speed, short-range wireless connectivity," *Proceedings of the IEEE*, vol. 92, no. 2, pp. 295–311, February 2004.

[282] A. J. Rustako, N. Amitay, G. J. Owens, and R. S. Roman, "Radio propagation at microwave frequencies for line-of-sight microcellular mobile and personal communications," *IEEE Transactions on Vehicular Technology*, vol. 40, no. 1, pp. 203–210, February 1991.

[283] A. Rustako, Y.-S. Yeh, and R. R. Murray, "Performance of feedback and switch space diversity 900 MHz FM mobile radio systems with Rayleigh fading," *IEEE Transactions on Communications*, vol. 21, no. 11, pp. 1257–1268, November 1973.

[284] D. Ryan, I. V. L. Clarkson, I. Collings, D. Guo, and M. Honig, "QAM and PSK codebooks for limited feedback MIMO beamforming," *IEEE Transactions on Communications*, vol. 57, no. 4, pp. 1184–1196, April 2009.

[285] A. Saleh and R. Valenzuela, "A statistical model for indoor multipath propagation," *IEEE Journal on Selected Areas in Communications*, vol. 5, no. 2, pp. 128–137, 1987.

[286] H. Sallam, T. Abdel-Nabi, and J. Soumagne, "A GEO satellite system for broadcast audio and multimedia services targeting mobile users in Europe," in *Advanced Satellite Mobile Systems Proceedings*, August 2008, pp. 134–139.

[287] S. Samejima, K. Enomoto, and Y. Watanabe, "Differential PSK system with nonredundant error correction," *IEEE Journal on Selected Areas in Communications*, vol. 1, no. 1, pp. 74–81, January 1983.

[288] S. Sanayei and A. Nosratinia, "Antenna selection in MIMO systems," *IEEE Communications Magazine*, vol. 42, no. 10, pp. 68–73, October 2004.

[289] S. Sandhu and M. Ho, "Analog combining of multiple receive antennas with OFDM," in *IEEE International Conference on Communications*, vol. 5, May 2003, pp. 3428–3432.

[290] A. Santamaria, *Wireless LAN Standards and Applications*. Artech House, 2006.

[291] H. Sari, G. Karam, and I. Jeanclaude, "Transmission techniques for digital terrestrial TV broadcasting," *IEEE Communications Magazine*, vol. 33, no. 2, pp. 100–109, February 1995.

[292] D. V. Sarwate and M. B. Pursley, "Crosscorrelation properties of pseudorandom and related sequences," *Proceedings of the IEEE*, vol. 68, no. 5, pp. 593–619, May 1980.

[293] T. Sato, D. M. Kammen, B. Duan, M. Macuha, Z. Zhou, J. Wu, M. Tariq, and S. A. Asfaw, *Smart Grid Standards: Specifications, Requirements, and Technologies*. Wiley, 2015.

[294] A. H. Sayed, *Adaptive Filters*. Wiley/IEEE, 2008.

[295] A. Sayeed, "Deconstructing multiantenna fading channels," *IEEE Transactions on Signal Processing*, vol. 50, no. 10, pp. 2563–2579, 2002.

[296] T. Schmidl and D. Cox, "Robust frequency and timing synchronization for OFDM," *IEEE Transactions on Communications*, vol. 45, no. 12, pp. 1613–1621, 1997.

[297] B. Schneier, *Applied Cryptography: Protocols, Algorithms, and Source Code in C, Second Edition*. John Wiley and Sons, 1996.

[298] L. Schumacher, K. Pedersen, and P. Mogensen, "From antenna spacings to theoretical capacities—Guidelines for simulating MIMO systems," in *13th IEEE International Symposium on Personal, Indoor and Mobile Radio Communications Proceedings*, vol. 2, pp. 587–592.

[299] S. Sesia, I. Toufik, and M. Baker, eds., *LTE: The UMTS Long Term Evolution*. John Wiley and Sons, 2009.

[300] A. Seyedi and D. Birru, "On the design of a multi-gigabit short-range communication system in the 60GHz band," in *4th IEEE Consumer Communications and Networking Conference Proceedings*, January 2007, pp. 1–6.

[301] C. Sgraja, J. Tao, and C. Xiao, "On discrete-time modeling of time-varying WSSUS fading channels," *IEEE Transactions on Vehicular Technology*, vol. 59, no. 7, pp. 3645–3651, September 2010.

[302] C. E. Shannon, "A mathematical theory of communication," *Bell System Technical Journal*, vol. 27, no. 3, pp. 379–423, 623–656, July, October 1948.

[303] ——, "Communication theory of secrecy systems," *Bell System Technical Journal*, vol. 28, pp. 656–715, 1949.

[304] P. Shelswell, "The COFDM modulation system: The heart of digital audio broadcasting," *Electronics and Communication Engineering Journal*, vol. 7, no. 3, pp. 127–136, June 1995.

[305] J. Shi, Q. Luo, and M. You, "An efficient method for enhancing TDD over the air reciprocity calibration," in *Proceedings of the IEEE Wireless Communications and Networking Conference*, March 2011, pp. 339–344.

[306] K. Shi and E. Serpedin, "Coarse frame and carrier synchronization of OFDM systems: A new metric and comparison," *IEEE Transactions on Wireless Communications*, vol. 3, no. 4, pp. 1271–1284, July 2004.

[307] H. Shirani-Mehr and G. Caire, "Channel state feedback schemes for multiuser MIMO-OFDM downlink," *IEEE Transactions on Communications*, vol. 57, no. 9, pp. 2713–2723, 2009.

[308] T. Siep, I. Gifford, R. Braley, and R. Heile, "Paving the way for personal area network standards: An overview of the IEEE P802.15 working group for wireless personal area networks," *IEEE Personal Communications*, vol. 7, no. 1, pp. 37–43, February 2000.

[309] M. Simon and J. Smith, "Alternate symbol inversion for improved symbol synchronization in convolutionally coded systems," *IEEE Transactions on Communications*, vol. 28, no. 2, pp. 228–237, February 1980.

[310] M. K. Simon and M.-S. Alouini, *Digital Communication over Fading Channels, Second Edition*. Wiley, 2004.

[311] A. Singer, J. Nelson, and S. Kozat, "Signal processing for underwater acoustic communications," *IEEE Communications Magazine*, vol. 47, no. 1, pp. 90–96, January 2009.

[312] E. N. Skomal, "The range and frequency dependence of VHF-UHF man-made radio noise in and above metropolitan areas," *IEEE Transactions on Vehicular Technology*, vol. 19, no. 2, pp. 213–221, May 1970.

[313] Q. Spencer, M. Rice, B. Jeffs, and M. Jensen, "A statistical model for angle of arrival in indoor multipath propagation," in *IEEE 47th Vehicular Technology Conference Proceedings*, vol. 3, May 1997, pp. 1415–1419.

[314] D. A. Spielman, "Linear-time encodable and decodable error-correcting codes," *IEEE Transactions on Information Theory*, vol. 42, no. 11, pp. 1723–1731, November 1996.

[315] M. Stojanovic and J. Preisig, "Underwater acoustic communication channels: Propagation models and statistical characterization," *IEEE Communications Magazine*, vol. 47, no. 1, pp. 84–89, January 2009.

[316] G. Strang, *Introduction to Linear Algebra*. Wellesley-Cambridge Press, 2003. Available at https://books.google.com/books?id=Gv4pCVyoUVYC.

[317] R. Struble, J. D'Angelo, J. McGannon, and D. Salemi, "AM and FM's digital conversion: How HD Radio^{TM} will spur innovative telematics services for the automotive industry," *IEEE Vehicular Technology Magazine*, vol. 1, no. 1, pp. 18–22, March 2006.

[318] G. Stuber, J. Barry, S. McLaughlin, Y. Li, M. Ingram, and T. Pratt, "Broadband MIMO-OFDM wireless communications," *Proceedings of the IEEE*, vol. 92, no. 2, pp. 271–294, 2004.

[319] W. Su, Z. Safar, and K. Liu, "Full-rate full-diversity space-frequency codes with optimum coding advantage," *IEEE Transactions on Information Theory*, vol. 51, no. 1, pp. 229–249, January 2005.

[320] N. Suehiro, C. Han, T. Imoto, and N. Kuroyanagi, "An information transmission method using Kronecker product," in *Proceedings of the IASTED International Conference on Communication Systems and Networks*, 2002, pp. 206–209.

[321] S. Sun, T. S. Rappaport, R. W. Heath, A. Nix, and S. Rangan, "MIMO for millimeter-wave wireless communications: Beamforming, spatial multiplexing, or both?" *IEEE Communications Magazine*, vol. 52, no. 12, pp. 110–121, December 2014.

[322] S. Sun, T. S. Rappaport, T. A. Thomas, A. Ghosh, H. C. Nguyen, I. Z. Kovács, I. Rodriguez, O. Koymen, and A. Partyka, "Investigation of prediction accuracy, sensitivity, and parameter stability of large-scale propagation path loss models for 5G wireless communications," *IEEE Transactions on Vehicular Technology*, vol. 65, no. 5, pp. 2843–2860, May 2016.

[323] Y. Sun, Z. Xiong, and X. Wang, "EM-based iterative receiver design with carrier-frequency offset estimation for MIMO OFDM systems," *IEEE Transactions on Communications*, vol. 53, no. 4, pp. 581–586, April 2005.

[324] T. Tang and R. W. Heath, Jr., "A space-time receiver with joint synchronization and interference cancellation in asynchronous MIMO-OFDM systems," *IEEE Transactions on Vehicular Technology*, vol. 57, no. 5, pp. 2991–3005, September 2008.

[325] V. Tarokh, A. Naguib, N. Seshadri, and A. R. Calderbank, "Space-time codes for high data rate wireless communication: Performance criteria in the presence of channel estimation errors, mobility, and multiple paths," *IEEE Transactions on Communications*, vol. 47, no. 2, pp. 199–207, February 1999.

[326] V. Tarokh, N. Seshadri, and A. R. Calderbank, "Space-time codes for high data rate wireless communication: Performance criterion and code construction," *IEEE Transactions on Information Theory*, vol. 44, no. 2, pp. 744–765, March 1998.

[327] V. Tarokh, H. Jafarkhani, and A. Calderbank, "Space-time block codes from orthogonal designs," *IEEE Transactions on Information Theory*, vol. 45, no. 5, pp. 1456–1467, July 1999.

[328] I. E. Telatar, "Capacity of multi-antenna Gaussian channels," *European Transactions on Telecommunications*, vol. 10, no. 6, pp. 585–595, 1999.

[329] Y. Toor, P. Muhlethaler, and A. Laouiti, "Vehicle ad hoc networks: Applications and related technical issues," *IEEE Communications Surveys Tutorials*, vol. 10, no. 3, pp. 74–88, October 2008.

[330] S. Tretter, "Estimating the frequency of a noisy sinusoid by linear regression (corresp.)," *IEEE Transactions on Information Theory*, vol. 31, no. 6, pp. 832–835, November 1985.

[331] TurboConcept, "WiMAX IEEE802.16e LDPC decoder." Available at www.turboconcept.com/ip_cores.php?p=tc4200-WiMAX-16e-LDPC-encoder-decoder.

[332] G. L. Turin, "The characteristic function of Hermitian quadratic forms in complex normal random variables," *Biometrika*, vol. 47, no. 1/2, pp. 199–201, June 1960.

[333] W. Tuttlebee and D. Hawkins, "Consumer digital radio: From concept to reality," *Electronics and Communication Engineering Journal*, vol. 10, no. 6, pp. 263–276, December 1998.

[334] G. Ungerboeck, "Channel coding with multilevel/phase signals," *IEEE Transactions on Information Theory*, vol. 28, no. 1, pp. 55–67, January 1982.

[335] ——, "Trellis-coded modulation with redundant signal sets, Part I: Introduction," *IEEE Communications Magazine*, vol. 25, no. 2, pp. 5–11, February 1987.

[336] "Urban transmission loss models for mobile radio in the 900 and 1800 MHz bands," European Cooperation in the Field of Scientific and Technical Research EURO-COST 231, Technical Report 2, September 1991.

[337] V. Va, T. Shimizu, G. Bansal, and R. W. Heath, Jr., "Millimeter wave vehicular communications: A survey," *Foundations and Trends in Networking*, vol. 10, no. 1, 2016.

[338] J. J. van de Beek, O. Edfors, M. Sandell, S. K. Wilson, and P. O. Borjesson, "On channel estimation in OFDM systems," in *IEEE 45th Vehicular Technology Conference Proceedings*, vol. 2, July 1995, pp. 815–819.

[339] H. Van Trees, *Detection, Estimation, and Modulation Theory: Nonlinear Modulation Theory*. Wiley, 2003.

[340] ——, *Detection, Estimation, and Modulation Theory: Detection, Estimation, and Filtering Theory*. Wiley, 2004.

[341] ——, *Detection, Estimation, and Modulation Theory: Optimum Array Processing*. Wiley, 2004.

[342] V. van Zelst and T. C. W. Schenk, "Implementation of a MIMO OFDM-based wireless LAN system," *IEEE Transactions on Signal Processing*, vol. 52, no. 2, pp. 483–494, February 2004.

[343] R. Velidi and C. N. Georghiades, "Frame synchronization for optical multi-pulse pulse position modulation," *IEEE Transactions on Communications*, vol. 43, no. 234, pp. 1838–1843, 1995.

[344] K. Venugopal and R. W. Heath, "Millimeter wave networked wearables in dense indoor environments," *IEEE Access*, vol. 4, pp. 1205–1221, 2016.

[345] S. Verdu, "Fifty years of Shannon theory," *IEEE Transactions on Information Theory*, vol. 44, no. 6, pp. 2057–2078, October 1998.

[346] A. Vielmon, Y. Li, and J. Barry, "Performance of Alamouti transmit diversity over time-varying Rayleigh-fading channels," *IEEE Transactions on Wireless Communications*, vol. 3, no. 5, pp. 1369–1373, September 2004.

[347] H. Vikalo and B. Hassibi, "On the sphere-decoding algorithm. II. Generalizations, second-order statistics, and applications to communications," *IEEE Transactions on Signal Processing*, vol. 53, no. 8, pp. 2819–2834, August 2005.

[348] A. Viterbi, "Error bounds for convolutional codes and an asymptotically optimum decoding algorithm," *IEEE Transactions on Information Theory*, vol. 13, no. 2, pp. 260–269, April 1967.

[349] J. Wallace and M. Jensen, "Modeling the indoor MIMO wireless channel," *IEEE Transactions on Antennas and Propagation*, vol. 50, no. 5, pp. 591–599, 2002.

[350] W. Weichselberger, M. Herdin, H. Ozcelik, and E. Bonek, "A stochastic MIMO channel model with joint correlation of both link ends," *IEEE Transactions on Wireless Communications*, vol. 5, no. 1, pp. 90–100, 2006.

[351] Q. Wen and J. Ritcey, "Spatial diversity equalization for underwater acoustic communications," in *Conference Record of the Twenty-sixth Asilomar Conference on Signals, Systems and Computers*, vol. 2, October 1992, pp. 1132–1136.

[352] S. Wicker, *Error Control Systems for Digital Communication and Storage*. Prentice Hall, 1995. Available at https://books.google.com/books?id=7_hSAAAAMAAJ&q.

[353] C. Williams, M. Beach, D. Neirynck, A. Nix, K. Chen, K. Morris, D. Kitchener, M. Presser, Y. Li, and S. Mclaughlin, "Personal area technologies for internetworked services," *IEEE Communications Magazine*, vol. 42, no. 12, pp. S15–S26, December 2004.

[354] G. Williams, *Linear Algebra with Applications: Alternate Edition*, Jones & Bartlett Learning Series in Mathematics. Jones & Bartlett Learning, 2012. Available at https://books.google.com/books?id=QDIn6WEByGQC.

[355] M. Williamson, "Satellites rock!" *IEEE Review*, vol. 49, no. 11, pp. 34–37, November 2003.

[356] WirelessHD, "WirelessHD specification version 1.1 overview," May 2010. Available at www.wirelesshd.org.

[357] A. Wittneben, "Basestation modulation diversity for digital simulcast," in *41st IEEE Vehicular Technology Conference, "Gateway to the Future Technology in Motion,"* St. Louis, MO, May 19–22, 1991, pp. 848–853.

[358] P. Xia and G. B. Giannakis, "Design and analysis of transmit-beamforming based on limited-rate feedback," *IEEE Transactions on Signal Processing*, vol. 54, no. 5, pp. 1853–1863, May 2006.

[359] X.-G. Xia, "Precoded and vector OFDM robust to channel spectral nulls and with reduced cyclic prefix length in single transmit antenna systems," *IEEE Transactions on Communications*, vol. 49, no. 8, pp. 1363–1374, August 2001.

[360] Y. Xiao, "IEEE 802.11n: Enhancements for higher throughput in wireless LANs," *IEEE Wireless Communications*, vol. 12, no. 6, pp. 82–91, December 2005.

[361] Y. Xiao, X. Shen, B. Sun, and L. Cai, "Security and privacy in RFID and applications in telemedicine," *IEEE Communications Magazine*, vol. 44, no. 4, pp. 64–72, April 2006.

[362] Xilinx, "3GPP LTE DL channel encoder." Available at www.xilinx.com/products/intellectual-property/do-di-chenc-lte.html.

[363] W. Yamada, K. Nishimori, Y. Takatori, and Y. Asai, "Statistical analysis and characterization of Doppler spectrum in large office environment," in *Proceedings of the 2009 International Symposium on Antennas and Propagation*, 2009, pp. 564–567.

[364] S. C. Yang, *3G CDMA2000: Wireless System Engineering*. Artech House, 2004.

[365] Y. Yao and G. Giannakis, "Blind carrier frequency offset estimation in SISO, MIMO, and multiuser OFDM systems," *IEEE Transactions on Communications*, vol. 53, no. 1, pp. 173–183, January 2005.

[366] W. R. Young, "Advanced mobile phone service: Introduction, background, and objectives," *The Bell System Technical Journal*, vol. 58, no. 1, pp. 1–14, January 1979.

[367] X. Zhang and S.-Y. Kung, "Capacity bound analysis for FIR Bézout equalizers in ISI MIMO channels," *IEEE Transactions on Signal Processing*, vol. 53, no. 6, pp. 2193–2204, June 2005.

[368] Y. R. Zheng and C. Xiao, "Simulation models with correct statistical properties for Rayleigh fading channels," *IEEE Transactions on Communications*, vol. 51, no. 6, pp. 920–928, June 2003.

[369] S. Zhou, B. Muquet, and G. B. Giannakis, "Subspace-based (semi-) blind channel estimation for block precoded space-time OFDM," *IEEE Transactions on Signal Processing*, vol. 50, no. 5, pp. 1215–1228, May 2002.

[370] H. Zhuang, L. Dai, S. Zhou, and Y. Yao, "Low complexity per-antenna rate and power control approach for closed-loop V-BLAST," *IEEE Transactions on Communications*, vol. 51, no. 11, pp. 1783–1787, November 2003.

Index

427

www.ingramcontent.com/pod-product-compliance
Lightning Source LLC
Chambersburg PA
CBHW080130220326
41598CB00032B/5015